Milk and Dairy Products as Functional Foods

The Society of Dairy Technology (SDT) has joined with Wiley-Blackwell to produce a series of technical dairy-related handbooks providing an invaluable resource for all those involved in the dairy industry, from practitioners to technologists, working in both traditional and modern large-scale dairy operations. For information regarding the SDT, please contact Maurice Walton, Executive Director, Society of Dairy Technology, PO Box 12, Appleby in Westmorland, CA16 6YJ, UK. email: execdirector@sdt.org

Other volumes in the Society of Dairy Technology book series:

Probiotic Dairy Products (ISBN 978 1 4051 2124 8)
Fermented Milks (ISBN 978 0 6320 6458 8)
Brined Cheeses (ISBN 978 1 4051 2460 7)
Structure of Dairy Products (ISBN 978 1 4051 2975 6)
Cleaning-in-Place (ISBN 978 1 4051 5503 8)
Milk Processing and Quality Management (ISBN 978 1 4051 4530 5)
Dairy Fats (ISBN 978 1 4051 5090 3)
Dairy Powders and Concentrated Products (ISBN 978 1 4051 5764 3)
Technology of Cheesemaking, Second Edition (ISBN 978 1 4051 8298 0)
Processed Cheese and Analogues (ISBN 978 1 4051 8642 1)

Milk and Dairy Products as Functional Foods

Edited by

A. Kanekanian
Cardiff Metropolitan University, UK

WILEY Blackwell

SDT

This edition first published 2014 © 2014 by John Wiley & Sons, Ltd

Registered office: John Wiley & Sons, Ltd, The Atrium, Southern Gate, Chichester, West Sussex, PO19 8SQ, UK

Editorial offices: 9600 Garsington Road, Oxford, OX4 2DQ, UK
The Atrium, Southern Gate, Chichester, West Sussex, PO19 8SQ, UK
111 River Street, Hoboken, NJ 07030-5774, USA

For details of our global editorial offices, for customer services and for information about how to apply for permission to reuse the copyright material in this book please see our website at www.wiley.com/wiley-blackwell.

Library of Congress Cataloging-in-Publication Data

Milk and dairy products as functional foods / [edited by] A. Kanekanian.
pages cm
Includes bibliographical references and index.
ISBN 978-1-4443-3683-2 (hardback)
1. Dairy products in human nutrition. 2. Milk – Analysis. 3. Functional foods.
I. Kanekanian, Ara, editor of compilation.
QP144.M54M534 2014
613.2'6–dc23

2013043865

A catalogue record for this book is available from the British Library.

Wiley also publishes its books in a variety of electronic formats. Some content that appears in print may not be available in electronic books.

Cover image: © iStock/ BeauSnyder
Cover design by www.hisandhersdesign.co.uk

Typeset in 10/12.5pt TimesLTStd by Laserwords Private Limited, Chennai, India
Printed and bound in Singapore by Markono Print Media Pte Ltd

1 2014

Contents

Preface to the Technical Series

For more than 60 years, the Society of Dairy Technology (SDT) has sought to provide education and training in the dairy field, disseminating knowledge and fostering personal development through symposia, conferences, residential courses, publications, and its journal, the *International Journal of Dairy Technology* (previously published as the *Journal of the Society of Dairy Technology*).

In recent years, there have been significant advances in our understanding of milk systems, probably the most complex natural food available to man. At the same time, improvements in process technology have been accompanied by massive changes in the scale of many milk processing operations, and the manufacture a wide range of dairy and other related products.

The Society has embarked on a project with Wiley-Blackwell to produce a Technical Series of dairy-related books to provide an invaluable source of information for practicing dairy scientists and technologists, covering the range from small enterprises to modern large-scale operation. This eleventh volume in this series, on *Membrane Processing – Dairy and Beverages Applications*, provides a timely and comprehensive update of the principles and practices involved in this technology. The commercial exploitation of membrane technology is developed further in its applications to the processing of milk and milk products, plus a wide range of non-dairy beverages, where membranes provide novel opportunities for separation, fractionation and concentration.

Andrew Wilbey
Chairman of the Publications Committee, SDT

Preface

Milk and dairy products provide all the necessary nutrients for healthy living. For the neonates, the colostrum produced by the mother during the 48–72 hours postpartum is the only source of all the necessary nutrients and the protection against microbial infection through the immunoglobulins. Babies obtain nourishment for several more months from milk with the additional benefits from the associated bacteria such as the bifidobacteria in their gut, especially breast-fed babies. For adults, milk and dairy products are consumed for their sensory qualities, nutritional value and for their many bioactive components, discussed throughout this textbook by the widely experienced contributing authors.

Although adverse reactions to some of the components in milk and dairy products are experienced by a very small sector of society, that does not make them unhealthy products to be avoided by the larger sector of society. The benefits of consuming dairy products far outweigh any negative health risk linked to dairy products. The bioactive compounds derived from protein for example, especially the bioactive peptides, have an important part to play in improving human health. They have been shown to act as antioxidants in reducing cholesterol and blood pressure, they have anticarcinogenic, anti-inflammatory, immune modulators, antimicrobial and wound healing properties, and also provide protection of tooth enamel from acid erosion. The health benefits of milk fat should not be underestimated, especially regarding some of the fatty acids such as conjugated linoleic acid and fat-soluble vitamins. The dairy industry has been very active in providing the consumer with a variety of products ranging from low-fat, low-lactose or lactose-free for those with lactose intolerance as well as some hypoallergenic milk formulations.

Although healthy living means providing the body with the necessary nutrients, it should be emphasised that many other factors need to be considered such as the prevalence of sedentary life style, smoking and genetics. The use of dairy products in sports and exercise is becoming popular with many athletes, especially whey protein and colostrum. Nonetheless, many more clinical trials should be conducted to provide evidence beyond any doubt of whether the consumption of milk and dairy products has any link to ill health, such as heart disease. On the contrary, everyone (especially children, the elderly, women and athletes) can benefit from the nutrients and bioactive compounds provided by milk and dairy products.

Fermented milk, such as yogurt and probiotic drinks based on lactobacilli and bifidobacteria, have gained popularity as functional foods and their consumption is on the increase throughout the world.

It is encouraging that advances in science have provided evidence of the health benefits of foods and their individual components. The food industry should however be accountable for the products they produce, in order to provide society with healthier alternatives to the consumption of unhealthy products which contribute to heart disease, obesity and diabetes.

The market for functional foods, where milk and dairy products play an important part, is growing fast and their popularity has exceeded expectation. Bioactive ingredients, including probiotics, are now being used in many other food applications. However, health claims must be based on real evidence and be carefully legislated.

Ara Kanekanian

List of Contributors

Editor

Dr. A. Kanekanian
Centre for Nutrition, Dietetics and Food Science
Cardiff Metropolitan University
Cardiff
Wales
United Kingdom
Tel: +44 2920416449
E-mail: akanekanian@cardiffmet.ac.uk

Contributors

Prof. A. G. D'Alessandro
Department of Engineering and Management of the Agricultural,
Livestock and Forest Systems
University of Bari
Italy
Tel: +393207980586
Email: dalex@agr.uniba.it

Dr. O. Alhaj
Department of Food Science & Nutrition
College of Food and Agricultural Sciences
King Saud University
Riyadh
Kingdom of Saudi Arabia
Tel: +966114693319
Email: oalhaj@ksu.edu.sa

Dr. D. E. Bauman
Department of Animal Science
Cornell University
Ithaca
New York
United States of America
Tel: +1-607-539-6562
E-mail: deb6@cornell.edu

Dr. G. Davison
School of Sport and Exercise Sciences
University of Kent
United Kingdom
Tel: 01634 888994
Email: G.Davison@kent.ac.uk

Prof. G. F. Fitzgerald
Department of Microbiology
University College Cork
Cork, Ireland
Tel: +353 21 490 2730
Email: g.fitzgerald@ucc.ie

Dr. D. I. Givens
Animal Science Research Group
School of Agriculture, Policy and Development
University of Reading
Reading
United Kingdom
Tel: +44 (0) 118 378 8491
Email: d.i.givens@reading.ac.uk

Dr. I. Gordon
1290 Versoix
Switzerland
Tel: +41793288908
Email: ianrgordon@btopenworld.com

Prof. P. Goulas
Department of Animal Production
Laboratory of Animal Production
Technological Educational Institute of Larissa
Larissa
Greece
Tel: +302410684300
Email: goulas@teilar.gr

Mr. M. Hickey
Charleville
Co. Cork
Ireland
Tel:+ 353 63 89392
Email: mfhickey@oceanfree.net

Mr. E. Jensen
Arla Foods Ingredients Group P/S
Aarhus
Denmark
Tel: +45 89 38 1000
Email: erik.jensen@arlafoods.com

Mrs. A. S. Kvistgaard
Arla Foods Ingredients Group P/S
Aarhus
Denmark
Tel: +45 89381196
Email: askv@arlafoods.com

Dr. A. L. Lock
Department of Animal Science
Michigan State University
United States of America
Tel: +1-517 353 8714
E-mail: allock@msu.edu

Dr. T. Marchbank
Office of the Vice Chancellor
Plymouth University
Plymouth
United Kingdom
Tel: + 44(0)20 7882 2260
Email: t.marchbank@qmul.ac.uk

Dr. R. McCarthy
Teagasc
Moorepark Food Research Centre
Fermoy, Co. Cork
Ireland
Tel: + 353-25-42686
Email: brian.mccarthy@teagasc.ie

Dr. S. Mills
Teagasc
Moorepark Food Research Centre
Fermoy, Co. Cork
Ireland
Tel: +3532542606
E-mail: susan.mills@teagasc.ie

Dr. K. Petrotos
Department of Biosystems Engineering
Laboratory of Food Process Engineering
Technological Educational Institute of Larissa
Larissa
Greece
Tel: 0030 69 44 191 500
Email: petrotos@teilar.gr

Prof. R. J. Playford
Office of the Vice Chancellor
Plymouth University
Plymouth
United Kingdom
Tel: +441752582002
Email: raymond.playford@plymouth.ac.uk

Prof. R. P. Ross
Teagasc
Moorepark Food Research Centre
Fermoy, Co. Cork
Ireland
Tel: +353 59 917 0200
Email: paul.ross@teagasc.ie

Ms. J. B. Schroder
Arla Foods Ingredients Group P/S
Aarhus
Denmark
Tel: +45 89 38 1000
Email: jscht@arlafoods.com

Dr. A. Setarehnejad
Centre for Nutrition, Dietetics and Food Science
Cardiff Metropolitan University
Cardiff
Wales
United Kingdom
Tel: +44 2920416455
E-mail: asetarehnejad@cardiffmet.ac.uk

Prof. C. Stanton
Teagasc
Moorepark Food Research Centre
Fermoy, Co. Cork
Ireland
Tel: + 353-25-42606
Email: Catherine.Stanton@teagasc.ie

Dr. E. Tsakali
Laboratory of Food Process Engineering
Department of Biosystems Engineering
Technological Educational Institute of Larissa
Larissa
Greece
Tel: +33 69 44 191 500
Email: efi.tsakali@gmail.com

1 The Health Benefits of Bioactive Compounds from Milk and Dairy Products

A. Kanekanian

Centre for Nutrition, Dietetics and Food Science, Cardiff Metropolitan University, Cardiff, Wales, United Kingdom

1.1 The importance of milk and dairy products

Milk and dairy products have been consumed for thousands of years as a wholesome staple diet in many cultures around the world. They provide the necessary nutritional, functional and physiological values to maintain health and wellbeing of the human body. Their nutritional value, due to the presence of the necessary macro and micronutrients for a healthy life style, has been known of for some time. The nutrients from milk and dairy products include proteins and peptides with their complete range of all the essential amino acids. Milk and dairy products also provide a number of physiological properties: the fat and the presence of both saturated and some unsaturated fatty acids, with the latter shown to have an important physiological role such as with the conjugated linoleic acid (CLA); lactose and other minor oligosaccharides; the major and essential minerals such as calcium, magnesium as well as phosphorous; and vitamins, especially the fat-soluble vitamins A, D, E and K as well as some of the water-soluble vitamins such as B12, riboflavin and C. Milk is the only source of nourishment for the neonates as nutritious food for adults. These nutrients provide the necessary elements for growth and maintenance of the human body. Infants can thrive just by consuming milk for several months, obtaining all the necessary nutrients for growth. Milk also provides the additional physiological properties of promoting the health of the baby and protecting against diseases through its immunoglobulins and other whey proteins.

Based on all these facts, milk and dairy products are considered an important component of functional foods and have seen an increase in their production in the last few decades. In this chapter the role of milk and dairy products, their nutritional importance and their role within the functional food sector are briefly described. Subsequent chapters provide a more detailed explanation of the role of the individual components.

Apart from their health benefits, dairy products are also recognised economically as an agricultural commodity and an essential commercial driver, especially in industrialised developed countries. The dairy industry is an important sector within the food industry, the most dynamic and with the highest gross output. Although the number of independent small dairy farms in most developed countries is in continuous decline due to economic pressure and international regulations, the dairy cooperatives and larger farms are on the increase

Milk and Dairy Products as Functional Foods, First Edition. Edited by A. Kanekanian.

and are able to reduce the overall cost of unit production and prosper in a demanding and competitive market.

The dairy industry, including the primary producers, has been at the frontline of innovation and diversification with emphasis on continuous progress to include animal breeding, quality of feed, new technologies and investment in research and development. Consequently, the development of new products with added value and the number of different varieties of dairy products have also been experiencing an increasing trend, especially the functional products which claim to improve the overall human health and wellbeing. Recently however, most research has focused on the additional health benefits provided by milk, its fermented products and the individual fractions, whether from the protein and its peptides, the fatty acids, the sugars, the minerals or the vitamins.

The importance of dairy products as part of a healthy daily dietary intake has also been emphasised by many government and other scientific bodies, not only for their economic value but also for their nutritional significance. This was clearly emphasised by the healthy balanced diet promoted by the Food Standards Agency (FSA 2007) and the nutrients of concern advocated by the United States Department of Agriculture (USDA 2005). However, it was found from the UK National Diet and Nutrition Surveys (Gregory 2000) that energy intake from fat consumption was almost 50% of total energy consumed; this is far greater than the recommended level, that of 30%. Prior to this recommendation, the dairy industry has been actively involved in introducing new products to satisfy the consumer demand for healthier products, especially those of low-fat (e.g. skimmed and semi-skimmed milk) and high-protein content.

1.2 Dairy products: Concerns and challenges

Although there have been numerous published research data indicating the positive role and benefits of consumption of milk and dairy products to human health, in the last few decades there has also been much adverse publicity regarding the major components of milk and dairy products: fat, protein and lactose. Concern has been raised regarding these three components as being either linked to the high fat content and saturated fatty acids level, protein allergy or lactose intolerance and any adverse symptoms associated with the consumption of these products. Although these three components do contribute to some undesirable side-effects, it is usually only a small segment of the population who are affected. It is however unfortunate that such unfavourable publicity has influenced a larger sector of the society, composed of normal healthy individuals with no adverse reaction to any of these major constituents in milk. Although some people have a genuine need to replace milk with other sources of protein such as soya, a bias against the consumption of milk should be based on real scientific facts and weighed up against the health benefits of consuming milk and dairy products.

Public misconceptions regarding the consumption of milk and dairy products must be addressed to restore the confidence of the public in the health benefits of milk and dairy products. There are huge challenges ahead facing all those concerned about the health and welfare of society – scientists, technologists, health practitioners, marketers and probably psychologists – in tackling this problem and changing consumers' concern or fear of dairy products, which is mostly based on saturated fat, protein, allergy and lactose intolerance. In tackling hostility towards milk, scientists and technologists have identified and explored methods to eliminate or minimise some of these adverse reactions through physical or chemical means. For example, although fat reduction in milk is still a controversial issue, this has been achieved by separating fat by centrifuging and introducing skimmed, semi-skimmed milk and low- to very-low-fat dairy products. Lactose content has also been reduced by physical means using membrane technology such as ultrafiltration. On the other hand, chemical means have been used to reduce or modify the protein content by partial enzymatic hydrolysis using proteolytic enzymes to produce a hypoallergenic milk powder especially for infant formula. However, the disadvantage of protein hydrolysis is the possibility of producing some bitter peptides which require further debittering processes (Minagawa *et al.* 1989; Gallagher *et al.* 1993, 1994). Removal of lactose can be achieved by its complete conversion to galactose and glucose by enzymatic hydrolysis using β-galactosidase or lactase. However, lactose hydrolysis increases the sweetness of milk, which might be unacceptable to those familiar with the low sweetness of milk. Nonetheless, this would be beneficial in fermented products.

The complete removal or reduction of the fat from milk might not be that challenging technologically, but the sensory quality is adversely affected and the resulting milk tastes quite watery and lacking in the body associated with full-fat milk. Fat reduction in dairy products also affects the texture and other functional properties, compensated for by the use of additives in some dairy products to improve consistency and texture. The nutritional quality of fat-soluble vitamins is also compromised, and external sources are needed to replace what has been removed to keep a reasonable amount of these important vitamins. Fat removal also reduces the concentration of the conjugated fatty acids (CLA), a vital component in preventing many diseases such as Type 2 diabetes, inflammation and cancer (Zulet *et al.* 2005). A comprehensive description of the importance of CLA in human diet with a wide range of examples is given in Chapter 6.

With all this negative and adverse publicity regarding the consumption of milk and dairy products, the only way to counteract this and increase their consumption is to develop new products in a similar trend to that of the emerging functional foods movement. The use of milk or the separation of certain ingredients from milk has also been at the frontline of the introduction and promotion of many dairy products ranging from probiotic fermented milk, colostrum, whey protein concentrates/isolates and bioactive peptides from the hydrolysis of casein and whey proteins. Some of these ingredients have been used as nutraceuticals and have, to some degree, compensated for the loss in revenue for some sectors within the dairy industry.

1.3 Dairy products and public health

The function of public health in any country is to protect the health and well-being of the people by preventing disease and ill health and promoting a healthy-living lifestyle. Diseases such as coronary heart disease (CHD), cardiovascular disease (CVD) and cancer and conditions such as diabetes, obesity and mental ill-health are a major concern and have seen an alarming surge in reported incidences. To maintain and improve public health, it is essential to provide the necessary information and tools to the public and make them aware of their role and responsibilities in achieving a healthier life. This could include quality of diet and associated nutrients as well as other important factors such as physical activity, education, environmental pollution and stress, which all have a role to play in overall welfare.

The concern about public health and the cost of healthcare is a major driver for many countries, especially in the western industrialised world. Expenditure on public health is mounting and in continuous increase year after year in most countries. In the UK, net expenditure is huge and has increased from £49 billion in 2001/02 to £104 billion in 2010/11, with mental health having the highest share. This total expenditure represents almost 8% of the net expenditure as a proportion of the gross domestic product (GDP; NHS 2011), with similar trends seen in other developed countries. The rise in the population of the elderly due to advances in medicine, nutrition and care provided by society has also added extra expenditure. In this section only one public health concerns is discussed.

1.3.1 *Heart disease*

Heart disease is a major cause of death, especially in the western world, and is the second-highest cost of health to the NHS in the UK after mental health (NHS 2011). It is well known that diet and lifestyle greatly contribute to the onset and the development of the disease, which could be easily managed. One of the important markers of heart disease is arterial stiffness. In recently published research, Crichton *et al.* (2012) investigated whether the consumption of dairy products could have any effect on the arterial stiffness. The study used pulse wave velocity (PWV) as an independent predictor of mortality due to cardiovascular damage. Their results demonstrated a negative linear relationship between the consumption of dairy products and the pulse wave velocity, independent of nutritional, demographic variables and other cardiovascular disease risk. Interestingly, they recommended further research studies to determine whether the consumption of dairy products would reduce arterial stiffness and cardiovascular disease risk in the elderly.

In a similar study to investigate the relationship between the consumption of dairy products and aortic PWV and arterial stiffness as a predictor of cardiovascular disease and mortality, Livingstone *et al.* (2012) used data from the Caerphilly Prospective Study. They concluded that consuming dairy products (excluding butter) does not adversely affect the arterial stiffness and metabolic markers.

In association within the Caerphilly Prospective Study, Elwood *et al.* (2007) examined the relationship between the consumption of dairy products and metabolic syndrome. The collected data from the metabolic syndrome were based on body mass index (BMI), blood pressure, high-density lipoprotein (HDL), the fasting blood glucose and insulin and fasting plasma triacylglycerol. Their investigation demonstrated a negative relationship between metabolic syndrome and the consumption of milk and dairy products, and no significant relationship between the incidence of diabetes and the consumption of dairy products.

Based on the research data described above, milk and dairy products should not be considered as unhealthy or the source of heart disease. However, further research is required to establish the link between diet and heart disease.

The health debate regarding the link between the consumption of saturated fatty acids and CVD, CHD and stroke, in isolation from the factors which cause heart disease, needs to be tackled in a more objective way based on more epidemiological studies rather than relying on existing data or using univalent analysis. The role of milk and dairy products should not be undermined considering that they provide protein, peptides, CLA, fat-soluble vitamins and important minerals such as calcium, potassium and zinc. All these could protect against many diseases including heart disease, diabetes, osteoporosis and stroke. Many long-term epidemiological studies involving a high number of subjects concluded that there is no direct link between consumption of a low-fat diet and a longer life expectancy, or whether a reduction in fat intake and its replacement with fruits and vegetables will reduce the incidence of CHD, CVD or stroke (Taubes 2001; Howard *et al.* 2006; Siri-Tarino *et al.* 2010).

It should be emphasised that milk and dairy products are a vital constituent of daily diet and consumers should be given a fair and impartial view about the importance of consuming a balanced diet. Indulgence in consuming any food can be unhealthy, which is true for many other food ingredients or additives. The challenge is to maintain both a healthy balanced diet and lifestyle.

1.4 Major nutrients in milk

The major constituent in milk which contributes towards improving human health includes protein and its bioactive peptides, fat and lactose and its derivatives. Each of these are discussed briefly in the following sections; for more details see Chapters 2–8.

1.4.1 *Proteins*

Caseins and whey proteins are present in milk at a concentration of 3.3–3.5% and have important nutritional value as well as many physiological functions after digestion. Milk proteins are important for the growth and maintenance of the body as well as providing bioactive peptides (BAP) and essential amino acids after their hydrolysis. Bioactive peptides derived from milk protein hydrolysis have an array of bioactive properties such as antihypercholesterolemic, hypotensive, anticarcinogenic and immunomodulators. More coverage is given in the sections below and in Chapters 2, 4, 7 and 8.

It appears that BAPs derived from milk protein offer better health benefits compared to other types of proteins such as soy protein, especially in building up skeletal protein. It has been shown that milk-derived peptides stimulate the uptake of amino acids after resistance exercise, consequently giving a larger increase in lean muscle which is beneficial to young athletes as well as strengthening muscles in the elderly (Phillips *et al.* 2005). Chapter 5 describes the importance of dairy products in sport nutrition.

Some whey proteins exhibit bioactivity without hydrolysis, such as immunoglobulins and lactoferrin. The immunoglobulins are considered the primary defence mechanism against any pathogenic micro-organisms in the gastrointestinal tract (GIT) of the infant, which is not yet fully developed to prevent infection. The immunoglobulins constitute the major protein in colostrum, which is the secreted milk by the mother postpartum. The highest concentration occurs in the first few hours at about 20% and drops to 0.5% after 48–72 hours, remaining at that level throughout the lactation period. Chapter 3 describes the health benefits of colostrum. Bovine colostrum and whey protein concentrate and isolates are available on the market for their health benefits, especially in modulating the immune system and muscle growth which is why they are mainly used by athletes (Phillips *et al.* 2005).

1.4.2 *Milk fat*

Milk fat provides energy, the essential fat-soluble vitamins A, D, E and K, some of the essential fatty acids including oleic acid and the conjugated linoleic acid (CLA) and its several isomers. These have all been studied by many scientists for their physiological functions such as anticancer and anti-atherogenic properties (Lock & Bauman 2004; Bauman *et al.* 2006). Of the isomers, *cis-9* and *trans-11*-octadecadienoic acid are prepredominantly found in milk fat in high concentration. They are mainly produced by anaerobic fermentation of bacteria in the rumen via the biohydrogenation of the polyunsaturated fatty acids. Some studies have linked CLA to its function as mutagenesis inhibitor and anticarcinogen (NRC 1996; Park *et al.* 2001), which might also be associated with its antioxidant property.

Although milk fat contains some saturated fatty acids like many other foods that does not mean that milk or dairy products are unhealthy products and should be avoided at all cost. A reduction in the consumption of saturated fatty acids (SFA) has been promoted by governments and several researchers in the last few decades. The link between milk fat and its saturated fatty acids and chronic heart disease has been under debate for several decades. Although many researchers have considered SFAs as one of the contributory factors in heart disease, so far there has been no real study or confirmed data to demonstrate conclusively a direct link between cardiovascular disease (CVD) and milk fat or to implicate dairy products in heart disease. Some useful data could be gathered from meta-analysis in order to provide a more comprehensive picture on this subject. Although there has been a lot of emphasis on reducing the intake of fat, SFAs and refined carbohydrates, the possible connection between obesity (and many other factors) and CVD should also been taken into account.

Although consumers should understand the importance of dietary fat for general health, heart disease is a multi-faceted problem and should not be solely linked to the consumption of SFAs or just a few foods or dietary habits. Other issues such as nutritional education, physical activity, smoking, stress, etc. all play a role in cardiovascular diseases. To emphasise this controversial issue, Taubes (2001) concluded that, after 50 years of research, there was little evidence to link a low consumption of saturated fat to prolonged life. Similarly, a recent meta-analysis of 21 prospective epidemiological studies by Siri-Tarino *et al.* (2010) concluded that there was no significant evidence to associate dietary saturated fat with an increased risk of CHD or CVD. This clearly shows that the relationship between fat, saturated fatty acids, cholesterol and CHD is a more complex matter than initially thought, and the risk of heart disease is multi-faceted (Parodi 2009). The impact of milk fat naturally enriched in *trans* fatty acids (TFA) has recently been examined in human clinical studies. It was found that even when TFA consumption was at the upper range of normal dietary intake, there were no negative effects on plasma biomarkers of CHD risk. It seems that there is little or no epidemiological, clinical or animal-model data indicating that consumption of natural milk fat, as a source of TFA, has a negative impact on human health (Motard-Belanger *et al.* 2008). Another recent study by Petyaev & Bashmakov (2012) highlighted the ongoing debate regarding the 'French Paradox' related to the consumption of dairy products and heart disease. It was observed that in France, despite high consumption of cheese and dairy products, CVD mortality is still low. Although this result was initially linked to the consumption of red wine and the presence of resveratrol, this link was not apparent in other countries. The link between milk fat and heart disease must therefore be proven beyond any doubt before a firm conclusion is drawn.

1.4.3 *Lactose and its derivatives*

Bovine milk contains about 4.6–4.8% lactose, which is lower than that found in human milk of about 7%. Lactose and its derivatives provide about 50% of the energy requirements of the infant (Vesa *et al.* 2000). The two components of lactose – glucose and galactose – play an important biological role in the body as the major source of energy supply and for brain development as part of the galactocerebroside (Kunz *et al.* 1999, 2002). In addition to lactose, about 15–18% of the total sugars in milk are present as lactose derivatives in the form of oligosaccharides in straight or branched chains. These vary in length over the range 3–8 monomers and contain mostly galactose, fucose and sialic acid (Fukuda *et al.* 2010).

It is of great interest to discover that human milk contains higher amounts of lactose and its derivatives compared to other mammals such as cows and goats; milk from the latter examples are also lacking in, for example, the fucosylated derivative (Newburg 1996; Urushima *et al.* 2001). It has been indicated that these oligosaccharides have physiological functions expressed mainly in the intestinal tract; they could act as prebiotics for the growth of the beneficial microflora in the gut such as *Bifidobacterium bifidum* which are naturally present in the gut of breast-fed babies, hence improving their general health (Venema 2012). Lactose derivatives have also been considered to improve health through the modulation of the defence system against pathogens, and could act as receptor analogues and prevent pathogenic micro-organism attachment to the epithelial membrane of the colon. Several of these oligosaccharides have also been isolated from Bactrian camel milk and colostrum in order to study their physiological activities (Fukuda *et al.* 2010).

One of the problems associated with milk consumption, experienced by small sector of the population, is lactose intolerance. It is well known that these people feel or perceive the adverse effect when consuming milk or dairy products, even when only a small amount of lactose is in the product such as in fermented milk products. Lactose is usually hydrolysed by the brush border enzyme β-galactosidase to galactose and glucose. With lactose-intolerant individuals, however, the lack of this enzyme leads to lactose fermentation inside the intestinal tract producing different acids (e.g. lactic and acetic) and gases (e.g. carbon dioxide) causing flatulence and discomfort. Lactose intolerance is more apparent in developing countries where milk consumption is usually low and the body produces a much smaller amount of the enzyme. This problem is less likely with fermented dairy products (e.g. yogurt), where the added bacterial culture utilises most of the lactose as the sole source of energy for their growth.

Lactose intolerance has been tackled by the dairy industry by the introduction of products containing low lactose or even lactose-free dairy products. Lactose is either physically removed using membrane technology such as ultrafiltration, or by direct enzymatic hydrolysis as indicated in Section 1.2.

1.4.4 *Vitamins and minerals*

Minerals in milk play an important role in the physiological functions of the body. It is well known that bioavailability of minerals is usually higher from animal sources, including milk, compared to that from plants. Milk and dairy products are a rich source of dietary calcium necessary for bone formation and, together with phosphate and vitamin D, they contribute to healthy bone structure. The lack of dietary calcium has been linked to osteoporosis for decades, a metabolic bone disease usually associated with a decrease in bone density and increased porosity, fragility and risk of fracture. Although osteoporosis has been linked to factors other than nutrition, such as genetic, environmental and lifestyle, it seems quite likely that calcium is the major factor in keeping and providing a supply of the lost calcium from the bone. Further investigation is however needed to establish the role of vitamin K in maintaining bone density (Vermeer *et al.* 1995; Szulc *et al.* 1996; Booth *et al.* 2000).

The importance of vitamins and their physiological role in improving health through the prevention of many diseases has been known of for a very long time. One of the few controversial issues related to the removal of fat from milk and dairy products is fat-soluble vitamin deficiency such as A, D, E and K. It has been shown that vitamin D has another physiological role regarding the onset of depression in older men and women. In a 6-year study of 531 women and 432 men aged 65 and over, Milaneschi *et al.* (2010) found that there was a link between 25-hydroxyvitamin D deficiency and depression. This was carried out by measuring the vitamin level in the serum and the signs of depression related to levels below 50 nmol L^{-1}. Chapter 9 describes the role of vitamins and minerals in human diet in more detail.

1.5 Dairy products as functional foods

Functional foods have been identified as any food which contains biologically active compounds that provide physiological functions, regardless of their normal nutritional value as a healthy food (Howlett 2008). Physiological functions include antioxidant activity, cholesterol reduction, immune modulation and lowering of blood pressure. Examples of functional foods include food and drink products enriched with vitamins or containing omega-3 fatty acid, probiotics, prebiotics, antioxidants, phytochemicals and many others. The most common and well-known functional ingredient which has also been used in many food formulations, including dairy products, is omega-3 fatty acid, especially the docosahexaenoic acid (DHA) and the eicosapentaenoic acid (EPA) which is mainly sourced from fish oil and, to some degree, from microalgae.

The functionality of these ingredients/foods has been known of for thousands of year throughout the world, especially in the Far East where plant infusions such as tea, herbs and mushrooms were used as a remedy for certain illnesses. The use of yogurt and honey is another example of food used for its tried-and-tested health remedies. The functionality is derived from the bioactive compounds already present in the food, either embedded or as part of one of the constituents within the food which exhibits the functionality only upon their release from the tissue or after hydrolysis e.g. peptides (Hartmann & Meisel 2007; Korhonen & Pihlanto 2007).

Nutraceuticals, a term used to combine nutrition with pharmaceuticals, have recently been recognised as any naturally occurring compounds extracted from food which could be used as supplements, similar to many available supplements in the market such as vitamins and minerals. The three terminologies – functional foods, bioactive compounds and nutraceuticals – have been used interchangeably in this chapter to describe food or some of its constituents for their physiological role in offering health benefits in addition to their normal nutritional value. The less-common name of 'pharmafood' has also been introduced to describe some functional foods with more specifically targeted physiological functions.

The consumption of functional foods is associated with wellbeing and health benefits to ameliorate symptoms and remedy or prevent certain diseases. A further development in functional foods was initiated by believing that diseases (especially chronic) cannot be cured just from the use of medication in the long term, but that other traditional and natural remedies are also required. This belief was also instigated by many published data indicating the health benefits of certain foods. However, the use of functional foods is yet to be proven and acknowledged clinically. Although the pharmaceutical industry offers a wide range of medication to cure many illnesses, the trend is shifting towards the use of food and some of its bioactive ingredients as an alternative or compliment to medicine. The consumption of certain fruits, vegetables, herbs and dairy products, especially the fermented dairy products, has been recommended as part of a healthy diet to provide essential nutrients and ease or cure many chronic diseases. Fermented dairy products containing probiotic bacteria are now a well-established functional food in most countries.

However, many organisations within Europe such as the European Food Safety Authority (EFSA), the UK's Food Standard Agency (FSA) and the American Food and Drug Administration (FDA) have not yet fully accepted all the claims presented to them by the food industry; they have however accepted, with some reservation, that a limited number of food and ingredients offer certain functionality. Many functional ingredients have been accepted to be used in food fortification or as supplements such as the omega-3 fatty acid and stanols from plants, non-starch polysaccharides/dietary fibres and many others, some with restrictions on label declarations. In some cases, there has been a heated debate between food manufacturers and legislative organisations regarding whether health claims can be substantiated.

Historically, some foods have long been used for their health benefits and functional properties. The recent increasing interest in functional foods is a result of their acknowledged effect on health, especially regarding heart disease associated with hypertension and cholesterol level in blood. The demand by the consumer for functional foods has been fuelled by the wealth of information, published literature and the continuous marketing and advertising of the food industry regarding the health benefits of functional foods, especially those linked to fermented milk and dairy products. The emphasis on the importance of these products and the different varieties of fermented milk drinks, yogurt, cheese and ice-cream has increased consumer interest; the market is increasing throughout the western world but not yet to the same extent as in Japan, for example. The estimated market value of functional foods in 2008 was $80 billion (Vergari *et al* 2010) could reach well over $100 billion with average annual growth of more than 5% per annum. This is true within developed countries as well as in the Far East and China.

Many of these functional foods have not yet been fully recognised, clinically proven or licensed in the western industrialised countries. Although there is no clear legislation to describe their use, so far they have been covered under other food legislation imposed by individual countries such as the FSA in the UK and the FDA in the US. In Japan, for example, clearer definition and legislation have been established by the Japanese authority where more than 200 functional food products are regulated by Food for Specified Health Use (or FOSHU; Shimizu 2003). In the past several decades, Japan has pioneered and spearheaded a functional food initiative; it is considered to be the largest-consuming country of these products with a well-matured market and willing and appreciative consumers (Arai *et al* 2002).

Some consumers are yet to be assured that functional foods offer the health benefits declared on their label. Such declarations need to be rigorously monitored follow labelling regulations and, where possible, backed by some known human interventions. In the European Union as well as the US and Canada, such declarations are strictly controlled under specific official regulations. Any claims made outside these regulations must be accompanied by a disclaimer indicating that the product is not intended to, for example, prevent or cure certain diseases. Apart from the typical health benefits associated with functional foods, some have also been considered or used as performance enhancers for athletes, immune modulation through gut health improvement, mental enhancement (described as 'brain' food) and relaxing agents (e.g. serotonin).

A tremendous amount of research has been published on the use of nutraceuticals, mostly based on *in vitro* studies. Fewer *in vivo* studies have been carried out on animals and in clinical trials. However, the approval of nutraceuticals by the authorities would be less problematic than that of pharmaceutical compounds as the former are originated from food with limited processing; the latter are associated with more stringent legislation and ethical documentation.

The emphasis in the last few decades has been on firmly establishing milk, dairy products and certain components of milk origin (such as whey protein) as an important segment of functional foods, nutraceuticals and bioactive compounds. Some of these functional foods of milk origin have included fermented milk products with probiotics, whey proteins (whether concentrate, isolate or as nutraceutical materials such as α-lactabumin, β-lactoglobulin, lactoferrin, lactoperoxidase and other minor constituents) and bioactive peptides from protein hydrolysis and oligonucleotides. Others milk constituents include the fatty acid conjugated linoleic acid (CLA), calcium and lactose derivatives. The list is increasing in number as consumers are becoming more aware of the importance of these ingredients that are either already present in the food or are added to the product. Based on this, milk and its individual components in their intact, modified or hydrolysed forms can be referred to as functional foods and their bioactive constituents as functional ingredients or nutraceuticals. This undoubtedly indicates the importance of milk as a major source of essential nutrients and bioactive compounds for healthy living.

Milk and dairy products are an important sector of functional foods; their individual components such as protein, peptides, fat and fatty acids, sugars, minerals and other minor constituents have been reviewed here as a source of nutrients and provider of physiological functions, reflecting the continuous advancement and progress in this field of science and technology. These findings have been exploited by the industry to produce either nutritional supplements or products that are healthy and tasty. In this context, the emphasis in the following section is on milk and dairy products as functional foods, concentrating mainly on the bioactive peptides derived from total milk proteins as well as the fermented probiotic yogurt drink and whey proteins. More detailed information is provided in most of the chapters and in Sections 1.6 and 1.7, highlighting the role of each product or component in providing functionality and as bioactive compounds.

The bioactivity and health benefits of some functional dairy products, such as probiotic fermented milk drinks and yogurts, are within the product itself. In some cases, the challenge might arise when functional ingredients are to be added to the product. This is true when a product is to be enriched with pure CLA, which could give an objectionable fishy flavour even when added at lower concentrations. To avoid this problem, manufacturers tend to use the micro- or nano-encapsulated form. This encapsulated form can also offer additional protection from environment processes such as oxidation.

1.6 Bioactive compounds from milk

Some intact proteins could provide physiological functions in addition to their nutritional value. During the digestion of most proteins, some of the released peptides offer additional physiological functions; these so-called 'bioactive peptides' are discussed along with other bioactive compounds in Section 1.6.1 and in more details in Chapters 2–9. The protein in this case is considered as the 'parent' or the precursors of these bioactive peptides which are derived either after digestion, as a result of fermentation or by hydrolysis using any source of proteolytic enzymes. Most of the basic nutrients released by hydrolysis such as amino acids are used as the building blocks for the growth and maintenance of the human body; bioactive peptides, which are needed in minute quantities, offer the additional benefits of protecting the human body from various diseases and dysfunctions of certain organs. The function of these bioactive peptides are similar to that of vitamins and minerals in that they are needed in minute quantities and can provide the body with additional health benefits and protection from certain diseases and malfunctions.

Bioactive compounds have been identified as either naturally occurring endogenous chemical compounds present in foods, such as lactoferrin and immunoglobulin, or produced as a result of the breakdown of the protein in food by the digestive enzymes in the human gastrointestinal tract (GIT). The bioactivities of these peptides are either demonstrated locally, such as antibacterial activity within the GIT, or induce some physiological effect by stimulating the immune system. They can also interact and bind to specific receptors on the outer surface of the epithelial cell membranes and be transported via the bloodstream to reach a specific target organ.

Fermented food can also be a vital contributor to bioactive compounds due to the presence of the beneficial micro-organisms such as the well-known lactic acid bacteria (LAB) or by the production of certain metabolites which provide their own specific health benefits. Most of the bioactive compounds are stable and do not undergo further breakdown; they can be absorbed through the epithelial wall of the GIT and be circulated to all the tissues in the body to provide health benefits and prevent DNA damage. Interest in bioactives or biologically active components (BAC) and their presence in or addition to foods has been growing at a fast rate in the last few decades due to their specific health benefits. Although knowledge of their health benefits has been widely available for some time, is it only recently that methods to isolate and identify these specific compounds have been developed; scientists have been studying these compounds using modern analytical techniques such as reverse-phase high-performance liquid chromatography (RP-HPLC) and liquid chromatography mass spectrometry (LC-MS), among others. *In vitro* and *in vivo* laboratory experiments on animals as well as several clinical trials have also been conducted to confirm their functionality and safety. Most of the biological activities have been tested *in vitro* adopting different chemical methods or using cell lines as well as *in vivo* and in some clinical trials.

The health benefits and the bioactivities derived from the consumption of milk and dairy products, their individual isolated constituents and derivatives as well as bioactive peptides derived from protein hydrolysis are briefly outlined in the following sections.

1.6.1 *Bioactive peptides*

During protein hydrolysis by proteolytic enzymes, whether by the digestive enzymes, those produced by the microbial cultures added to milk or a combination of both, the nutritional value of the resulting peptides is associated with the amino acids they contain. In addition to their nutritional value, some of these peptides have been shown to have biological or functional activities which could have physiological properties such as antihypertensive (FitzGerald *et al.* 2004; Hernandez-Ledesma *et al.* 2007; Bader & Ganten 2008). The activities of these peptides are exhibited only when they are released from their original precursor or 'parent' protein, where they exist in a dormant or inactive state. These bioactive peptides can vary in length between 3 and 20 amino acids with one or more physiological effects. Several other macropeptides, up to 100 amino acids in length, have also been identified to have physiological function. It has been shown that the short-length peptides can be absorbed into the bloodstream through a specific peptide transporter mechanism in the epithelial wall of the intestine, and undergo no further hydrolysis by the blood serum peptidases. They eventually reach their particular target organ in the body where they perform a specific function. Large-scale production of such peptides is normally carried out using exogenous enzymes to hydrolyse the protein under certain conditions, followed by extraction and purification. Peptides have also been synthesised with good purity, quality and different chain length on a small- or large-scale operation using liquid- or solid-phase peptide synthesis. Such techniques could synthesise up to 200 amino acid peptides.

It has been shown that conjugation or interaction of some peptides with other molecules such as a simple sugar e.g. glucose, xylose or arabinose or an oligosaccharide could induce bioactivity or even enhance it. For example, it was found that glycation of β-lactoglobulin by attaching arabinose or glucose could increase the antioxidant activity (Chevalier *et al.* 2001). Most of the research on the bioactivity of peptides has been on the antioxidant, hypotensive and hypocholesterolemic effects.

All the above-mentioned physiological activities and functions related to the presence of BAC, whether from peptides or other food constituents, have been assessed either *in vitro* using different chemical assays, using cell lines or a model digestive system. A dynamic gastric model (DGM) has been utilised to investigate the exposure of food to different pH values and enzymes, for example: using different amylases and proteinases at pH 7.0 to resemble the action in the mouth; or using pepsin at pH 2.0 (the condition in the stomach) followed by pH 6.0 using pancreatic juice containing enzymes such as those of proteolytic, lypolytic and amylolytic nature. *In vivo* assessment, on the other hand, has been based on animal experiments, clinical trials and human interventions. The physiological activities provided by the various bioactive compounds can be summarised as follows.

Antihypertensive/hypotensive properties

The mechanism of hypertension has been linked to the angiotensin-I-converting enzyme (ACE) which is needed for the rennin angiotensin system. The conversion of angiotensin I to angiotensin II causes the restriction of blood flow, hence the increase in blood pressure. It is known that the inhibition of the ACE action help to normalise blood pressure. Some of these bioactive peptides are derived from milk protein and other protein sources as short peptides. These peptides have been shown to lower the blood pressure *in vitro*, *in vivo*, through animal experiments as in spontaneous hypertensive rats (SHR) and in a few clinical trials (Pihlanto-Leppälä *et al.* 1998; Li *et al.* 2004).

The antihypertensive property of the peptides has been studied by many researchers *in vivo* and *in vitro* on crude casein hydrolysates and whey protein digest (Pihlanto-Leppälä *et al.* 2000; FitzGerald *et al.* 2004). The main peptides which have been linked to this property are the short di- and tripeptides such as those having the amino acid sequence of VY, IPP and VPP. These peptides are mainly produced from the hydrolysis of β-casein and κ-casein producing, for example, fraction 84–86 and fraction 74–76, respectively. However, peptides containing 16–19 amino acids have also been shown to induce an antihypertensive property. The production of such short di- and tripeptides can be achieved by enzymatic hydrolysis or as a result of fermentation. In Japan, the fermented sour milk 'Calpis' has been marketed as containing the tripeptide IPP and VPP. In Europe, a Finnish company also produces fermented milk called 'Evolus' which is marketed as containing ACE inhibitor peptides (Korhonen 2009).

To find out the real effect of these peptides on blood pressure, a meta-analysis of many clinical trials was undertaken by Pripp (2008). In this study, it was concluded that there was a significant reduction in blood pressure when peptides from food proteins were used. In cases of mild to medium hypertension, it was suggested that these peptides could be used as part of a diet, a supplement and/or an alternative to existing pharmaceutical drugs.

The blood-pressure-lowering effect has also been associated with fermented milk products that contain γ-aminobutyric acid (GABA). Inoue *et al.* (2003) showed that feeding fermented milk containing GABA to a group of 39 patients with mild hypertension for 12 weeks lowered the blood pressure significantly within 2–4 weeks.

Antihypercholesterolemic/hypocholesterolemic peptides

Although cholesterol plays an important physiological role in humans, it can also be one of the major factors in heart disease. Its exogenous source is that of food and the rest is produced in the liver. It is present in the blood as free cholesterol or is mostly bound with protein in different forms, the most important forms being the low- and high-density lipoproteins (LDL and HDL). High cholesterol levels in the blood (especially LDL) is one of the factors, among many others, which cause heart disease through atherosclerosis (Stein *et al.* 2005). The pharmaceutical industry has introduced several medications as antihypertensive drugs such as statins which inhibit 3-hydroxy-3-methylglutaryl hydroxymethylglutaryl-CoA

(HMG-CoA) reductase, an enzyme necessary for the formation of endogenous cholesterol. Statins are usually prescribed to people with high cholesterol and at risk from cardiovascular disease.

There are other ways to reduce levels of cholesterol; one of these is through diet including some of the constituents in foods such as plant protein, soluble fibres, plant sterols, probiotics and prebiotics (Taylor & Williams 1998). It has also been shown that some milk protein hydrolysates as well as the small peptides isolated from the hydrolysate, mostly the di- and tripeptides, could also inhibit 3-hydroxy-3-methylglutaryl coenzyme A (HMG-CoA) reductase to exhibit a similar effect to that of statins. Several *in vivo* studies on human subjects demonstrated the hypocholesterolemic effect of fermented milk products (Lewis & Burmeister 2005). It has also been reported that tryptic hydrolysate of β-lactoglobulin prevented cholesterol absorption through direct interaction with the cholesterol (Nagaoka *et al.* 2001).

Antioxidants

The oxidation process in the human body is a natural and important metabolic mechanism within the cell which also results in the formation of reactive oxygen species (ROS) as free radicals. When in excess, these free radicals, such as the hydroxyl (OH*) and peroxyl (OOR*) radicals, could damage and destroy the cell by affecting the lipids, protein and nucleic acid causing disease and pathogenesis. The damage and the apoptosis of the cells could have a major role in the manifestation of some diseases such as Alzheimer's disease, atherosclerosis and inflammation. The body has its own defence system to counteract the effect of these free radicals, by stabilising them through its own endogenous antioxidant system through catalase, glutathione peroxidase and superoxide dismutase. A similar stabilising effect could also take place by other non-enzymatic actions such as: ascorbic acid and α-tocopherol; by some bioactive peptides such as caseinophosphopeptides (Diaz & Decker 2005; Kim *et al.* 2007) or those derived from the hydrolysates of e.g. β-lactoglobulin (Hernandez-Ledesma *et al.* 2007), glutathione and carnosine (Babizhayev *et al.* 1994); and by the presence of some amino acids such as cysteine, histidine and tyrosine.

In some cases, oxidative stress could take place when natural endogenous antioxidants cannot cope with the surge of free radicals as a result of intensive exercise by endurance athletes or any other reason. During any imbalance due to the high level of free radicals and low level of endogenous or exogenous antioxidants, pathogenesis of some diseases will materialise, especially in vulnerable groups such as the elderly showing signs of some diseases such as Alzheimer's, atherosclerosis, cancer, inflammation and Type 2 diabetes (Stadtman 2006). In these circumstances, the intake of supplements or consumption of foods rich in antioxidants could help to minimise such adverse reactions. Regarding dairy products, some casein and whey proteins and their short peptides from their hydrolysates could maintain the correct balance.

Anti-inflammatory effect

Inflammation, whether acute or chronic, has been recognised as the basis of pathogenesis of many diseases such as Alzheimer's, cancer, cardiovascular, diabetes and osteoporosis. It has also been shown that symptoms as diverse as depression, muscle fatigue, tiredness and few more might also be linked, to some degree, to inflammation. Although inflammation is a defensive mechanism against foreign agents entering the body or as a result of internal injury, this action needs to be controlled before causing further damage to the tissues.

The majority of people tend to use the general over-the-counter medication based on non-steroidal anti-inflammatory drugs (NSAID). However, some people rely on diet and/or nutritional supplements as a means to minimise inflammation rather than taking medication. The best-known supplement is the omega-3 fatty acids such as docosahexaenoic acid (DHA) or conjugated linoleic acid (CLA), especially in inflammatory bowel disease (Simopoulos 2002). Other supplements which been demonstrated to have an effect on inflammation include: vitamins such as A, C, D and E; minerals such as Zn, Se and Cr; fruits, vegetables and their extracts such as tomato lycopene; probiotics; and some well-known peptides.

Milk proteins and peptides have been used *in vivo* and *in vitro* to assess their effect as anti-inflammatory agents. The non-heme and iron-binding glycoprotein lactoferrin, along with immunoglobulins, are vital protective proteins in an innate immune mechanism. The role of lactoferrin in regulating and controlling the production of cytokines such as tumour necrosis factor alpha (TNF-α) and interleukin-1β, IL-1β and IL-6 (Crouch *et al.* 1992; Conneely 2001) has also been demonstrated. Glycomacropeptides, fraction 106–169 from κ-casein, and other casein tryptic hydrolysates and proteose peptone-3 also contribute to anti-inflammatory action. There are also some commercially available synthetic anti-inflammatory oligopeptides of about 7–9 amino acids with molecular weight around 1 kDa. Most of these rely on a specific sequence of a few amino acids in a certain position on the peptide chain.

Immunomodulation

The immune system protects the body from infective and foreign agents by destroying them through an effective mechanism based on immunoglobulins and other agents. It was found that certain milk proteins and their hydrolysates could be effective immunomodulators. For example, lactoferrin and whey protein hydrolysate enhances the production of immunoglobulins IgA, IgM and IgG (Miyauchi *et al.* 1997). Similarly, other whey proteins such as α-lactalbumin and β-lactoglobulin hydrolysates containing peptide fractions with 2–10 amino acids could enhance activity and lymphocyte proliferation (Biziulevicius *et al.* 2006; Jacquot *et al.* 2010). The effect of partially hydrolysed β-Casein, fraction 1–28, on the immune modulation was examined for its effect on human T, B and monocytes cell lines proliferation. The study showed that the proliferation of these cell lines was improved. It was also shown that after 96 hours of incubation, IgA production was stimulated (Kawahara *et al.* 2004). In an *in vivo* study, Otani *et al.* (2000) demonstrated that oral administration of diets containing fractions extracted from caseinophosphopeptides, CPP-I hydrolysates such as fraction 1–32 from α-casein and 1–28 from β-casein, enhanced intestinal IgA levels in animals. Similarly, Kitamura & Otani (2002) showed that oral administration of a diet containing CPP-III enhanced fecal IgA level in healthy human subjects.

Most researchers have shown that the physiological effects of the specific peptide fractions are usually associated with a known chain length and amino acid sequence. However, it is interesting to examine the synergistic effect of a mixture of several peptide fractions which could yield better results than the individual fractions.

1.6.2 *Other bioactivities*

Several other bioactivities related to protein and peptides which affect human health are briefly discussed in the following sections.

Ageing

The proportion of the ageing population is increasing due to better health care, nutrition and the improved understanding of the importance of a balanced diet to enhance health. The process of ageing within the human body is still not yet fully understood, and is usually linked to the gradual diminishing of the physiological function of the cell as well as genetics. However, it is still linked to the presence of reactive oxygen species (ROS) and their effect on cellular function which could lead to their malfunction and the manifestation of several diseases such as Alzheimer's, arthritis, cancer, diabetes and Parkinson's. Ageing has been linked to many other factors, one of which is metabolic rate as well as the effect of oxidative stress and lack of antioxidants. It seems clear that taking some supplements such as vitamins, minerals, omega-3, antioxidants, etc. may, to some degree, reduce the rate of ageing. Other factors such as exercise, nutrition and some social factors also have an impact on ageing.

Much research has been published regarding the process of ageing, linked to the shortening of the telomeres which could be limited by the presence of carnosine, a dipeptide of β-alanine and histidine (Nagasawa *et al.* 2001). Some even demonstrated a rejuvenating effect in the cell. More research regarding the assessment and severity of ageing has focused on biomarkers such as telomerase deficiency or overexpression and its link to telomere length homeostasis and cell viability (Bianchi & Shore 2008).

Opioids and satiety

Opioids are bioactive molecules which bind to opioid receptors in the nervous system and gastrointestinal tract and can affect mood, emotion and behaviour. They might have an indirect effect on food intake by inducing satiety. Many opioids can pass through the epithelial membranes of the gastrointestinal tract to the bloodstream and reach the brain and the central nervous system. Peptide-based opioids such as casomorphines are released after enzymatic hydrolysis of the casein. Gauthier & Pouliot (2003) have found that β-lg releases a hydrophobic opioid tetra peptide, fraction 102–105, which could also undergo further interaction with the central cavity of β-lg. Interestingly, β-lg interaction with peptide could also be used as a carrier to deliver peptides which might undergo further hydrolysis by gastric enzymes. Other opioid peptides from protein hydrolysis have been described by Meisel (1996) and Teschemacher *et al.* (1997).

Anticarcinogenic peptides

It appears that the anticarcinogenic properties of physiologically active compounds are mainly related to their indirect effect of enhancing the immune system and, in a few cases, direct effect in shrinking the tumour. Some of these bioactive compounds have been linked to the intact protein such as the role of lactoferrin from whey; it is thought that its binding to the iron in the tissue and reducing its oxidative role has an effect on oral carcinoma cells

(Mohan *et al.* 2006). It has also been found that some synthesised peptides containing 21 and 26 amino acids with two functional domains could facilitate the internalisation of these peptides into the targeted cancerous cells to induce apoptosis (Ellerby *et al.* 1999).

Wound healing

There are many bioactive components in food which could enhance wound healing. Some of these studies have been based on using the entire or whole food such as honey, and observing the rate of wound recovery. Milk protein and especially peptides derived from whey protein have also been shown to have a similar effect in wound healing. In a study by Wang *et al.* (2010), they found that intragastric administration of whey protein to rats improved wound healing compared to the control.

Antimicrobial properties

Some of the naturally released peptides by the gastrointestinal mucosa or the epithelial cells of the intestine, including defensins, bind to the negatively charged phospholipids on the microbial cell membrane and cause their destruction (Ramasundara *et al.* 2009). However, there is a fine balance between these and the probiotics present in the GIT (Jager *et al.* 2010). As the resistance of some bacteria to antimicrobial agents increases, including those based on naturally produced peptides, the emphasis is now on the synthesis of some peptides as anti-infective agents.

1.7 Probiotics and dairy products

The human gut contains a huge number (estimated to be over 100 billion) of micro-organisms, referred to as microbiota/microflora. These micro-organisms are normally divided into three categories: the beneficial, termed as probiotic bacteria; the harmful pathogenic; and the commensal bacteria with no harmful effect. The balance is naturally kept under control in the human gut in healthy individuals. During illness and infection however, as in the case of diarrhoea, the balanced is tipped where the number of the harmful bacteria exceeds those of the beneficial bacteria. The use of antibiotics is then necessary to halt the effect of the harmful bacteria to restore the balance. It has been recognised however that the use of the antibiotic itself can inflict diarrhoea unrelated to pathogens, referred to as 'antibiotic-associated diarrhoea' (Hampel *et al.* 2012). Some of the probiotics such as *Lactobacillus rhamnosus* GG, *Lb acidophilus* and *Lb reuteri* have been used in the case of travellers or antibiotics-related diarrhoea (Katelaris *et al.* 1995; Vanderhoof *et al.* 1999).

The main probiotic bacteria that have been studied and used in fermented dairy products and other foods are members of *Lactobacillus* spp and *Bifidobacterium* spp (Sanders 2000). To derive their health benefits, it was suggested they should be present in these products at concentrations of more than 10^6 cfu g^{-1} or 10^6 cfu mL^{-1} of a product (Shah 2000). There are several factors that affect the viability and survival of the probiotics, some of which have been attributed to the storage temperature, pH and the type of acids added or produced such as acetic acid or ascorbic acid and the level of oxygen (Dave & Shah 1997; Kailasapathy & Chin 2000).

The beneficial effects of probiotics taken in fermented foods materialise after their ingestion and survival within the gastrointestinal tract of the host, where they colonise the GIT and prevent the adherence of the pathogenic bacteria. They can also have an indirect effect through the production of some bioactive compounds such as antimicrobial and immunomodulating agents as described in Section 1.6.1. The antimicrobial agents can be isolated and applied to many dairy products, such as cheese, to improve their shelf life.

Fermented dairy products such as drinking yogurt has dominated the probiotic market and been regarded as the main source of probiotic intake (Tamime *et al.* 1995). Their inclusion into other products has been limited to a few other dairy products, for example, ice-cream and cheese. Micro-encapsulation technology has made their use is more varied products possible, where some heat processing or unfavourable conditions exist for example (Shah & Ravula 2000).

1.8 Summary and future trends

Milk and dairy products are an essential part of a healthy diet. They have nutritional as well as physiological functions which need to be explored further to improve human health. In the past few decades there has been considerable progress in the understanding of many biochemical and analytical techniques which has enabled researchers to better interpret the research data. Advances in bioinformatics and the emergence of 'omics' technology has created a huge interest in the field; progress in genomics, proteomics and peptidomics has created many discoveries that could link all these technologies together to understand the role of food or its bioactive components to improve human health. Computer-based docking has been used for the discovery of new interactions such as in the creation of novel peptides with specific amino acid sequences to perform certain physiological functions, which could be confirmed experimentally. More research should be undertaken to verify and confirm the physiological functions in human intervention trials, important to minimise the use of drugs (especially those with unpleasant side-effects). Further research is also needed on the role of bioactive compounds, especially the bioactive peptides, and how to transfer the technology to produce supplements or new products for the consumer.

References

Arai, S., Moringa, Y., Yoshikawa, T., Ichiishi, E., Kiso, Y., Yamazaki, M., Morotimo, M., Shimizu, M., Kuwata, T., & Kaminogawa, S. (2002). Recent trends in functional food science and industry in Japan. *Bioscience Biotechnology & Biochemistry* **66**, 2017–2029.

Babizhayev, M.A., Seguin, M.C., Gueyne, J., Evstigneeva, R.P., Ageyeva, E.A. & Zheltukhina, G.A. (1994) L-Carnosine (β-alanyl-L-hystidine and carcinine (β-alanylhistamine) act as natural antioxidants with hydroxylradical-scavenging and lipid-peroxidaseactivities. *Biochemical Journal*, **304**, 509–516.

Bader, M. & Ganten, D. (2008) Update on tissue rennin-angiotensinsystems. *Journal of Molecular Medicine*, **86**, 615–621.

Bauman, D.E., Lock, A.L., Corl, B.A., Salter, A.M. & Parodi, P.W. (2006) Milk fatty acids and human health: potential role of conjugated linoleic acid and *trans* fatty acids. *Ruminant Physiology: Digestion, Metabolism and Impact of Nutrition on Gene Expression, Immunology and Stress* (eds K. Sejrsen, T. Hvelplund & M.O. Nielsen), pp. 529–561, Wageningen Academic, The Netherlands.

Bianchi, A. & Shore, D. (2008) How telomerase reachesits end: mechanism of telomerase regulation by the telomeric complex. *Molecular Cell*, **31**, 153–165.

Biziulevicius, G.A., Kislukhina, O.V., Kazlauskaite, J. & Zukaite, V. (2006) Food-protein enzymatic hydrolysates possess both antimicrobial and immunomodulatory activities: a 'cause and effect' theory of bifunctionality. *FEMS Immunology and Medical Microbiology*, **46**, 131–138.

Booth, S., Tucker, K., Chen, H., Hannan, M., Gagnon, D., Cupples, L., Wilson, P., Ordovas, J., Schaefer, E., Dawson-Hughes, B. & Kiel, D. (2000) Dietary vitamin K intakes are associated with hip fracture but not with bone mineral density in elderly men and women. *American Journal of Clinical Nutrition*, **71**, 1201–1208.

Chevalier, F., Chobert, J.M., Genot, C. & Haertle, T. (2001) Scavenging of free radicals, antimicrobial, and cytotoxic activities of the Maillard reaction products of beta lactoglobulin glycated with several sugars. *Journal of Agricultural and Food Chemistry*, **49**(10), 5031–5038.

Conneely, O.M. (2001) Antiinflammatory activities of lactoferrin. *Journal of the American College of Nutrition*, **20**(5), 389S–395S.

Crichton, G.E., Elias, M.F., Dore, G.A., Abhayaratna, W.P. & Robbins, M.A. (2012) Relations between dairy food intake and arterial stiffness: pulse wave velocity and pulse pressure. *Hypertension*, **59**(5), 1044–1050.

Crouch, S.P., Slater, K.J. & Fletcher, J. (1992) Regulation of cytokine release from mononuclear cells by the iron binding protein lactoferrin. *Blood*, **80**(1), 235–240.

Dave, R.I. & Shah, N.P. (1997) Viavility of yogurt and probiotic bacteria in yogurts made from commercial starter cultures. *International Dairy Journal*, **7**, 31–41.

Diaz, M. & Decker, E.A. (2005) Antioxidant mechanism of caseinophosphopeptides and casein hydrolysates and their application in ground beef. *Journal of Agricultural and Food Chemistry*, **52**, 8208–8213.

Ellerby, H.M., Arap, W., Ellerby, L.M., Kain, R., Andrusiak, R., Rio, G.D., Krajewski, S., Lombardo, C.R., Rao, R., Ruoslahti, E., Brendesen, D.E. & Pasqualini, R. (1999) Anticancer activity of targeted pro-apoptotic peptides. *Nature Medicine*, **5**(9), 1032–1038.

Elwood, P.C., Pickering, J.E. & Fehily, A.M. (2007) Evidence-based public health policy and practice. Milk and dairy consumption, diabetes and the metabolic syndrome: the Caerphilly prospective study. *Journal of Epidemiology Community Health*, **61**, 695–698.

FitzGerald, J.R., Murray, A.B. & Walsh, J.D. (2004) Hypotensive peptides from milk proteins. *Journal of Nutrition*, **134**, 980–988.

FSA (2007) FSA nutrient and food based guidelines for UK institutions. Available at www.food.gov.uk (accessed October 2013).

Fukuda, K., Yamamoto, A., Ganzorig, K., Khukhenbattar, J., Senda, A., Saito, T. & Urashima, T. (2010) Chemical characterization of the oligosaccharides in bacterian camel (Camelus bacterianus) milk and colostrum. *Journal of Dairy Science*, **93**(12), 5572–5587.

Gallagher, J., Kanekanian, A.D. & Evans, P.E. (1993) Debittering of alpha casein hydrolysates by fungal peptidase. In *Biochemistry of Milk Products* (eds. Andrews, A.T. & Varley, J.M.), pp. 143–151. Proceedings of the Royal Society of Chemistry, Reading University, Reading, 143–151.

Gallagher, J., Kanekanian, A.D. & Evans, P.E. (1994) Hydrolysis of casein: A comparative of two proteases and their peptides. *International Journal of Food Science and Technology*, **29**(3), 279–285.

Gauthier, S.F. & Pouliot, Y. (2003) Functional and biological properties of peptides obtained by enzymatic hydrolysis of whey proteins. *Journal of Dairy Science*, **86**, E78–E87.

Gregory, J. (2000) *National Diet and Nutrition Survey: Young People Aged 4–18 Years*. The Stationary Office, London.

Hampel, S., Newberry, S.J., Maher, A.R., Wang, Z., Miles, J.N., Shanman, R., Johnsen, B. & Shekelle, P.G. (2012) Probiotics for the prevention and treatment of antibiotic-associated diarrhoea: a systematic review and meta-analysis. *JAMA Journal*, **307**(18), 1959–69.

Hartmann, R. & Meisel, H. (2007) Food–derived peptides with biological activity: From research to food applications. *Current Opinion in Biotechnology*, **18**(2), 136–169.

Hernandez-Ledesma, B., Amigo, L., Recio, I. & Bartolome, B. (2007) ACE-inhibitory and radical-scavenging activity of peptidesderived from β-lactoglobulin f(19–25). Interactions with ascorbic acid. *Journal of Agricultural and Food Chemistry*, **55**, 3392–3397.

Howard, B.V., van Horn, L., Hsia, J., Manson, J.E. & Stefanick, M.L. (2006) Low fat dietary pattern and risk of cardiovascular disease: The women's health initiative randomised controlled dietary modification trial. *Journal of American Medical Association*, **295**, 655–666.

Howlett, J. (2008) *Functional Foods from Science to Health and Claims*. Washington DC: International Life Sciences Institute. Europe Concise Monograph Series.

Inoue, K., Shirai, T., Ochiai, M., Hayakawa, K. & Sansawa, H. (2003) Blood pressure lowering effect of a novel fermented milk containing γ-aminobutyric acid (GABA) in mild hypertensives. *European Journal of Clinical Nutrition*, **57**, 490–495.

Jacquot, A., Gauthier, S.F., Drouin, R. & Boutin, Y. (2010) Proliferative effects of synthetic peptides from β-lactoglobulin and α-lactalbumin on murine splenocytes. *International Dairy Journal*, **20**, 514–521.

Jager, S., Stange, E.F. & Wehkamp, J (2010) Antimicrobial peptides in gastrointestinal inflammation. *International Journal of Inflammation*, 2010, article ID 910283.

Kailasapathy, K. & Chin, J. (2000) Survival and therapeutic potential of probioticorganisms with reference *to Lactobacillus acidophilus* and *Bifidobacterium* spp. *Immunology and Cell Biology*, **78**, 80–88.

Katelaris, P.H., Salam, I. & Farthing, M.J. (1995) Lactobacilli to prevent traveler's diarrhoea? *New England Journal of Medicine*, **333**(20), 1360–1361.

Kawahara, T., Katayama, D. & Otani, H. (2004) Effect of β-Casein (1–28) on prolifrative responses and secretory functions of human immunocompetent cell lines. *Bioscience, Biotechnology and Biochemistry*, **68**(10), 2091–2095.

Kim, G.N., Jang, H.D. & Kim, C.I. (2007) Antioxidant capacity of caseinophosphopeptides prepared from sodium caseinate using alcalase. *Food Chemistry*, **104**, 1359–1565.

Kitamura, H. & Otani, H. (2002) Fecal IgA levels in healthy persons who ingested cakes with or without bovine casein phosphopeptides. *Milchwissenschaft*, **57**(11), 611–614.

Korhonen, H. (2009) Milk-derived bioactive peptides: from science to applications. *Journal of Functional Foods*, **1**, 177–187.

Korhonen, H. & Pihlanto, A. (2007) Technological options for the production of health promoting proteins and peptides derived from milk and colostrum. *Current Pharmaceutical Design*, **13**(8), 829–843.

Kunz, C., Rudloff, S., Schad, W. & Braun, D. (1999) Lactose-derived oligosaccharidesin milk in elephant: comparison to human milk. *British Journal of Nutrition*, **82**, 391–399.

Kunz, C., Rudloff, S. & Michaelson, K (2002) Concentration of human milk oligosaccharides during late lactation. *FASEB Journal*, **15**, A986.

Lewis, S.J. & Burmeister, S. (2005) A double blind placebo controlled study of the effects of *Lactobacillus acidophilus* on plasma lipids. *European Journal of Clinical Nutrition*, **59**, 776–780.

Li, G.H., Le, G.W., Shi, Y.H. & Shrestha, S. (2004) Angiotensin I converting enzyme inhibitory peptides derived from food proteins and their physiological and pharmacological effects. *Nutrition Research*, **24**, 469–486.

Livingstone, K.M., Lovegrove, J.A., Cockcroft, J.R., Elwood, P.C., Pickering, J.E. & Givens, I.G. (2012) Does dairy food intake predict arterial stiffness and blood pressure in men? Evidence from the Caerphilly Prospective Study. *Hypertension*, **61**, 42–47.

Lock, A.L. & Bauman, D.E. (2004) Modifying milk fat composition of dairy cows to enhance fattyacids beneficial to human health. *Lipids*, **39**, 1197–1206.

Meisel, H. (1996) Chemical characterization and opioid activity of an oxorphin isolated from in vivo digest of casein. *FEBS Letters*, **399**, 196–223.

Milaneschi, Y., Shardell, M., Corsi, A.M., Vazzana, R., Bandinelli, S., Guralnik, J.M. & Ferrucci, L. (2010) Serum 25-hydroxyvitamin D and depressive symptoms in older women and men. *Journal of Clinical Endocrinology and Metabolism*, **95**(7), 3225.

Minagawa, E., Kaminogawa, S., Tsukasaki, F. & Yamauchi, K. (1989) Debittering mechanism in bitter peptides of enzymatic hydrolysates from milk casein by aminopeptidase T. *Journal of Food Science*, **54**(5), 1225–1229.

Miyauchi, H., Kaino, A., Shinoda, I., Fukuwatari, Y. & Hayasawa, H. (1997) Immunomodulatory effect of bovine hydrolysate on murine splenocytes and Peyer's cells. *Journal of Dairy Science*, **80**, 2330–2339.

Mohan, K.V., Gunasekaran, P., Varalakshmi, E., Hara, Y. & Nagini, S. (2006) In vitro evaluation of the anticancer effect of lactoferrin and tea polyphenol combination on oral carcinoma cells. *Cell Biology International*, **31**(6), 599–608.

Motard-Belanger, A., Charest, A., Grenier, G., Paquin, P. & Chouinard, Y. (2008) Study of the effect of *trans* fatty acids from ruminants on blood lipids and other risk factors for cardiovascular disease. *American Journal of Clinical Nutrition*, **87**, 593–599.

Nagaoka, S., Futamura, Y., Miwa, K., Awano, T., Yamauchi, K., Kanamaru, Y., Tadashi, K. & Kuwata, T. (2001) Identification of novel hypocholesterolemic peptides derived from bovine milk β-lactoglobulin. *Biochemical and Biophysical Research Communications*, **281**, 11–17.

Nagasawa, T., Yonekura, T., Nishizawa, N. & Kitts, D.D. (2001) In vitro and in vivo inhibition of musclelipid and protein oxidation by carnosine. *Molecular and Cellular Biochemistry*, **225**, 29–34.

Newburg, D.S. (1996) Oligosaccharides and glycoconjugates in human milk: their role in host defense. *Journal of Mammary Gland Biology*, **1**, 271–283.

NHS (2011) Key statistics on the NHS. Available at www.nhsconfed.org (accessed October 2013).

NRC (1996) *Carcinogenesis and Anti-Carcinogenesis in the Human Diet*. The National Academy of Sciences, National Academy Press, Washington DC.

Otani, H., Kihara, T. & Park, M. (2000) The immunoenhancing property of dietary casein phosphopeptide preparation in mice. *Food and Agricultural Immunology*, **12**, 165–173.

Park, H.S., Ryu, J.H., Ha, Y.L. & Park, J.H. (2001) Dietary conjugated linoleic acid (CLA) induces apoptosis of colonic mucosa in 1,2-dimethylhydrazine-treated rats: a possible mechanism of the anticarcinogenic effect by CLA. *British Journal of Nutrition*, **86**(5), 549–555.

Parodi, P.W. (2009) Has the association between saturated fatty acids, serum cholesterol and coronary heart disease been over emphasised? *International Dairy Journal*, **19**, 345–361.

Petyaev, A.A. & Bashmakov, Y.K. (2012) Could cheese be the missing piece in the French paradox puzzle? *Medical Hypotheses*, **79**, 746–749.

Phillips, S.M., Hartman, J.W. & Wilkinson, S.B. (2005) Dietary protein to support anabolism with resistance exercise in young men. *Journal of American College of Nutrition*, **24**(2), 134S–139S.

Pihlanto-Leppälä, A., Rokka, T. & Korhonen, H. (1998) Angiotensin I converting enzyme inhihitory peptides derived from bovine milk proteins. *International Dairy Journal*, **8**, 325–331.

Pihlanto-Leppälä, A., Koskinien, P., Piilola, K., Tupasela, T. & Korhonen, H. (2000) Angiotensin I converting enzyme inhihitory properties of whey protein digest:concentration and characterization of active peptides. *Journal of Dairy Research*, **67**, 53–64.

Pripp, A.P. (2008) Effect of peptides from food proteins on blood pressure: a meta-analysis of randomized controlled trials. *Food and Nutrition*, doi: 10.3402/fnr.v52i0.1641.

Ramasundara, M., Leach, S.T., Lemberg, D.A. & Day, A.S. (2009) Defensins and inflammation: the role of defensins in inflammatory bowel disease. *Journal of Gastroenterology and Hepatology*, **24**(2), 202–208.

Sanders, M.E. (2000) Consideration for use of probiotic bacteria to modulate human health. *Journal of Nutrition*, **130**, S384–S390.

Shah, N.P. (2000) Probiotic bacteria: selective enumeration and survival in dairy foods. *Journal of Dairy Science*, **83**, 894–907.

Shah, N.P. & Ravula, R. (2000) Microencapsulation of probiotic bacteriaand their survival in frozen fermented dairy desserts. *Australian Journal of Dairy Technology*, **55**, 139–144.

Shimizu, T. (2003) Health claims on functional foods: the Japanese regulation and an international comparison. *Nutrition Research Reviews*, **16**, 241–252.

Simopoulos, A.P. (2002) Omega-3 fatty acids in inflammation and autoimmune diseases. *Journal of American College of Nutrition*, **21**(6), 495–505.

Siri-Tarino, P.W., Sun, Q., Hu, F.B. & Krauss, R.M. (2010) Meta-analysis of prospective cohort studies evaluating the association of saturated fat with cardiovascular disease. *American Journal of Clinical Nutrition*, **91**, 535–546.

Stadtman, E.R. (2006) Protein oxidation and aging. *Free Radical Research*, **40**, 1250–1258.

Stein, E.A., Sniederman, A. & Laskkarzewski, P. (2005) Assessment of reaching goal in patients with combined hyperlipidemia: low density lipoprotein cholesterol, non high density lipoprotein cholesterol or apolipoprotein A. *American Journal of Cardiology*, **96**, 36K–43K.

Szulc, P., Chapuy, M.C., Meunier, P.J. & Delmas, P.D. (1996) Serum undercarboxylated osteocalcin is a marker of risk of hip fracture: A three follow-up study. *Bone*, **18**(5), 487–488.

Tamime, A.Y., Marshall, V.M.E. & Robinson, R.K. (1995) Microbiological and technological aspects of fermented milks by bifidobacteria. *Journal of Dairy Research*, **62**, 151–187.

Taubes, G. (2001) The soft science of dietary fat. *Science*, **291**, 2535–2541.

Taylor, G.R. & Williams, C.M. (1998) Effects of probiotics and prebiotics on blood lipids. *British Journal of Nutrition*, **80**, S225–S230.

Teschemacher, H., Kock, G. & Brantl, V. (1997) Milk protein derived opioid receptor ligands. *Peptide Science*, **43**(2), 99–117.

Urushima, T., Saito, T., Nakamura, T. & Messer, M (2001) Oligosaccharides of milk and colostrum in non-human mammals. *Glycocongugate Journal*, **18**, 357–371.

USDA (2005) *Dietary Guidelines for Americans*. USDA, Washington DC. Available at: http://www.health.gov/dietaryguidelines/ (accessed October 2013).

Vanderhoof, J.A., Whitney, D.B., Antonson, D.L., Hanner, T.L., Lupo, J.V. & Young, R.J. (1999) *Lactobacillus* GG in the prevention of antibiotic-associated diarrhoea in children. *Journal of Pediatrics*, **135**, 564–568.

Veermer, C., Jie, K. & Knapen, M. (1995) Role of vitamin K in bone metabolism. *Annual Review of Nutrition*, **15**, 1–22.

Venema, K. (2012) Intestinal fermentation of lactose and prebiotic lactose derivatives, including human milk oligosaccharides. *International Dairy Journal*, **22**, 123–140.

Vergari, A., Tibuzzi, A. & Basile, G. (2010) An overview of the functional food market: from marketing issues & commercial players to future demand from life in space. *Exp Med Biol* **698**, 308–21.

Vesa, T.H., Marteau, P. & Korpela, R. (2000) Lactose intolerance. *Journal of American College of Nutrition*, **19**(2), 165S–175S.

Wang, J., Zhao, M., Liang, R., Zhang, Z., Zhao, H., Zhang, J., Li, T. & Li, Y. (2010) Whey peptides improve wound healing following caesarean section in rats. *British Journal of Nutrition*, **104**(11), 1621–1627

Zulet, M.A., Marti, A., Parra, M.D. & Martinez, J.A. (2005) Inflammation and conjugated linoleic acid: mechanism of action and implications for human health. *Journal of Physiology and Biochemistry*, **61**(3), 483–94.

2 Bioactive Peptides from Casein and Whey Proteins

R. McCarthy[1], S. Mills[1], R.P. Ross[1], G.F. Fitzgerald[2] and C. Stanton[1]

[1]Teagasc, Moorepark Food Research Centre, Fermoy, Co. Cork, Ireland
[2]Department of Microbiology, University College Cork, Cork, Ireland

2.1 Introduction

Milk is a highly complex and nutritious food which has evolved to meet the nutritional demands of the young mammal. While the macromolecules of milk have received much attention, science has revealed that milk also encodes various other molecules which exhibit a wide array of health effects. This is particularly evident for the protein fraction of milk. Bovine milk is composed of approximately 5% lactose, 3.2% protein, 4% lipid and 0.7% mineral salts (Séverin & Wenshui 2005). Casein and whey proteins constitute the protein fraction, where casein (αs1 , αs2-, β- and κ-casein) accounts for approximately 78% and whey proteins (β-lactoglobulin, α-lactalbumin, serum albumin and immunoglobulins), lactoferrin, lactoperoxidase and lysozyme comprise *c*. 17% of the total protein. While casein and whey proteins themselves serve specific biological requirements, these proteins also contain discrete peptide sequences which, when released through proteolysis, can exhibit unique biological activities (Benkerroum 2010; Hartmann & Meisel 2007; Madureira *et al*. 2010). Indeed, it has been reported that bovine milk, cheese and dairy products are among the greatest sources of bioactive proteins and peptides from any food source (Moller *et al*. 2008). In many cases, the released bioactive peptides interact with appropriate mammalian receptors and hence exhibit hormone-like activity (Dziuba & Darewicz 2007), while others interact directly with micro-organisms, exhibiting antimicrobial activity. In general, bioactive peptides range in size from 3 to 20 amino acids (Pihlanto 2001), although they can be larger in some cases. The activity of each bioactive peptide is dependent on the amino acid composition and sequence (Shahidi & Zhong 2008). Release of the peptides from the native protein can be achieved through the action of digestive proteases (Clare & Swaisgood 2000), or through fermentation with proteolytic bacteria (Hayes *et al*. 2006; Mills *et al*. 2009). These peptides may also be released during transit through the human gastrointestinal tract (GIT) by human proteolytic enzymes.

Milk and Dairy Products as Functional Foods, First Edition. Edited by A. Kanekanian.

Perhaps the most renowned bioactive peptides are those which exhibit antihypertensive activity by inhibiting a key enzyme involved in the regulation of blood pressure, namely angiotensin-I-converting enzyme (ACE). Indeed, ACE-inhibitory peptides from casein are already commercially available in fermented food products such as Calpis® (Calpis, Co. Ltd., Tokyo) and Evolus® (Valio, Finland). In both cases, the bioactive peptides consist of the casein-derived amino acid sequences V-P-P (Hamel *et al.* 1985; Juillard *et al.* 1995; Clare *et al.* 2003) and I-P-P (Drouet *et al.* 1990; Chabance *et al.* 1995). Milk proteins also contain a suite of peptide sequences which interfere with the formation of thrombi, known as the antithrombotic peptides. Interestingly, the most potent food-derived antithrombotic peptides have been released from κ-casein; Jolles *et al.* (1978) hypothesised that human fibrinogen γ-chain (involved in blood clotting) and κ-casein may have evolved from a common ancestor since they share a degree of structural similarity. A number of other bioactivities are also inscribed in the peptide sequences of casein and whey. The casein phosphopeptides (CPPs) are phosphorylated peptides which have a high content of negative charge and can therefore bind divalent cations. CPPs have been shown to prevent enamel demineralisation and exhibit anticariogenic activity (Grenby *et al.* 2001; Aimutis 2004) and are available commercially in products such as mineral drinks, nutritional supplements for children, confectionary and products for dental care (Cross *et al.* 2007; Luo & Wong 2007). Cytomodulatory peptides have also been identified, particularly from a variety of fermented dairy products (Hayes *et al.* 2007; Parodi 2007), suggesting that they are released by bacterial proteolytic digestion. These peptides exert activity by affecting the viability of cancer cells. For example, a significant antiproliferative effect was observed on Caco-2 cells following treatment with whey peptides derived from the waste stream of Mozzarella di Bufala Campana cheese production (De Simone *et al.* 2009). Casein and whey proteins have also proven to be a rich source of opioid peptides which exhibit pharmacological properties similar to opium. The major opioid peptides are derived from β-casein, and are hence called β-casomorphins (Teschemacher *et al.* 1997; Clare & Swaisgood 2000). Whey proteins such as lactoferrin, β-lactoglobulin and bovine serum albumin also contain opioid peptides encrypted within the primary sequence (Rokka *et al.* 1997; Belem *et al.* 1999), which have been associated with several biological activities ranging from antihypertensive to immunomodulatory, antidepressant to antisecretory and antidiarrhoeal (Pihlanto 2001).

Bioactive peptides derived from milk proteins have the potential to impact on the immune response through both direct and indirect mechanisms. Considering that correct development of the immune system is perhaps one of the most fundamental biological elements to long-term health, such peptides may have a crucial role to play in staving off chronic diseases including allergy, inflammation, autoimmune diseases and even cancer. Dietert & Zelikoff (2009) reported that in some developed countries, immune-based diseases affect as many as 25% of children. As a consequence, strategies with the potential to alleviate or prevent immune dysfunction are worthy of considerable attention. The aim of this review is to highlight the accumulating evidence on bioactive peptides with potential to impact positively on the immune system, derived from casein and whey proteins. Many of these peptides have been shown to interact directly with cells of the immune system and are referred to as immunomodulatory peptides; others act indirectly, for example by exhibiting antimicrobial activities, stimulating the growth of specific beneficial gut bacteria and regulating gut homeostasis through mucin production, all of which have dramatic consequences for the immune response. In this review, we also consider strategies for the successful delivery of bioactive peptides to the gut, particularly the infant whose sole source of nutrition in the earliest stages of life is milk.

2.2 Direct effects of casein and whey peptides on host immunity

Certain peptides derived from the casein and whey fractions of milk have demonstrated a capacity to interact with various components of the immune system, suggesting positive consequences following ingestion. Many of these peptides have been released through the action of commercial proteases or through fermentation with proteolytic bacteria, but others have been released via the action of gastric enzymes, simulating conditions in the human GIT. Collectively, such immunomodulatory effects have been identified in both the innate or non-specific immune system and also the acquired or specific immune system, and include lymphocyte activity/function, cytokine production, antibody-forming cell counts (AFCs), maturation of the immune system, phagocytic/granulocyte activity and allergic tolerances, among others (Hachimura *et al.* 1993; Brody 2000; Li & Mine 2004; Fischer *et al.* 2006; Jacquot *et al.* 2010). In many studies, the exact peptides exhibiting such immunomodulatory effects have been identified (Table 2.1) and these range in size from 2 to 64 amino acids. In other studies however, the hydrolysates and peptide mixes extracted from hydrolysates of milk proteins, but not the intact proteins themselves, have been associated with immunomodulatory activity. The examples cited in the following sections demonstrate how specific peptides or hydrolysates from casein- and whey can influence immune responses.

2.2.1 *Immunomodulatory peptides derived from casein*

β-casein

β-casein accounts for *c.* 42 % of total casein in bovine milk (Ginger & Grigor 1999). Peptides derived from β-casein have been shown to influence a wide array of immune functions including lymphocyte proliferation, phagocytic activity of macrophages, cytokine levels and antibody synthesis. For example, the β-casein-derived peptide fragment (f) (193–209) has been shown to affect macrophage activity following co-incubation with bone-marrow-derived macrophages from germ-free and human gut microbiota-associated mice (Sandre *et al.* 2001). This peptide was also shown to increase major histocompatabilty complex II (MHC II) expression and increase phagocytotic activity in murine bone-marrow-derived macrophages. Indeed, a greater increase in MHC II expression and phagocytic activity was observed in cells following stimulation with the f(193–209) peptide compared with lipopolysaccharide- (LPS-) stimulated cells (Sandre *et al.* 2001). A phosphopeptide derived from β-casein f(1–28) has also been shown to alter spleen lymphocyte status by direct interaction with cell surface proteins. Indeed, Tobita *et al.* (2006) demonstrated that this phosphopeptide could stimulate the proliferation of CD19+ spleenocytes by interacting with toll-like receptor (TLR-) 4 and, upon stimulation via TLR-4, cytokine production was also affected. Furthermore, following stimulation with the

Table 2.1 Immunomodulatory peptides derived from casein and whey proteins.

Bioactive peptide	Source	Immunomodulatory effect	Reference	Sequence (where available)
Immunomodulatory peptides derived from casein				
f(193–209)	β-casein	Stimulate macrophage activity; up-regulate MHC II expression	Sandre *et al.* (2001)	YGEGEPVLGPVRGPFPIIV
f(1–28)	β-casein	Interacts with TLR-4; stimulate CD19+ spleenocytes	Tobita *et al.* (2006)	MKVLILACLVALALARELELNVPGEIV
f(145–160) f(145–154) f(143–154) f(192–202)	β-casein	Stimulate TNF-α and IL-6	Tellez *et al.* (2010)	HQPHQPLPPTVMFPPQ HQPHQPLPPT WMHQPHQPLPPT LYQEPVLGPVR
f(54–59) f(54–59 with modified N terminus	β-casein β-casein	Immunostimulant Immunosuppressive	Puri *et al.* (2009)	VEPIPY VDEGIPYVEDPIPY
β-casomorphin-7 β-casokinin-10	β-casein	Low concentrations: suppress lymphocyte proliferation; high concentrations: stimulate lymphocyte proliferation	Kayser & Meisel (1996)	YPFPGPIYQQPVLGPVR
κ-caseino glycomacropeptide (GMP)	κ-casein	Up-regulate IL-1ra: prevention of IL-1 induction of IL-1 induction of Lymphocyte proliferation; down-regulate IL-17 and TNF; anti-inflammatory	Daddaoua *et al.* (2005); Brody (2000); Monnai & Otani (1997); Otani & Hata (1995); Requena *et al.* (2008)	64-aa structure
CMP hydrolysates	κ-casein	Induce phagocytic activity in U937 cells		
Isracidin (1–23)	αs1-casein	Increase antibody forming cell numbers; immunostimulant; up-regulate inflammatory responces (non-anaphylactic)	Lahov & Regelson (1996)	RPKHPIKHQGLPQEVLNEN LLRF

Table 2.1 *(Continued)*

Bioactive peptide	Source	Immunomodulatory effect	Reference	Sequence (where available)
α-casein hydrolysates	α-casein	Oral tolerance Stress relieving	Peng *et al.* (2005) Kim *et al.* (2006)	
Immunomodulatory peptides derived from whey				
f(15-20) f(55-60) f(139–148)	β-lactoglobulin	Induce spleenocyte proliferation, alter cytokine expression	Jacquot *et al.* (2010)	(K)bVAGTWY (L)EILLQK (K)ALKALPMHIR
f(1–8)	B-lactoglobulin	Cytotoxic	Jacquot *et al.* (2010)	LIVTQTMK
α-lactalbumin hydrolysates	WPI	Enhance macrophage activity	Biziulevicius *et al.* (2006)	
f(18–19) f(50–51) f(18–20) f(10–16)	α-lactalbumin	Enhance human lymphocyte proliferation	Kayser & Meisel (1996); Jacquot *et al.* (2010); Berthou *et al.* (1987)	YG YG YGG RELKDLK
f(51–53)	α-lactalbumin	Up-regulate phagocyte activity	Berthou *et al.* (1987); Gauthier *et al.* (2006)	GLL
f(104–108)	α-lactalbumin	Cytotoxic	Jacquot *et al.* (2010)	WLAHK
Bovine lactoferrin hydrolysate	Bovine lactoferrin	Increase spleenocyte proliferation; increased antibody production (IgA IgM and IgG)	Miyauchi *et al.* (1997)	
Lactoferricin f(17–42)	Bovine lactoferrin	Up-regulate IL-8 production	Shinoda *et al.* (1996)	FKCRRWQWRMKK- LGAPSITCVRRAF

peptide, up-regulations of interleukin- (IL-) 6, IL-10 and tumour necrosis factor (TNF)-α (which in turn serve to modulate the immune system further) were also observed. In contrast to these findings, bioactive peptides/hydrolysates resulting from the digestion of β-caseins with pepsin and trypsin or after fermentation with *Lactobacillus casei* GG (*Lactobacillus rhamnosus* GG) were shown to consistently infer suppressive effects on the proliferation of lymphocytes at varying concentrations (Sutas *et al.* 1996).

Recently, β-casein-derived peptides from *Lactobacillus helveticus*-fermented milk have also been shown to alter cytokine expression in human mononuclear cells and murine macrophages (Laffineur *et al.* 1996; Tellez *et al.* 2010). Specifically, peptide fraction (F5) identified as containing β-casein f(145–160), f(145–154), f(143–154) and f(192–202) was shown to induce TNF-α and IL-6 (Tellez *et al.* 2010). Interestingly, TNF-α levels following exposures equalled those observed after LPS stimulation (Tellez *et al.* 2010). In contrast, modification at the N-terminus sequence of β-casein peptide f(54–59) (a previously identified potent immunostimulant) has been shown to infer significant immunosuppressive activity in murine models (Puri *et al.* 2009). The modifications in amino acid sequence included the insertion of an aspartic acid residue and the substitution of a proline for a glycine residue, resulting in two immunosuppressive derivatives. In animal studies, the modified β-casein f(54–59) peptides suppressed both cell-mediated immune responses and also humoral immune responses following challenge with sheep erythrocytes. These peptides were also shown to inhibit interferon gamma production (IFN-γ), a cytokine which affects a broad range of immune and indeed other physiological functions (such as up-regulation of pathogen recognition and reduction in cell proliferation; Schroder *et al.* 2004) and increases IL-4 and IL-10 production in skin graft studies in mice (Puri *et al.* 2009). Such immunomodulatory fractions have been proposed as potential alternatives to immunosuppressive compounds such as cycloporins (Puri *et al.* 2009).

Many peptides derived from β-casein, such as ACE-inhibitory and opioid peptides have demonstrated multifunctional properties. Indeed, *in vitro* studies with the opioid peptide β-casomorphin-7 and the ACE inhibitor β-casokinin-10 demonstrated that these peptides also have the capacity to influence lymphocyte proliferation in human-derived peripheral blood mononuclear cells (PBMC; Kayser & Meisel 1996). Specifically, at lower concentrations of both β-casomorphin-7 and β-casokinin-10, a reduction in lymphocyte proliferation was observed; an increase in proliferation was however observed following exposure with higher concentrations (Kayser & Meisel 1996).

κ-casein

κ-Caseins constitute *c.* 12% of total casein in bovine milk (Ginger & Grigor 1999) and have been identified as a significant source of bioactive peptides exhibiting immunomodulatory potential (Brody 2000). Caseinomacropeptide (CMP), also known as glycomacropeptide (GMP), is a 64 amino acid peptide derived from κ-casein and contains N-acetylneuraminic acid (sialic acid) at varying concentrations (0–5 U) (Daddaoua *et al.* 2005). Due to the action of chymosin on κ-casein, CMP is produced in the neonatal stomach (Brody 2000) and is also produced during cheese manufacture (10–15% of milk whey) from the milk clotting process. In addition to its bifidogenic properties in the human gut (as discussed in Section 2.4), bovine CMP has been shown to potentially alter the status of the immune system (Brody 2000; Li &

Mine 2004; Daddaoua *et al.* 2005; Requena *et al.* 2008). For example, CMP has been shown to exert a wide array of effects on many areas of the immune system, including lymphocyte proliferation and cytokine production (Otani & Hata 1995; Otani *et al.* 1995; Monnai & Otani 1997). Indeed, CMP has been shown to alleviate LPS and other mitogenic compound-induced effects on lymphocytes (Otani *et al.* 1995). Monnai & Otani (1997) concluded that this action of CMP was brought about through its interaction with lymphocyte cell surface proteins. In response, lymphocyte up-regulation of the IL-1 receptor antagonist results in the prevention of IL-1 induction of lymphocyte proliferation (Monnai & Otani 1997).

CMP has been shown to infer a dose-dependant anti-inflammatory effect on a murine colitis model (Daddaoua *et al.* 2005). Daddaoua *et al.* (2005) concluded that the mechanism by which CMP reduced colitis symptoms in the animal model may well be similar to those elucidated by Monnai & Otani (1997), that is, through up-regulation of the IL-1 receptor antagonist, thus blocking the action of IL-1β. Furthermore, the action of CMP on the immune system can also extend to 'proximal regions' of the GIT (Requena *et al.* 2008). For example, in a rat-based ileitis model, CMP had a significant anti-inflammatory effect when fed at 500 mg kg^{-1} day^{-1}. Similar to the observations of Monnai & Otani (1997) and Daddaoua *et al.* (2005), the mechanism was partly attributed to the blocking of IL-1. Requena *et al.* (2008) speculated that the anti-inflammatory effects inferred by dietary supplementation with CMP may also be linked to the down-regulation of TNF-α and IL-17. Interestingly, further *in vitro* studies using the U937 cell line demonstrated that CMP (and CMP digested with pepsin) had the ability to stimulate the immune system in terms of cell proliferation and phagocytic activity (Li & Mine 2004). These studies would suggest that the immunomodulatory potential of CMP can vary significantly, depending on the composition and arrangement of attached molecules e.g. sialic acids and carbohydrates (Li & Mine 2004).

α-casein

αs1- and αs2-caseins together make up the majority of casein in bovine milk (34% and 12%, respectively; Ginger & Grigor 1999). One of the first antimicrobial peptides identified from bovine casein, referred to as isracidin, was released from αs1-casein B following digestion with chymosin. Isracidin has a molecular weight of 2770 Da and consists of amino acids 1–23 of αs1-casein B. As well as exhibiting antimicrobial properties (discussed in Section 2.3), isracidin has been shown to exert immunomodulatory effects both *in vitro* and *in vivo* (Lahov & Regelson 1996). Indeed, isracidin altered both specific and innate immune responses in murine studies (Lahov & Regelson 1996). Along with the injection of 10^8 sheep erythrocytes, isracidin was administered intraperitoneally to mice at varying concentrations (1–40 mg kg^{-1}) and AFC numbers were assessed on day 10. Isaracidin was shown to impact maximally on AFC numbers for IgG and also IgM at 10 mg kg^{-1} and 40 mg kg^{-1}, respectively. For both AFC populations, a two-fold increase was observed (Lahov & Regelson 1996).

Casein and casein digests have also been shown to aid in prevention of milk protein intolerances. Tryptic digests of whole casein fractions fed to mice resulted in a poor humoral response to subsequent immunisation with casein, suggesting that the inclusion of the hydrolysed protein fraction in the diet inferred oral tolerance to casein (Hachimura *et al.* 1993). Such findings suggest that casein contains peptide sequences which, when included in the diet, are capable of preventing the occurrence of milk allergy (Hachimura *et al.* 1993). Examples of such findings have been observed for hydrolysed α-caseins (Peng *et al.* 2005). Indeed, mice fed casein hydrolysates prior to sensitisation with α-casein showed a significantly suppressed immune response (measured by specific IgE) (Peng *et al.* 2005). However, ongoing sensitisation was not found to be suppressed when α-casein hydrolysates were included in the diet following sensitisation. Peng *et al.* (2005) concluded that oral tolerance in mice to α-casein only occurs when the hydrolysates are included in the diet prior to sensitisation.

Interestingly, αs1-casein hydrolysates have been linked to the alleviation of symptoms associated with stress in women (Kim *et al.* 2006). Indeed, in a double-blind placebo-controlled trial, a significant improvement in self-reported stress-related symptoms was observed following consumption of αs1-casein hydrolysate. Research is now ongoing to evaluate its effect on the general population, including males and non-clinical subjects and also to assess its long-term efficacy. As stress has been proven to impact extensively on the efficacy of the immune system (Segerstrom & Miller 2004), such properties of dietary milk caseins could in practice play beneficial roles in immune system maintenance.

2.2.2 *Immunomodulatory peptides derived from whey proteins*

Whey proteins account for 20% of total bovine milk protein and are made up of globular and more water-soluble proteins when compared with casein. The whey fraction of milk protein largely comprises α-lactalbumin, β-lactoglobulin, bovine lactoferrin, bovine serum albumin (BSA) and immunoglobulins. CMP can also be found in whey protein preparations following release from κ-casein (Gauthier *et al.* 2006; Haug *et al.* 2007; Madureira *et al.* 2010).

Peptides and hydrolysates derived from whey proteins have demonstrated abilities to impact on many components of the immune system. Micro-filtered whey protein isolate (WPI) (consisting of mainly α-lactalbumin, β-lactoglobulin and CMP) was shown to significantly increase lymphocyte proliferation (Mercier *et al.* 2004). Digests of WPI generated with trypsin and chymosin were shown to infer less lymphocyte stimulation, which was reportedly due to peptide–peptide interaction in the complex peptide mix. However, fractions isolated from these digests were shown to reproduce the stimulating effects of the WPI on lymphocyte at far less concentrations (0.5–500 μg mL^{-1} as opposed to 2000 μg mL^{-1}), suggesting that immunomodulation was due to peptide sequences contained within the WPI (Mercier *et al.* 2004). These fractions (named F1–F4) were made up of short-chain peptides, mostly in the region of 5 kDa (Mercier *et al.* 2004).

In further studies, these fractions demonstrated immunomodulatory capabilities in infected and non-infected mice. Indeed, expressions of IFN-γ and TGF-β1 were altered differentially depending on the fraction (F1–F4) tested (Saint-Sauveur *et al.* 2009). Specifically, the F1 (acidic) fraction resulted in a decrease in IFN-γ production and resulted in significant weight loss in infected mice, the F2 (neutral) fraction showed a significant up-regulation in IFN-γ and the F3 (basic) fraction up-regulated IFN-γ as well as increasing IgA and TGF-β1. These data demonstrate the diversity of immunomodulatory potential of whey protein and whey protein hydrolysates (Saint-Sauveur *et al.* 2009).

β-lactoglobulin and α-lactalbumin

β-lactoglobulin and α-lactalbumin have been shown to contain encrypted peptide sequences with potential to influence immune responses. Peptides released from the hydrolysis of β-lactoglobulin, following fermentation with *Lactobacillus paracasei* NCC2461, have the ability to stimulate IL-10, down-regulate spleenocyte proliferation and down-regulate INF-γ and IL-4 secretion in murine models (Prioult *et al.* 2004). Synthetic peptides, theoretically released from β-lactoglobulin (and α-lactalbumin) upon digestion with trypsin and chymotrypsin, were found to have immunomodulatory properties when used to stimulate murine spleenocytes (Jacquot *et al.* 2010). In this case, the majority of peptides were shown to induce proliferation in murine spleenocytes (resting and/or following ConA stimulation). In particular, f(15–20), f(55–60) and β-LG f(139–148) were not only shown to induce spleenocyte proliferation but also to induce significant changes in cytokine expression in both resting spleenocytes and con-A stimulated spleenocytes. In addition, some fractions were shown to have cytotoxic properties e.g. β-LG f(1–8) which could potentially exert a change in immune status (Jacquot *et al.* 2010).

In a study investigating immunomodulatory and antimicrobial activity of milk protein hydrolysates, α-lactalbumin hydrolysates were the most effective at stimulating the immune system of all the whey hydrolysates tested (Biziulevicius *et al.* 2006). Indeed, the hydrolysates were found to enhance macrophage activity *in vivo* using murine models and induced autolysis of microbial preparations (Biziulevicius *et al.* 2006). Bioactive peptides derived from α-lactalbumin, namely f(18–19), f(50–51) and f(18–20), demonstrated proliferating effects on human lymphocytes (Kayser & Meisel 1996). While β-casein-derived peptides yielded similar results, the α-lactalbumin peptides also enhanced total cellular protein synthesis. The peptide f(51–53) of α-lactalbumin was synthesised and used in studies evaluating macrophage activity (Berthou *et al.* 1987). Results demonstrated a significant up-regulation of phagocytic activity in murine macrophages.

The α-lactalbumin-derived peptide f(10–16) was also shown to have immunomodulatory effects on spleenocytes (either resting or stimulated with Con A) (Jacquot *et al.* 2010). In the same study, a peptide from α-lactalbumin f(104–108) showed cytotoxic effects on spleenocytes indicating the diversity of bioactivity for such peptides (Jacquot *et al.* 2010). Matin *et al.* (2001) investigated cytotoxic effects of peptides derived from α-lactoglobulin and showed that acidification of certain whey peptides produced multimeric forms which were shown to be cytotoxic to mouse spleenocytes. This would suggest that these cytotoxic peptides may be generated in the acidic environment of the stomach with consequences for the local immune response (Matin *et al.* 2001).

Bovine lactoferrin

Bovine lactoferrin is recognised for its antimicrobial and immunomodulatory properties. Interestingly, while purified lactoferrin has been identified as an inhibitor of spleenocyte proliferation, trypic/peptic-digested lactoferrin increases spleenocyte proliferation (Miyauchi *et al.* 1997). In the same investigation, bovine lactoferrin digests were shown to influence antibody production in murine spleenocytes, specifically the antibodies IgM, IgG and IgA (Miyauchi *et al.* 1997). In a follow-up study, lactoferrin and lactoferrin hydrolysates were both shown to up-regulate phagocyte activity (Miyauchi *et al.* 1998). Furthermore, a lactoferrin hydrolysate preparation with lactoferricin (a peptide derived from the N-terminus region of lactoferrin) removed was shown to infer no up-regulation of phagocyte activity, suggesting that the lactoferricin peptide was responsible for these immunomodulatory effects. In addition, retention of the immunomodulated state was observed following washing of the co-incubated cells with the lactoferrin and lactoferrin hydrolysates, suggesting a direct interaction between the bioactive sequences in the lactoferrin/lactoferrin hydrolysates and the neutrophils used (Miyauchi *et al.* 1998).

Indeed, Vogel *et al.* (2002) speculated that the positive charge of lactoferricin in an aqueous solution (elucidated by Hwang & Vogel 1998) may resemble chemokine structures, allowing it to bind to and activate chemokine receptors on immunomodulating cells. It has also been hypothesised that lactoferrin is transported to the nucleus and has a direct effect on DNA expression (He & Furmanski 1995), which would contribute to the immunomodulatory activity of lactoferricin. Such interactions may be responsible for the induction of IL-8 production in human polymorphonuclear leukocytes, following co-incubation with lactoferricin (released from lactoferrin by pepsin hydrolysis) *in vitro* (Shinoda *et al.* 1996).

Lactoferricin has also been shown to significantly modulate cytokine production in bovine leukocytes and monocytes (Prgomet *et al.* 2006). This study demonstrated the up-regulation (using quantitative PCR) of TNF-α, IL-1β, IL-6 and IL-10 mRNA levels. Interestingly, in the same study it was shown that lactoferricin (which has been shown to bind bacterial LPS) was unable to alter LPS-induced cytokine patterns in cells at any LPS concentration (Prgomet *et al.* 2006). A synthetic peptide whose sequence was derived from lactoferrin has been shown to stimulate PBMCs (Ueta *et al.* 2001). Upon co-incubation with this peptide (P2), significant changes occurred in the status of the PBMCs including up-regulation of superoxide production and up-regulation of protein kinase C, which in turn affect many signalling cascades.

2.2.3 *Indirect effects of casein and whey peptides on host immunity*

Casein and whey proteins are a rich reservoir of antimicrobial and growth-stimulating peptides. In all cases, these peptides have dramatic consequences for the viability of their target bacteria and hence for host immunity through two indirect mechanisms. Primarily, antimicrobial peptides provide a level of bioprotection by directly inhibiting the proliferation of target bacteria. Secondly, growth-stimulating peptides have the potential to modulate the gut microbiota by promoting the growth of beneficial micro-organisms, which is now known to significantly impact on host immunity (O'Hara & Shanahan 2007).

The examples described in the following section (Section 2.3) demonstrate the range of inhibitory activities encoded by antimicrobial peptides from casein and whey proteins. While many of these peptides were isolated with a view to therapeutic or perhaps food safety applications, their existence within the native protein sequence suggest that these peptides may be released *in vivo* via gastrointestinal digestion. Furthermore, examples are provided of milk-derived peptides which have been shown to directly promote the growth of certain species of bacteria, especially of the genus *Bifidobacterium*. In this sense, peptides derived from milk proteins, whether antimicrobial or bifidogenic, have a fundamental role to play in the correct development of host immunity.

Finally, certain milk-derived peptides have been shown to stimulate mucin production in the gut either *in vivo* or *in vitro*. While these peptides are not involved in direct killing of pathogenic micro-organisms, such peptides may have a critical role to play in maintaining gut homeostasis by preventing the adherence of pathogenic bacteria. In this sense, such peptides provide a third indirect mechanism towards maintaining a healthy immune response as discussed in the following section.

2.3 Antimicrobial peptides

2.3.1 *Antimicrobial peptides derived from casein*

α-casein

As early as 1966, the first of a series of patents relating to chymosin digests of casein with antibiotic properties were filed by Katzir Katchalsky and colleagues (Katzir Katchalsky *et al.* 1973; Lahov & Regelson 1996). Shortly after, Lahov *et al.* (1971) were the first to describe glycopeptides called casecidins which were released from bovine milk following heat treatment or chymosin digestion and inhibited the growth of pathogenic bacteria and lactobacilli.

To improve the process, the concentration of enzyme was increased for the digestion of αs1-casein B which resulted in the isolation of isracidin (Hill *et al.* 1974; Table 2.2.). Remarkably, high concentrations of isracidin (in the range 0.1–1 mg mL^{-1}) were required for the *in vitro* inhibition of a variety of Gram-positive bacteria. On the contrary however, much smaller quantities were required to exhibit an effect *in vivo* and a single injection protected mice for up to 5 months, presumably as a result of its immunomodulatory capabilities. Moreover, a strong protective effect was observed if the peptide was administered prior to bacterial challenge (Lahov & Regelson 1996). For example, an intramuscular dose of 1 mg of isracidin to mice infected with staphylococci resulted in significant survival compared to control mice, depending on the time of bacterial challenge after administration of the peptide (Lahov & Regelson 1996). In another experiment, *Staphylococcus aureus*-infected mice were protected from death with a concentration as low as 10 μg, although the effect was found to be strain-dependent.

Isracidin also afforded protection to guinea pigs and rabbits using up to 20 mg per animal (Lahov & Regelson 1996). The *in vivo* antimicrobial activities could also be extended to the pathogens *Listeria monocytogenes* and *Streptococcus pyogenes*. Moreover, isracidin also afforded protection to cows and sheep against *S. aureus*, *Streptococcus uberous* and/or *Streptococcus disagalactia* and isracidin-treated animals developed long-term immunity (Lahov & Regelson 1996). More recently, Hayes *et al.* (2006) demonstrated that isracidin is effective against pathogens such as *Enterobacter sakazakii* and *Escherichia coli* 0157:H7.

Indeed, *E. sakazakii* (*Cronobacter*) is a problematic contaminant of milk powder and hence represents a significant health risk to neonates (Chenu & Cox 2009). Moreover, Hayes *et al.* (2006) isolated three peptides (Table 2.2) from bovine αs1-casein after fermentation which exhibited antimicrobial activities against *E. sakazakii* and *E. coli*. The authors suggest

Table 2.2 Antimicrobial peptides derived from casein and whey proteins.

Antimicrobial peptide	Source	Amino acid sequence	Examples of target micro-organisms	Reference
Antimicrobial peptides derived from casein				
Isracidin f(1–23)	Bovine αs1-casein B	RPKHPIKHQGLPQEVLNENLLRF	*Staphylococcus aureus*, *Enterobacter sakazakii*, *Escherichia coli* 0157: H7, *Listeria monocytogenes*, yeast	Hayes *et al*. 2006; Lahov & Regelson 1996
f(21–29)	Bovine αs1-casein	IKHQGLPQE	*Escherichia coli*, *Enterobacter sakazakii*, *Lactobacilllus bulgaricus*, *Streptococcus mutans*	Hayes *et al*. 2006
f(30–37)	Bovine αs1-casein	VLNENLLR		
f(195–208)	Bovine αs1-casein	SDIPNPIGSENSEK	*Listeria innocua*	
Casocidin-1 f(165–203)	Bovine αs2-casein	KTKLTEEEKNRLNFLKKISQ-RYQKFALPQYLKTVYQHQK	*Escherichia coli*, *Staphyococcus carnosus*, *Bacillus subtilis*. *Staphylococcus epidermidis*, *Enterococcus faecium*, Rhodotorula (yeast)	Forssmann *et al*. 2003; Zucht *et al*. 1995
f(183–207)	Bovine αs2-casein	VYQHQKAMKPWIQPKTKVIPYVRYL	*Escherichia coli*, *Listeria innocua*, *Bacillus cereus*, *Micrococcus flavus*, *Streptococcus thermophilus*	Recio & Visser 1999a
f(164–179)	Bovine αs2-casein	LKKISQRYQKFALPQY		
N-23-T (Lactophoricin) f(113–135)	Component 3 of proteose peptone	NTVKETIKYLKSLFSHAFEVVKT	*Streptococcus thermophilus*, *Salmonella* St. Paul, *Pseudomas aeroginosa*	Campagna *et al*. 2004
Cr1 f(181–207)	Bovine αs2-casein	KTVYQHQKAMKPWIQPKTKV-IPYVRYL	*Bacillus cereus*, *Bacillus subtilis*, *Listeria monocytogenes*, *Escherichia coli*, *Salmonella* Enteriditis, *Salmonella* Typhimurium	McCann *et al*. 2005
Cr4 f(175–207)		ALPQYLKTVYQHQKAMKPWIQ-PKTKVIPYVRYL	*Bacillus subtilis*, *Listeria monocytogenes*, *Escherichia coli*, *Salmonella* Typhimurium	

Table 2.2 *(Continued)*

Antimicrobial peptide	Source	Amino acid sequence	Examples of target micro-organisms	Reference
Cr5/6 f(164–207)		LKKISQRYQKFALPQYLKTVY-QHQKAMKPWIQPKTKVIPYVRYL	*Bacillus subtilis*, *Listeria monocytogenes*, *Escherichia coli*	
f(165–170)	Ovine αs2-casein	LKKISQ	*Escherichia coli*, *Staphyococcus carnosus*, *Staphylococcus epidermidis*, *Listeria innocua*	Lopez-Exposito *et al.* 2006a
f(165–181)		LKKISQYYYQKFAWPQYL	*Escherichia coli*, *Staphyococcus marcescens*, *Staphyococcus carnosus*, *Staphylococcus epidermidis*, *Enterococcus faecalis*, *Listeria innocua*	
f(184–208)		VDQHQKAMKPWTQPKTNAIPYVRYL	*Escherichia coli*, *Staphyococcus carnosus*, *Staphylococcus epidermidis*, *Listeria innocua*	
f(203–208)		PYVRYL		
f(184–210)	Human β-casein	QELLLNPTHQYPVTQPLAPVHNFISV	*Enterococcus faecium*, *Bacillus megaterium*, *Listeria innocua*, *Salmonella* species, *Yersinia enterocolitica*, *Staphylococcus aureus*, *Escherichia coli*	
Casecidin 17 f(193–209)	Bovine β-casein	YQEPVLGPVRGPFPIIV	*Escherichia coli*	Birkemo *et al.* 2009
Casecidin 15 f(193–207)		YQEPVLGPVRGPFPI		
f(63–117)	Human κ-casein	YQRRPAIAINNPYVPRTYYANPA-VVRPHAQIPQRQYLPNSHPFTVVR-RPNLHPSF	Gram-positive bacteria, Gram-negative bacteria, yeasts	Liepke *et al.* 2001
f(18–24)	Bovine κ-casein	FSDKIAK	*Escherichia coli*, *Listeria innocua*, *Staphylococcus carnosus*	Lopez-Exposito *et al.* 2006b

(continued)

Table 2.2 *(Continued)*

Antimicrobial peptide	Source	Amino acid sequence	Examples of target micro-organisms	Reference
f(30–32)		YVL	*Staphylococcus marcescens*, *Listeria innocua*, *Staphylococcus carnosus*	
f(139–146)		VESTVATL	*Escherichia coli*, *Staphylococcus carnosus*	
Kappacin A f(106–169)	Bovine κ-casein	AIPPKKNQDKTEIPTINTIASGEPT-STPTTEAVESTVATLEDΣPEVIES-PPEINTVQVTSTAV (Σ=phosphoserine)	*Streptococcus mutans*	Malkoski *et al.* 2001
Ser(*P*)^{149} κ-caseinA(138–158)	Bovine κ-casein	AVESTVATLEDΣPEVIESPPE (Σ=phosphoserine)		
Antimicrobial peptides derived from whey proteins				
Lactoferricin B f(17–41/42)	Bovine lactoferrin	FKCRRWQWRMKKLGAPSICVRRAF/A	*Escherichia coli*, *Salmonella enteritidis*, *Klebsiella pneumoniae*, *Proteus vulgaris*, *Yersinia enterocolitica*, *Pseudomonas aeruginosa*, *Campylobacter jejuni*, *Staphylococcus aureus*, *Streptococcus mutans*, *Corynebacterium diphtheriae*, *Listeria monocytogenes*, *Clostridium perfringens*, *Pseudomonas fluorescens*, *Bacillus cereus*	Bellamy *et al.* 1992a; Hoek *et al.* 1997
Lactoferricin B f(1–16)S-S(17–48)	Bovine lactoferrin	APRKNVRWCTISQPEWFKCR-RWQWRMKKGAPSITCVRRAFALE-CIRA	*Escherichia coli*	Hoek *et al.* 1997
Lactoferricin B f(1–16)S-S(45–48)	Bovine lactoferrin	APRKNVRWCTISQPEWCIRA	*Micrococcus flavus*	Recio & Visser 1999b
Lactoferricin B f(1–11)S-S(17–47)		APRKNVRWCTIFKCRRWQWR-MKKLGAPSITCVRRAFALECIRA	*Micrococcus flavus*	

Table 2.2 *(Continued)*

Antimicrobial peptide	Source	Amino acid sequence	Examples of target micro-organisms	Reference
Lactoferricin B f(17–30)	Bovine lactoferrin	FKCRRWQWRMKKLG	Oral pathogenic bacteria	Groenink *et al.* 1999
Lactoferrampin f(268–284)	Bovine lactoferrin	WKLLSKAQEKFGKNKSR	*Candida albicans*, *Escherichia coli*, *Bacillus subtilis*, *Pseudomonas seroginosa*	van der Kraan *et al.* 2004
Lactoferricin H f(1–47)	Human lactoferrin	GRRRRSVQWCAVSQPEATKC-FQWQRNMRKVRGPPVSCIKRD-SPIQCI	*Escherichia coli*	Bellamy *et al.* 1992b
Lfpep f(118–140)	Human lactoferrin	TKCFQWQRNMRKVRGPPVSCIKR	*Candida albicans*	Viejo-Diaz *et al.* 2005
Kaliocin I f(153–183)		FFSASCVPGADKGQFPNLCRL-CAGTGENKCA		
f(1–5)	α-lactalbumin	EQLTK	Several Gram-positive bacteria	Pellegrini *et al.* 1999
f(17–31)S-S(109–114)		GYGGVSLPEWVCTTFALCSEK		
f(61–68)S-S(75–80)		CKDDQNPH ISCDKF		
f(15–20)	β-lactoglobulin	VAGTWY	Several Gram-positive bacteria	Pellegrini *et al.* 2001
f(25–40)		AASDISLLDAQSAPLR		
f(78–83)		IPAVFK		
f(92–100)		VLVLDTDYK		

that such peptides have potential bio-protective applications in milk-based formula. Indeed, the sodium caseinate fermentate afforded protection against *E. sakazakii* growth in reconstituted infant formula and at 3.33% wt/vol numbers of *E. sakazakii* were reduced to 0 within one hour (Hayes *et al.* 2009). The fermentate, also afforded protection against *E. coli* 0157:H7 and *Listeria innocua*.

Casocidin-I was one of the first peptides released from αs2-casein to be characterised (Zucht *et al.* 1995). This peptide, which was released after heat treatment under acidic conditions, contains 39 amino acids corresponding to f(165–203) of αs2-casein (Table 2.2). Antibacterial activity was observed against *E. coli* and *Staphylococcus carnosus*. The authors suggested that the findings could explain the differences in the gut microbiota profile between formula-fed and breast-fed infants, considering that αs2-casein is absent from human milk. More recently, Forssmann and co-workers obtained a patent on casocidin-I (Forssmann *et al.* 2003) and demonstrated that the peptide has a wider spectrum of inhibition with antimicrobial activity against the following micro-organisms (in order of sensitivity): *Bacillus subtilis* > *S. carnosus* = *Staphylococcus epidermidis* > Rhodotorula (yeast) ≫ *E. coli* > *Enterococcus faecium*, at concentrations within the range 1–180 μg mL^{-1}. The authors suggested that the peptide could be used to treat diarrhoea or diseases related to disturbances of the intestinal microflora, mucosa of the vagina or mouth or could be used to treat diseases of the skin (in the form of an ointment) as well as fungal diseases.

Most recently, the structural integrities of casocidin-I and isracidin were tested in the presence of yogurt starters *Streptococcus thermophilus* and *Lactobacillus delbreuckii* subsp. *bulgaricus* (Somkuti & Paul 2010). At pH 4.5, both peptides remained largely unchanged but at pH 7.0 casocidin was degraded. The authors suggested that both peptides may have potential as food-grade antimicrobial supplements if added near the end of fermentation.

Pepsin hydrolysis of αs2-casein resulted in the release of two active cationic fractions, f(183–207) and f(164–179) (Recio & Visser 1999a). Both peptides contain a high proportion and asymmetric clustering of basic amino acid residues. The peptides exhibited activity against both Gram-positive and Gram-negative bacteria (Table 2.2) with minimum inhibitory concentrations (MICs) in the region of 25–99 μM for f(164–179) and 8–16 μM for f(183–207). Interestingly, a synergistic effect was also observed between f(183–207) and the lactococcal bacteriocin nisin against the food-borne pathogen *Listeria monocytogenes* (Lopez-Exposito *et al.* 2008a). Further studies with f(183–207) indicated that this peptide binds initially to lipotechoic acid in Gram-positive bacteria and lipopolysaccharide in Gram-negative bacteria (Lopez-Exposito *et al.* 2008b). The peptide was shown to permeabilise both inner and outer membranes, generating pores in the outer membrane of Gram-negative bacteria and in the cell wall of Gram-positive bacteria. In the case of the latter, the cytoplasmic contents leaked from the cell whereas the peptide caused cytoplasm condensation and, in many cases, leakage in Gram-negative bacteria.

Barzyk *et al.* (2009) also demonstrated that the biological activities of two antimicrobial peptides derived from milk may be related to their electrostatic interactions with cell membranes based on studies using lipid membranes. Moreover, one peptide (N-23-T) underwent conformational rearrangements, adopting an α-helical conformation when inserted into the monolayer. It has been proposed that the ability of peptides to form α-helical channels in lipid bilayers is linked to the longer length of the peptide chain (Campagna *et al.* 1999, 2001). In addition, the α-helical content of amphipathic antimicrobial peptides has been linked to their potency (Uteng *et al.* 2003; Lee *et al.* 2006; Park *et al.* 2008). However, for a more detailed account of the mode of action of milk-derived antimicrobial peptides, the recent review by Benkerroum (2010) is recommended.

Five antibacterial peptides (referred to as Cr1, Cr3, Cr4, Cr5 and Cr7) were isolated from a chymosin digest of bovine sodium caseinate corresponding to amino acid residues 180–207 of αs2-casein (Table 2.2) (McCann *et al.* 2005). While all five were different to the sequence of casocidin-I reported by Zucht *et al.* (1995), Cr1 corresponded to the antibacterial peptide f(181–207) previously reported (Recio & Visser 1999a). The remaining peptides shared 27 amino acids with f(181–207) but differed in that they contained between 1 and 17 additional amino acids at the N-terminus. Interestingly, the extra amino acids served to increase potency. Indeed, peptide Cr4 which had an additional 6 amino acids compared to Cr1 displayed MIC values 2–3 times lower than CR1 against Gram-positive bacteria. Cr5, which had an extra 17 amino acids, displayed MIC values that were seven times lower than CR1. A partially purified chymosin preparation which contained peptides Cr1, Cr4 and Cr5 as well as other peptides, termed Cr_{MIX}, was shown to be highly heat stable. While Cr_{MIX} exhibited antimicrobial activity against *Salmonella* Typhimurium in peptone medium, the effect was lost in skim milk; this was attributed to the cationic nature of the peptides. Indeed, the presence of metal cations have been linked to the reduced activity of cationic peptides (Chantaysakorn & Richter 2000) where it has been proposed that they stabilise negatively charged peptide target sites on the outer membranes of Gram-negative bacteria (Ellison & Giehl 1991; Stevens *et al.* 1992).

Pepsin digestion of ovine milk resulted in the isolation of ten cationic peptides from αs2 casein, four of which were chemically synthesised and demonstrated antimicrobial activities against several Gram-positive and Gram-negative bacteria (Table 2.2) (Lopez-Exposito *et al.* 2006a). Interestingly, although one peptide f(184–208) differed in only four amino acids from the bovine αs2-casein f(183–207), its activity was at least 200 times less potent.

β-casein

Minervini *et al.* (2003) tested proteinase-hydrolysed caseins from six different species for both ACE-inhibitory and antimicrobial peptides using a partially purified proteinase from *Lactobacillus helveticus*. Although hydrolysed sodium caseinates were prepared from bovine, ovine, caprine, porcine, buffalo and human milk, only the latter generated an antimicrobial peptide. Indeed, the peptide was isolated from human β-casein f(184–210), and exhibited a wide spectrum of activity against Gram-positive and Gram-negative bacteria including species of *Enterococcus faecium*, *Bacillus megaterium*, *Listeria innocua*, *Salmonella* species, *Yersinia enterocolitica*, *S. aureus* and *E. coli* at a concentration of *c.* 100 μg mL^{-1}. However, *Lb. helveticus* and *Lactobacillus plantarum* were not inhibited by the peptide. While human β-casein f(184–210) shares a similar length (26 amino acids) with isracidin (Lahov & Regelson 1996) and lactoferricin (Tomita *et al.* 1993), it displayed a lower positive charge, a higher content of non-polar hydrophobic residues (15 of the 26 amino acids) and contained some proline residues towards the C-terminal end which could act to provide a level of protection from degradation by peptidases. Such antimicrobial peptides generated in the human gut would be protected from enzymatic digestion and therefore available to exert antimicrobial activities *in vivo*. Two antimicrobial peptides, casecidin 15 and casecidin 17 were isolated from bovine β-casein of colostrum (Table 2.2; Birkemo *et al.* 2009). Both peptides had identical MICs against *E. coli* at 0.4 mg mL^{-1}.

κ-casein

In an experiment to simulate digestion in the infant stomach, human milk was acidified and hydrolysed with pepsin (Liepke *et al.* 2001), which resulted in the release of a κ-casein fragment f(63–117) which inhibited the growth of Gram-positive and Gram-negative bacteria (Table 2.2). The results of the study confirmed that antimicrobial peptides could be liberated from human milk during passage through the GIT, which undoubtedly has a significant impact on the host immune response. Hydrolysis of bovine casein with gastric enzymes resulted in the release of several peptides which enabled the synthesis of identical synthetic derivatives (Lopez-Exposito *et al.* 2006b). Fractions f(18–24), f(30–32) and f(139–146) were the most active against the bacteria tested (Table 2.2).

CMP, the heterogenous C-terminal fragment of bovine κ-casein, is composed of glycosylated and phosphorylated forms of different genetic variants and has demonstrated antimicrobial activity against the oral pathogens *Streptococcus mutans*, *Porphyromonas gingivalis* and *E. coli* (Malkoski *et al.* 2001). The non-glycosylated, phosphorylated form of the molecule, termed kappacin f(106–169), was deemed responsible for the observed antimicrobial activity. Further hydrolysis of kappacin released a range of peptides, but only one phosphorylated peptide exhibited antimicrobial activity, Ser(*P*)^{149} κ-caseinA(138–158) (Malkoski *et al.* 2001). Chemical synthesis of Ser(*P*)^{149} κ-casein(138–158) and its non-phosphorylated derivative revealed that phosphorylation is essential for antimicrobial activity as the non-phosphorylated form did not inhibit the growth of *S. mutans*.

Kappacin A and kappacin B are the most common variants; however, kappacin B is less active than variant A. Substitution of the hydrophilic residue Asp^{148} in variant A for the hydrophobic residue Ala^{148} in variant B dramatically reduced the antibacterial activity of kappacin A against *S. mutans* (Dashper *et al.* 2005). Moreover, kappacin was shown to have membranolytic activity at acidic pH but not so less at neutral pH. However, the activity of kappacin was markedly enhanced at neutral pH in the presence of the divalent cations zinc or calcium. Indeed, in the presence of excess divalent metal ions and the membrane solvent trifluoroethanol, the kappacin fragment κ-casein-A(138–158) underwent a conformational change, which the authors propose is responsible for the increased antibacterial activity of the peptide in the presence of excess cations. Kappacin is now available commercially as a dental care product by the Cooperative Research Centre for Oral Health Science (Australia) as Kappacin™ or as KappaZin™, the latter of which consists of a 1:1 combination of kappacin and zinc for enhanced antimicrobial activity (Benkerroum 2010).

2.3.2 *Antimicrobial peptides derived from whey proteins*

Many whey proteins such as lysozyme and lactoferrin have well-documented antimicrobial properties. However, beyond the bioactivities of the native whey proteins themselves, antimicrobial peptide sequences are also encrypted within these proteins (Lopez-Exposito & Recio 2008; Benkerroum 2010; Madureira *et al.* 2010). The antimicrobial peptides of lactoferrin are referred to as lactoferricins and have been released from both bovine and human lactoferrin (Table 2.2) (Wakabayashi *et al.* 2003; Madureira *et al.* 2010). While bovine lactoferricin has been shown to possess antimicrobial activity against Gram-positive and Gram-negative bacteria and yeasts as reviewed by Madureira *et al.* (2010), it was found to be inactive against strains of *Streptococcus lactis* and *Lactobacillus casei* (Korhonen 2001). Moreover, the combination of lactoferrin with lactoferricin B was found to exert a synergistic effect against *E. coli*, a phenomenon that could have a protective effect in the infant gut (Lopez-Exposito *et al.* 2008a).

The mode of activity of lactoferricin has been linked to cell surface binding and membrane disruption, thus following a different antimicrobial mechanism to the native lactoferrin protein (Madureira *et al.* 2010). Interestingly, the residues tryptophan and arginine were demonstrated to be especially important for the antimicrobial and antifungal activity of bovine lactoferricin, being prevalent in 'penetrations' which are described as regions of proteins or peptides that can spontaneously cross membranes (Vogel *et al.* 2002). Lactoferrampin f(268–284) represents another antimicrobial peptide derived from the N1 domain of lactoferrin (van der Kraan *et al.* 2004; Table 2.2); as well as anti-*Candida* activity, the peptide was effective against *B. subtilis*, *E. coli* and *Pseudomonas aeruginosa*.

The first reported human lactoferricin, lactoferricin H f(1–47), was released following gastric pepsin cleavage (Table 2.2; Bellamy *et al.* 1992b). Several other shorter synthetic antimicrobial peptides have since been derived from lactoferricin H. Indeed, human lactoferrin and its peptides derived from the N-terminal domain (amino acids 1-11) were found to be effective against antibiotic resistant *S. aureus* and *Klebsiella pneumoniae* (Nibbering *et al.* 2001). However, peptide fragments lacking the first 3 N-terminal amino acids were less effective than the peptide designated hLF(1-11) which contains all 11 amino acids from the N-terminal domain. The synthetic peptides Lfpep (derived from residues 118–140 of lactoferricin H) and kaliocin I (residues 153–183 of human lactoferrin) were shown to exhibit antifungal activity against *Candida* species including fluconazole- and amphotericin B-resistant isolates (Viejo-Diaz *et al.* 2005). While the inhibitory activity of Lfpep was related to its ability to permeabilise the *Candida albicans* cells, kaliocin I was unable to disrupt the cytoplasmic membrane.

Both α-lactalbumin and β-lactoglobulin hydrolysates have also yielded peptides with potent antimicrobial activities. For example, *E. coli* growth was dramatically altered in the presence of bovine α-lactalbumin or β-lactoglobulin hydrolysates generated via digestion with pepsin and trypsin, but not in the presence of the intact proteins (Pihlanto-Leppälä *et al.* 1999). Indeed, in the presence of 25 mg mL^{-1} of either hydrolysate, the activity of *E. coli* was only 21% that of the control after incubation for 6 hours. Digestion of α-lactalbumin with trypsin and chymotrypsin generated three peptide sequences of varying lengths (f(1–5), f(17–31)S-S(109–114), f(61–68)S-S(75–80); Table 2.2), two of which were disulphide-linked heterodimers (Pellegrini *et al.* 1999). Synthetic derivatives of the peptides revealed antimicrobial activities against Gram-positive and Gram-negative bacteria, although the latter group was only poorly susceptible (Pellegrini *et al.* 1999). The heterodimers containing the disulphide bridges were the most potent.

Interestingly, disulphide bridges have been associated with increased potency for class IIa bacteriocins due to improved structural stability (Drider *et al.* 2006), but this has not been proven for the antimicrobial peptides described above. Tryptic digestion of β-lactoglobulin revealed four peptide fragments covering f(15–20), f(25–40), f(78–83) and f(92–100) (Table 2.2; Pellegrini *et al.* 2001). Synthetic derivatives of the four peptides demonstrated antimicrobial activity against Gram-positive bacteria only. Interestingly, modifying the peptide VLVLDTDYK, by replacing aspartic acid with arginine and adding an extra lysine to generate the peptide VLVLDTRYKK, enabled the peptide to target the Gram-negatives *E. coli* and *Bordatella bronchiseptica* but reduced its activity against *B. subtilis*.

2.4 Peptides that stimulate the gut microbiota

The healthy gut microbiota should contain a balanced composition of many bacterial classes including symbionts (organisms with known health-promoting functions), commensals (permanent residents providing no benefit or detriment to the host) and pathobionts (permanent residents with potential to induce pathology) (Round & Mazmanian 2009). The gut microbiota exerts a significant effect on numerous host processes including gastrointestinal immunity. Imbalances in microbial composition have been linked with immune-dysfunction and the various pathologies which accompany it, including inflammatory bowel disease (IBD), Crohn's disease and even cancer.

Bifidobacteria are naturally present amongst the gut microbiota and can be categorised as symbionts, on account of the many beneficial health effects which have been associated with the genus (Turroni *et al.* 2009). In this respect, dietary strategies involving the consumption of probiotics and bifidogenic prebiotics, to increase the content of bifidobacteria in the gut, have received extensive attention. Human milk is a rich source of oligosaccharides, which are known to stimulate the growth of bifidobacteria (Mills *et al.* 2011a). As a consequence, infant formulae derived from bovine milk has been supplemented with prebiotic mixes such as fructo- and galacto-oligosaccharides (FOS and GOS) to promote the growth of bifidobacteria and decrease the numbers of pathogenic micro-organisms (Miniello *et al.* 2003; Boehm *et al.* 2005; Fanaro *et al.* 2005; Moro & Arslanoglu 2005; Bruzzese *et al.* 2006). However, bovine casein and whey proteins have also been linked to the growth promotion of beneficial gut bacteria.

Interestingly, this prebiotic effect has been associated with hydrolysed fractions of casein and whey and the peptides therein. Over 30 years ago, a bovine casein hydrolysate termed CBF was identified as being responsible for promoting the growth of *Bifidobacterium bifidum* var. *pennsylvanicus* (Kehagias *et al.* 1977). Growth stimulation by CBF was deemed as effective as that obtained by the bifidogenic monosaccharide N-acetylglucosamine. The CBF fraction was found to contain 0.15% phosphorous, 7% lactose and various peptides. Casein and whey constituents of human milk and whey constituents of bovine milk were also previously shown to enhance the growth of bifidobacteria isolated from infant stools (Petschow & Talbott 1990). While the whey-associated effects were not identified or characterised in either bovine or human milk, the lactose content was ruled out as having a stimulatory effect on growth. Growth of *Bifidobacterium bifidum* and *B. longum* isolated from infant stools was found to be enhanced in the presence of trypsin-digested κ-casein (Poch & Bezkorovainy 1991). The carbohydrate portion was ruled out as having any effect while the disulphide or sulphhydryl residues were deemed essential for the growth-promoting activity. However, glutathione or other cysteine/cystine peptides were unable to replicate the effect. Growth of *Bifidobacterium lactis* Bo and *Lactobacillus acidophilus* Ki was improved in milk by the addition of milk hydrolysates, particularly proteinase-mediated hydrolysate. The latter had a greater array of peptides, especially peptides <2000 Da, as opposed to Neutrase™-mediated hydrolysate (Gomes *et al.* 1998). The authors postulated that the poor growth of bifidobacteria in milk is related to the lack of small peptides and free amino acids which are provided in the hydrolysates.

As well as its immunomodulatory functions, CMP has been associated with the anti-adhesion of various pathogenic micro-organisms including *Helicobacter pylori*, *E. coli* and viruses as well as toxins (Brody 2000). However, as discussed by Brody (2000) CMP is also bifidogenic. For example, supplementation of milk with 2% CMP (isolated from cow's or combined ewe's and goat's sweet cheese whey) was shown to increase counts of *B. lactis* in milk by 1.5 logs after 24 hours at 37°C (Janer *et al.* 2004). However, 2% whey protein concentrate (WPC) had the most dramatic effect on bifidobacterial growth in milk (Janer *et al.* 2004). The authors postulated that as well as CMP, WPC may also contain other bifidobacterial growth factors. Supplementation of formula (spiked with either enteropathogenic *E. coli* or *Salmonella* serotype Typhimurium) with CMP and α-lactalbumin inoculated with infant faecal specimens resulted in stable bifidobacterial counts comparable to breast milk (Bruck *et al.* 2003). Moreover, *Salmonella* counts decreased significantly in vessels containing breast milk, CMP and α-lactalbumin.

The efficacy of CMP and α-lactalbumin for infant formula supplementation was further investigated by feeding test formulae enriched with α-lactalbumin, with higher or lower CMP, to 6-week-old healthy term infants (4–8 weeks) for 6 months (Bruck *et al.* 2006). However, there was no statistically significant difference between control (standard formula- and breast-fed infants) and test groups. All infants in the study had been breast fed before recruitment such that a threshold level of bifidobacteria may have already been established (Bruck *et al.* 2006). The authors therefore postulated that the prebiotic effect of both CMP and α-lactalbumin may only be observed with low starting populations of beneficial bacteria.

Bifidogenic peptides were isolated from human milk following digestion with pepsin and demonstrated for the first time that the bifidogenic effect of human milk is not solely based on its carbohydrate content (Liepke *et al.* 2002). The prebiotic peptides were derived from the secretory component of the soluble polyimmunoglobulin receptor and lactoferrin, both of which exhibit antimicrobial activities when intact. The stimulatory effect was even observed in extraordinarily rich growth medium and the peptides were found to be 100 times more effective on a molar basis than the known bifidogenic monosaccharide N-acetylglucosamine and stimulated a wider set of different bifidobacterial strains. As no other bacterial species were stimulated in a similar manner, the peptides are specifically bifidogenic. A common structural motif was identified across the bifidogenic peptides consisting of a pair of cysteine residues forming a disulphide bond and two small hydrophobic domains located C-terminally to the two cysteines. Based on this observation, a synthetic peptide was generated with the sequence CAVGGGCIAL which was as bifidogenic as the isolated peptides. Interestingly, as observed by Poch & Bezkorovainy (1991) and Ibrahim & Bezkorovainy (1994), the disulphide bridge is essential for the bifidogenic effect.

A peptidic construct composed of bovine lactoferricin and lactoferrampicin fed to piglets administered enterotoxigenic *E. coli* was found to be as effective as the antibiotic colistin sulphate for enhancing growth performance (Tang *et al.* 2009). Moreover, bifidobacteria and lactobacilli occurred at an increased concentration in the ileum, caecum and colon in the lactoferricin-lactoferrampicin-fed piglets compared to controls.

The identification of an intracellular endopeptidase in *Bifidobacterium animalis* subsp. *lactis*, termed PepO, may provide some insight into the growth-stimulating properties of certain milk-derived peptides (Janer *et al.* 2005). Indeed, PepO is a member of the M13 peptidase family of zinc metallopeptidases. The enzyme exhibited activity with oligopeptide substrates of at least 5 amino acids and with the 23 amino acid peptide αs1-casein f(1–23) and mainly cleaved bonds on the N-terminal side of phenylalanine residues. The authors postulated that the enzyme may play a role in the enhanced growth rate of bifidobacteria in milk when supplemented with peptide fractions from whey (Janer *et al.* 2004; Martın-Diana *et al.* 2003).

Bacteriocin production has been deemed a probiotic trait (Ross *et al.* 2010) by providing direct protection from pathogen infection (Corr *et al.* 2007) and by also ensuring microbial survival and dominance of the producing strain in a given niche (O'Shea *et al.* 2009). This latter trait was particularly apparent following an intensive screening program for antimicrobial activity from a variety of intestinal sources, involving the screening of over 40,000 LAB colonies and resulting in the repeated isolation of the same strains within individual intestinal sources (O'Shea *et al.* 2009). Factors which induce bacteriocin production indirectly ensure the survival and dominance of the producing strain. Biosynthesis of the bacteriocin macedocin by *Streptococcus macedonicus* in milk was recently linked to an induction-active fraction from milk containing several peptides of low mass corresponding to fragments of αs1- and β-casein (Georgalaki *et al.* 2010). A chemically synthesised derivative of αs1-casein f(37–55) was itself able to induce synthesis. It is therefore possible that milk peptides also serve as induction factors for other bacteriocin-producing bacteria. Such a phenomenon would have dramatic consequences for gut microbial profiles and may be worth further investigation to determine if this is a common mechanism among gut microbial isolates.

2.5 Peptides that regulate gut homeostasis

It has been suggested that mucins, which are high-molecular-weight glycoproteins produced by epithelial cells and secreted onto the intestinal surface, play a role in limiting the pathogenesis of microbial infections in the gut (Bergstrom *et al.* 2010). For example, the membrane-bound mucin (MUC)1 produced by all intestinal cells has been shown to protect against *Campylobacter jejuni* infection *in vivo* (McAuley *et al.* 2007). *In vitro* studies have demonstrated that MUC3 expression decreases colonisation of enteropathogenic *E. coli* (EPEC) (Mack *et al.* 1999). More recently, MUC2 was linked to host defence *in vivo* by providing protection against attaching and effacing infections (Bergstrom *et al.* 2010). However, as with MUC2, the biological efficacy of these mucins is open to modulation at various levels including differences in gene expression, post-translational modification or even at the level of secretion into the intestinal lumen (Bergstrom *et al.* 2010). Interestingly, certain milk-derived peptides have been shown to modulate the expression of specific mucin-encoding genes. Mammary-associated serum amyloid A3 (M-SAA3) is highly abundant in the colostrum of mammals (McDonald *et al.* 2001). Peptides derived from the N-terminal region of bovine M-SAA3 were shown to enhance MUC3 gene expression in human intestinal cells by a factor of *c.* 2.5 (Mack *et al.* 2003). Likewise, an N-terminal peptide derived from human M-SAA3 was shown to enhance MUC3 gene expression by a factor of almost 4.3 in the same cells (Larson *et al.* 2003). Moreover, pre-treatment of the cells with the peptide reduced the adherence of EPEC by 73% compared to untreated controls. A casein-derived peptide, β-casomorphin 7, was shown to increase the expression of mucin-encoding genes in both rat and human mucin-producing cell lines (Zoghbi *et al.* 2006). While this is an indirect form of protection, peptides with mucin-stimulating abilities could provide a primary line of defence against infection, ensuring that pathogens are unable to colonise the GIT.

2.6 Potential for peptide bioactivities *in vivo* and outlook for the future

While the scientific evidence regarding the functionalities of encrypted bioactive peptides from casein and whey proteins is compelling, the question remains as to whether these peptides are actually released during gastrointestinal digestion and if they reach their target site in a biologically intact manner. This is particularly important for the newborn infant, considering that milk is the sole food in the earliest stages of life at a time when the immune system is developing. The scientific evidence suggests that intact bioactive peptides from casein and whey proteins can be liberated during gastrointestinal digestion, although few of these studies were performed *in vivo*. For example, bioavailability studies have demonstrated that the bioactive peptide I-P-P can escape intestinal degradation and reach the circulatory system intact (Foltz *et al.* 2007).

Antithrombotic peptides have been identified in the plasma of 5-day-old infants consuming a cow-milk-based formula (Chabance *et al.* 1995). Hydrolysis of caprine whey with human gastric juice at pH 2.5 and human duodenal juice at pH 8.0 resulted in the degradation of proteins such as serum albumin, lactoferrin and immunoglobulins which strongly inhibited *L. monocytogenes* (Almaas *et al.* 2008). However, digestion of the caprine whey with human gastric juice alone did not produce a significant antilisterial effect. The authors suggested that during digestion the antibacterial effects of caprine whey hydrolysates are most effective in the duodenum. Likewise, human gastric and duodenal juices successfully released antilisterial peptides from donkey milk samples (Tidona *et al.* 2010).

However, not all studies have produced such convincing evidence. Recently, Picariello *et al.* (2010) developed an *in vitro* multi-step static model to simulate as closely as possible complete digestion of bovine milk proteins through the GIT and to identify peptides which were biologically active. This involved the sequential use of pepsin, pancreatic proteases and extracts of human intestinal brush border membranes in simulated gastric, duodenal and jejuneal environments. Both casein and whey proteins were heavily degraded by digestive proteases suggesting that most of the bioactive peptides would not survive intact, an observation previously reported by Schmelzer *et al.* (2007) for β-casein. However, the model served to simulate the GIT of adults and hence the peptide profile of infants may be much different due to the lack of or limited concentration of specific proteases in the gastrointestinal tract (Picariello *et al.* 2010). Interestingly, specific regions of β-lactoglobulin were found to be particularly resistant to proteolysis and especially β-lactoglobulin f(125–135), and the high resistance of β-lactoglobulin to hydrolysis could have implications for the antigenicity associated with bovine β-lactoglobulin and cow's milk allergy.

Engineering bioactive peptides which can withstand gastrointestinal digestion is a feasible strategy, and has been successfully executed to increase the proteolytic resistance of bacteriocins for example (O'Shea *et al.* 2010). Moreover, genetic engineering approaches can be used to increase the bioavailability of peptide sequences in certain foods through the science of biopharming, as reviewed by Mills *et al.* (2011b). Despite the strides which have been made in the laboratory, strict regulations regarding molecular genetics in certain countries as well as consumer attitudes will undoubtedly delay the exploitation of such approaches. This suggests that determining the susceptibility of milk-derived bioactive peptides to gastrointestinal enzymes is of vital importance if the peptide is to exert a biological effect after gastrointestinal digestion.

Interestingly, recent evidence suggests that the consumption of unprocessed bovine milk may be inversely associated with the development of asthma, hay fever and atopic sensitisation (Braun-Fahrlander & von Mutius 2011). While the underlying mechanisms for such an association are not yet clear, it could be speculated that the ability of the host to produce bioactive peptides with immunomodulatory capacities *in vivo* may be a contributing factor. Moreover, it has been suggested that early exposure to cow's milk protein as a supplement to breast-feeding may promote tolerance to cow's milk protein based on large-scale clinical evidence (Katz *et al.* 2010). However, a more recent analysis of available data did not confirm this (Koletzko *et al.* 2011). One could speculate that the presence of immunomodulatory peptides in hydrolysed milk protein preparations may at least be partly responsible for the anti-allergenic effects associated with hypoallergenic infant formulas, again suggesting that the native peptides can exert an effect *in vivo*, although this has yet to be confirmed.

Acknowledgements

This work was funded by FP7 EU project EFRAIM (KBBE-2007-2-2-06), Enterprise Ireland, Irish Government under the National Development Plan, 2000–2006 and Science Foundation Ireland (SFI). R. McCarthy was in receipt of a Teagasc Walsh Fellowship.

References

Aimutis, W.R. (2004) Bioactive properties of milk proteins with particular focus on anticariogenesis. *Journal of Nutrition*, **134**, 989S–995S.

Almaas, H., Berner, V., Holm, H., Langsrud, T. & Vegarud, G.E. (2008) Degradation of whey from caprine milk by human proteolytic enzymes, and the resulting antibacterial effect against *Listeria monocytogenes*. *Small Ruminant Research*, **79**, 11–15.

Barzyk, W., Campagna, S., Wieclaw, K., Korchowiec, B. & Rogalska, E. (2009) The affinity of two antimicrobial peptides derived from bovine milk proteins for model lipid membranes. *Colloids Surf A: Physiochemical Engineering Aspects*, **343**, 104–110.

Belem, M.A.F., Gibbs, B.F. & Lee, B.H. (1999) Proposing sequences for peptides derived from whey fermentation with potential bioactive sites. *Journal of Dairy Science*, **82**, 486–493.

Bellamy, W., Takase, M., Wakabayashi, H., Kawase, K. & Tomita, M. (1992a) Antibacterial spectrum of lactoferricin B, a potent bactericidal peptide derived from the N-terminal region of bovine lactoferrin. *Journal of Applied Bacteriology*, **73**, 472–479.

Bellamy, W., Takase, M., Yamauchi, K., Wakabayashi, H., Kawase, K. & Tomita, M. (1992b) Identification of the bactericidal domain of lactoferrin. *Biochimica et Biophysica Acta*, **1121**, 130–136.

Benkerroum, N. (2010) Antimicrobial peptides generated from milk proteins: a survey and prospects for application in the food industry. *International Dairy Journal*, **63**, 320–338.

Bergstrom, K.S., Kissoon-Singh, V., Gibson, D.L. *et al.* (2010) Muc2 protects against lethal infectious colitis by disassociating pathogenic and commensal bacteria from the colonic mucosa. *PLoS Pathogens*, **6**, e1000902.

Berthou, J., Miglioresamour, D., Lifchitz, A., Delettre, J., Floch, F. & Jolles, P. (1987) Immunostimulating properties and 3-dimensional structure of 2 tripeptides from human and cow caseins. *FEBS Letters*, **218**, 55–58.

Birkemo, G.A., O'Sullivan, O., Ross, R.P. & Hill, C. (2009) Antimicrobial activity of two peptides casecidin 15 and 17, found naturally in bovine colostrum. *Journal of Applied Microbiology*, **106**, 233–240.

Biziulevicius, G.A., Kislukhina, O.V., Kazlauskaite, J. & Zukaite, V. (2006) Food-protein enzymatic hydrolysates possess both antimicrobial and immunostimulatory activities: a 'cause and effect' theory of bifunctionality. *FEMS Immunology and Medical Microbiology*, **46**, 131–138.

Boehm, G., Stahl, B., Jelinek, J., Knol, J., Miniello, V. & Moro, G.E. (2005) Prebiotic carbohydrates in human milk and formulas. *Acta Pediatrica*, (Suppl) **94**, 18–21.

Braun-Fahrlander, C. & von Mutius, E. (2011) Can farm milk consumption prevent allergic diseases? *Clinical and Experimental Allergy*, **41**, 29–35.

Brody, E.P. (2000) Biological activities of bovine glycomacropeptide. *British Journal of Nutrition*, **84** Suppl 1, S39–46.

Bruck, W.M., Graverholt, G. & Gibson, G.R. (2003) A two-stage continuous culture system to study the effect of supplemental α-lactalbumin and glycomacropeptide on mixed cultures of human gut bacteria challenged with enteropathogenic *Escherichia coli* and *Salmonella* serotype Typhimurium. *Journal of Applied Microbiology*, **95**, 44–53.

Bruck, W.M., Redgrave, M., Tuohy, K.M., Lonnerdal, B., Graverholt, G., Hernell, O. & Gibson, G.R. (2006) Effects of bovine α-lactalbumin and casein glycomacropeptide-enriched infant formulae on faecal microbiota in healthy term infants. *Journal of Pediatric Gastroenterology and Nutrition*, **43**, 673–679.

Bruzzese, E., Volpicelli, M., Squaglia, M., Tartaglione, A. & Guarino, A. (2006) Impact of prebiotics on human health. *Digestive and Liver Disease*, **38**, S283–S287.

Campagna, S., Mau, N.V., Heitz, F., Humbert, G. & Gaillard, J.L. (1999) Specific interaction between anionic phosphplipids and milk bovine component PP3 and its 119-135C-terminal fragment. *Colloids Surf B Biointerfaces*, **13**, 299–309.

Campagna, S., Cosette, P., Molle, G. & Gaillard, J.L. (2001) Evidence for membrane affinity of the C-terminal domain of bovine milk PP3 component. *Biochimica et Biophysica Acta*, **1513**, 217–222.

Campagna, S., Mathot, A.G., Fleury, Y., Girardet, J.M. & Gaillard, J.L. (2004) Antibacterial activity of lactophoricin, a synthetic 23-residues peptide derived from the sequence of bovine milk component-3 of proteose peptone. *Journal of Dairy Science*, **87**, 1621–1626.

Chabance, B., Jolles, P., Izquierdo, C., Mazoyer, E., Francoual, C., Drouet, L. & Fiat, A.M. (1995) Characterisation of an antithrombotic peptide from kappa-casein in newborn plasma after milk ingestion. *British Journal of Nutrition*, **73**, 583–590.

Chantaysakorn, P. & Richter, R.L. (2000) Antimicrobial properties of pepsin-digested lactoferrin added to carrot juice and filtrates of carrot juice. *Journal of Food Protection*, **63**, 376–380.

Chenu, J.W. & Cox, J.M. (2009) *Cronobacter* ('*Enterobacter sakazakii*'): current status and future prospects. *Letters in Applied Microbiology*, **49**, 153–159.

Clare, D.A. & Swaisgood, H.E. (2000) Bioactive milk peptides: a prospectus. *Journal of Dairy Science*, **83**, 1187–1195.

Clare, D.A., Catignani, G.L. & Swaisgood, H.E. (2003) Biodefense properties of milk: the role of antimicrobial proteins and peptides. *Current Pharmaceutical Design*, **9**, 1239–1255.

Corr, S.C., Li, Y., Riedel, C.U., O'Toole, P.W., Hill, C. & Gahan, C.G. (2007) Bacteriocin production as a mechanism for the antiinfective activity of *Lactobacillus salivarius* UCC118. *Proceedings of National Academy of Science, USA*, **104**, 7617–7621.

Cross, K.J., Huq, N.L. & Reynolds, E.C. (2007) Casein phosphopeptides in oral health: chemistry and clinical applications. *Current Pharmaceutical Design*, **13**, 793–800.

Daddaoua, A., Puerta, V., Zarzuelo, A., Suarez, M.D., de Medina, F.S. & Martinez-Augustin, O. (2005) Bovine glycomacropeptide is anti-inflammatory in rats with hapten-induced colitis. *Journal of Nutrition*, **135**, 1164–1170.

Dashper, S.G., O'Brien-Simpson, N.M., Cross, K.J., Paolini, R.A., Hoffmann, B., Catmull, D.V., Malkoski, M. & Reynolds, E.C. (2005) Divalent metal cations increase the activity of the antimicrobial Peptide kappacin. *Antimicrobial Agents and Chemotherapy*, **49**, 2322–2328.

De Simone, C., Picariello, G., Mamone, G., Stiuso, P., Dicitore, A., Vanacore, D., Chianese, L., Addeo, F. & Ferranti, P. (2009) Characterisation and cytomodulatory properties of peptides from Mozzarella di Bufala Campana cheese whey. *Journal of Peptide Science*, **15**, 251–258.

Dietert, R.R. & Zelikoff, J.T. (2009) Pediatric immune dysfunction and health risks following early life immune insult. *Current Pediatric Reviews*, **5**, 36–51.

Drider, D., FimLand, G., Hechard, Y., McMullen, L.M. & Prevost, H. (2006) The continuing story of class IIa bacteriocins. *Microbiology and Molecular Biology Reviews*, **70**, 564–582.

Drouet, L., Bal dit Sollier, C., Cisse, M., Pignaud, G., Mazoyer, E., Fiat, A.M., Jolles, P. & Caen, J.P. (1990) The antithrombotic effect of KRDS, a lactotransferrin peptide, compared with RGDS. *Nouvelle Revue Françoise d'Haematologie*, **32**, 59–62.

Dziuba, M. & Darewicz, M. (2007) Food proteins as precursors of bioactive peptides: classification into families. *Food Science and Technology International*, **13**, 393–404.

Ellison, R.T. & Giehl, T.J. (1991) Killing of gram-negative bacteria by lactoferrin and lysozyme. *Journal of Clinical Investigations*, **88**, 1080–1091.

Fanaro, S., Boehm, G., Garssen, J., Knol, J., Mosca, F., Stahl, B. & Vigo, V. (2005) Galacto-oligosaccharides and long-chain fructooligosaccharides as prebiotics in infant formulas: a review. *Acta Paediatrica*, **94** (suppl), S22–S26.

Fischer, R., Debbabi, H., Dubarry, M., Boyaka, P. & Tome, D. (2006) Regulation of physiological and pathological Th1 and Th2 responses by lactoferrin. *Biochemistry and Cell Biology: Biochimie et Biologie Cellulaire*, **84**, 303–311.

Foltz, M., Meynen, E.E., Bianco, V., van Platerink, C., Koning, T.M. & Kloek, J. (2007) Angiotensin converting enzyme inhibitory peptides from a lactotripeptide-enriched milk beverage are absorbed intact into the circulation. *Journal of Nutrition*, **137**, 953–958.

Forssmann, W.G., Zucht, H.D., Raida, M., Adermann, K. & Magert, H.J. (2003) Antibiotic peptides from bovine milk. US Patent no. 657984917.

Gauthier, S.F., Pouliot, Y. & Saint-Sauveur, D. (2006) Immunomodulatory peptides obtained by the enzymatic hydrolysis of whey proteins. *International Dairy Journal*, **16**, 1315–1323.

Georgalaki, M., Papadelli, M., Chassioti, E., Anastasiou, R., Aktypis, A., De Vuyst, L., Van Driessche, G., Devreese, B. & Tsakalidou, E. (2010) Milk protein fragments induce the biosynthesis of macedocin, the lantibiotic produced by *Streptococcus macedonicus* ACA-DC 198. *Applied Environmental Microbiology*, **76**, 1143–1151.

Ginger, M.R. & Grigor, M.R. (1999) Comparative aspects of milk caseins. *Comparative Biochemistry and Physiology Part B: Biochemistry and Molecular Biology*, **124**, 133–145.

Gomes, A.M., Malcata, F.X. & Klaver, F.A. (1998) Growth enhancement of *Bifidobacterium lactis* Bo and *Lactobacillus acidophilus* Ki by milk hydrolyzates. *Journal of Dairy Science*, **81**, 2817–2825.

Grenby, T.H., Andrews, A.T., Mistry, M. & Williams, R.J.H. (2001) Dental caries-protective agents in milk products: Investigations *in vitro*. *Journal of Dentistry*, **29**, 83–92.

Groenink, J., Walgreen-Weterings, E., van't Hof, W., Veerman, E.C. & Nieuw Amerongen, A.V. (1999) Cationic amphipathic peptides, derived from bovine and human lactoferrins, with antimicrobial activity against oral pathogens. *FEMS Microbiology Letters*, **179**, 217–222.

Hachimura, S., Takahashi, Y., Fujikawa, Y., Tsumori, C., Enomoto, A., Yoshino, U. & Kaminogawa, S. (1993) Suppression of the systemic immune-response to casein by oral-administration of a tryptic digest of casein. *Bioscience Biotechnology and Biochemistry*, **57**, 1674–1677.

Hamel, U., Kielwein, G. & Teschemacher, H. (1985) β-Casomorphin immunoreactive materials in cows' milk incubated with various bacterial species. *Journal of Dairy Research*, **52**, 139–148.

Hartmann, R. & Meisel, H. (2007) Food-derived peptides with biological activity: from research to food applications. *Current Opinion in Biotechnology*, **18**, 163–169.

Haug, A., Hostmark, A. & Harstad, O. (2007) Bovine milk in human nutrition - a review. *Lipids in Health and Disease*, **6**, 25.

Hayes, M., Ross, R.P., Fitzgerald, G.F., Hill, C. & Stanton, C. (2006) Casein-derived antimicrobial peptides generated by Lactobacillus acidophilus DPC6026. *Applied and Environmental Microbiology*, **72**, 2260–2264.

Hayes, M., Stanton, C., Fitzgerald, G.F. & Ross, R.P. (2007) Putting microbes to work: dairy fermentation, cell factories and bioactive peptides. Part II: bioactive peptide functions. *Biotechnology Journal*, **2**, 435–449.

Hayes, M., Barrett, E., Ross, R.P., Fitzgerald, G.F., Hill, C. & Stanton, C. (2009) Evaluation of an antimicrobial ingredient prepared from a *Lactobacillus acidophilus* casein fermentate against *Enterobacter sakazakii*. *Journal of Food Protection*, **72**, 340–346.

He, J. & Furmanski, P. (1995) Sequence specificity and transcriptional activation in the binding of lactoferrin to DNA. *Nature*, **373**, 721–724.

Hill, R.D., Lahov, E. & Givol, D. (1974) A rennin-sensitive bond in alpha and beta casein. *Journal of Dairy Research*, **41**, 147–153.

Hoek, K.S., Milne, J.M., Grieve, P.A., Dionysius, D.A. & Smith, R. (1997) Antibacterial activity in bovine lactoferrin-derived peptides. *Antimicrobial Agents and Chemotherapy*, **41**, 54–59.

Hwang, P.M. & Vogel, H.J. (1998) Structure-function relationships of antimicrobial peptides. *Biochemistry and Cell Biology*, **76**, 235–246.

Ibrahim, S.A. & Bezkorovainy, A. (1994) Growth-promoting factors for *Bifidobacterium longum*. *Journal of Food Science*, **59**, 189–191.

Jacquot, A., Gauthier, S.F., Drouin, R. & Boutin, Y. (2010) Proliferative effects of synthetic peptides from β-lactoglobulin and α-lactalbumin on murine splenocytes. *International Dairy Journal*, **20**, 514–521.

Janer, C., Pelaez, C. & Requena, T. (2004) Caseinomacropeptide and whey protein concentrate enhance *Bifidobacterium lactis* growth in milk. *Food Chemistry*, **86**, 263–267.

Janer, C., Arigoni, F., Lee, B.H., Pelaez, C. & Requena, T. (2005) Enzymatic ability of *Bifidobacterium animalis* subsp. *lactis* to hydrolyze milk proteins: identification and characterization of endopeptidase O. *Applied and Environmental Microbiology*, **71**, 8460–8465.

Jolles, P., Loucheux-Lefebvre, M.H. & Henschen, A. (1978) Structural relatedness of κ-casein and fibrinogen g-chain. *Journal of Molecular Evolution*, **11**, 271–277.

Juillard, V., Laan, H., Kunji, E.R., Jeronimus-Stratingh, C.M., Bruins, A.P. & Konings, W.N. (1995) The extracellular PI-type proteinase of *Lactococcus lactis* hydrolyzes beta-casein into more than one hundred different oligopeptides. *Journal of Bacteriology*, **177**, 3472–3478.

Katz, Y., Rajuan, N., Goldberg, M.R., Eisenberg, E., Heyman, E., Cohen, A. & Leshno, M. (2010) Early exposure to cow's milk protein is protective against IgE-mediated cow's milk protein allergy. *Journal of Allergy and Clinical Immunology*, **126**, 77–82 e71.

Katzir Katchalsky, A., Lahov, E. & Sode-Mogensen, M.T. (1973) Poly-peptidic antibiotic substances derived from casein. US Patent no. 3764670 A.

Kayser, H. & Meisel, H. (1996) Stimulation of human peripheral blood lymphocytes by bioactive peptides derived from bovine milk proteins. *FEBS Letters*, **383**, 18–20.

Kehagias, C., Jao, Y.C., Mikolajcik, E.M. & Hansen, P.M.T. (1977) Growth response of *Bifidobacterium bifidum* to a hydrolytic product isolated from bovine casein. *Journal of Food Science*, **42**, 146–150.

Kim, J.H., Desor, D., Kim, Y.T., Yoon, W.J., Kim, K.S., Jun, J.S., Pyun, K.H. & Shim, I. (2006) Efficacy of [alpha]s1-casein hydrolysate on stress-related symptoms in women. *European Journal of Clinical Nutrition*, **61**, 536–541.

Koletzko, S., Filipiak-Pittroff, B., von Berg, A. *et al.* (2011) Supplementation with cow's milk at birth is not recommended. *Journal of Allergy and Clinical Immunology*, doi: 10.1016/j.jaci.2010.11.051.

Korhonen, H. (2001) Antibacterial and antiviral activities of whey proteins, the importance of whey and whey components in food and nutrition. In *Proceedings of 3rd International Whey Conference*, pp. 303–321. Munich, Germany.

Laffineur, E., Genetet, N. & Leonil, J. (1996) Immunomodulatory activity of beta-casein permeate medium fermented by lactic acid bacteria. *Journal of Dairy Science*, **79**, 2112–2120.

Lahov, E. & Regelson, W. (1996) Antibacterial and immunostimulating casein-derived substances from milk: Casecidin, isracidin peptides. *Food and Chemical Toxicology*, **34**, 131–145.

Lahov, E., Edelsten, D., Sode-Mogensen, M.T. & Sofer, E. (1971) Properties of basic glycopeptides from cow milk protein by heat. *Milchwissenschaft*, **26**, 489–495.

Larson, M.A., Wei, S.H., Weber, A., Mack, D.R. & McDonald, T.L. (2003) Human serum amyloid A3 peptide enhances intestinal MUC3 expression and inhibits EPEC adherence. *Biochemical and Biophysical Research Communications*, **300**, 531–540.

Lee, K.H., Lee, D.G., Park, Y., Kang, D.I., Shin, S.Y., Hahm, K.S. & Kim, Y. (2006) Interactions between the plasma membrane and the antimicrobial peptide HP (2-20) and its analogues derived from *Helicobacter pylori*. *Biochemical Journal*, **394**, 105–114.

Li, E.W.Y. & Mine, Y. (2004) Immunoenhancing effects of bovine glycomacropeptide and its derivatives on the proliferative response and phagocytic activities of human macrophagelike cells, U937. *Journal of Agricultural and Food Chemistry*, **52**, 2704–2708.

Liepke, C., Zucht, H.D., Forssmann, W.G. & Standker, L. (2001) Purification of novel peptide antibiotics from human milk. *Journal of Chromatography B: Biomedical Sciences and Application*, **752**, 369–377.

Liepke, C., Adermann, K., Raida, M., Magert, H.J., Forssmann, W.G. & Zucht, H.D. (2002) Human milk provides peptides highly stimulating the growth of bifidobacteria. *European Journal of Biochemistry*, **269**, 712–718.

Lopez-Exposito, I. & Recio, I. (2008) Protective effect of milk peptides: antibacterial and antitumor properties. *Advances in Experimental Medicine and Biology*, **606**, 271–293.

Lopez-Exposito, I., Gomez-Ruiz, J.A., Amigo, L. & Recio, I. (2006a) Identification of antibacterial peptides from ovine a_{s2}-casein. *International Dairy Journal*, **16**, 1072–1080.

Lopez-Exposito, I., Minervini, F., Amigo, L. & Recio, I. (2006b) Identification of antibacterial peptides from bovine kappa-casein. *Journal of Food Protection*, **69**, 2992–2997.

Lopez-Exposito, I., Pellegrini, A., Amigo, L. & Recio, I. (2008a) Synergistic effect between different milk-derived peptides and proteins. *Journal of Dairy Science*, **91**, 2184–2189.

Lopez-Exposito, I., Amigo, L. & Recio, I. (2008b) Identification of the initial binding sites of alphas2-casein f(183-207) and effect on bacterial membranes and cell morphology. *Biochimica et Biophysica Acta*, **1778**, 2444–2449.

Luo, S. J. & Wong, L. L. (2007) Oral care chewing gums and confections. US Patent no. 20020071858.

Mack, D.R., Michail, S., Wei, S., McDougall, L. & Hollingsworth, M.A. (1999) Probiotics inhibit enteropathogenic *E. coli* adherence in vitro by inducing intestinal mucin gene expression. *American Journal of Physiology*, **276**, G941–950.

Mack, D.R., McDonald, T.L., Larson, M.A., Wei, S. & Weber, A. (2003) The conserved TFLK motif of mammary-associated serum amyloid A3 is responsible for up-regulation of intestinal MUC3 mucin expression in vitro. *Pediatric Research*, **53**, 137–142.

Madureira, A.R., Tavares, T., Gomes, A.M., Pintado, M.E. & Malcata, F.X. (2010) Invited review: physiological properties of bioactive peptides obtained from whey proteins. *Journal of Dairy Science*, **93**, 437–455.

Malkoski, M., Dashper, S.G., O'Brien-Simpson, N.M., Talbo, G.H., Macris, M., Cross, K.J. & Reynolds, E.C. (2001) Kappacin, a novel antibacterial peptide from bovine milk. *Antimicrobial Agents and Chemotherapy*, **45**, 2309–2315.

Martın-Diana, A.B., Janer, C., Pelaez, C. & Requena, T. (2003) Development of a fermented goat's milk containing probiotic bacteria. *International Dairy Journal*, **13**, 827–833.

Matin, M.A., Mizumoto, G. & Otani, H. (2001) Cytotoxic properties of alpha-lactalbumins toward mammalian cells. *Milchwissenschaft-Milk Science International*, **56**, 430–433.

McAuley, J.L., Linden, S.K., Png, C.W. *et al.* (2007) MUC1 cell surface mucin is a critical element of the mucosal barrier to infection. *Journal of Clinical Investigations*, **117**, 2313–2324.

McCann, K.B., Shiell, B.J., Michalski, W.P., Lee, A., Wan, J., Roginski, H. & Coventry, M.J. (2005) Isolation and characterisation of antibacterial peptides derived from the f(164-207) region of bovine a_{s2}-casein. *International Dairy Journal*, **15**, 133–143.

McDonald, T.L., Larson, M.A., Mack, D.R. & Weber, A. (2001) Elevated extrahepatic expression and secretion of mammary-associated serum amyloid A 3 (M-SAA3) into colostrum. *Vetinary Immunology and Immunopathology*, **83**, 203–211.

Mercier, A., Gauthier, S.F. & Fliss, L. (2004) Immunomodulating effects of whey proteins and their enzymatic digests. *International Dairy Journal*, **14**, 175–183.

Mills, S., Stanton, C. & Ross, R.P. (2009) Microbial production of bioactives: from fermented Functional Foods to probiotic mechanisms. *Australian Journal of Dairy Technology*, **64**, 41–49.

Mills, S., Ross, R.P., Hill, C., Fitzgerald, G.F. & Stanton, C. (2011a) Milk intelligence: Mining milk for bioactive substances associated with human health. *International Dairy Journal*, **21**, 377–401.

Mills, S., Stanton, C., Hill, C. & Ross, R.P. (2011b) New Developments and applications of bacteriocins and peptides in foods. *Annual Review of Food Science and Technology*, **2**, 299–329.

Minervini, F., Algaron, F., Rizzello, C.G., Fox, P.F., Monnet, V. & Gobbetti, M. (2003) Angiotensin I-converting-enzyme-inhibitory and antibacterial peptides from *Lactobacillus helveticus* PR4 proteinase-hydrolyzed caseins of milk from six species. *Applied and Environmental Microbiology*, **69**, 5297–5305.

Miniello, V.L., Moro, G.E. & Armenio, L. (2003) Prebiotics in infant formulas: new perspectives. *Acta Pediatrica*, **9**, S68–S76.

Miyauchi, H., Kaino, A., Shinoda, I., Fukuwatari, Y. & Hayasawa, H. (1997) Immunomodulatory effect of bovine lactoferrin pepsin hydrolysate on murine splenocytes and Peyer's patch cells. *Journal of Dairy Science*, **80**, 2330–2339.

Miyauchi, H., Hashimoto, S., Nakajima, M., Shinoda, I., Fukuwatari, Y. & Hayasawa, H. (1998) Bovine lactoferrin stimulates the phagocytic activity of human neutrophils: Identification of its active domain. *Cellular Immunology*, **187**, 34–37.

Moller, N.P., Scholz-Ahrens, K.E., Roos, N. & Schrezenmeir, J. (2008) Bioactive peptides and proteins from foods: indication for health effects. *European Journal of Nutrition*, **47**, 171–182.

Monnai, M. & Otani, H. (1997) Effect of bovine kappa-caseinoglycopeptide on secretion of interleukin-1 family cytokines by P388D1 cells, a line derived from mouse monocyte/macrophage. *Milchwissenschaft-Milk Science International*, **52**, 192–196.

Moro, G.E. & Arslanoglu, S. (2005) Reproducing the bifidogenic effect of human milk in formula-fed infants: why and how? *Acta Pediatrica*, **94**, S14–S17.

Nibbering, P.H., Ravensbergen, E., Welling, M.M., van Berkel, L.A., van Berkel, P.H., Pauwels, E.K. & Nuijens, J.H. (2001) Human lactoferrin and peptides derived from its N terminus are highly effective against infections with antibiotic-resistant bacteria. *Infection and Immunity*, **69**, 1469–1476.

O'Hara, A. & Shanahan, F. (2007) Mechanisms of action of probiotics in intestinal diseases. *Scientific World Journal*, **7**, 31–46.

O'Shea, E.F., Gardiner, G.E., O'Connor, P.M., Mills, S., Ross, R.P. & Hill, C. (2009) Characterization of enterocin- and salivaricin-producing lactic acid bacteria from the mammalian gastrointestinal tract. *FEMS Microbiology Letters*, **291**, 24–34.

O'Shea, E.F., O'Connor, P.M., Cotter, P.D., Ross, R.P. & Hill, C. (2010) Synthesis of trypsin-resistant variants of the *Listeria*-active bacteriocin salivaricin P. *Applied and Environmental Microbiology*, PMID: 20581174.

Otani, H. & Hata, I. (1995) Inhibition of proliferative responses of mouse spleen lymphocytes and rabbit peyers patch cells by bovine-milk caseins and their digests. *Journal of Dairy Research*, **62**, 339–348.

Otani, H., Monnai, M., Kawasaki, Y., Kawakami, H. & Tanimoto, M. (1995) Inhibition of mitogen-induced proliferative responses of lymphocytes by bovine kappa-caseinoglycopeptides having different carbohydrate chains. *Journal of Dairy Research*, **62**, 349–357.

Park, S.C., Kim, M.H., Hossain, M.A., Shin, S.Y., Kim, Y., Stella, L., Wade, J.D., Park, Y. & Hahm, K.S. (2008) Amphipathic a-helical peptide, HP (2-20), and its analogues derived from *Helicobacter pylori*: pore formation mechanism in various lipid compositions. *Biochimica et Biophysica Acta*, **1778**, 229–241.

Parodi, P.W. (2007) A role for milk proteins and their peptides in cancer prevention. *Current Pharmaceutical Design*, **13**, 813–828.

Pellegrini, A., Thomas, U., Bramaz, N., Hunziker, P. & von Fellenberg, R. (1999) Isolation and identification of three bactericidal domains in the bovine α-lactalbumin molecule. *Biochimica et Biophysica Acta*, **1426**, 439–448.

Pellegrini, A., Dettling, C., Thomas, U. & Hunziker, P. (2001) Isolation and characterization of four bactericidal domains in the bovine beta-lactoglobulin. *Biochimica et Biophysica Acta*, **1526**, 131–140.

Peng, T.J., Su, S.N., Wu, T.G., Ho, C.K. & Kuo, H.L. (2005) Effect of ingestion of cow's milk protein hydrolysate formulas on alpha-casein-specific immunoglobulin E and G1 antibody responses in naive and sensitized mice. *Journal of Pediatric Gastroenterology and Nutrition*, **41**, 438–444.

Petschow, B.W. & Talbott, R.D. (1990) Growth promotion of *Bifidobacterium* species by whey and casein fractions from human and bovine milk. *Journal of Clinical Microbiology*, **28**, 287–292.

Picariello, G., Ferranti, P., Fierro, O., Mamone, G., Caira, S., Di Luccia, A., Monica, S. & Addeo, F. (2010) Peptides surviving the simulated gastrointestinal digestion of milk proteins: biological and toxicological implications. *Journal of Chromatography B: Analytical Technologies in the Biomedical and Life Sciences*, **878**, 295–308.

Pihlanto, A. (2001) Bioactive peptides derived from bovine whey proteins: opioid and ace-inhibitory peptides. *Trends in Food Science and Technology*, **11**, 347–356.

Pihlanto-Leppälä, A., Marnila, P., Hubert, L., Rokka, T., Korhonen, H. J. & Karp, M. (1999) The effect of α-lactalbumin and β-lactoglobulin hydrolysates on the metabolic activity of *Escherichia coli* JM103. *Journal of Applied Microbiology*, **87**, 540–545.

Poch, M. & Bezkorovainy, A. (1991) Bovine milk κ-casein trypsin digest is a growth enhancer for the genus *Bifidobacterium*. *Journal of Agriculture and Food Chemistry*, **39**, 73–77.

Prgomet, C., Peters, S. & Pfaffl, M.W. (2006) Influence of bovine lactoferrin and lactoferricin on cytokine expression in LPS-treated cultivated bovine blood cells. *Journal of the Science of Food and Agriculture*, **86**, 640–647.

Prioult, G., Pecquet, S. & Fliss, I. (2004) Stimulation of interleukin-10 production by acidic beta-lactoglobulin-derived peptides hydrolyzed with Lactobacillus paracasei NCC2461 peptidases. *Clinical and Diagnostic Laboratory Immunology*, **11**, 266–271.

Puri, A., Bhattacharya, M., Tripathi, L.M. & Haq, W. (2009) Derivatives of human beta-Casein fragments (54-59) exhibit highly potent immunosuppressant activity. *International Immunopharmacology*, **9**, 1092–1096.

Recio, I. & Visser, S. (1999a) Identification of two distinct antibacterial domains within the sequence of bovine a_{s2}-casein. *Biochimica et Biophysica Acta*, **1428**, 314–326.

Recio, I. & Visser, S. (1999b) Two ion-exchange chromatographic methods for the isolation of antibacterial peptides from lactoferrin. *In situ* enzymatic hydrolysis on an ion-exchange membrane. *Journal of Chromatography A*, **831**, 191–201.

Requena, P., Daddaoua, A., Martinez-Plata, E., Gonzalez, M., Zarzuelo, A., Suarez, M.D., de Medina, F.S. & Martinez-Augustin, O. (2008) Bovine glycomacropeptide ameliorates experimental rat ileitis

by mechanisms involving downregulation of interleukin 17. *British Journal of Pharmacology*, **154**, 825–832.
Rokka, T., Syvaoja, E.L., Tuominen, J. & Korhonen, H. (1997) Release of bioactive peptides by enzy matic proteolysis of *Lactobacillus* GG fermented UHT milk. *Milchwissenschaft*, **52**, 675–677.
Ross, R.P., Mills, S., Hill, C., Fitzgerald, G.F. & Stanton, C. (2010) Specific metabolite production by gut microbiota as a basis for probiotic function. *International Dairy Journal*, **20**, 269–276.
Round, J.L. & Mazmanian, S.K. (2009) The gut microbiota shapes intestinal immune responses during health and disease. *Nature Reviews Immunology*, **9**, 313–323.
Saint-Sauveur, D., Gauthier, S.F., Boutin, Y., Montoni, A. & Fliss, I. (2009) Effect of feeding whey peptide fractions on the immune response in healthy and Escherichia coli infected mice. *International Dairy Journal*, **19**, 537–544.
Sandre, C., Gleizes, A., Forestier, F., Gorges-Kergot, R., Chilmonczyk, S., Leonil, J., Moreau, M.C. & Labarre, C. (2001) A peptide derived from bovine beta-casein modulates functional properties of bone marrow-derived macrophages from germfree and human flora-associated mice. *Journal of Nutrition*, **131**, 2936–2942.
Schmelzer, C.E., Schops, R., Reynell, L., Ulbrich-Hofmann, R., Neubert, R.H. & Raith, K. (2007) Peptic digestion of beta-casein. Time course and fate of possible bioactive peptides. *Journal of Chromatography A*, **1166**, 108–115.
Schroder, K., Hertzog, P.J., Ravasi, T. & Hume, D.A. (2004) Interferon-{gamma}: an overview of signals, mechanisms and functions. *Journal of Leukocyte Biology*, **75**, 163.
Segerstrom, S.C. & Miller, G.E. (2004) Psychological stress and the human immune system: A meta-analytic study of 30 years of inquiry. *Psychological Bulletin*, **130**, 601–630.
Séverin, S. & Wenshui, X. (2005) Milk biologically active components as neutraceuticals. *Critical Reviews in Food Science and Nutrition*, **45**, 645–656.
Shahidi, F. & Zhong, Y. (2008) Bioactive peptides. *Journal of AOAC International*, **91**, 914–931.
Shinoda, I., Takase, M., Fukuwatari, Y., Shimamura, S., Koller, M. & Konig, W. (1996) Effects of lactoferrin and lactoferricin(R) on the release of interleukin 8 from human polymorphonuclear leukocytes *Bioscience Biotechnology and Biochemistry*, **60**, 521–523.
Somkuti, G.A. & Paul, M. (2010) Enzymatic fragmentation of the antimicrobial peptides casocidin and isracidin by *Streptococcus thermophilus* and *Lactobacillus delbrueckii* ssp. *bulgaricus*. *Applied Microbiology and Biotechnology*, **87**, 235–242.
Stevens, K.A., Klapes, N.A., Sheldon, B.W. & Klaenhammer, T.R. (1992) Antimicrobial action of nisin against *Salmonella typhimurium* lipopolysaccharide mutants. *Applied and Environmental Microbiology*, **58**, 1786–1788.
Sutas, Y., Soppi, E., Korhonen, H., Syvaoja, E.L., Saxelin, M., Rokka, T. & Isolauri, E. (1996) Suppression of lymphocyte proliferation in vitro by bovine caseins hydrolyzed with Lactobacillus casei GG-derived enzymes. *Journal of Allergy and Clinical Immunology*, **98**, 216–224.
Tang, Z., Yin, Y., Zhang, Y. *et al.* (2009) Effects of dietary supplementation with an expressed fusion peptide bovine lactoferricin-lactoferrampin on performance, immune function and intestinal mucosal morphology in piglets weaned at age 21 d. *British Journal of Nutrition*, **101**, 998–1005.
Tellez, A., Corredig, M., Brovko, L.Y. & Griffiths, M.W. (2010) Characterization of immune-active peptides obtained from milk fermented by *Lactobacillus helveticus*. *Journal of Dairy Research*, **77**, 129–136.
Teschemacher, H., Koch, G. & Brantl, V. (1997) Milk protein-derived opioid receptor ligands. *Biopolymers*, **43**, 99–117.
Tidona, F., Sekse, C., Criscione, A., Jacobsen, M., Bordonaro, S., Marletta, D. & Vegarud, G.E. (2010) Antimicrobial effect of donkeys' milk digested in vitro with human gastrointestinal enzymes. *International Dairy Journal*, doi: 10.1016/j.idairyj.2010.10.008.
Tobita, K., Kawahara, T. & Otani, H. (2006) Bovine beta-casein (1-28), a casein phosphopeptide, enhances proliferation and IL-6 expression of mouse CD19(+) cells via toll-like receptor 4. *Journal of Agricultural and Food Chemistry*, **54**, 8013–8017.

Tomita, Y.K.M., Giehl, T.J. & Ellison, R.T. (1993) Antibacterial activity of lactoferrin and a pepsin-derived lactoferrin peptide fragment. *Infection and Immunity*, **61**, 719–728.

Turroni, F., van Sinderen, D. & Ventura, M. (2009) Bifidobacteria: from ecology to genomics. *Frontiers in Bioscience*, **14**, 4673–4684.

Ueta, E., Tanida, T. & Osaki, T. (2001) A novel bovine lactoferrin peptide, FKCRRWQWRM, suppresses Candida cell growth and activates neutrophils. *Journal of Peptide Research*, **57**, 240–249.

Uteng, M., Hauge, H.H., Markwick, P.R., FimLand, G., Mantzilas, D., Nissen-Meyer, J. & Muhle-Goll, C. (2003) Three-dimensional structure in lipid micelles of the pediocin-like antimicrobial peptide sakacin P and a sakacin P variant that is structurally stabilized by an inserted C-terminal disulfide bridge. *Biochemistry*, **42**, 11417–11426.

van der Kraan, M.I., Groenink, J., Nazmi, K., Veerman, E.C., Bolscher, J.G. & Nieuw Amerongen, A.V. (2004) Lactoferrampin: a novel antimicrobial peptide in the N1-domain of bovine lactoferrin. *Peptides*, **25**, 177–183.

Viejo-Diaz, M., Andres, M.T. & Fierro, J.F. (2005) Different anti-Candida activities of two human lactoferrin-derived peptides, Lfpep and kaliocin-1. *Antimicrobial Agents and Chemotherapy*, **49**, 2583–2588.

Vogel, H.J., Schibli, D.J., Jing, W., Lohmeier-Vogel, E.M., Epand, R.F. & Epand, R.M. (2002) Towards a structure-function analysis of bovine lactoferricin and related tryptophan- and arginine-containing peptides. *Biochemistry and Cell Biology*, **80**, 49–63.

Wakabayashi, H., Takase, M. & Tomita, M. (2003) Lactoferricin derived from milk protein lactoferrin. *Current Pharmaceutical Design*, **9**, 1277–1287.

Zoghbi, S., Trompette, A., Claustre, J., El Homsi, M., Garzon, J., Jourdan, G., Scoazec, J.Y. & Plaisancie, P. (2006) β-Casomorphin-7 regulates the secretion and expression of gastrointestinal mucins through a mu-opioid pathway. *American Journal of Physiology: Gastrointestinal and Liver Physiology*, **290**, G1105–1113.

Zucht, H.D., Raida, M., Adermann, K., Magert, H.J. & Forssmann, W.G. (1995) Casocidin-I: a casein-alpha s2 derived peptide exhibits antibacterial activity. *FEBS Letters*, **372**, 185–188.

3 Colostrum: Its Health Benefits

T. Marchbank and R.J. Playford
Office of the Vice Chancellor, Plymouth University, Plymouth, United Kingdom

3.1 Introduction

Bovine colostrum is the first milk produced after birth and is a rich natural source of nutrients, particularly immunoglobulins, antimicrobial peptides (e.g. lactoferrin and lactoperoxidase) and other bioactive peptides including growth factors. In combination with the milk that is subsequently produced, it is important for the nutrition, growth, development and immunological defence of the newborn infant. It is produced as a by-product of the milk industry and is commercially sold as a health food product to promote general health. There is however increasing evidence that it may be useful for the specific treatment of both neonatal and adult gastrointestinal disease (Ghosh & Playford 2003). There are also a number of studies which have shown that colostrum may act in stimulating and modulating the immune system. Bovine colostrum has been shown to have positive effects when used in treating side-effects of Type 2 diabetes, on bone growth and development, in surgery and in respiratory infections. Bovine colostrum is also widely taken as a supplement by athletes to improve exercise performance and recovery. There is also increasing scientific evidence to show that the many biological factors present in bovine colostrum have beneficial effects on human health.

The identity of the factor(s) in colostrum responsible for many of the biological actions is unclear. Multiple peptide growth factors are present in colostrum, including transforming growth factor (TGF) -α and -β, insulin-like growth factor (IGF) -1 and -2, epidermal growth factor (EGF) and granulocyte colony-stimulating factor (G-CSF) (Playford *et al.* 2000). More than one peptide is likely to be important, as synergistic activity between various factors in colostrum have been demonstrated. For example, co-administration of bovine lactoferrin with EGF resulted in synergistic activity in stimulating growth of the rat intestinal epithelial cell line IEC-18 (Hagiwara *et al.* 1995). However, two factors present in colostrum that are of particular interest for the treatment of gastrointestinal disease include TGF-β (Donnet-Hughes *et al.* 2000) and EGF (Shing & Klagsbrun 1987; Shing *et al.* 1987). It is also important to note that there is a marked variation in the relative abundance of the growth factor constituents across species (e.g. human colostrum has much higher concentrations of EGF than bovine colostrum) and when the colostrum changes to milk.

Colostrum contains multiple specific (antibody) and non-specific (e.g. lactoferrin) antimicrobial factors in addition to several cytokines, including interleukin (IL)-1β, IL-6, IL-10, tumour necrosis factor (TNF)-α and granulocyte-, macrophage- and granulocyte/macrophage

Milk and Dairy Products as Functional Foods, First Edition. Edited by A. Kanekanian.

colony-stimulating factors. Natural bioactive substances in colostrum also include nucleosides and nucleotides (Schlimme *et al.* 2000) which have been shown to be gastroprotective (Belo *et al.* 2006). Both the specific and non-specific constituents of colostrum may have relevance to immune modulatory and antimicrobial activity when given to neonates or adults. Several reports support the potential of colostrum for the prevention and treatment of microbial infections (Seifert *et al.* 2002) acting on the host's general and specific immune function (Cross & Gill 1999) as well as the microbe itself. For example, healthy volunteers given bovine colostrum responded to attenuated oral *Salmonella typhi* Ty21a vaccine administration with a greater increase in circulating specific IgA compared to controls (He *et al.* 2001). However, not all trials have confirmed beneficial effects, for example, adjuvant hyperimmune bovine colostrum was ineffective in decreasing the stool frequency, duration or severity of childhood shigellosis (Ashraf *et al.* 2001).

The main area of research involves studies that suggest that administration of (usually bovine) colostrum, growth-factor-enriched colostrum or purified peptides derived from colostrum may be useful to treat a variety of gastrointestinal diseases. Administration of commercial colostral preparations have been shown to decrease non-steroidal anti-inflammatory-induced gut injury in rats, mice (Fig. 3.1) and humans (Playford *et al.* 1999, 2001) and to stimulate mucosal healing of patients with inflammatory bowel disease (Khan *et al.* 2002). Other gastrointestinal conditions for which colostrum-derived growth factors may be useful include chemotherapy-induced mucositis, necrotising enterocolitis and short bowel syndrome (Playford *et al.* 2000). Other diseases and areas have also been studied for the beneficial use of colostrum. Various fractions of colostrum or purified components of colostrum have been investigated for use in treatment of Type 2 diabetes, influenza, bone development and metabolism in humans and rats, asthma, athletic performance and Alzheimer's disease.

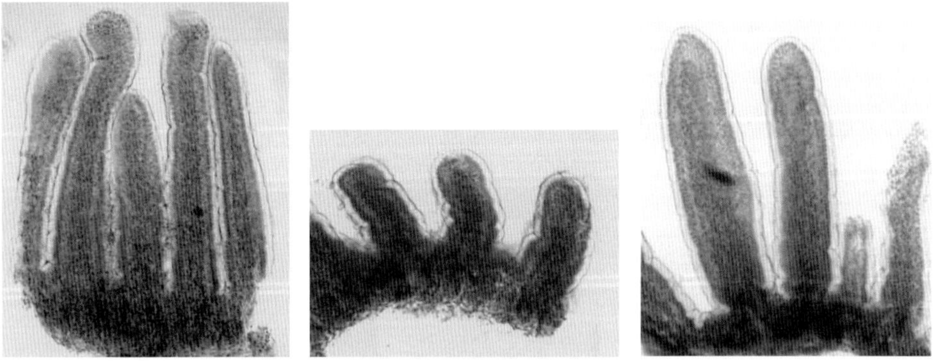

(a) (b) (c)

Fig. 3.1 Effect of bovine colostrum on non-steroidal anti-inflammatory drug-induced small intestinal injury in mice. Mice received placebo or colostrum supplementation in their drinking water for 14 days. Twenty-four hours before being killed, some animals also received 85 mg indomethacin kg^{-1} subcutaneously. The morphology of microdissected villi was determined throughout the small intestine (200x magnification). (a) Control mice did not receive indomethacin or colostrum and had long, slightly tapering, villi. (b) Mice that received indomethacin alone had markedly shortened villi with bulbous expansion of the tips. (c) Mice that received indomethacin and colostrum showed much less marked changes to the villi. Playford, Macdonald & Johnson (2000). Reproduced with permission of the American Society for Nutrition.

This chapter will cover the bioactive constituents in bovine colostrum and the research that has been performed using bovine colostrum in its various forms or constituents, for the treatment of both gastrointestinal and non-gastointestinal diseases and for its use in enhancing athletic performance.

3.2 Bioactive constituents of colostrums

Colostrum contains many factors that can influence cell growth, differentiation and function. Some of the major constituents of colostrum that can interact with peptide growth factors, non-peptide growth factors and immunological factors are discussed in the following sections.

3.2.1 *Peptide growth factors*

Epidermal growth factor receptor ligand family

This group of polypeptides with the common property of binding to the EGF receptor (also known as the c-erb1 receptor) includes EGF itself, TGF-α, mammary-derived growth factor II (MDGF-II) and human milk growth factor III (HMGF-III), which might be the same molecule as EGF (see following section). Other related polypeptides with these binding characteristics but that are not present in significant concentrations in colostrum are amphiregulin, betacellulin and heparin-binding EGF (for a more comprehensive review of these peptides see Barnard *et al.* 1995).

Epidermal growth factor

EGF is a 53 amino acid peptide produced by the salivary glands and the Brunner's glands of the duodenum in adults. EGF is present in human colostrum (200 mg L^{-1}) and milk (30–50 mg L^{-1}) and in many other species, but is not found in significant amounts in bovine secretions (Read *et al.* 1985) although related molecules have been identified and characterised. *In vitro* experiments using gastric juice from preterm infants indicate that milk-borne EGF is not deactivated under typical gastric proteolytic conditions (Kodolvsky *et al.* 1991). In contrast, our group has shown that adult gastric juice digests EGF_{1-53} to an EGF_{1-49} form that has only 25% of the biological activity of the intact EGF molecule (Playford *et al.* 1995). Once EGF enters the small intestine, it is susceptible to proteolytic digestion under fasting conditions but is preserved in the presence of ingested food proteins (Playford *et al.* 1993).

There is controversy over the physiological function of EGF in the gastrointestinal lumen under normal (non-damaged) conditions. Most studies examining the distribution of the EGF receptor in normal adult human gastrointestinal tract showed it to be present only on basolateral membranes and not on the apical (luminal) surfaces (Playford *et al.* 1996a). The distribution of the EGF receptors might, however, vary between species as autoradiographic studies identified apical receptors in the pig intestine (Kelly *et al.* 1992). If EGF receptors are distributed only on the basolateral membranes of the normal adult human gut, then EGF in the intestinal lumen is unlikely to exert any biological activity, except at sites of injury. Evidence in favour of this role for EGF include the finding that rats that have had their salivary glands removed do not develop spontaneous ulcers or atrophy of the gut. However, compared with control animals, they do develop more extensive ulceration with diminished repair if artificial ulcers are induced (Skov-Olsen *et al.* 1986). This has led to the suggestion that EGF acts as a 'luminal surveillance peptide' in the adult gut, readily available to stimulate the repair process at sites of injury (Playford *et al.* 1995).

It is important to note, however, that luminal EGF might gain access to basolateral receptors in the immature neonatal gut (Thompson *et al.* 1994) because of its increased permeability. The EGF in colostrum and milk may therefore play a role in preventing bacterial translocation (Okuyama *et al.* 1998) and stimulating gut growth in suckling neonates. *Lactococcus lactis* overexpessing EGF has been shown to be beneficial in enhancing intestinal development in early weaned pigs by stimulating proliferation, increasing duodenal villus height and increasing intestinal length (Kang *et al.* 2010).

In areas other than the gastrointestinal tract recombinant EGF has been used in the treatment of diabetic foot ulcers. Two groups have seen positive results on the healing of moderate-to-severe foot ulcers by topical application (Tuyet *et al.* 2009) or intra-lesion injections (Fernández-Montequín *et al.* 2009) of recombinant EGF preparations.

Transforming growth factor α

TGF-α is a 50 amino acid molecule that is present in human colostrum and milk at much lower concentrations (2.2–7.2 mg L^{-1}) than EGF (Okada *et al.* 1991). In contrast with EGF, TGF-α is produced within the mucosa throughout the gastrointestinal tract (Cartlidge & Elder 1989). Systemic administration of TGF-α stimulates gastrointestinal growth and repair, inhibits acid secretion, stimulates mucosal restitution after injury and increases gastric mucin concentrations (Barnard *et al.* 1995). Within the small intestine and colon, TGF-α expression occurs mainly in the upper (non-proliferative) zones which suggests that its physiological role may be to influence differentiation and cell migration rather than cell proliferation. TGF-α may therefore play a complementary role to that of TGF-β (see the following section) in controlling the balance between proliferation and differentiation in the intestinal epithelium (Koyama & Podolsky1989).

Up-regulation of TGF-α expression has been shown to occur in the gastrointestinal mucosa at sites of injury as well as in the liver after partial hepatectomy, supporting a role for TGF-α in mucosal growth and repair (Coffey *et al.* 1995). Further evidence for this role comes from research in mice that have had the TGF-α gene 'knocked out' by homologous recombination. These rats have a relatively normal phenotype under control conditions but an increased sensitivity to colonic (Egger *et al.* 1997), although not small intestinal (MacDonald *et al.* 1998), injury. These findings support the role of TGF-α in maintaining epithelial continuity but suggest that the relative importance of peptides such as this might vary from one region of the gut to another. Taken together, most studies suggest that the major physiological role of TGF-α is to act as a mucosal-integrity peptide, maintaining normal epithelial function in the non-damaged mucosa (Playford 1995).

Other peptides within this family are MDGF-II (Zwiebel *et al.* 1986) and HMGF-III. HMGF-III has a molecular mass of about 6 kDa and is the predominant growth factor in human milk, accounting for about 75% of total mitotic activity (Shing & Klagsbrun 1987; Shing *et al.* 1987). There is uncertainty as to whether HMGF-III is a distinct molecule or is, in fact, the same as EGF.

Transforming growth factor β family

This family of molecules is structurally distinct from TGF-α and, in most systems, actually inhibits proliferation. There are 35 different isoforms of TGF-β and their major site of expression in the normal gastrointestinal tract is in the superficial zones, where they may inhibit proliferation once the cells have left the crypt region (Koyama & Podolsky 1989). TGF-β has many diverse functions; it is a potent chemoattractant for neutrophils and stimulates epithelial cell migration at wound sites (Dignas & Podolsky 1993). It is therefore likely to be a key player in stimulating restitution, the early phase of the repair process during which surviving cells from the edge of a wound migrate over the denuded area to re-establish epithelial continuity.

TGF-β and TGF-β-like molecules are present in high concentrations in both bovine milk (1–2 mg L^{-1}) and colostrum (20–40 mg L^{-1}). These concentrations are sufficient to prevent indomethacin-induced gastric injury in rats (Playford *et al.* 1999), suggesting that the TGF-β in colostrum may be a key component in mediating its ability to maintain gastrointestinal integrity in suckling neonates (Walker 2010). A TGF-β-like milk growth factor has been described as being associated with the casein fraction of cow milk; this has since been shown to be a mixture of TGF-β1 and TGF-β2, predominantly the β2 form (85%) (Jin *et al.* 1991).

Insulin-like growth factors (somatomedins) and their binding proteins

IGF-I and IGF-II promote cell proliferation and differentiation (Daughaday & Rotwein 1989). They are similar in structure to pro-insulin and it is possible that they also exert insulin-like effects at high concentrations. The liver is a major site of IGF synthesis (Lund *et al.* 1986); IGF-I and IGF-II are also both expressed in particularly high amounts in the developing human foetal stomach and small intestine, with expression reaching a maximum soon after birth (Han *et al.* 1987). Bovine colostrum contains much higher concentrations of IGF-I than human colostrum (500 mg L^{-1} compared to 18 mg L^{-1}) (Baxter *et al.* 1984; Vacher & Blum 1993), with lower concentrations in mature bovine milk (10 mg L^{-1}) (Collier *et al.* 1991).

These growth factors are relatively stable to both heat and acidic conditions. They therefore survive the harsh conditions of both commercial milk processing and gastric acid to maintain their biological activity (Lowe 1991). IGF-I is known to promote protein accretion, i.e. it is an anabolic agent (Lo *et al.* 1996), and is at least partly responsible for mediating the growth-promoting activity of growth hormone (GH). Administration of IGF-1 following brain injury has been shown to be protective in both grey and white matter and leads to improved somatic function (Guan *et al.* 2003). IGF-II is present in bovine milk and colostrum at much lower concentrations than is IGF-I but, like IGF-I, it has anabolic activity and has been shown to reduce the catabolic state in starved animals (Koea *et al.* 1992).

IGFs in bovine and human colostrum and milk are present in both free and bound forms. The amount of free IGF varies during the perinatal period, with most of the IGF-I in bovine colostrum being present in the free form (i.e. not associated with its binding protein), whereas the reverse is true in the antepartum period and in mature milk (Schams & Einspanier 1991). Six IGF binding proteins (IGFBPs) have been identified and cloned. It was initially thought that the main function of IGFBPs was to act as carrier proteins, reducing the proteolytic digestion of IGF and limiting its biological activity because only the free forms of IGF are thought to have any major proliferative activity. Additional roles for IGFBPs have been suggested because it has been shown that different IGFBPs have distinct patterns of distribution in different tissues and their concentrations are altered in response to hormonal or nutrient status.

Examples include the findings that administration of dexamethasone to rats increases hepatic production of IGFBP-1 (Suh & Rechler 1997) and that malnutrition of neonatal rats decreases serum IGF-I and IGF-II but increases serum IGFBP-2 (Donovan *et al.* 1991). The detailed functions of IGFBPs are unclear, although it is probable that one of the roles of secreted or soluble IGFBP is to inhibit IGF-mediated proliferation or amino acid uptake by limiting the availability of free IGF to bind to its receptors. Conversely, cell surface- and cell matrix-associated IGFBPs may potentiate the actions of IGF by increasing local concentrations of IGF-I and IGF-II next to their receptors. A review of the clinical relevance of IGF-1 was published by Yakar *et al.* (2005). A detailed review of IGFBPs was published by Rechler (1993) and a general review of the role of IGFs and IGFBPs was published by Lund & Zimmermann (1996).

Platelet-derived growth factor

Platelet-derived growth factor (PDGF) is an acid-stable molecule that was originally identified from platelets but is also synthesised and secreted by macrophages. It consists of 2 disulphide linked polypeptides: chain A (14 kDa) and chain B (17 kDa). The dimer, therefore, exists in 3 isoforms (AA, AB and BB) that bind to tyrosine kinase-type receptors. PDGF is a potent mitogen for fibroblasts and arterial smooth muscle cells, and administration of exogenous PDGF has been shown to facilitate ulcer healing when administered orally to animals. Although PDGF is present in human and bovine milk and colostrum, most of the PDGF-like mitogenic activity in bovine milk is actually derived from bovine colostral growth factor, which shares sequence homology with PDGF (Shing & Klagsbrun 1984, 1987; Shing *et al.* 1987). A general review of the effects of PDGF were published by Szabo & Sandor (1996).

Vascular endothelial growth factor

Vascular endothelial growth factor (VEGF) is a homodimeric 34–42 kD heparin-binding glycoprotein with potent angiogenic, mitogenic and vascular permeability-enhancing factors that is related to PDGF (Keck *et al.* 1989). VEGF is present in human breast milk at a concentration of about 75 mg L^{-1} during the first week of lactation, and concentrations fall to about 25 mg L^{-1} during the second postnatal week (Siafakas *et al.* 1999). Specific receptors for VEGF have been identified on the apical membranes of the human colonic cell line Caco-2 (Siafakas *et al.* 1999) and also on the human cell line H-4. Although VEGF bound to these cell lines, it did not induce a proliferative response (Siafakas *et al.* 1999). The pathophysiological role of VEGF is therefore unclear, although its angiogenic activity may play an important role in the healing of conditions such as peptic ulceration.

Growth hormone and its releasing factor

Growth hormone (GH), along with its releasing factor (GHRF) and binding protein (Amit *et al.* 1997), is present in human and bovine colostrum and milk. Human GHRF concentrations have been reported to be about 41 ng L^{-1} in colostrum, falling to about 23 mg L^{-1} in mature milk (Werner *et al.* 1988). Suckling neonates have high circulating concentrations of GH, probably because of a combination of GH and GHRF ingestion, which stimulates the neonate to release GH from the pituitary gland (Grosvenor *et al.* 1992). Many of the growth-promoting effects of GH are mediated by release of IGF-1 (Van Wyk *et al.* 1988), although GH may also have direct mitogenic effects (Ulshen *et al.* 1993).

There is increasing evidence that systemic GH plays important modulatory roles in gut growth and function. GH receptors have been reported to be present throughout the human gastrointestinal tract (Delehate-Zervas *et al.* 1994) and transgenic mice that overexpressed GH had higher total body weights and heavier small intestines than that of control (non-transgenic) mice (Ulshen *et al.* 1993). The importance of GH in the lumen is however unclear. It is not known whether GH receptors are present on the apical membranes of enterocytes. Further studies examining the effect of GH in adults and neonates, when given via the lumen, are required to determine the pathophysiologic significance of GH in milk and colostrum. In a recent pilot study in juvenile Crohn's disease patients, treatment with an exclusion diet with nutraceutical therapy (including bovine colostrum) and recombinant growth hormone resulted in long-term remission with good growth response (Slonim *et al.* 2009).

Casein

Casein is the predominant phosphoprotein (αS2, β, κ) that accounts for nearly 80% of proteins in cow milk and cheese, with αs1-casein being the major protein fraction of bovine milk (Ginger & Grigor 1999). Casein contains opiod-type peptides that can influence gut motility, decreasing gastric emptying in rats (Daniel *et al.* 1990) in addition to possessing immune modulatory activity using *in vitro* (Pessi *et al.* 2001) and *ex vivo* studies (Meister *et al.* 2002).

In addition to any specific activities present in casein, an important effect of casein ingestion may be to preserve activity and aid adsorption of other biologically active peptides by acting as a competitive substrate for pancreatic proteolytic enzymes. Studies have demonstrated that EGF is protected from digestion in humans (Playford *et al.* 1993) and stability and adsorption of IGF-1 (Kimura *et al.* 1997) has been shown to be improved in the presence of casein. Casein has been shown to be protective in experimental bacteremia by eliciting myelopoiesis (Kanwar *et al.* 2009) and casein hydrolysates were also protective in diabetic animals, reduced the tumor growth and diminished colicky symptoms in infants.

The effect of casein (and whey) in human health are discussed further in Chapter 4 of this book.

Other less-well-defined peptide growth factors

Bovine and human milk contain several other peptides whose structure and function are less clearly defined, including: MDGF-I, a 62-kDa peptide that has been shown to stimulate the growth of mammary cells and enhance collagen production (Bano *et al.* 1992); HMGF-I and -II, acidic polypeptides that are poorly characterised (Kidwell & Solomon 1989); bovine colostral growth factor, a 35-kDa molecule responsible for most of the mitogenic activity of bovine colostrum that appears to be biochemically similar to HMGF-II and possibly to PDGF (Shing & Klagsbrun 1984, 1987; Shing *et al.* 1987); and other bovine MDGFs such as b-MDGF-I, which has a molecular mass of about 30 kDa and exhibits EGF-like activity and b-MDGF-II, which is larger (50–150 kDa) (Talhouk *et al.* 1996). Several other peptides reportedly exist; however, some of these were shown subsequently to be highly homologous with known existing molecules; the details of structure and function have not been elucidated for others. It is however likely that, over the next few years, additional novel potent growth factors with clinical potential will be identified within colostrum and milk (Belford *et al.* 1997).

3.2.2 *Non-peptide growth factors*

Several non-peptide constituents of colostrum have stimulated proliferation when added to cells *in vitro* or when infused into animal models. These factors include glutamine, polyamines and nucleotides. It is debatable whether these factors should be considered growth factors *per se* because the increased proliferation is not mediated by the classic receptor-ligand secondary messenger system. Factors such as glutamine are therefore often referred to as preferred substrates. Nevertheless, these factors play an important role in maintaining gastrointestinal mucosal mass and modulating the immune system via multiple mechanisms, e.g. altering intestinal flora and influencing the actions of growth factors. For example, the trophic response of EGF on the rat small intestinal cell line IE6 requires the presence of glutamine within the medium (Ko *et al.* 1993) and nucleotides have been shown to be gastroprotective (Belo *et al.* 2006).

Hormones

It is well established that milk and colostrum contain many hormones which, when infused systemically, influence a wide variety of end-organ systems. These systems include the hypothalamic–hypophyseal system (because colostrum contains prolactin, somatostatin, oxytocin and luteinising hormone-releasing hormone), thyroid gland (because colostrum contains thyroid-stimulating hormone, thyroxine and calcitonin), sexual glands (because colostrum contains estrogen and progesterone) and adrenal and pancreatic glands. It is probable that at least some of these hormones (e.g. luteinising hormone-releasing hormone) influence plasma concentrations and the development of various end organs of suckling neonates (Baram *et al.* 1977) because of the passage of the hormones through the bowel wall into the systemic circulation. These hormones are likely to be less influential in adults because the lower permeability of the adult bowel is likely to restrict passage of most of these factors. However, it is important to appreciate that when these factors are administered to adult patients with a damaged bowel, e.g. those with celiac or Crohn's disease or following strenuous exercise, the increased bowel permeability associated with these conditions might allow these hormones to reach their receptors and mediate pathophysiologic effects.

Leptin

Leptin is a 16-kDa 167 amino acid protein hormone that plays a key role in regulating energy intake and energy expenditure, including appetite and metabolism. Leptin acts on receptors in the hypothalamus of the brain where it inhibits appetite by counteracting the effects of the potent feeding stimulators neuropeptide Y and anandamide and by stimulating the synthesis of the appetite suppressant α-melanocyte-stimulating hormones. Leptin has been shown to be present in bovine colostrum at low concentrations (13.9 mg L^{-1}; Pinotti & Rosi 2006). However, a recent study in a diabetic mouse (ob/ob) strain showed that 25 ng hr^{-1} of leptin reduces plasma glucose and insulin levels to normal without significantly reducing body weight (Hedbacker *et al.* 2010), showing that very low concentrations of leptin exert most potent effects on glucose metabolism by up-regulation of IGF Binding Protein 2 (IGFBP2).

3.2.3 *Immunological factors*

Immunoglobulins

In mammals immunoglobulins are important in passing passive immunity from the mother to its young. In bovine colostrum the major immunoglobulin present is IgG (30–87 g L^{-1}); it makes up nearly 80% of the total immunoglobulins found in colostrum (Rona 1998; Solomons *et al.* 2002) in contrast to human milk where IgA is the major immunoglobulin present. In bovine colostrum IgA (3.2–6.2 g L^{-1}) and IgM (3.7–6.1 g L^{-1}) each make up 10–15% of total immunoglobulins whereas IgE and IgD are present in small amounts (Stelwagen *et al.* 2009). IgG is important in systemic immunity and IgG_1, the major class present (Barrington *et al.* 1997), is essential for the identification of and protection against a range of bacterial toxins (McConnell *et al.* 2001). IgA is mainly involved with local immunity and secretory IgA, which is present in colostrum, has a major role in protecting against gastrointestinal infections. IgM is the first antibody produced in response to an antigen, IgE is involved in allergic responses to antigens and IgD acts as an antigen receptor on B cells.

The possibility that the levels and types of antibodies can be altered and enhanced in bovine colostrum has allowed for the production of hyperimmune colostrum which can be used in many areas for the benefit of human health. Hyperimmune colostrum from cows vaccinated with a multivalent killed bacterial antigen has been shown to have anti-inflammatory activity mediated by tight junction-regulated suppression of neutrophil emigration (Ormrod & Miller 1992; Stelwagen & Ormrod 1998). A concentrate of hyperimmune milk has also been shown to prevent adherence of the yeast *Candida Albicans* in a tooth surface model (Hodgkinson *et al.* 2007) and hyperimmune colostral preparation reduced dental plaque (Loimaranta *et al.* 1999). Eye infections and dryness also have responded well to enriched bovine colostrum lactoserum (Chaumeil *et al.* 1994). Bovine colostrum containing anti-measles lactoglobulin given orally (100 mL per day for 30 days) to patients with multiple sclerosis resulted in a significant improvement in their condition compared to patients receiving normal bovine colostrum (Ebina *et al.* 1984). Further more detailed examples of specific uses of hyperimmune colostrum are discussed in Section 3.3.1.

Lactoferrin

Lactoferrin is an iron binding glycoprotein (80 kDa) that is present in human colostrum at a concentration of about 7 g L^{-1}, with mature milk having a lower concentration (about 1 g L^{-1}). Bovine colostrum also contains lactoferrin which has a high degree of homology with human lactoferrin (77%), but the concentration is only about 10% of that of human milk (<0.1 g L^{-1}) (Masson & Heremans 1971; Sanchez *et al.* 1988). Lactoferrin exerts multiple effects, including facilitating iron absorption and acting as an antimicrobial agent (Aisen & Litowsky 1980; Arnold *et al.* 1980). In addition, lactoferrin has been shown to stimulate the growth of various cell lines *in vitro*, including fibroblasts and intestinal epithelial cells (Hagiwara *et al.* 1995), suggesting that its presence in milk may be important in regulating gut growth in developing neonates. Bovine lactoferrin has been shown to decrease respiratory tract infections (King *et al.* 2007) in bottle-fed infants, reduced pollen antigen-induced allergic airway inflammation in a mouse model of asthma (Kruzel *et al.* 2006) and reduced *Giardia lamblia* colonisation in infants (Ochoa *et al.* 2008). Manzoni *et al.* (2009) recently reported a small placebo-controlled study in very low birth weight pre-term neonates treated with bovine lactoferrin (100 mg) with or without Lactobacillus GG for 35–45 days. Bovine lactoferrin treatment resulted in a reduction in incidence rate of sepsis. Based on this promising result, larger trials are ongoing (Manzoni *et al.* 2010). For a full review of lactoferrin for the prevention of sepsis and NEC see Venkatesh & Abrams (2010).

Cytokines

Cytokines are any of a number of small proteins, peptides or glycoprotein that are secreted by specific cells of the immune system and glial cells and which carry signals locally between cells, thus having an effect on other cells. Cytokines have a broad range of cellular function and are active in picomolar to nanomolar concentrations. In general, cytokines do not regulate normal cellular homeostasis but alter cellular metabolism during times of perturbation such as in response to inflammation (Dinarello 1994). Cytokines trigger acute cellular responses such as chemotaxis, protein synthesis and cellular differentiation.

Although cytokines and growth factors are often considered to be separate entities, it is important to appreciate that the distinction between them is sometimes blurred. For example, IL-8 has been shown to stimulate migration of the human colonic epithelial cell line LIM 1215 (Wilson *et al.* 1999), an effect that is usually attributed to growth factors such as EGF and TGF-β. In addition, some studies have shown 'cross-talk' between cytokines and growth factors. For example, Yasunaga *et al.* (1996) examined the molecular mechanisms underlying *Helicobacter pylori (H. pylori)* -induced gastric hyperproliferation in patients with large-fold gastritis. The presence of *H. pylori* caused the gastric mucosa to release the cytokine IL-1β, which in turn resulted in the local production of hepatocyte growth factor.

Human colostrum has been shown to contain a wide range of cytokines, however the content of bovine colostrum has not been as extensively investigated. Bovine colostrum has been shown to contain a number of cytokines that are important in stimulating the immune system such as IL-1β, IL-6, TNF-α, Interferon (IFN)-y and IL-1 receptor agonists

(Hagiwara *et al.* 2000). These pro-inflammatory cytokines from bovine colostrum have been shown to regulate the mitogenic response in peripheral blood mononuclear cells from newborn calves (Yamanaka *et al.* 2003). Bovine colostrum is likely to contain other facors that influence the human immune system by stimulating cytokine production. In cells from intestinal Peyer's patches in weaned piglets, bovine colostrum (1–5 g per day for 3 weeks) increased cytokine messenger RNA (Boudry *et al.* 2007) and in human peripheral blood mononuclear cells bovine colostrum stimulated IFN-ψ, IL-1 and IL-2 secretion under resting conditions whereas secretion of TNF, IL-6 and IL-4 were reduced under inflammatory conditions (Shing *et al.* 2009a), suggesting that bovine colostrum may reduce pro-inflammatory cytokine production. A study in human colonic carcinoma cells (HT-29) showed that treatment with bovine colostrum inhibited IL-1β-induced IL-8 expression, suppressed IL-1β-induced nuclear factor kappa beta (NFκβ) activation and inhibited the inhibitor protein of NFκβ degradation (An *et al.* 2009), suggesting that bovine colostrum may protect against intestinal inflammation by inhibiting the NFκβ pathway.

Colostrinin

Colostrinin (also known as CLN, proline-rich polypeptides or PRP) is a naturally occurring mixture of proline-rich polypeptides derived from colostrum. The amino acid compositions of colostrinin from ovine, bovine and human colostrum are very similar (Kruzel *et al.* 2001). Ovine colostrinin was first characterised in animal and *in vitro* studies as a substance that generally stimulates the immune response (Janusz & Lisowski 1993). Colostrinin has not been completely defined chemically; it was identified as a single protein with a molecular weight of 17–18 kDa but subsequent studies have shown it to consist of a mixture of at least 32 peptides ranging in size from 0.5 to 3 kDa (Rattray 2005). Most of the peptides appear to be derived from proteolytic processing of β casein and a β-casein homologue from colostrum. The importance of non-peptide components of the mixture is unknown but, since its effects cannot be completely mimicked by any one of the peptide components tested so far, the biological activity of colostrinin appears to be due to more than one of the components.

Preliminary research has been performed on the possible efficacy of colostrinin in Alzheimer's disease (Leszek *et al.* 1999), the most common form of dementia. The disease is characterised by extracellular senile plaques consisting mainly of aggregated (Aß) and intracellular neurofibrillary tangles, containing the cytoskeletal protein tau. There have been several small placebo controlled trials examining the value of colostrinin in Alzheimer's disease. The results are too preliminary to form any firm conclusions but are of interesting regards to performing further larger clinical trials. The potential mechanisms by which this product may work have also been investigated using microarray experiments and unsurprisingly have shown multiple actions on various signalling pathways (Szaniszlo *et al.* 2009b). A further study using human cell lines (Froud *et al.* 2010) has reported that colostrinin might influence ameloid-beta (Aβ)-induced cytotoxicity and SOD1 levels.

Other *in vitro* studies using colostrinin have shown it can increase the lifespan of cells isolated from inbred mice predisposed to premature aging and death (Boldogh & Kruzel 2008), induce neurite outgrowth of pheochromocytoma cells and inhibit beta amyloid-induced apoptosis (Bacsi *et al.* 2007). In animal studies 1-day-old domestic chicks showed enhancement of long-term memory retention (Stewart & Banks 2006) following treatment with colostrinin. In a recent microarray experiment using a human buccal mucosal cell line (TR146) treatment with colostrinin resulted in modulation of gene networks involved in cellular oxidative state, chemokine and cytokine production and cell differentiation, as well as on genes that are involved in the development, maintenance and regeneration of neurons in the central nervous system. This may also explain the improvements observed in Alzheimer's patients with mild-to-moderate dementia during treatment with colostrinin (Szaniszlo *et al.* 2009b).

In addition to studies related to Alzheimer's disease, the effects of colostrinin have also been investigated in other diseases. Colostrinin prevented allergic inflammation due to common indoor and outdoor allergens (Boldogh *et al.* 2008) in a well-characterised mouse model of allergic airway inflammation. In a study in rats that had been immunocompromised and infected with enterotoxigenic *Escherichia coli* (*E.coli*) as a model of sepsis, treatment with colostrinin alone could reduce endotoxin levels and infection of lymph nodes and interperitoneal lavage. However, when the animals were treated with colostrinin in addition to bovine colostrums, the effect was greater than either colostrinin or bovine colostrum alone (Ragab & Ibrahim 2004).

β-lactoglobulin

β-lactoglobulin (β-lg) is the major whey protein of cow's milk (*c.* 3 g L^{-1}) and is also present in many other mammalian species apart from humans. Bovine β-lg is a relatively small protein of 162 residues, with a molecular weight of 18.4 kDa. No clear function has been identified for β-lg although it binds to several hydrophobic molecules, suggesting its role in their transport. The strong suggestion is that the molecule exists primarily as a food source.

α-lactalbumin

α-lactalbumin (α-la) is an important whey protein in cow's milk (*c.* 1 g L^{-1}); it has a molecular weight of 14178 Da and forms the regulatory subunit of the lactose synthase. As a monomer, α-la strongly binds calcium and zinc ions and may possess bactericidal or antitumor activity. Recent work by Jaini *et al.* (2010) used α-la as a target vaccine autoantigen and showed that the vaccine can protect against growth of tumours in transgenic mouse models of breast cancer.

Lysozyme

Lysozyme, also known as muramidase, is a family of enzymes that damage bacterial cell walls. Lysozyme is abundant in a number of secretions such as tears, saliva, human milk and mucus. Lysozyme is part of the innate immune system. Reduced lysozyme levels have been associated with bronchopulmonary dysplasia in newborns (Revenis & Kaliner 1992). Children fed infant formula lacking lysozyme in their diet have three times the rate of diarrhoeal disease (Lönnerdal 2003). Lysozyme is effective in treatment of periodentitis and prevention of tooth decay.

Lactoperoxidase

Lactoperoxidase is a peroxidase enzyme secreted from mammary gland into colostrum and is a member of the heme peroxidase family of enzymes. Lactoperoxidase is an effective antimicrobial agent and, as such, is already used as a preservative in a range of food products, cosmetics, opthalmic and oral hygiene products.

For a full and more detailed review of other less-defined components in colostrum see Gagnaire *et al.* (2009).

3.3 Bovine colostrum use in human health

3.3.1 *Gastrointestinal diseases*

Esophagitis and H. pylori-related disease

Colostrum has been shown to have bacteriacidal activity against *H. pylori in vitro* (Korhonen *et al.* 1995; Bitzan *et al.* 1998). In addition, studies using gastric adenocarcinoma (AGS) cells have shown that hyperimmune colostrum could reduce *H. pylori* adhesion to AGS cells. The mechanism for this was not due to specific *H. pylori* antibodies but more likely due to lactoferrin or other colostral constituents, as the control colostral preparations had the same effect (Rokka *et al.* 2008). These results have not translated to clinical studies as of yet.

Studies using hyperimmune colostrum in infants in Bangladesh did not eradicate *H. pylori* (Casswall *et al.* 1998), so colostrum is unlikely to be of major clinical value for the treatment of reflux esophagitis or *H. pylori*-induced peptic ulceration. This is also because acid suppressant therapies, particularly proton-pump inhibitors, are highly efficacious and cheap (compared to recombinant peptides). Furthermore, standard *H. pylori*-eradication regimens, usually consisting of a proton-pump inhibitor and 2 antibiotics for 7d, have an eradication success rate of >90% and effectively provide a life-long cure for *H. pylori*-induced peptic ulceration.

Short-bowel syndrome

Some patients have an insufficient length of bowel to digest and absorb food adequately, usually as a result of massive intestinal resection for vascular insufficiency or after repeated operations for inflammatory bowel disease. Current therapeutic options are unpleasant and associated with a high risk of morbidity or mortality, e.g. long-term parenteral (intravenous) feeding and small-bowel transplantation. Strategies to optimise the function of residual bowel and ultimately wean patients off total parenteral nutrition would therefore be of great benefit.

There is evidence that growth factors could be instrumental in achieving this goal; for example, systemic administration of individual growth factors such as EGF have been shown to stimulate bowel growth in rats receiving total parenteral nutrition (Playford *et al.* 1996b) and oral administration of EGF helped restore glucose transport and phlorizin binding in rabbit intestines after jejunal resection (O'Loughlin *et al.* 1994). In a clinical pilot study in pediatric short bowel syndrome, EGF treatment improved nutrient absorption, increased tolerance of enteral feeds and reduced the infection rate (Sigalet *et al.* 2005).

Colostrum supplementation of piglet feeding regimens resulted in a significant increase in intestinal proliferation (Kelly *et al.* 1993). Using a pig model of infant short bowel syndrome has shown some varied results. One study showed that there was no effect on growth or gut adaptation when piglets were fed colostrum-based formula compared to normal formula (Heemskerk *et al.* 2002). In later studies however, supplementation with colostrum protein concentrate resulted in resected animals having normal weight gain and there was evidence of enhanced adaptation of the jejenum and ileum (Nagy *et al.* 2004). Levels of glucagon-like peptide-2 were shown to be increased in animals supplemented with bovine colostrum protein concentrate, indicating a possible role in the adaptive response in the intestine following resection (Paris *et al.* 2004). In a more recent study using the same model, piglets underwent a 75% proximal small bowel resection and were supplemented with $13\,g\ kg^{-1}$ of bovine colostrum protein concentrate. In the group receiving the colostrum supplementation there was a significant increase in circulating IGF-1, IGFBP-2 and IGFBP-3 and muscular hypertrophy, suggesting that these factors in bovine colostrum may be of use in increasing muscular adaptation after resection (Pereira-Fantini *et al.* 2008). Colostrum supplementation may be of particular value in young children who have undergone intestinal resection, as gut adaptation is more likely during early childhood than it is in adulthood.

Nonsteroidal anti-inflammatory drug-induced gut injury

Nonsteroidal anti-inflammatory drugs (NSAIDs) are widely prescribed and are effective in the treatment of musculoskeletal injury and chronic arthritic conditions. Nevertheless, about 2% of subjects taking NSAIDs for 1 year suffer from gastrointestinal adverse effects including bleeding, perforation and stricture formation of the stomach and intestine (MacDonald *et al.* 1997). Acid suppressants and prostaglandin analogues have been shown to be effective in reducing gastric injury induced by NSAIDs, but are less effective in preventing small intestinal injury. Novel therapeutic approaches to deal with these problems, such as the use of recombinant peptides, are therefore still required. A recent series of *in vivo* and *in vitro* studies support this idea; EGF (Playford *et al.* 1995) and TGF-α and TGF-β (Playford *et al.* 1999) have all been shown to reduce NSAID-induced gastric injury.

The beneficial effects of recombinant growth factors on NSAID-induced small and large intestinal injury are however less well documented. It was shown recently that a defatted colostrum preparation, which is rich in the growth factors discussed in Section 3.2, reduced NSAID-induced gastric and intestinal injury in rats and mice (Fig. 3.2; Playford *et al.* 1999). This material was also shown to effectively reduce gastric erosions in human volunteers taking NSAIDs (J Hunter, personal communication, 1998). Further support for this approach comes from the finding that this defatted colostrum preparation (125 mL three times daily for 7 days) reduced small intestinal permeability, which was used as a marker of intestinal damage in human volunteers taking clinically relevant doses of the drug indomethacin

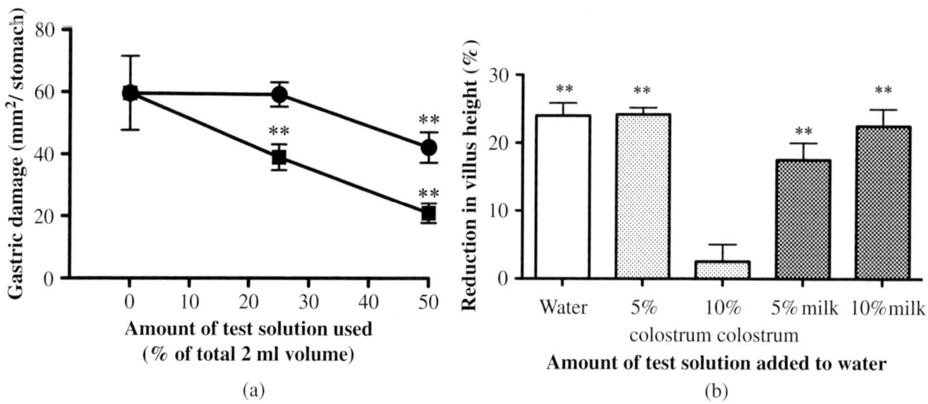

Fig. 3.2 Effect of colostrum on indomethacin-induced gastric and small intestinal injury. (a) Rats were gavaged with 2 mL of saline, defatted colostrum (■) or milk (●), all solutions also contained 2% hydroxymethylpropylcellulose to reduce the rate of gastric emptying. Thirty minutes later all rats received indomethacin (20 mg kg^{-1} subcutaneously) and were placed in Bollman-type restraint cages. Animals were killed three hours later and their stomachs removed and inflated with 4 mL of 10% formalin. The stomachs were randomly coded and macroscopic and microscopic assessment of injury were assessed in a blinded fashion. Macroscopic injury was assessed using a dissecting microscope (×10) with the aid of a reference square grid and reported as the total area of ulceration per stomach (mm^2/stomach). Results expressed as mean (standard error of the mean or SEM) of six rats per group. $^{**}p<0.01$ versus amount of injury seen in animals receiving control solution. (b) Mice received placebo or colostrum supplementation in their drinking water for 14 days. Twenty-four hours before being killed, some animals also received 85 mg indomethacin kg^{-1} subcutaneously. Microdissected samples were assessed for villus height and width by tracing the outline of the crypts and villi using a precalibrated drawing tube. The tracings were then used to measure the height of the villi. Twenty individual villi were assessed in each animal at each site and the mean value from these 20 measurements was used in the subsequent ANOVA. Data are expressed as the percentage reduction in villus height of indomethacin-treated animals compared with animals given the equivalent supplementation but no indomethacin. Results shown are those obtained from the jejunum and are expressed as mean (SD). $^{**}p<0.01$ versus villus height in animals given the equivalent supplementation but no indomethacin. Playford *et al.* (1999). Reproduced with permission of the BMJ Publishing Group.

(Fig. 3.3; Playford *et al.* 2001). However, a small clinical trial involving patients taking NSAIDs long term showed that these patients had very low intestinal permeability and this was not influenced by co-administration of bovine colostrum, possibly due to small intestinal adaptation (Playford *et al.* 2001). Further studies appear to be warranted.

Chemotherapy-induced mucositis

Current regimens for the treatment of cancers require patients to take much higher doses of chemotherapeutic agents than were used previously. As a result of these higher doses, toxic adverse effects on the bone marrow and gastrointestinal tract can be the factor limiting the dose or duration of treatment. Strategies to protect these tissues and encourage their recovery may facilitate the use of higher doses of chemotherapy, with greater potential for cure. For example, EGF enhances the repair of rat intestinal mucosa damaged by methotrexate (Hirano *et al.* 1995), TGF-β ameliorates chemotherapy-induced mucositis (Sonis *et al.* 1994)

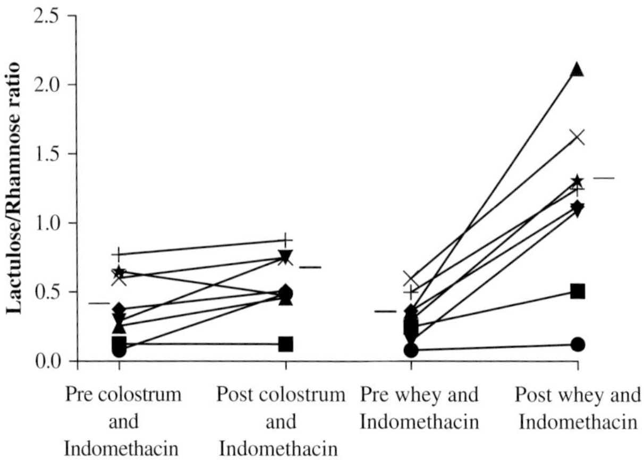

Fig. 3.3 Influence of colostrum on indomethacin-induced increases in permeability. Seven volunteers participated in this double-blinded crossover study. Following an initial baseline permeability assessment they received, in random order, the colostral preparation (125 mL 3 times daily or tsd) or control solution for 7 days. For the last 5 days of each study arm, they also took indomethacin 50 mg, tds. At the end of the test period intestinal permeability was reassessed. A 2 week 'washout' period was left between the two stages of the study. Each subject's individual results are shown by a different symbol. The group means from each stage are also marked. Baseline permeability values were similar prior to each arm of the study. In the control (whey) arm, 5 days of indomethacin caused a 3-fold rise in lactulose/rhamnose ratios ($p < 0.01$). Co-administration of the colostral preparation prevented this increase. Playford *et al.* (2001). Reproduced with permission of the Biochemical Society.

and administration of a cheese-whey-derived preparation reduced methotrexate-induced gut injury in mice (Howarth *et al.* 1996). Administration of IGF-1 improved small bowel integrity following abdominal irradiation in rats (Howarth *et al.* 1997) and IGF-1 accelerated epithelial repair following methotrexate treatment of rats (Howarth *et al.* 1998a).

Not all studies have shown favourable results; EGF had only a minor beneficial effect in reducing mouth ulceration in a phase I clinical study of patients undergoing chemotherapy (Gordler *et al.* 1995). However, in a more recent Phase 2 trial by Wu *et al.* (2009), patients undergoing radiotherapy for head and neck cancer did see significant reduction in the incidence of severe oral mucositis.

If peptides with growth stimulatory or inhibitory effects are to be used, the timing of administration is likely to be critical; growth-arresting factors might protect bone marrow or gut from the damaging effects of chemotherapy, which tend to affect areas with the highest cell turnover, if given before chemotherapy. In contrast, growth-stimulating factors might 'rescue' recovery of injured areas if administered after chemotherapy. This latter approach is already being used clinically, e.g. colony-stimulating growth factor is being used to stimulate bone marrow recovery after chemotherapy.

Inflammatory bowel disease

The etiology of ulcerative colitis and Crohn's disease is unknown; current treatment of these severe and incapacitating conditions therefore has to be on an empiric basis. Milk-derived products are already in clinical use for the treatment of inflammatory bowel disease;

casein-based enteral feeds are used for the treatment of Crohn's disease and their efficacy might be partly due to the presence of MDGFs in the preparation, which are preserved during the processing of the milk protein. Studies examining the effect of administration of EGF, PDGF, TGF-β or IGF-I in animal models of colitis have had encouraging results (Procaccino *et al.* 1994), and a cheese whey growth factor extract containing several of these growth factors had positive results in a similar model (Porter *et al.* 1998). IGF-1 has been shown to reduce edema in the mucosa and submucosa in a dextran sodium sulphate (DSS) model of colitis in rats (Howarth *et al.* 1998b).

Other peptides, not present in milk or colostrum in significant concentrations, under study as potential therapeutic agents for these conditions include keratinocyte growth factor (Zeeh *et al.* 1996) and trefoil factor family (TFF) peptides (Mashimo *et al.* 1996). Treatment with EGF and TFF1 by subcutaneous injection resulted in an improvement in histological ulcer score and myoloperoidase activity in a rat model of colitis (FitzGerald *et al.* 2004). The treatment with both peptides was significantly better than either peptide alone, suggesting this combined therapy could be useful in the treatment of colitis patients with disease extending beyond the reach of enema therapy.

To date, randomised placebo-controlled clinical trials have been performed with both EGF enemas (Sinha *et al.* 2003) and TFF3 enemas (Mahmood *et al.* 2005) for the treatment of mild-to-moderate left-sided colitis. Both trials had to be performed, including treatment with mesalazine (1.2 g per day), for ethical reasons. EGF enema treatment (5 mg in 100 mL carrier, daily for 2 weeks) resulted in a significant improvement in disease activity-, simoidoscopy- and histological-score compared to the placebo group. In contrast, TFF-3 enema treatment (10 mg mL^{-1} in 75 mL carrier, daily for 2 weeks) did not result in an additional benefit above standard treatment alone.

The effect of colostrum on inflammatory bowel disease has also been investigated in a double-blind randomised placebo-controlled clinical trial. In this trial all patients were also treated with 1.6 g mesalazine for ethical reasons. Patients with mild to moderately severe distal colitis were treated with a colostrum enema (100 mL of 10% solution twice daily) for 4 weeks. The treatment resulted in a reduction in symptom severity score, symptom duration and an improvement in histological score of muscle biopsies in the colostrum group (Fig. 3.4; Khan *et al.* 2002). The proposed mechanism by which the colostrum acts is stimulating cell migration and proliferation to re-establish continuity as seen in animal experiments (Playford *et al.* 1999). The presence of colostrum also acts to stabilise the mucosa to further damage by acting as a cytoprotective agent. The results from this are clinically relevant as the bovine colostrum enemas were given to patients who already had disease activity, demonstrating the potential as new therapy for left-sided colitis with additional benefits as compared to standard (mesalazine) treatments alone.

Necrotising enterocolitis

Necrotising enterocolitis (NEC) is a severe life-threatening illness of young children that causes severe ulceration of the small and large bowel. Its etiology is unclear, although there are many possible risk factors including prematurity, enteric infections, intestinal ischemia and abnormal immune responses. Although many proinflammatory molecules are likely to be involved in the etiology of NEC, there is currently interest in the role of the phospholipid-mediator platelet-activating factor (PAF), which is produced by intestinal flora

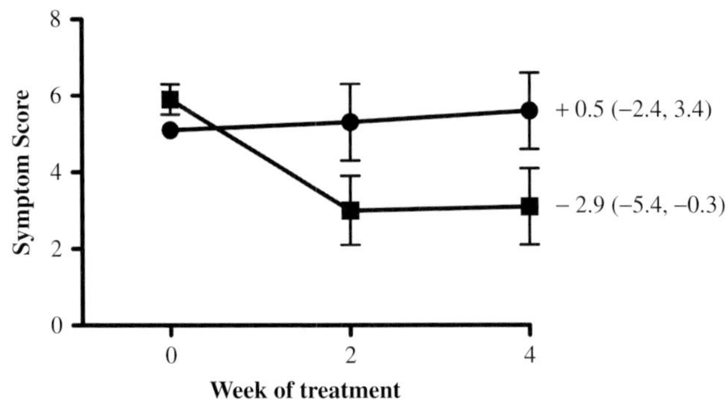

Fig. 3.4 Colostrum therapy for the treatment of inflammatory bowel disease. Patients with ulcerative colitis restricted to the left side of the colon were randomised to receive enemas containing either the colostral preparation (100 mL, 10% solution) or control solution, twice daily for 4 weeks. Patients in both arms of the trial were also given mesalazine at a dose of 1.6 g day^{-1} or, if already taking mesalazine, the dose was increased by a further 1.6 g. Colostral enemas (■) were found to be superior to control (●) in reducing symptom score. Mean change and 95% confidence interval for the change at 4 weeks are shown, illustrating that only colostrum caused a significant reduction in score. The symptom score is a well-validated scoring system score based on the assessment of patient well-being (0–3), abdominal pain (0–2), rectal bleeding (0–3), temperature (0–2), anorexia/nausea (0–1), bowel frequency (0–3), stool consistency (0–2), abdominal tenderness (0–3) and presence of extra intestinal manifestations (0–1). All patients were required to have a minimum Powell–Tuck score of 5 to qualify for entry into the study. Data are presented as the group mean ± SEM. Khan *et al.* (2002). Reproduced with permission of John Wiley & Sons.

and inflammatory cells during the development of NEC. The finding that human colostrum contains the enzyme PAF acetylhydrolase (Moya *et al.* 1994), which degrades PAF, might therefore be relevant in explaining why human milk feeds protect against the development of NEC.

Although the molecular mechanisms underlying the development of NEC are unclear there is no doubt that, once established, it is associated with a very high mortality rate. Current treatment consists of general supportive measures consisting of fluid-replacement and antibiotic therapy, although intestinal resection is often required. There is therefore a need for novel therapeutic approaches, e.g. the use of peptides to stimulate the repair process. Support for this idea comes from an *in vivo* study in neonates with micro-villous atrophy; treatment with EGF demonstrated a significant increase in gastrointestinal villus hight and depth (Walker-Smith *et al.* 1985). In a recent case study a continuous infusion of EGF resulted in a remarkable restorative effect on gut histology in a child with NEC (Sullivan *et al.* 1991). A subsequent clinical trial in neonates with NEC treated with recombinant EGF(1–48) produced positive and measurable remodelling trophic effects on the gastrointestinal mucosa (Sullivan *et al.* 2007). No clinical studies have been performed using colostrum for the treatment of NEC; however, in *in vitro* studies spray-dried colostrum significantly decreased the attachment of a range of bacteria, isolated from NEC-positive infants, to HT29 colonic carcinoma cells (Brooks *et al.* 2006). Colostrum increased the inflammatory mediator IL-8, and the authors of the study suggest that the isolation of the antibacterial (possibly lactoferrin, see above) component of colostrum could be useful for the treatment of NEC (Venkatesh & Abrams 2010).

Infective diarrhoea

Most cases of infective diarrhoea resolve spontaneously and only occasionally require a short course of antibiotics. For immunocompromised subjects, such as those with HIV infection, prophylaxis against the unusual organisms that they are susceptible to, e.g. *Cryptosporidium*, may be beneficial. Patients with AIDS with diarrhoea caused by *C. parvum* were treated with colostrum-derived bovine immunoglobulin concentrate for 21 days. Patients treated with a powdered form of the medication had a significant decrease in stool output and frequency; those taking it in a capsule form and those with *C.parvum* independent diarrhoea saw no improvement (Greenberg & Cello 1996). This has been confirmed by a recent clinical trial in Nigeria where patients with HIV-associated diarrhoea were treated with colostrum for 4 weeks. The study showed that a colostrums-based product could significantly decrease stool evacuations, decrease fatigue, increase body weight and increase CD4+ count (Florén *et al.* 2006).

Hyperimmune colostrum, colostrum from cows vaccinated with specific viruses or bacteria, has been shown to be beneficial in the prevention and treatment of enteropathic infections. Several trials have been performed in healthy volunteers treated with specific immunoglobulin concentrates extracted from immunised cow colostrum and then challenged with the organism to which the cow was immunised. A trial in healthy volunteers given buffered colostral immunoglobulin concentrate against enterotoxic *E. coli* (3 times per day) had significantly less diarrhoea after being challenged with enterotoxic *E.co*li than the volunteers given placebo (Tacket *et al.* 1988). Healthy volunteers given bovine hyperimmune anti *Cryptosporidium* colostrum immunoglobin (10 g 3 times daily for 5 days) had significantly reduced diarrhoea and oocyst secretion after challenge with *C. parvum* (Okhuysen *et al.* 1998). A similar trial was performed in healthy volunteers given hyperimmune immunoglobulin concentrate extracted from colostrum from *Shigella flexneri*-immunised cows (3 times/day for a week; Tacket *et al.* 1992). Following a challenge with *S. flexneri*, significantly less of the volunteers in the treatment group became ill and the organism was shed far faster compared to the placebo group.

In other trials, hyperimmune colostrum has been used to treat infection in patients who were already ill. A trial in Iraq using bovine anti-*E. coli* hyperimmune colostrum concentrate in infants (7 days) resulted in a lower incidence of diarrhoea and also significantly increased weight gain in the 8 months following treatment (Tawfeek *et al.* 2003). In Bangladesh, children with rotavirus diarrhoea were treated for 4 days with 10 g per day immunoglobulins extracted from bovine colostrum from rotavirus-immunised cows. Children in the treatment group had significantly less daily and total stool output and frequency, they required less oral hydration and the rotavirus cleared quicker than those taking placebo (Sarker *et al.* 1998).

The use of whole hyperimmune colostrum rather than specific antibodies purified from milk or other sources has the added value of potentially stimulating the repair process (due to the presence of growth factors) as well as facilitating the eradication of the infection by mechanisms involving non-specific antibacterial factors in colostrum and milk. In a small study using whole hyperimmune colostrum to human rotavirus given to infants, there were significantly less episodes of diarrhoea caused by rotavirus compared to those infants given milk (Ebina *et al.* 1985, 1992). In a more recent study by Mitra *et al.* (1995), infants infected with rotavirus diarrhoea were treated for 3 days with 100 mL of hyperimmune colostrum from cows immunised with four types of human rotavirus. The patients in the treatment group had a significant reduction in stool output and a significantly shorter duration of diarrhoea.

However, in another trial in Bangladesh, children with bloody mucoid diarrhoea and found to be infected with *Shigella dysentriae* were treated with either hyperimmune colostrum containing *S. dysentriae* antigen or bovine colostrum (100 mL 3 times per day for 3 days) alongside pivmecillinam. Treatment with either the hyperimmune colostrum or normal bovine colostrum had no significant beneficial effect (Ashraf *et al.* 2001).

3.3.2 *Non-gastrointestinal diseases*

Type 2 diabetes

In a clinical trial performed by Kim *et al.* (2009) Type 2 diabetic patients were treated with 5 g of bovine colostrum twice daily for 4 weeks. This treatment resulted in a significant decrease in levels of blood glucose and ketone bodies as well as reduction in the levels of cholesterol and triglycerides. The reduction in all these factors appears to be mediated by the presence of IGF-1, which lowers blood glucose and increases insulin sensitivity (Frystyk 2004), and leptin, which works in concert with IGF-1 to lower cholesterol and triglycerides and reduces plasma glucose levels (Shimomura *et al.* 2000) and EGF which restores glucose transport (O'Loughlin *et al.* 1994). This result may have relevance in reducing the numerous complications in Type 2 diabetic patients such as cardiovascular disease (Grundy *et al.* 1999), blindness, kidney damage, endothelial dysfunction, impaired microcirculatory flow and increased thrombotic risk (Rosenson *et al.* 2001).

In addition to the prevention of complications, bovine colostrum may also have a therapeutic potential as an anti-Type 2 diabetes treatment. This is confirmed by recent research by Szaniszlo *et al.* (2009b) who found that colostrinin had an effect on adipocytokine gene networks that play a crucial role in the pathobiology of diabetes.

Bone development

In a study in juvenile rats (Lee *et al.* 2008) a 1–30 kDa fraction of bovine colostrum (GP-C) comprising 5% of the diet was fed to rats over a period of 7 weeks. In animals that had received the colostral fraction there was a significant increase in body weight, femur length, femur weight and bone mineral density compared to animals fed a diet not including the colostral fraction. The animals fed the colostral preparation had increased serum levels of growth hormone, IGF-1 and osteocalcin. The mechanism by which this preparation has such an effect is unclear but the various factors present either directly or indirectly raise serum GH and IGF-1 levels which results in an increase in proliferation in chondrocytes, resulting in bone growth and an increase in serum osteocalcin. This is confirmed by the finding that GP-C also stimulated proliferation *in vitro* in a mouse osteoblast-like cell line (MC3T3-E1).

How these results translate to human health has yet to be investigated, but human clinical trials using dietary supplementation with milk basic protein (MBP) from milk whey are ongoing. MBP was reported to produce a significant increase in bone mineral density in adult women (Aoe *et al.* 2001), prevented bone loss in menopausal woman (Aoe *et al.* 2005) and promoted bone formation and repressed bone resorption in healthy adult men and young women (Toba *et al.* 2001; Uenishi *et al.* 2007). In another study, however, no effect was seen on bone density and metabolism in young women when given MBP (Zou *et al.* 2009).

As colostrum contains higher levels of growth-promoting protein compared to milk, this could potentially be more effective in bone development. In a recent study, ovariectomised rats treated with bovine colostrum acidic proteins (50 mg per day for 12 weeks) increased bone mineral content and density of the femur (Du *et al.* 2010). The effects were attributed to the presence of osteopontin, lactoferrin, EGF and IGF-2. Osteoprotegerin is another peptide found in bovine colostrum that has been shown to contribute to anti-resorptive activity on bone (Vidal *et al.* 2004).

Respiratory infections

Brinkworth & Buckley (2003) performed a double-blind placebo-controlled randomised study investigating the effect of concentrated bovine colostrum protein (CBC, 60 g per day, for 8 weeks) on adult males. A significantly smaller proportion of volunteers in the CBC arm reported symptoms of upper respiratory tract infection; however, treatment had no effect on duration of symptoms. In a study in children with recurrent upper respiratory tract infections, treatment with bovine colostrum for 12 weeks resulted in a reduction in episodes of upper respiratory tract infection and resulting hospitalisation (Patel & Rana 2006). The mechanism by which the colostrum increased resistance to infection is unclear, however recent studies have reported that colostrum supplementation increases salivary IgA (Mero *et al.* 2002; Crooks *et al.* 2006). Increased salivary IgA levels have been shown to reduce the incidence of upper respiratory tract infection in a separate study (MacKinnon 1996).

Bovine colostrum (a 400 mg capsule once daily for 8 weeks) has also been investigated for its prophylactic use against influenza in both healthy volunteers and in high-risk cardiovascular subjects (Cesarone *et al.* 2007). In the colostrum group in healthy volunteers, the number of flu episodes and the number of days of illness was significantly less compared to both the non-treatment group and the vaccination group. In the cardiovascular patients colostrum alone, or in combination with vaccination, significantly reduced the number of complications and hospital admissions compared to the vaccination alone group, suggesting colostrum might have value in preventing influenza.

Surgery

Bölke *et al.* (2002a) performed a placebo-controlled pilot study investigating the effect of pre-operative oral application of a bovine colostrum preparation (56 g per day for 3 days) in patients undergoing abdominal surgery. Prophylactic oral application of bovine colostrum resulted in a significant reduction in endotoxemia and prevented reduction in endotoxin-neutralising capacity. The authors suggest that the presence of bovine colostrum stabilises the gut barrier during abdominal surgery. However, in a similar trial by the same group where patients were given enteral bovine colostrum (42 g per day for 2 days) prior to undergoing coronary bypass operations, there was no significant effect on perioperative endotoxemia (Bölke *et al.* 2002b).

In a pilot study bone marrow transplant recipients were treated with 10 g of bovine-anti-*Candida* antibody concentrated from bovine colostrum from cows immunised with *C. albicans* (3 doses per day from 4 days before to 28 days after transplant) as a mouthwash. Allthough this was a small suboptimal study, the treatment resulted in a reduction in *Candida* organisms in the oral cavity (Tollemar *et al.* 1999).

Failure to thrive

In a recent study, children with non-organic failure-to-thrive condition were treated with 40 mg kg^{-1} day^{-1} of bovine colostrum. The children that received the bovine colostrum supplement significantly increased weight compared to the standard treatment group (Panahi *et al.* 2010).

3.3.3 *Colostrum and athletic performance*

Colostrum contains growth factors such as EGF, IGF, TGF-α and -β, immunoglobulins, cytokines, lactoferrin and lysozyme, as well as hormones such as growth hormone, gonadotrophin-releasing hormone, luteinising hormone and glucocorticoids. This mixture of constituents suggests that supplementation with colostrum could be beneficial to improve immune function, gastrointestinal integrity and the neuroendocrine system, all areas that may be affected by intensive training and exercise. Athletes have used bovine colostrum as a nutritional supplement to enhance immune function, improve exercise performance and recovery and increase lean muscle mass long before there was any scientific evidence to support these effects. There have now been a range of studies investigating the impact of bovine colostrum supplementation on exercise performance with varying results (for a full review see Shing *et al.* 2009b).

Mero *et al.* (1997) were the first group to investigate the effect of bovine colostrum supplementation on IGF-1 levels in athletes in a randomised double-blind crossover trial. The athletes (n=9) were given 25 mL of bovine colostrum mixed with placebo (containing 13.5 mg IGF-1), 125 mL of bovine colostrum (containing 67.7 mg IGF-1) or placebo (milk whey) for 8 days. The 125 mL bovine colostrum resulted in an increase in serum IGF-1; this increase was not significant compared to the other treatments, however. There was also no significant effect on secretory IgA, serum IgG or jumping performance. Another trial where athletes (n=22) received supplementation with 20 g bovine colostrum for 8 weeks showed no significant effect on strength exercise performance; however, the colostrum group did have a significant increase in bone-free lean body mass (Antonio *et al.* 2001). Kerksick *et al.*(2007) also reported an increase in fat-free mass after 12 weeks supplementation with colostrum (60 g per day) with or without creatine in resistance-trained subjects (n=49).

Other trials have looked at the effect of colostrum supplementation on exercise performance. Elite field hockey players (n=35) supplemented with bovine colostrum for 8 weeks (Hofman *et al.* 2002) showed a significant improvement in sprint performance compared to performance following whey supplementation. However, there was no change in vertical jump performance, body composition or in an endurance test. Competitive cyclists (n=42) supplemented with 20 g and 60 g of bovine colostrum per day for 8 weeks showed an increase in timed trial performances (Coombes *et al.* 2002).

Buckley *et al.* (2002, 2003) performed two randomised double-blind placebo-controlled trials supplementing athletes (n=30 and 51, respectively) with 60 g per day bovine colostrum for 8 weeks. They looked at various exercise parameters, but found no significant change in plasma IGF-1 levels in either study. In the first study after 8 weeks of supplementation however, running performance (effective peak running speed) was significantly increased, possibly due to enhancement of recovery. In the second study, peak vertical jump power and peak cycle power were also significantly increased. The same group (Brinkworth *et al.* 2002) looked at the effect of 60 g bovine colostrum supplementation in elite female rowers (n=13) during 9 weeks of pre-competition training. They found that bovine colostrum enhanced blood buffer capacity but not performance. In a more recent study using the same supplementation parameters and athletes, Brinkworth & Buckley (2004) went on to show that resting haemoglobin concentration and plasma buffer capacity were unaffected by colostrum supplemenation. The reported enhancement to buffer capacity was due to adaptations in the tissue buffer system. In another more recent study by this group (Brinkworth *et al.* 2004) the effect of 60 g per day bovine colostrum for 8 weeks on tissue composition of resistance trained and untrained limbs (elbow flexors) was investigated. The circumference and cross-sectional area of the trained limbs in the bovine colostrum group significantly increased compared to the placebo group. There was no change in any of these measures in the untrained limb, suggesting that the increases were in skin cross-sectional area and not fat.

In a later study, Mero *et al.* (2005) investigated the effect of supplementation with 20 g bovine colostrum for 2 weeks on protein metabolism and strength performance in non-competitive athletes (n=12). They showed that bovine colostrum supplementation resulted in a significant increase in serum essential and branched chain amino acids. There was no effect on protein net balance, strength performance or serum IGF-1 concentration.

Shing *et al.* (2006) looked at the effects of 10 g per day bovine colostrum for 10 weeks in highly trained cyclists (n=29). The supplementation with bovine colostrum had no effect on 40 km time trial performance during normal training. At the end of 5 days of high-intensity training there was however significant improvement in 40 km time trial performance, a reduction in fatigue and a decrease in ventilatory threshold. The same group (Shing *et al.* 2007) went on to investigate the effects of 10 g per day bovine colostrum supplementation for 6 weeks in highly trained cyclists (n=29) on immune variables following a 40 km time trial during normal training and at the end of 5 days of high-intensity training. They showed that bovine colostrum supplementation significantly increased pre-exercise serum-soluble TNF receptor 1, suppressed post-exercise decrease in cytotoxic/suppressor T cells, prevented post-exercise decrease in serum IgG_2 concentration and resulted in a trend towards reduced incidence of upper respiratory illness symptoms.

Davison & Diment (2010) investigated the effect of bovine colostrum supplementation to counter exercise-induced immunodepression. Supplementation of 20 g per day of bovine colostrum for 4 weeks in active men resulted in a faster recovery of neutrophil function and prevented the decrease in salivary lysozyme release, but not IgA, after prolonged exercise, suggesting that bovine colostrum supplementation may be beneficial to host defence. In another study in elite swimmers (n=25) supplementation with 25 g per day of bovine colostrum for 10 weeks had no effect on salivary or plasma immunoglobulin levels; there was however a reduction in upper respiratory tract symptoms in the colostrum group compared to the placebo group, although this did not reach significance (Crooks *et al.* 2010).

Heavy exercise causes gut symptoms and 'heat stroke' in extreme cases, partially due to increased intestinal permeability of luminal toxins. Gastrointestinal symptoms including cramps, diarrhoea, nausea and bleeding are commonly reported by long-distance runners (Pals *et al.* 1997). These symptoms are likely due to a combination of reduced splanchnic blood flow, hormonal changes, altered gut permeability and increased body temperature. There is evidence that bovine colostrum supplementation can reduce heat-stress-induced hyperpermeability and endotoxin concentration in rats (Prosser *et al.* 2004).

Our group has recently performed a double-blind placebo-controlled crossover study in active volunteers (n=12) supplemented with 20 g per day of colostrum for 14 days (Marchbank *et al.* 2010) and the effect on exercise-induced intestinal permeability. We have shown that bovine colostrum supplementation did not affect core temperature, cardiovascular parameters and gut hormone profiles in response to exercise at the assessments at the end of the 14-day study arms. However, gut permeability increased by a factor of *c.* 2.5 in response to exercise in the placebo arm, but subjects who had ingested colostrum rather than placebo showed a marked reduction in the increase in gut permeability following exercise (Fig. 3.5). The possible mechanisms include reducing temperature-induced apoptosis, reducing the temperature-induced reduction in cellular resistance and induction of HSP70 expression (Fig. 3.6).

The effects of bovine colostrum supplementation on exercise performance and recovery appear to be variable depending on the parameters measured; results are equivocal. There are as many studies suggesting improvements as there are showing no significant changes. More research is necessary to investigate the mechanisms by which bovine colostrum constituents

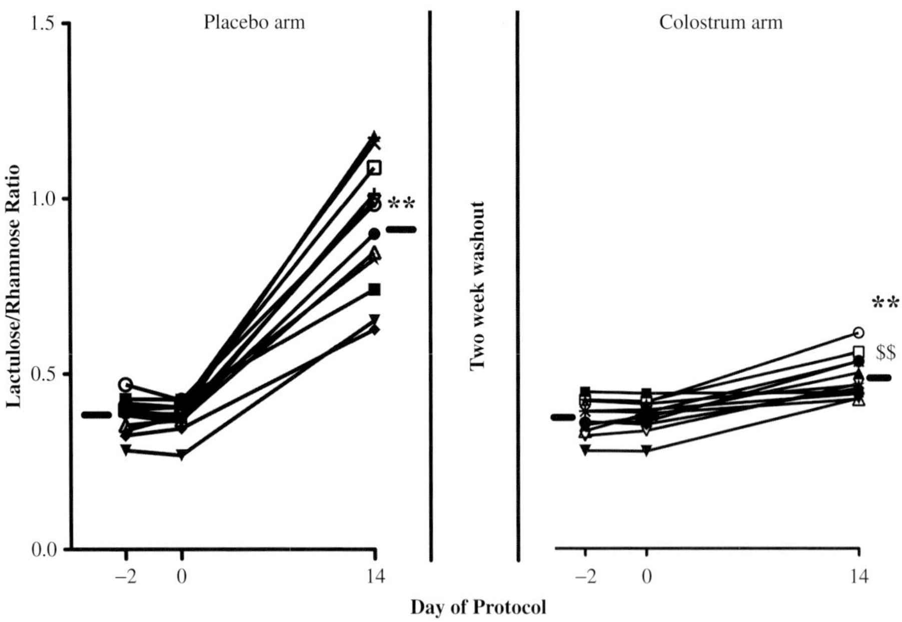

Fig. 3.5 Effect of colostrum on exercise-induced increases in intestinal permeability. Two baseline assessments (no exercise) were performed before each arm and at the end of 14 days ingestion of placebo or colostrum. The third value for each arm of the study was performed immediately after the subject had followed a 20 min 80% $\dot{V}O_2$max protocol. Bars show mean value for each time period. **, $$ signify p<0.01 different from baseline for that arm and p<0.01 compared to equivalent period in opposite arm. Marchbank, Davison, Oakes, Ghatei, Patterson & Playford (2010). Reproduced with permission of Elsevier.

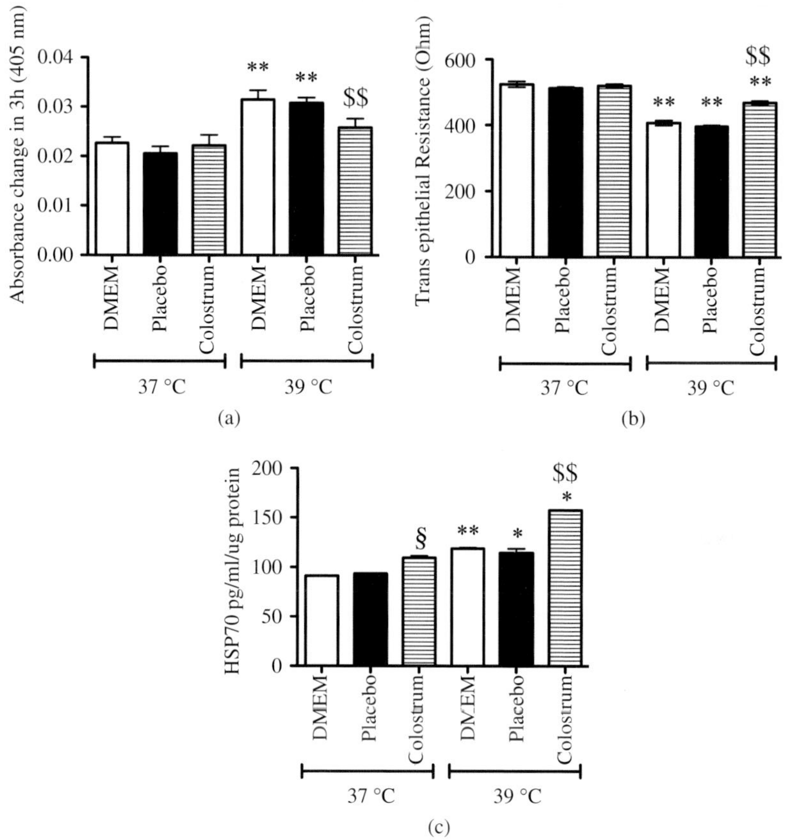

Fig. 3.6 Effect of colostrum on temperature induced changes *in vitro*. (a) Effect of colostrum on temperature-induced apoptosis. T84, polarising human colonic carcinoma cells were incubated at 37 or 39°C in medium alone, with placebo (4%) or with colostrum (4%). Changes in apoptosis were determined using active caspase-3 assay kits, following changes in absorbance at 405 nM. (b) Effect of colostrum on temperature-induced changes in cellular resistance. T84 cells were grown as confluent polarised monolayers on transwell permeable supports until resistance >500 Ohms. Cells were incubated under same conditions as for (a). Resistance was measured using a Milicell-electrical resistance system at 4 hours. (c) Effect of colostrum on heat shock protein expression (HSP70). Cells were incubated in similar conditions to (a). Presence of colostrum increased HSP expression at both 37 and 39°C. **,* represents $p<0.01$ and $p<0.05$ different from 37°C equivalent; & represents $p<0.05$ compared to medium alone at 37°C; $$ represents $p<0.01$ compared to medium alone at 39°C. Marchbank *et al.* (2011). Reproduced with permission of the American Physiological Society.

exert their effect, and larger-scale trials using well-defined bovine colostrum and a standard dose and length of supplementation are required for a definitive answer.

3.4 Considerations

Nearly all of the colostral preparations available are taken as oral preparations. It is therefore important to consider what happens to the biological activity that has been shown to be present *in vitro*; it may not be present once ingested. Many peptide growth factors are likely to be degraded to inactive forms by acid/pepsin and pancreatic proteolytic digestion soon after ingestion (Playford *et al.* 1993). In addition, many of the receptor-binding sites such as the EGF-receptor are not accessible from the apical (luminal) surface (Playford *et al.* 1996a); even if growth factors reach the small intestine intact, this may not result in receptor activation.

Until recently, most research has focused on the use of a single peptide for the treatment of a particular condition. There is now increasing evidence that administration of a combination of many peptides, whether purified or recombinantly produced, can result in additive or synergistic activity. For example, the co-administration of GH and IGF-I stimulates anabolism (Kupfer *et al.* 1993), the co-administration of bovine lactoferrin and EGF stimulates the growth of the rat intestinal epithelial cell line IEC-18 (Hagiwara *et al.* 1995) and administration of TFF1 and EGF reduces DSS colitis in rats (FitzGerald *et al.* 2004). Orally administered colostrum-derived preparations therefore appear to be an attractive therapeutic option because they contain many different growth factors in a formulation that provides inherent protection against proteolytic digestion.

Much of the appeal of colostral bioactive products is the idea that they are naturally occurring products and therefore seen as a 'healthy option'. If a product is seen as a food supplement rather than a drug, regulation is much less stringent even although some of these compounds contain poorly identified factors with major biological activity. Batch variation and quality control issues are also aspects that are relevant if the active ingredients of the products are poorly defined. In the UK colostrum is regulated by standard food hygiene regulatory mechanisms, but the Medicines Control Agency has recognised the need for vigilance by incorporating adverse reactions reporting on unlicensed remedies into their drug reaction monitoring function. So far, two studies have reported side-effects including occasional minor gastrointestinal complaints such as nausea and flatulence (Rump *et al.* 1992; Plettenberg *et al.* 1993); however, most clinical studies performed report no side-effects from the supplementation with bovine colostrum.

Other approaches currently under scrutiny include altering the volume and nature of the components of colostrum, e.g. GH (Gunn *et al.* 1996), prolactin and colony-stimulating factor-1 (Sapi & Kacinski 1999) or using genetic modification technology to improve the healing and protective properties of milk. With the use of recombinant technology the production of the required peptides, including human homologues, can be specifically targeted to the breast tissue of the animal by using specific promoters such as the β-lg gene (Simons *et al.* 1998). This approach therefore provides the potential to specifically modify bovine colostrum to increase its content of beneficial peptides, including human homologues such as human lactoferrin (Hyvönen *et al.* 2006) and casein (Laible *et al.* 2007). These products could then be used in a similar way to that of colostrum for the prevention and treatment of gut injury and other diseases. Hyperimmune colostrum and colostral products are already being studied in clinical trials and their results are promising against a range of pathogens including *C. parvum* (Florén *et al.* 2006), enterotoxic *E. coli* (Tawfeek *et al.* 2003) and rotavirus (Sarker *et al.* 1998).

Many bovine colostral preparations are already available in health-food shops and, as for any other milk product for human consumption, their manufacture is regulated by food hygiene standards. All of these colostral preparations are pasteurised, microfiltered or otherwise treated to prevent the risk of contamination with enteropathogens; the concentration of endotoxins in these preparations is similar to that of standard commercial milk. If colostrum or modified milk products are to be used in clinical practice, several issues regarding their safety will however need to be addressed. It is unlikely that human colostrum or milk will find a major role in clinical practice because of its limited supply and because of concerns regarding the transmission of infectious agents such as cytomegalovirus. It is therefore likely that further research into the commercial aspects of using purified peptides to treat human diseases will focus on colostrum derived from ruminants. Regulatory authorities require bovine herds to be certified free from bovine spongiform encephalopathy and require sheep, which are being used in several studies to produce recombinant peptides in milk (Simons *et al.* 1998; Dalrymple & Garner 1998), to be free from the ovine equivalent of bovine spongiform encephalopathy, scrapie. An additional area of research concerns the use of recombinant hormones, such as bovine somatotropin, to increase milk yields. Although approval for the use of bovine somatotropin was granted by the US Food and Drug Administration in 1993, the European Union banned its use until at least the end of 1999 and there is continuing controversy regarding the safety of its use. For further discussion of the use of bovine somatotropin, readers are referred to the article by Morris (1999).

Current farming methods allow the production of large amounts of bovine colostrum for clinical use. It is important that batch variations during production be kept to a minimum to ensure consistency of the product produced and that processing methods be developed to prevent deactivation. Such preparations have the advantage of being perceived as 'natural' products, which might result in greater patient acceptance and compliance. Further therapeutic advantages might also be gained by developing formulations specifically tailored for individual conditions, e.g. the use of a hyperimmune milk or colostrum formulation for the treatment of immunocompromised patients.

3.5 Conclusion

Looking at the results of all the clinical trials taken together, colostrum-derived proteins and peptides are bio-available and appear safe for the prevention and treatment of various disorders in humans. Bovine colostrum or colostral preparations may play a complementary role (natural agents) rather than a substitutional role to the synthetic pharmacological drugs. Continuing research is necessary to investigate the mechanisms by which bovine colostrum constituents exert their effect and larger-scale trials, using well-defined bovine colostrum, a standard dose and length of supplementation, are required for a definitive answer.

References

Aisen, P. & Listowsky, I. (1980) Iron transport and storage proteins. *Annual Review of Biochemistry*, **49**, 357–393.

Amit, T., Dibner, C. & Barkey, R.J. (1997) Characterization of prolactin- and growth hormone-binding proteins in milk and their diversity among species. *Molecular and Cellular Endocrinology*, **130**, 167–180.

An, M.J., Cheon, J.H., Kim, S.W., Park, J.J., Moon, C.M., Han, S.Y., Kim, E.S., Kim, T.I. & Kim, W.H. (2009) Bovine colostrum inhibits nuclear factor kappa-mediated proinflammatory cytokine expression in intestinal epithelial cells. *Nutritional Research*, **29**, 275–280.

Antonio, J., Sanders, M.S. & Van Gammeren, D. (2001) The effects of bovine colostrum supplementation on body composition and exercise performance in active men and women. *Nutrition*, **17**, 243–247.

Aoe, S., Toba, Y., Yamamura, J., Kawakami, H., Yahiro, M., Kumegawa, M., Itabashi, A. & Takada, Y. (2001) Controlled trial of the effects of milk basic protein (MBP) supplementation on bone metabolism in healthy adult women. *Biosciecne Biotechnology and Biochemistry*, **65**, 913–918.

Aoe, S., Koyama, T., Toba, Y., Itabashi, A. & Takada, Y. (2005) A controlled trial of the effect of milk basic protein (MBP) supplementation on bone metabolism in healthy menopausal women. *Osteoporosis International*, **16**, 2123–2128.

Arnold, R.R., Brewer, M. & Gauthier, J.J. (1980) Bactericidal activity of human lactoferrin, sensitivity of a variety of micro-organisms. *Infection and Immunology*, **28**, 893–898.

Ashraf, H., Mahalanabis, D., Mitra, A.K., Tzipori, S. & Fuchs, G.J. (2001) Hyperimmune bovine colostrum in the treatment of shigellosis in children, a double-blind, randomized, controlled trial. *Acta Paediatrica*, **90**, 1373–1378.

Bacsi, A., Woodbury, M., Kruzel, M.L. & Boldogh, I. (2007) Colostrinin delays the onset of proliferative senescence of diploid murine fibroblast cells. *Journal of Neuropeptides*, **41**, 93–101.

Bano, M.,Worland, P., Kidwell, W.R., Lippman, M.E. & Dickson, R.B. (1992) Receptor induced phosphorylation by mammary derived growth factor 1 in mammary epithelial cell lines. *Journal of Biological Chemistry*, **267**, 10389–10392.

Baram, T., Koch, Y., Hazum, E. & Friedkin, M. (1977) Gonadotropin-releasing hormone in milk. *Science*, **198**, 300–302.

Barnard, J.A., Beauchamp, R.D., Russell, W.E., Dubois, R.N. & Coffey, R.J. (1995) Epidermal growth factor-related peptides and their relevance to gastrointestinal pathophysiology. *Gastroenterology*, **108**, 564–180.

Barrington, G.M., Besser, T.E., Davis, W.C., Gay, C.C., Reeves, J.J. & McFadden, T.B. (1997) Expression of immunoglobulin G1 receptors by bovine mammary epithelial cells and mammary leukocytes. *Journal of Dairy Science*, **80**, 86–93.

Baxter, R.C., Zaltsman, Z. & Turtle, J.R. (1984) Immunoreactive somatomedin-C/insulin-like growth factor I and its binding protein in human milk. *Journal of Clinical Endocrinology and Metabolism*, **58**, 955–959.

Belford, D.A., Rogers, M.L., Francis, G.L., Payne, C., Ballard, F.J. & Goddard, C. (1997) Platelet-derived growth factor, insulin-like growth factors, fibroblast growth factors and transforming growth factor b do not account for the cell growth activity present in bovine milk. *Journal of Endocrinology*, **154**, 45–55.

Belo, A., Marchbank, T., Fitzgerald, A., Ghosh, S. & Playford, R.J. (2006) Gastroprotective effects of oral nucleotide administration. *Gut*, **55**, 165–171.

Bitzan, M.M., Gold, B.D., Philpott, D.J., Huesca, M., Sherman, P.M., Karch, H., Lissner, R., Lingwood, C.A. & Karmali, M.A. (1998) Inhibition of Helicobacter pylori and Helicobacter mustelae binding to lipid receptors by bovine colostrum. *Journal of Infectious Diseases*, **177**, 955–961.

Boldogh, I. & Kruzel, M.L. (2008) Colostrinin, an oxidative stress modulator for prevention and treatment of age-related disorders. *Journal of Alzheimers Disease*, **13**, 303–321.

Boldogh, I., Aguilera-Aguirre, L., Bacsi, A., Choudhury, B.K., Saavedra-Molina, A. & Kruzel, M.L. (2008) Colostrinin decreases hypersensitivity and allergic responses to common allergens. *International Archives of Allergy and Immunology*, **146**, 298–306.

Bölke, E., Jehle, P.M., Hausmann, F., Däubler, A., Wiedeck, H., Steinbach, G., Storck, M. & Orth, K. (2002) Preoperative oral application of immunoglobulin-enriched colostrum milk and mediator response during abdominal surgery. *Shock*, **17**, 9–12.

Bölke, E., Orth, K., Jehle, P.M., Schwarz, A., Steinbach, G., Schleich, S., Ulmer, C., Storck, M. & Hannekum, A. (2002) Enteral application of an immunoglobulin-enriched colostrum milk preparation for reducing endotoxin translocation and acute phase response in patients undergoing coronary bypass surgery: a randomized placebo-controlled pilot trial. *Wiener Klinische Wochenschrift*, **114**, 923–928.

Boudry, C., Buldgen, A., Portetelle, D., Collard, A., Théwis, A. & Dehoux, J.P. (2007) Effects of oral supplementation with bovine colostrum on the immune system of weaned piglets. *Research in Vetinary Science*, **83**, 91–101.

Brinkworth, G.D. & Buckley, J.D. (2003) Concentrated bovine colostrum protein supplementation reduces the incidence of self-reported symptoms of upper respiratory tract infection in adult males. *European Journal of Nutrition*, **42**, 228–232.

Brinkworth, G.D. & Buckley, J.D. (2004) Bovine colostrum supplementation does not affect plasma buffer capacity or haemoglobin content in elite female rowers. *European Journal of Applied Physiology*, **91**, 353–356.

Brinkworth, G.D., Buckley, J.D., Bourdon, P.C., Gulbin, J.P. & David, A. (2002) Oral bovine colostrum supplementation enhances buffer capacity but not rowing performance in elite female rowers. *International Journal of Sport Nutrition and Exercise Metabolism*, **12**, 349–365.

Brinkworth, G.D., Buckley, J.D., Slavotinek, J.P. & Kurmis, A.P. (2004) Effect of bovine colostrum supplementation on the composition of resistance trained and untrained limbs in healthy young men. *European Journal of Applied Physiology*, **91**, 53–60.

Brooks, H.J., McConnell, M.A., Corbett, J., Buchan, G.S., Fitzpatrick, C.E. & Broadbent, R.S. (2006) Potential prophylactic value of bovine colostrum in necrotizing enterocolitis in neonates, an in vitro study on bacterial attachment, antibody levels and cytokine production. *FEMS Immunology and Medical Microbiology*, **48**, 347–354.

Buckley, J.D., Abbott, M.J., Brinkworth, G.D. & Whyte, P.B. (2002) Bovine colostrum supplementation during endurance running training improves recovery, but not performance. *Journal of Science and Medicine in Sport*, **5**, 65–79.

Buckley, J.D., Brinkworth, G.D. & Abbott, M.J. (2003) Effect of bovine colostrum on anaerobic exercise performance and plasma insulin-like growth factor I. *Journal of Sports Science*, **21**, 577–588.

Cartlidge, S.A. & Elder, J.B. (1989) Transforming growth factor a and EGF levels in normal human gastrointestinal mucosa. *British Journal of Cancer*, **60**, 657–660.

Casswall, T.H., Sarker, S.A., Albert, M.J., Fuchs, G.J., Bergström, M., Björck, L. & Hammarström, L. (1998) Treatment of Helicobacter pylori infection in infants in rural Bangladesh with oral immunoglobulins from hyperimmune bovine colostrum. *Alimentary Pharmacology and Therapeutics*, **12**, 563–568.

Cesarone, M.R., Belcaro, G., Di Renzo, A. *et al.* (2007) Prevention of influenza episodes with colostrum compared with vaccination in healthy and high-risk cardiovascular subjects, the epidemiologic study in San Valentino. *Clinical and Applied Thrombosis/Hemostasis*, **13**, 130–136.

Chaumeil, C., Liotet, S. & Kogbe, O. (1994) Treatment of severe eye dryness and problematic eye lesions with enriched bovine colostrum lactoserum. *Advances in Experimental Medicine and Biology*, **350**, 595–599.

Coffey, R.J., Romano, M. & Goldenring, J. (1995) Roles for transforming growth factor-α in the stomach. *Journal of Clinical Gastroenterology*, **21**(suppl), S36–S39.

Collier, R.J., Miller, M.A., Hidebrant, J.R., Torkelson, A.R., White, T.C., Madsen, K.S., Vicini, J.L., Eppard, P.J. & Lanza, G.M. (1991) Factors affecting insulin-like growth factor I concentration in bovine colostrum. *Journal of Dairy Science*, **74**, 2905–2911.

Coombes, J.S., Conacher, M., Austen, S.K. & Marshall, P.A. (2002) Dose effects of oral bovine colostrum on physical work capacity in cyclists. *Medicine and Science in Sports and Exercise*, **34**, 1184–1188.

Crooks, C.V., Wall, C.R., Cross, M.L. & Rutherfurd-Markwick, K.J. (2006) The effect of bovine colostrum supplementation on salivary IgA in distance runners. *International Journal of Sport Nutrition and Exercise Metabolism*, **16**, 47–64.

Crooks, C., Cross, M.L., Wall, C. & Ali, A. (2010) Effect of bovine colostrum supplementation on respiratory tract mucosal defenses in swimmers. *International Journal of Sport Nutrition and Exercise Metabolism*, **20**, 224–235.

Cross, M.L. & Gill, H.S. (1999) Modulation of immune function by a modified bovine whey protein concentrate. *Immunology and Cell Biology*, **77**, 345–350.

Dalrymple, M.A. & Garner, I. (1998) Genetically modified livestock for the production of human proteins in milk. *Biotechnology and Genetic Engineering Reviews*, **15**, 33–49.

Daniel, H., Vohwinkel, M. & Rehner, G. (1990) Effect of casein and beta-casomorphins on gastrointestinal motility in rats. *Journal of Nutrition*, **120**, 252–257.

Daughaday, W.H. & Rotwein, P. (1989) Insulin-like growth factors I & II. Peptide messenger RNA-like structures, serum and tissue concentrations. *Endocrinology Review*, **10**, 68–91.

Davison, G. & Diment, B.C. (2010) Bovine colostrum supplementation attenuates the decrease of salivary lysozyme and enhances the recovery of neutrophil function after prolonged exercise. *British Journal of Nutrition*, **103**, 1425–1432.

Delehate-Zervas, M.C., Mertani, H., Martini, J.F., Nihoul-Fekete, C., Morel, G. & Postel-Vinay, M.C. (1994) Expression of the growth hormone receptor gene in human digestive tissues. *Journal of Clinical Endocrinology and Metabolism*, **78**, 1473–1480.

Dignas, A.U. & Podolsky, D.K. (1993) Cytokine modulation of intestinal epithelial cell restitution, central role of transforming growth factor-β. *Gastroenterology*, **105**, 1323–1332.

Dinarello, C.A. (1994) The interleukin-1 family, 10 years of discovery. *FASEB Journal*, **8**, 1314–1325.

Donnet-Hughes, A., Duc, N., Serrant, P., Vidal, K. & Schiffrin, E.J. (2000) Bioactive molecules in milk and their role in health and disease, the role of transforming growth factor-beta. *Immunology and Cell Biology*, **78**, 74–79.

Donovan, S.M., Atilano, L.C., Hintz, R.L., Wilson, D.M. & Rosenfeld, R.G. (1991) Differential regulation of the insulin-like growth factors (IGF-I and -II) and IGF binding proteins during malnutrition in the neonatal rat. *Endocrinology*, **129**, 149–157.

Du, M., Xu, W., Yi, H., Han, X., Wang, C. & Zhang, L. (2010) Protective effects of bovine colostrum acid proteins on bone loss of ovariectomized rats and the ingredients identification. *Molecular Nutrition and Food Research*, **55**, 220–228.

Ebina, T., Sato, A., Umezu, K., Ishida, N., Seki, H., Tsukamoto, T., Takase, S., Hoshi, S. & Ohta, M. (1984) Treatment of multiple sclerosis with anti-measle cow colostrum. *Medical Microbiology and Immunology*, **173**, 87–93.

Ebina, T., Sato, A., Umezu, K., Ishida, N., Ohyama, S., Oizumi, A., Aikawa, K., Katagiri, S., Katsushima, N. & Imai, A. (1985) Prevention of rotavirus infection by oral administration of cow colostrum containing antihumanrotavirus antibody. *Medical Microbiology and Immunology*, **174**, 177–185.

Ebina, T., Ohta, M., Kanamaru, Y., Yamamoto-Osumi, Y. & Baba, K. (1992) Passive immunizations of suckling mice and infants with bovine colostrum containing antibodies to human rotavirus. *Journal of Medical Virology*, **38**, 117–123.

Egger, B., Procaccino, F., Lakshmanan, J., Reinshagen, M., Hoffmann, P., Patel, A., Reuben, W., Gnanakkan, S., Liu, L., Barajas, L. & Eysselein, V.E. (1997) Mice lacking transforming growth factor b have an increased susceptibility to dextrin sulphate-induced colitis. *Gastroenterology*, **113**, 825–832.

Fernández-Montequín, J.I., Valenzuela-Silva, C.M., Díaz, O.G. *et al.* (2009) Intra-lesional injections of recombinant human epidermal growth factor promote granulation and healing in advanced diabetic foot ulcers, multicenter, randomised, placebo-controlled, double-blind study. *International Wound Journal*, **6**, 432–443.

FitzGerald, A.J., Pu, M., Marchbank, T., Westley, B.R., May, F.E., Boyle, J., Yadollahi-Farsani, M., Ghosh, S. & Playford, R.J. (2004) Synergistic effects of systemic trefoil factor family 1 (TFF1) peptide and epidermal growth factor in a rat model of colitis. *Peptides*, **25**, 793–801.

Florén, C.H., Chinenye, S., Elfstrand, L., Hagman, C. & Ihse, I. (2006) ColoPlus, a new product based on bovine colostrum, alleviates HIV-associated diarrhoea. *Scandinavian Journal of Gastroenterology*, **41**, 682–686.

Froud, K.E., Wardhaugh, T., Banks, D., Saffrey, M.J. & Stewart, M.G. (2010) Colostrinin alleviates amyloid-β induced toxicity in rat primary hippocampal cultures. *Journal of Alzheimer's Disease*, **20**, 423–426.

Frystyk, J. (2004) Free insulin-like growth factors – measurements and relationships to growth hormone secretion and glucose homeostasis. *Growth Hormone and IGF Research*, **14**, 337–375.

Gagnaire, V., Jardin, J., Jan, G. & Lortal, S. (2009) Invited review. *Proteomics of milk and bacteria used in fermented dairy products: from qualitative to quantitative advances. Journal of Dairy Science*, **92**, 811–825.

Ghosh, S. & Playford, R.J. (2003) Bioactive natural compounds for the treatment of gastrointestinal disorders. *Clinical Science (London)*, **104**, 547–556.

Ginger, M.R. & Grigor, M.R. (1999) Comparative aspects of milk caseins. *Comparative Biochemistry and Physiology Part B: Biochemistry and Molecular Biology*, **124**, 133–145.

Gordler, N.M., McGurk, M., Aqual, S. & Prince, M. (1995) The effect of EGF mouthwash on cytotoxic-induced oral ulceration. *American Journal of Clinical Oncology*, **18**, 403–406.

Greenberg, P.D., Cello, J.P. (1996) Treatment of severe diarrhea caused by Cryptosporidium parvum with oral bovine immunoglobulin concentrate in patients with AIDS. *Journal of Acquired Immune Deficiency Syndrome and Human Retrovirology*, **13**, 348–354.

Grosvenor, C.E., Picciano, M.F. & Baumrucker, C.R. (1992) Hormones and growth factors in milk. *Endocrinology Review*, **14**, 710–728.

Grundy, S.M., Benjamin, I.J. & Burke, J.L. (1999) Diabetes and cardiovascular disease, a statement for healthcare professionals from the American Heart Association. *Circulation*, **100**, 1134–1146.

Guan, J., Bennet, L., Gluckman, P.D. & Gunn, A.J. (2003) Insulin-like growth factor-1 and post-ischemic brain injury. *Progress in Neurobiology*, **70**, 443–462.

Gunn, J., Gunn, T.R., Rabone, D.L., Breier, B.H., Blum, W.F. & Gluckman, P.D. (1996) Growth hormone increases breast milk volumes in mothers of preterm infants. *Pediatrics*, **98**, 279–282.

Hagiwara, T., Shinoda, I., Fukuwatari, Y. & Shimamura, S. (1995) Effects of lactoferrin and its peptides on proliferation of rat intestinal epithelial cell line, IEC-18, in the presence of epidermal growth factor. *Bioscience, Biotechnology and Biochemistry*, **59**, 1875–1881.

Hagiwara, K., Kataoka, S., Yamanaka, H., Kirisawa, R. & Iwai, H. (2000) Detection of cytokines in bovine colostrum. *Vetinary Immunology and Immunopathology*, **76**, 183–190.

Han, V.K.M., D'Ercole, A.J. & Lund, P.K. (1987) Cellular localization of somatomedin (insulin-like growth factor) messenger RNA in the human fetus. *Science*, **236**, 193–197.

He, F., Tuomola, E., Arvilommi, H. & Salminen, S. (2001) Modulation of human humoral immune response through orally administered bovine colostrum. *FEMS Immunology and Medical Microbiology*, **31**, 93–96.

Hedbacker, K., Birsoy, K., Wysocki, R.W., Asilmaz, E., Ahima, R.S., Farooqi, I.S. & Friedman, J.M. (2010) Antidiabetic effects of IGFBP2, a leptin-regulated gene. *Cellular Metabolism*, **11**, 11–22.

Heemskerk, V.H., van Heurn, L.W., Farla, P., Buurman, W.A., Piersma, F., ter Riet, G. & Heineman, E. (2002) Effect of IGF-rich colostrum on bowel adaptation in neonatal piglets with short bowel syndrome. *Journal of Pediatric Gastroenterology and Nutrition*, **34**, 47–51.

Hirano, M., Iweakiri, R., Fujimoto, K. *et al.* (1995) Epidermal growth factor enhances repair of rat intestinal mucosa damaged after oral administration of methotrexate. *Journal of Gastroenterology*, **30**, 169–176.

Hodgkinson, A.J., Cannon, R.D., Holmes, A.R., Fischer, F.J. & Willix-Payne, D.J. (2007) Production from dairy cows of semi-industrial quantities of milk-protein concentrate (MPC) containing efficacious anti-Candida albicans IgA antibodies. *Journal of Dairy Research*, **74**, 269–275.

Hofman, Z., Smeets, R., Verlaan, G., Lugt, R. & Verstappen, P.A. (2002) The effect of bovine colostrum supplementation on exercise performance in elite field hockey players. *International Journal of Sport Nutrition and Exercise Metabolism*, **12**, 461–469.

Howarth, G.S., Francis, G.L., Cool, J.C., Ballard, R.W. & Read, L.C. (1996) Milk growth factors enriched from cheese whey ameliorate intestinal damage by methotrexate when administered orally to rats. *Journal of Nutrition*, **126**, 2519–2530.

Howarth, G.S., Fraser, R., Frisby, C.L., Schirmer, M.B. & Yeoh, E.K. (1997) Effects of insulin-like growth factor-I administration on radiation enteritis in rats. *Scandinavian Journal of Gastroenterology*, **32**, 1118–1124.

Howarth, G.S., Cool, J.C., Bourne, A.J., Ballard, F.J. & Read, L.C. (1998a) Insulin-like growth factor-I (IGF-I) stimulates regrowth of the damaged intestine in rats, when administered following, but not concurrent with, methotrexate. *Growth Factors*, **15**, 279–292.

Howarth, G.S., Xian, C.J. & Read, L.C. (1998) Insulin-like growth factor-I partially attenuates colonic damage in rats with experimental colitis induced by oral dextran sulphate sodium. *Scandinavian Journal of Gastroenterology*, **33**, 180–190.

Hyvönen, P., Suojala, L., Haaranen, J., von Wright, A. & Pyörälä, S. (2006) Human and bovine lactoferrins in the milk of recombinant human lactoferrin-transgenic dairy cows during lactation. *Biotechnol Journal*, **1**, 410–412.

Jaini, R., Kesaraju, P., Johnson, J.M., Altuntas, C.Z., Jane-Wit, D. & Tuohy, V.K. (2010) An autoimmune-mediated strategy for prophylactic breast cancer vaccination. *Natural Medicine*, **16**, 799–803.

Janusz, M. & Lisowski, J. (1993) Proline-rich polypeptide (PRP) an immunomodulatory peptide from ovine colostrums. *Archivum Immunologiae et Therapiae Experimentalis*, **41**, 275–279.

Jin, Y., Cox, D.A., Knecht, R., Raschdorf, F. & Cerletti, N. (1991) Separation, purification and sequence identification of TGF-β1 and TGF-β2 from bovine milk. *Journal of Protein Chemistry*, **10**, 565–575.

Kang, P., Toms, D., Yin, Y., Cheung, Q., Gong, J., De Lange, K. & Li, J. (2010) Epidermal growth factor-expressing Lactococcus lactis enhances intestinal development of early-weaned pigs. *Journal of Nutrition*, **140**, 806–811.

Kanwar, J.R., Kanwar, R.K., Sun, X., Punj, V., Matta, H., Morley, S.M., Parratt, A., Puri, M. & Sehgal, R. (2009) Molecular and biotechnological advances in milk proteins in relation to human health. *Current Protein and Peptide Science*, **10**, 308–338.

Keck, P.J., Hauser, S.D., Krivi, G., Sanzo, K., Warren, T., Feder, J. & Connolly, D.T. (1989) Vascular permeability factor, an endothelial cell mitogen related to PDGF. *Science*, **246**, 1309–1312.

Kelly, D., McFadyen, M., King, T.P. & Morgan, P.J. (1992) Characterization and autoradiographic localization of the epidermal growth factor receptor in the jejunum of neonatal and weaned pigs. *Reproduction, Fertility and Development*, **4**, 183–191.

Kelly, D., King, T.P., McFadyen, M. & Coutts, A.G.P. (1993) Effect of preclosure colostrum intake on the development of the intestinal epithelium of artificially reared piglets. *Biology of the Neonate*, **64**, 235–244.

Kerksick, C.M., Rasmussen, C., Lancaster, S., Starks, M., Smith, P., Melton, C., Greenwood, M., Almada, A. & Kreider, R. (2007) Impact of differing protein sources and a creatine containing nutritional formula after 12 weeks of resistance training. *Nutrition*, **23**, 647–656.

Khan, Z., Macdonald, C., Wicks, A.C., Holt, M.P., Floyd, D., Ghosh, S., Wright, N.A. & Playford, R.J. (2002) Use of the 'nutriceutical', bovine colostrum, for the treatment of distal colitis, results from an initial study. *Alimentary Pharmacology and Therapeautics*, **16**, 1917–1922.

Kidwell, W.R. & Salomon, D.S. (1989) Growth factors in human milk, sources and potential physiological roles. In *Protein and Non-Protein and Nitrogen in Human Milk* (eds Atkinson, S.A. & Lonnerdal, B.), pp. 77–91. CRC Press, Boca Raton, FL.

Kim, J.H., Jung, W.S., Choi, N.J., Kim, D.O., Shin, D.H. & Kim, Y.J. (2009) Health-promoting effects of bovine colostrum in Type 2 diabetic patients can reduce blood glucose, cholesterol, triglyceride and ketones. *Journal of Nutritional Biochemistry*, **20**, 298–303.

Kimura, T., Murakawa, Y., Ohno, M., Ohtani, S. & Higaki, K. (1997) Gastrointestinal absorption of recombinant human insulin like growth factor-I in rats. *Journal of Pharmacology and Experimental Therapeutics*, **283**, 611–618.

King, J.C. Jr,, Cummings, G.E., Guo, N., Trivedi, L., Readmond, B.X., Keane, V., Feigelman, S. & de Waard, R. (2007) A double-blind, placebo-controlled, pilot study of bovine lactoferrin supplementation in bottle-fed infants. *Journal of Pediatric Gastroenterology and Nutrition*, **44**, 245–251.

Ko, T.C., Beauchamp, R.D., Townsend, C.M. Jr, & Thompson, J.C. (1993) Glutamine is essential for epidermal growth factor-stimulated intestinal cell proliferation. *Surgery*, **114**, 147–153.

Koea, J.B., Gallaher, B.W., Breier, B.H., Douglas, R.G., Hodgkinson, S., Shaw, J.H. & Gluckman, P.D. (1992) Passive immunization against circulating insulin-like growth factor-I (IGF-I) increases protein catabolism in lambs, evidence for a physiological role for circulating IGF-I. *Journal of Endocrinology*, **135**, 279–284.

Koldovsky, O., Britton, J., Davis, D., Grimes, J., Kong, W., Rao, R. & Schaudies, P. (1991) The developing gastrointestinaltract and milk-borne epidermal growth factor. In *Immunology of Milk and the Neonate* (ed. Mestecky, J.), pp. 99–105. Plenum Press, New York.

Korhonen, H., Syvaoja, E.L., Ahola-Luttila, H., Sivela, S., Kopola, S. & Kosunen, T.U. (1995) Bactericidal effect of bovine normal and immune serum, colostrum and milk against Helicobacter pylori. *Journal of Applied Bacteriology*, **78**, 655–662.

Koyama, S. & Podolsky, D.K. (1989) Differential expression of transforming growth factors a and b in rat intestinal epithelial cells. *Journal of Clinical Investigation*, **83**, 1768–1773.

Kruzel, M., Janusz, M., Lisowski, J., Fischleigh, R. & Georgiades, J. (2001) Towards an understanding of the biological role of colostrinin peptides. *Journal of Molecular Neuroscience*, **17**, 115–125.

Kruzel, M.L., Bacsi, A., Choudhury, B., Sur, S. & Boldogh, I. (2006) Lactoferrin decreases pollen antigen-induced allergic airway inflammation in a murine model of asthma. *Immunology*, **119**, 159–166.

Kupfer, S.R., Underwood, L.E., Baxter, R.C. & Clemmons, D.R. (1993) Enhancement of the anabolic effects of growth hormone and insulin-like growth factor I by use of both agents simultaneously. *Journal of Clinical Investgation*, **91**, 391–396.

Laible, G., Brophy, B., Knighton, D. & Wells, D.N. (2007) Compositional analysis of dairy products derived from clones and cloned transgenic cattle. *Theriogenology*, **67**, 166–177.

Lee, J., Kwon, S.H., Kim, H.M., Fahey, S.N., Knighton, D.R. & Sansom, A. (2008) Effect of a growth protein-colostrum fraction on bone development in juvenile rats. *Bioscience, Biotechnology and Biochemistry*, **72**, 1–6.

Leszek, J., Inglot, A.D., Janusz, M., Lisowski, J., Krukowska, K. & Georgiades, J.A. (1999) Colostrinin, a praline-rich polypeptide (PRP) complex isolated from ovine colostrum for the treatment of Alzheimer's disease: a double-blind, placebo-controlled study. *Archivum Immunologiae et Therapie Experimentalis*, **47**, 377–385.

Leszek, J., Inglot, A.D., Janusz, M., Byczkiewicz, F., Kiejna, A., Georgiades, J. & Lisowski, J. (2002) Colostrinin praline-rich polypeptide complex from ovine colostrums: a long-term study of its efficacy in Alzheimer's disease. *Medical Science Monitor*, **8**, PI93–PI96.

Lo, H.-C., Hinton, P.S.,Yang, H., *et al.* (1996) Insulin-like growth factor-I but not growth hormone attenuates dexamethasone-induced catabolism in parenterally fed rats. *JPEN Journal of Parenteral and Enteral Nutrition*, **20**, 171–177.

Loimaranta, V., Laine, M., Söderling, E., Vasara, E., Rokka, S., Marnila, P., Korhonen, H., Tossavainen, O. & Tenovuo, J. (1999) Effects of bovine immune and non-immune whey preparations on the composition and pH response of human dental plaque. *European Journal of Oral Science*, **107**, 244–250.

Lönnerdal, B. (2003) Nutritional and physiologic significance of human milk proteins. *American Journal of Clinical Nutrition*, **77**, 1537S–1543S.

Lowe, W.L. (1991) Biological actions of the insulin-like growth factors. In *Insulin-like Growth Factors, Molecular and Cellular Aspects* (ed. LeRoith, D.), pp. 49–85. CRC Press, Boca Raton, FL.

Lund, P.K. & Zimmermann, E.M. (1996) Insulin-like growth factors and inflammatory bowel disease. *Bailliere's Clinical Gastroenterology*, **10**, 83–96.

Lund, P.K., Moats-Staats, B.M., Hynes, M.A., Simmons, J.G., Jansen, M., D'Ercole, A.J. & Van Wyk, J.J. (1986) Somatomedin-C/insulin-like growth factor-I and insulin-like growth factor-II mRNAs in rat fetal and adult tissues. *Journal of Biological Chemistry*, **261**, 14539–14544.

MacDonald, C.E., Playford, R.J., Khatri, M. & Goodlad, R.A. (1998) Transforming growth factor a knockout mice have smaller small intestines, larger large intestines, but no increased sensitivity to NSAID induced small intestinal injury. *Gut*, **42**(suppl), A3 (abstr).

MacDonald, T.M., Morant, S.V., Robinson, G.C., Shield, M.J., McGilchrist, M.M., Murray, F.E. & McDevitt, D.G. (1997) Association of upper gastrointestinal toxicity of non-steroidal anti-inflammatory drugs with continued exposure, cohort study. *British Medical Journal*, **315**, 1333–1337.

MacKinnon, L.T. (1996) Immunoglobulin, antibody, and exercise. *Exercise Immunology Review*, **2**, 1–35.

Mahmood, A., Melley, L., Fitzgerald, A.J., Ghosh, S. & Playford, R.J. (2005) Trial of trefoil factor 3 enemas, in combination with oral 5-aminosalicylic acid, for the treatment of mild-to-moderate left-sided ulcerative colitis. *Alimentary Pharmacology and Therapeutics*, **21**, 1357–1364.

Manzoni, P., Rinaldi, M., Cattani, S. *et al.* (2009) Bovine lactoferrin supplementation for prevention of late-onset sepsis in very low-birth-weight neonates, a randomized trial. *Journal of the American Medical Association*, **302**, 1421–1428.

Manzoni, P., Decembrino, L., Stolfi, I. *et al.* (2010) Lactoferrin and prevention of late-onset sepsis in the pre-term neonates. *Early Human Development*, **86**(Suppl), 159–161.

Marchbank, T., Davison, G., Oakes, J.R., Ghatei, M.A., Patterson, M. & Playford, R.J. (2010) Clinical trial, influence of bovine colostrum on intestinal permeability in healthy athletes after heavy exercise. *Gastroenterology*, **138**, S725.

Marchbank, T., Davison, G., Oakes, J.R., Ghatei, M.A., Patterson, M., Moyer, M.P. & Playford, R.J. (2011) The nutriceutical bovine colostrum truncates the increase in gut permeability caused by heavy exercise in athletes. *American Journal of Physiology: Gastrointestinal and Liver Physiology*, **300**, G477–484.

Mashimo, H., Wu, D.C., Podolsky, D.K. & Fishman, M.C. (1996) Impaired defense of intestinal mucosa in mice lacking intestinal trefoil factor. *Science*, **274**, 262–265.

Masson, P.L. & Heremans, J.F. (1971) Lactoferrin in milk from different species. *Comperative Biochemistry and Physiology*, **39**, 119–129.

McConnell, M.A., Buchan, G.A.A., Borissnko, M.V. & Brooks, H.J.L. (2001) A comparison of IgG and IgG1 in an early milk concentrate from non-immunized cows and milk from hyperimmunized animals. *Food Research International*, **34**, 255–261.

Meister, D., Bode, J., Shand, A. & Ghosh, S. (2002) Anti-inflammatory effects of enteral diet components on Crohn's disease-affected tissues in vitro. *Digestive and Liver Disease*, **34**, 430–438.

Mero, A., Miikkulainen, H., Riski, J., Pakkanen, R., Aalto, J. & Takala, T. (1997) Effects of bovine colostrum supplementation on serum IGF-I, IgG, hormone, and saliva IgA during training. *Journal of Applied Physiology*, **83**, 1144–1151.

Mero, A., Kähkönen, J., Nykänen, T., Parviainen, T., Jokinen, I., Takala, T., Nikula, T., Rasi, S. & Leppäluoto, J. (2002) IGF-I, IgA, and IgG responses to bovine colostrum supplementation during training. *Journal of Applied Physiology*, **93**, 732–739.

Mero, A., Nykänen, T., Keinänen, O., Knuutinen, J., Lahti, K., Alen, M., Rasi, S. & Leppäluoto, J. (2005) Protein metabolism and strength performance after bovine colostrum supplementation. *Amino Acids*, **28**, 327–335.

Mitra, A.K., Mahalanabis, D., Ashraf, H., Unicomb, L., Eeckels, R. & Tzipori, S. (1995) Hyperimmune cow colostrum reduces diarrhoea due to rotavirus, a double-blind, controlled clinical trial. *Acta Paediatrica*, **84**, 996–1001.

Morris, K. (1999) Bovine somatotropin: who's crying over spilt milk? *Lancet*, **353**, 306.

Moya, F.R., Eguchi, H., Zhao, B., *et al.* (1994) Platelet-activating factor acetylhydrolase in term and preterm human milk, a preliminary report. *Journal of Pediatric Gastroenterology and Nutrition*, **19**, 236–239.

Nagy, E.S., Paris, M.C., Taylor, R.G., Fuller, P.J., Sourial, M., Justice, F. & Bines, J.E. (2004) Colostrum protein concentrate enhances intestinal adaptation after massive small bowel resection in juvenile pigs. *Journal of Pediatric Gastroenterology and Nutrition*, **39**, 487–492.

Ochoa, T.J., Chea-Woo, E., Campos, M., Pecho, I., Prada, A., McMahon, R.J. & Cleary, T.G. (2008) Impact of lactoferrin supplementation on growth and prevalence of Giardia colonization in children. *Clinical Infectious Diseases*, **46**, 1881–1883.

Okada, M., Ohmura, E., Kamiya, Y., Murakami, H., Onoda, N., Iwashita, M., Wakai, K., Tsushima, T. & Shizume, K. (1991) Transforming growth factor (TGF)-α in human milk. *Life Sciences*, **48**, 1151–1156.

Okhuysen, P.C., Chappell, C.L., Crabb, J., Valdez, L.M., Douglass, E.T. & DuPont, H.L. (1998) Prophylactic effect of bovine anti-Cryptosporidium hyperimmune colostrum immunoglobulin in healthy volunteers challenged with Cryptosporidium parvum. *Clinical Infectious Diseases*, **26**, 1324–1329.

Okuyama, H., Urao, M., Lee, D., Drongowski, R.A. & Coran, A.G. (1998) The effect of epidermal growth factor on bacterial translocation in newborn rabbits. *Journal of Pediatric Surgery*, **33**, 225–228.

O'Loughlin, W., Winter, M. & Shun, A. (1994) Structural and functional adaptation following jejunal resection in rabbits, effect of epidermal growth factor. *Gastroenterology*, **107**, 87–93.

Ormrod, D.J. & Miller, T.E. (1992) A low molecular weight component derived from the milk of hyperimmunised cows suppresses inflammation by inhibiting neutrophil emigration. *Agents Actions*, **37**, 70–79.

Pals, K.L., Chang, R.T., Ryan, A.J. & Gisolfi, C.V. (1997) Effect of running intensity on intestinal permeability. *Journal of Applied Physiology*, **82**, 571–576.

Panahi, Y., Falahi, G., Falahpour, M., Moharamzad, Y., Khorasgani, M.R., Beiraghdar, F. & Naghizadeh, M.M. (2010) Bovine colostrum in the management of nonorganic failure to thrive, a randomized clinical trial. *Journal of Pediatric Gastroenterology and Nutrition*, **50**, 551–554.

Paris, M.C., Fuller, P.J., Carstensen, B., Nagy, E., Taylor, R.G., Sourial, M., Holst, J.J., Hartmann, B. & Binesm, J.E. (2004) Plasma GLP-2 levels and intestinal markers in the juvenile pig during intestinal adaptation, effects of different diet regimens. *Digestive Diseases and Sciences*, **49**, 1688–1695.

Patel, K. & Rana, R. (2006) Pedimune in recurrent respiratory infection and diarrhoea: the Indian experience: the pride study. *Indian Journal of Pediatrics*, **73**, 585–591.

Pereira-Fantini, P.M., Thomas, S.L., Taylor, R.G., Nagy, E., Sourial, M., Fuller, P.J. & Bines, J.E. (2008) Colostrum supplementation restores insulin-like growth factor-1 levels and alters muscle morphology following massive small bowel resection. *JPEN Journal of Parenteral and Enteral Nutrition*, **32**, 266–275.

Pessi, T., Isolauri, E., Sütas, Y., Kankaanranta, H., Moilanen, E. & Hurme, M. (2001) Suppression of T-cell activation by Lactobacillus rhamnosus GG-degraded bovine casein. *International Immunopharmacology*, **1**, 211–218.

Pinotti, L. & Rosi, F. (2006) Leptin in bovine colostrum and milk. *Hormone and Metabolic Research*, **38**, 89–93.

Playford, R.J. (1995) Leading article, peptides and gastrointestinal mucosal integrity. *Gut*, **37**, 595–597.

Playford, R.J., Woodman, A.C., Clark, P., Watanapa, P., Vesey, D., Deprez, P.H., Williamson, R.C. & Calam, J. (1993) Effect of luminal growth factor preservation on intestinal growth. *Lancet*, **341**, 843–848.

Playford, R.J., Marchbank, T., Calam, J. & Hansen, F.H. (1995) EGF is digested to smaller, less active, forms in acidic gastric juice. *Gastroenterology*, **108**, 92–101.

Playford, R.J., Hanby, A., Gschmeissner, S., Peiffer, L.P., McGarrity, T. & Wright, N.A. (1996a) The epidermal growth factor receptor (EGF-R) is present on the basolateral, but not the apical, surface of enterocytes in the human gastrointestinal tract. *Gut*, **39**, 262–266.

Playford, R.J., Boulton, R., Ghatei, M.A., Bloom, S.R., Wright, N.A. & Goodlad, R.A. (1996b) Comparison of the effects of TGFa and EGF on gastrointestinal proliferation and hormone release. *Digestion*, **57**, 362–367.

Playford, R.J., Floyd, D.N., Macdonald, C.E., Calnan, D.P., Adenekan, R.O., Johnson, W., Goodlad, R.A. & Marchbank, T. (1999) Bovine colostrum is a health food supplement which prevents NSAID induced gut damage. *Gut*, **44**, 653–658.

Playford, R.J., Macdonald, C.E. & Johnson, W.S. (2000) Colostrum and milk-derived peptide growth factors for the treatment of gastrointestinal disorders. *American Journal of Clinical Nutrition*, **72**, 5–14.

Playford, R.J., MacDonald, C.E., Calnan, D.P., Floyd, D.N., Podas, T., Johnson, W., Wicks, A.C., Bashir, O. & Marchbank, T. (2001) Co-administration of the health food supplement, bovine colostrum, reduces the acute non-steroidal anti-inflammatory drug-induced increase in intestinal permeability. *Clinical Science (London)*, **100**, 627–633.

Plettenberg, A., Stoehr, A., Stellbrink, H.J., Albrecht, H. & Meigel, W. (1993) A preparation from bovine colostrum in the treatment of HIV-positive patients with chronic diarrhea. *Clinical Investigations*, **71**(1), 42–45.

Porter, S.N., Howarth, G.S. & Butler, R.N. (1998) An orally administered growth factor extract derived from bovine whey suppresses breath ethane in colitic rats. *Scandinavian Journal of Gastroenterology*, **33**, 967–974.

Procaccino, F., Reinshagen, M., Hoffman, P., Zeeh, J.M., Lakshmanan, J., McRoberts, J.A., Patel, A., French, S. & Eysselein, V.E. (1994) Protective effect of epidermal growth factor in an experimental model of colitis. *Gastroenterology*, **107**, 12–17.

Prosser, C., Stelwagen, K., Cummins, R., Guerin, P., Gill, N. & Milne, C. (2004) Reduction in heat-induced gastrointestinal hyperpermeability in rats by bovine colostrum and goat milk powders. *Journal of Applied Physiology*, **96**, 650–654.

Ragab, S.S. & Ibrahim, M.K. (2004) Role of bovine colostrum and its biofunctional fraction PRP in oral treatment of enterogenic endotoxaemia in rats. *International Journal of Agriculture and Biology*, **6**, 576–580.

Rattray, M. (2005) Technology evaluation, Colostrinin. *Current Opinion in Molecular Therapeutics*, **7**, 78–84.

Read, L.C., Francis, G.L., Wallace, J.C. & Ballard, F.J. (1985) Growth factor concentrations and growth-promoting activity in human milk following premature birth. *Journal of Development Physiology*, **7**, 135–145.

Rechler, M.M. (1993) Insulin-like growth factor binding proteins. *Vitamins and Hormones*, **47**, 1–114.

Revenis, M.E. & Kaliner, M.A. (1992) Lactoferrin and lysozyme deficiency in airway secretions, association with the development of bronchopulmonary dysplasia. *Journal of Pediatrics*, **121**, 262–270

Rokka, S., Myllykangas, S. & Joutsjoki, V. (2008) Effect of specific colostral antibodies and selected lactobacilli on the adhesion of Helicobacter pylori on AGS cells and the Helicobacter-induced IL-8 production. *Scandinavian Journal of Immunology*, **68**, 280–286.

Rona, Z. (1998) Bovine colostrum emerges as immunity modulator. *American Journal of Natural Medicine*, **5**, 19–23.

Rosenson, R.S., Shott, S. & Lu, L. (2001) Hypertriglyderidemia and other factors associated with plasma viscosity. *American Journal of Medicine*, **110**, 488–492.

Rump, J.A., Arndt, R., Arnold, A. *et al.* (1992) Treatment of diarrhoea in human immunodeficiency virus-infected patients with immunoglobulins from bovine colostrum. *Clinical Investigations*, **70**(7), 588–594.

Sanchez, L., Aranda, P., Perez, M.D. & Calvo, M. (1988) Concentration of lactoferrin and transferrin throughout lactation in cow's colostrum and milk. *Biological Chemistry Hoppe-Seyler*, **369**, 1005–1008.

Sapi, E. & Kacinski, B.M. (1999) The role of CSF-1 in normal and neoplastic breast physiology. *Proceedings of the Society for Experimental Biology and Medicine*, **220**, 1–8.

Sarker, S.A., Casswall, T.H., Mahalanabis, D., Alam, N.H., Albert, M.J., Brüssow, H., Fuchs, G.J. & Hammerström, L. (1998) Successful treatment of rotavirus diarrhea in children with immunoglobulin from immunized bovine colostrum. *Pediatric Infectious Diseases Journal*, **17**, 1149–1154.

Schams, D. & Einspanier, R. (1991) Growth hormone, IGF-I and insulin in mammary gland secretion before and after parturition and possibility of their transfer into the calf. *Endocrine Regulations*, **25**, 139–143.

Schlimme, E., Martin, D. & Meisel, H. (2000) Nucleosides and nucleotides, natural bioactive substances in milk and colostrum. *British Journal of Nutrition*, **84**, S59–68.

Seifert, J., Molkewehrum, M., Oesser, S., Nebermann, L. & Schulze, C. (2002) Endotoxin inactivation by enterally applied colostrum of different composition. *European Surgery Research*, **34**, 68–72.

Shimomura, I., Matsuda, M., Hammer, R.E., Bashmakov, Y., Brown, M.S. & Goldstein, J.L. (2000) Down regulation of IRS-2 and up-regulation of SREBP-1c leads to mixed insulin resistance and sensitivity in livers of lipodystrophic and ob/ob mice. *Molecular Cell*, **6**, 77–86.

Shing, Y.W. & Klagsbrun, M. (1984) Human and bovine milk contain different sets of growth factors. *Endocrinology*, **115**, 273–282.

Shing, Y.W. & Klagsbrun, M. (1987) Purification and characterization of a bovine colostrums-derived growth factor. *Molecular Endocrinology*, **1**, 335–338.

Shing, Y., Davidson, S. & Klagsbrun, M. (1987) Purification of polypeptide growth factors from milk. *Methods in Enzymology*, **146**, 42–48.

Shing, C.M., Jenkins, D.G., Stevenson, L. & Coombes, J.S. (2006) The influence of bovine colostrum supplementation on exercise performance in highly trained cyclists. *British Journal of Sports Medicine*, **40**, 797–801.

Shing, C.M., Peake, J., Suzuki, K., Okutsu, M., Pereira, R., Stevenson, L., Jenkins, D.G. & Coombes, J.S. (2007) Effects of bovine colostrum supplementation on immune variables in highly trained cyclists. *Journal of Applied Physiology*, **102**, 1113–1122.

Shing, C.M., Peake, J.M., Suzuki, K., Jenkins, D.G. & Coombes, J.S. (2009a) Bovine colostrum modulates cytokine production in human peripheral blood mononuclear cells stimulated with lipopolysaccharide and phytohemagglutinin. *Journal of Interferon and Cytokine Research*, **29**, 37–44.

Shing, C.M., Hunter, D.C. & Stevenson, L.M. (2009b) Bovine colostrum supplementation and exercise performance, potential mechanisms. *Sports Medicine*, **39**, 1033–1054.

Siafakas, C.G., Anatolitou, F., Fusunyan, R.D., Walker, W.A. & Sanderson, I.R. (1999) Vascular endothelial growth factor (VEGF) is present in human breast milk and its receptor is present on intestinal epithelial cells. *Pediatric Research*, **45**, 652–657.

Sigalet, D.L., Martin, G.R., Butzner, J.D., Buret, A. & Meddings, J.B. (2005) A pilot study of the use of epidermal growth factor in pediatric short bowel syndrome. *Journal of Pediatric Surgery*, **40**, 763–768.

Simons, J.P., Wilmut, I., Clark, A.J., Archibald, A.L., Bishop, J.O. & Lathe, R. (1998) Gene transfer into sheep. *Biotechnology*, **6**, 179–183.

Sinha, A., Nightingale, J., West, K.P., Berlanga-Acosta, J. & Playford, R.J. (2003) Epidermal growth factor enemas with oral mesalamine for mild-to-moderate left-sided ulcerative colitis or proctitis. *New England Journal of Medicine*, **349**, 350–357.

Skov-Olsen, P., Poulsen, S.S., Therkelsen, K. & Nexo, E. (1986) Effect of sialoadenectomy and synthetic human urogastrone on healing of chronic gastric ulcers in rats. *Gut*, **27**, 1443–1449.

Slonim, A.E., Grovit, M. & Bulone, L. (2009) Effect of exclusion diet with nutraceutical therapy in juvenile Crohn's disease. *Journal of American College of Nutrition*, **28**, 277–285.

Solomons, N.W. (2002) Modulation of the immune system and the response against pathogens with bovine colostrum concentrates. *European Journal of Clinical Nutrition*, **56**, S24–28.

Sonis, S.T., Lindquist, L., Van Vugt, A. *et al.* (1994) Prevention of chemotherapy-induced ulcerative mucositis by transforming growth factor beta 3. *Cancer Research*, **54**, 1135–1138.

Stelwagen, K. & Ormrod, D.J. (1998) An anti-inflammatory component derived from milk of hyperimmunised cows reduces tight junction permeability in vitro. *Inflammation Research*, **47**, 384–388.

Stelwagen, K., Carpenter, E., Haigh, B., Hodgkinson, A. & Wheeler, T.T. (2009) Immune components of bovine colostrums and milk. *Journal of Animal Science*, **87**, 3–9.

Stewart, M.G. & Banks, D. (2006) Enhancement of long-term memory retention by Colostrinin in one-day-old chicks trained on a weak passive avoidance learning paradigm. *Neurobiology of Learning and Memory*, **86**, 66–71.

Suh, D.S. & Rechler, M.M. (1997) Hepatocyte nuclear factor 1 and the glucocorticoid receptor synergistical activate transcription of the rat insulin-like growth factor binding protein-1 gene. *Molecular Endocrinology*, **11**, 1822–1831.

Sullivan, P.B., Brueton, M.J., Tabara, Z., Goodlad, R.A., Lee, C.Y. & Wright, N.A. (1991) Epidermal growth factor in necrotizing enterocolitis. *Lancet*, **338**, 53–54.

Sullivan, P.B., Lewindon, P.J., Cheng, C., Lenehan, P.F., Kuo, B.S., Haskins, J.R., Goodlad, R.A., Wright, N.A. & de la Iglesia, F.A. (2007) Intestinal mucosa remodeling by recombinant human epidermal growth factor(1–48) in neonates with severe necrotizing enterocolitis. *Journal of Pediatric Surgery*, **42**, 462–469.

Szabo, S. & Sandor, Z. (1996) Basic fibroblast growth factor and PDGF in GI diseases. *Bailliere's Clinical Gastroenterology*, **10**, 97–112.

Szaniszlo, P., German, P., Hajas, G., Saenz, D.N., Kruzel, M. & Boldogh, I. (2009a) New insights into clinical trial for Colostrinin in Alzheimer's Disease. *Journal of Nutrition Health and Aging*, **13**, 235–241.

Szaniszlo, P., German, P., Hajas, G., Saenz, D.N., Woodberry, M.W., Kruzel, M.L. & Boldogh, I. (2009b) Effects of Colostrinin on gene expression-transcriptomal network analysis. *International Immunopharmacology*, **9**, 181–198.

Tacket, C.O., Losonsky, G., Link, H., Hoang, Y., Guesry, P., Hilpert, H. & Levine, M.M. (1988) Protection by milk immunoglobulin concentrate against oral challenge with enterotoxigenic Escherichia coli. *New England Journal of Medicine*, **318**, 1240–1243.

Tacket, C.O., Binion, S.B., Bostwick, E., Losonsky, G., Roy, M.J. & Edelman, R. (1992) Efficacy of bovine milk immunoglobulin concentrate in preventing illness after Shigella flexneri challenge. *American Journal of Tropical Medicine and Hygiene*, **47**, 276–283.

Talhouk, R.S., Neiswander, R.L. & Schanbacher, F.L. (1996) Developmental regulation and partial characterization of growth factors in the bovine mammary gland. *Journal of Reproduction and Fertility*, **106**, 221–230.

Tawfeek, H.I., Najim, N.H. & Al-Mashikhi, S. (2003) Efficacy of an infant formula containing anti-Escherichia coli colostral antibodies from hyperimmunized cows in preventing diarrhea in infants and children, a field trial. *International Journal of Infectious Diseases*, **7**, 120–128.

Thompson, J.F., Van Den Berg, M. & Stokkers, P.C.F. (1994) Developmental regulation of epidermal growth factor receptor kinase in rat intestine. *Gastroenterology*, **107**, 1278–1287.

Toba, Y., Takada, Y., Matsuoka, Y. *et al.* (2001) Milk basic protein promotes bone formation and suppresses bone resorption in healthy adult men. *Bioscience, Biotechnology and Biochemistry*, **65**, 1353–1357.

Tollemar, J., Gross, N., Dolgiras, N., Jarstrand, C., Ringdén, O. & Hammarström, L. (1999) Fungal prophylaxis by reduction of fungal colonization by oral administration of bovine anti-Candida antibodies in bone marrow transplant recipients. *Bone Marrow Transplantation*, **23**, 283–290.

Tuyet, H.L., Nguyen Quynh, T.T., Vo Hoang Minh, H., Thi Bich, D.N., Do Dinh, T., Le Tan, D., Van, H.L., Le Huy, T., Doan Huu, H. & Tran Trong, T.N. (2009) The efficacy and safety of epidermal growth factor in treatment of diabetic foot ulcers, the preliminary results. *International Wound Journal*, **6**, 159–166.

Uenishi, K., Ishida, H., Toba, Y., Aoe, S., Itabashi, A. & Takada, Y. (2007) Milk basic protein increases bone mineral density and improves bone metabolism in healthy young women. *Osteoporos International*, **18**, 385–390.

Ulshen, M.H., Dowling, R.H., Fuller, C.R., Zimmermann, E.M. & Lund, P.K. (1993) Enhanced growth of small bowel in transgenic mice overexpressing bovine growth hormone. *Gastroenterology*, **104**, 973–980.

Vacher, P.Y. & Blum, J.W. (1993) Age dependency of insulin like growth factor 1, insulin protein and immunoglobulin concentrations and gamma glutamyl transferase activity in first colostrum of dairy cows. *Milchwissenschaft*, **48**, 423–425.

Van Wyk, J.J., Casella, S.J., Hynes, M.A. & Lund, P.K. (1988) Indirect actions of growth hormone. In *Human Growth Hormone, Progress and Challenges* (ed. *Underwood, L.E.)*, pp. 26–61. Marcel Dekker, New York.

Venkatesh, M.P. & Abrams, S.A. (2010) Oral lactoferrin for the prevention of sepsis and necrotizing enterolcolitis in preterm infants. *Cochrane Database of Systematic Reviews*, **12**, 1–27.

Vidal, K., van den Broek, P., Lorget, F. & Donnet-Hughes, A. (2004) Osteoprotegerin in human milk, a potential role in the regulation of bone metabolism and immune development. *Pediatric Research*, **55**, 1001–1008.

Walker, A. (2010) Breast milk as the gold standard for protective nutrients. *Journal of Pediatrics*, **156**, S3–7.

Walker-Smith, J.A., Phillips, A.D., Walford, N., Gregory, H., Fitzgerald, J.D., MacCullagh, K. & Wright, N.A. (1985) Intravenous epidermal growth factor/urogastrone increases small-intestinal cell proliferation in congenital microvillous atrophy. *Lancet*, **2**, 1239–1240.

Werner, H., Katz, P., Fridkin, M., Koch, Y. & Levine, S. (1988) Growth hormone releasing factor and somatostatin concentrations in the milk of lactating women. *European Journal of Pediatrics*, **147**, 252–256.

Wilson, A.J., Byron, K. & Gibson, P.R. (1999) Interleukin 8 stimulates the migration of human colonic epithelial cells in vitro. *Clinical Science*, **97**, 385–390.

Wu, H.G., Song, S.Y., Kim, Y.S., Oh, Y.T., Lee, C.G., Keum, K.C., Ahn, Y.C. & Lee, S.W. (2009) Therapeutic effect of recombinant human epidermal growth factor (RhEGF) on mucositis in patients undergoing radiotherapy, with or without chemotherapy, for head and neck cancer, a double-blind placebo-controlled prospective phase 2 multi-institutional clinical trial. *Cancer*, **115**, 3699–3708.

Yakar, S., Pennisi, P., Wu, Y., Zhao, H. & LeRoith, D. (2005) Clinical relevance of systemic and local IGF-I. *Endocrinology Development*, **9**, 11–16.

Yamanaka, H., Hagiwara, K., Kirisawa, R. & Iwai, H. (2003) Proinflammatory cytokines in bovine colostrum potentiate the mitogenic response of peripheral blood mononuclear cells from newborn calves through IL-2 and CD25 expression. *Microbiology and Immunology*, **47**, 461–468.

Yasunaga, Y., Shimomura, Y., Kanayama, S. *et al.* (1996) Increased production of interleukin 1b and hepatocyte growth factor may contribute to foveolar hyperplasia in enlarged fold gastritis. *Gut*, **39**, 787–794.

Zeeh, J.M., Procaccino, F., Hoffmann, P., Aukerman, S.L., McRoberts, J.A., Soltani, S., Pierce, G.F., Lakshmanan, J., Lacey, D. & Eysselein, V.E. (1996) Keratinocyte growth factor ameliorates mucosal injury in an experimental model of colitis in rats. *Gastroenterology*, **110**, 1077–1083.

Zou, Z.Y., Lin, X.M., Xu, X.R., Xu, R., Ma, L., Li, Y. & Wang, M.F. (2009) Evaluation of milk basic protein supplementation on bone density and bone metabolism in Chinese young women. *European Journal of Nutrition*, **48**, 301–306.

Zwiebel, J.A., Baho, M., Nexo, E., Salomon, D.S. & Kidwell, W.R. (1986) Partial purification of transforming growth factors from human milk. *Cancer Research*, **46**, 933–939.

4 Casein and Whey Proteins in Human Health

K. Petrotos[1], E. Tsakali[1], P. Goulas[2] and A.G. D'Alessandro[3]

[1]Department of Biosystems Engineering, Laboratory of Food Process Engineering, Technological Educational Institute of Larissa, Larissa, Greece
[2]Department of Animal Production, Laboratory of Animal Production, Technological Educational Institute of Larissa, Larissa, Greece
[3]Department of Engineering and Management of the Agricultural, Livestock and Forest Systems, University of Bari, Italy

4.1 Introduction

Milk is a complex physiological liquid that simultaneously provides nutrients and bioactive components that facilitate the successful postnatal adaptation of the newborn infant by stimulating cellular growth and digestive maturation, the establishment of symbiotic microflora and the development of gut-associated lymphoid tissues. The health benefits of milk and fermented milk products have been known of since medieval times. Drinking milk has taken the advantage of the extensive nutritional value not only to the child, but also to the adult and the elderly. In recent years, milk constituents have become recognised as functional foods, suggesting their use has a direct and measurable effect on health outcomes (Gill *et al.* 2000a).

Milk is composed of water, fat, proteins (mainly casein micelles and whey proteins), carbohydrates (mainly lactose) and quantitatively minor bioactive components: minerals, vitamins and enzymes (Jensen 1995). It contains two major protein types: casein and whey protein. Milk proteins have received increasing attention as potential ingredients of health-promoting functional foods targeted at diet-related chronic diseases such as cardiovascular disease, Type 2 diabetes and obesity. To this end, growing interest has been focused on physiologically active peptides derived from milk proteins. These peptides are inactive within the sequence of the parent protein molecule and can be liberated by gastrointestinal digestion of milk, fermentation of milk with proteolytic starter cultures or hydrolysis by proteolytic enzymes. Milk protein-derived peptides have been shown, under *in vitro* and *in vivo* conditions, to exert a number of activities affecting the digestive, endocrine, cardiovascular, immune and nervous systems.

A great variety of naturally formed bioactive peptides have been found in fermented dairy products such as yogurt, sour milk and cheese. Recently, industrial-scale technologies suitable for the industrial production of bioactive milk peptides have been developed. In addition, a few commercial food products supplemented with milk protein-derived bioactive peptides have been launched on limited markets. Some of these products carry clinically documented

Milk and Dairy Products as Functional Foods, First Edition. Edited by A. Kanekanian.

benefits, in particular for reduction of mild hypertension. The multifunctional properties of milk peptides appear to offer considerable potential for the development of many similar products in the near future.

Important research projects to: investigate the bioavailability and safety of the above-mentioned bioactive peptides; assess and elucidate the mechanisms by which bioactive peptides exert their activities via molecular studies; and measure the impact of proteins and peptides on the expression of genes and hence optimise the nutritional and health effects of these compounds are all underway. Further challenges with functional food products containing novel bioactive peptides are associated with regulatory issues if any health-related claims are attached to the products. Currently, health claim regulations seem to vary greatly from one country to another but regional and continental harmonisation processes are in progress. The aim of this contribution is to provide a comprehensive overview of the effects of casein and whey protein on human health.

4.2 Casein-derived proteins and their effect in human health

Casein (from the Latin word *caseus* for cheese) is the main proteinaceous component of milk, where it accounts for *c.* 80% of the total protein inventory. The amount of casein in cow's whole milk varies according to the breed of cow and stage of lactation, but is generally in the range 24–29 g L^{-1}. Casein contains 0.7–0.9% phosphorus, covalently bound to the protein by a serine ester linkage. Casein is consequently known as a phospho-protein. All the amino acids that are essential to humans are present in casein in high proportions, with the possible exception of cysteine. Casein may therefore be considered as a highly nutritious protein

Casein exists in milk in complex groups of molecules (sometimes referred to as calcium phospho-caseinate) that are called micelles. The micelles consist of casein molecules, calcium, inorganic phosphate and citrate ions, and have a typical molecular weight of several hundred million. In terms of physical chemistry, the casein micelles may be considered to exist in milk as a very stable colloidal dispersion. The caseins, as proteins, are made up of many hundreds of individual amino acids, each of which may have a positive or a negative charge depending on the pH of the (milk) system. At some pH value, all the positive charges and all the negative charges on the (casein) protein will be in balance, so that the net charge on the protein will be zero. That pH value is known as the isoelectric point (IEP) of the protein and is generally the pH at which the protein is least soluble. For casein, the IEP is approximately 4.6 and it is the pH value at which acid casein is precipitated. In milk, which has a pH of about 6.6, the casein micelles have a net negative charge and are quite stable.

Casein has been shown to consist of several individual casein components, referred to as αs1-, αs2-, β- and κ-casein, with slightly different properties which are caused by small variations in their amino acid content. The four different types are known to occur in at least 10 genetic variants (A1–A3 and B–H) from which the A2, A1 and B forms are the most prevalent. It is composed of 209 amino acids (AA) and the molecular mass for variant A2 is 23.9 kDa. CN contains five serine-phosphate residues located in the N-terminal region. In the past, the main physiological role of casein in the milk system was widely accepted to be a source of amino acids required for the growth of the neonate. However, the dominant physiological feature of the casein micelle system has been proven a long time ago to be the prevention of pathological calcification of the mammary gland (Holt 1997).

While no specific physiological property has been proposed for the whole casein system (or its individual fractions, for that matter), various peptides hidden (or inactive) in the amino acid sequence have been the subject of increasingly intense studies. Much work regarding those peptides, which are known to possess bioactivities, is currently underway regarding their release via selective enzymatic hydrolysis. Until now, such release has been demonstrated primarily *in vitro* and, to a lower extent, *in vivo* using animal models. Experimental verification of the effectiveness of such bioactive peptides involving human subjects is still scarce, however.

Due to the increasing awareness of the effect of food ingredients upon promotion of health, several publications that cover recent advances in milk peptide research can be studied (Wal 2002; Clare *et al.* 2003; Florisa *et al.* 2003; Kilara & Panyam 2003; Kitts & Weiler 2003; Korhonen & Pihlanto 2003a; Lonnerdal 2003; Meisel & FitzGerald 2003; Pellegrini 2003; Pihlanto & Korhonen 2003b; Teschemacher 2003; Walker-Smith 2003; Janecka *et al.* 2004).This chapter provides (1) an overview of the major classes of bioactive peptides putatively derived from caseins which play a role in the nervous, cardiovascular, digestive and immune systems (Fig. 4.1), and (2) a brief critical discussion of the most recent data made available in the literature regarding that growing family of molecules.

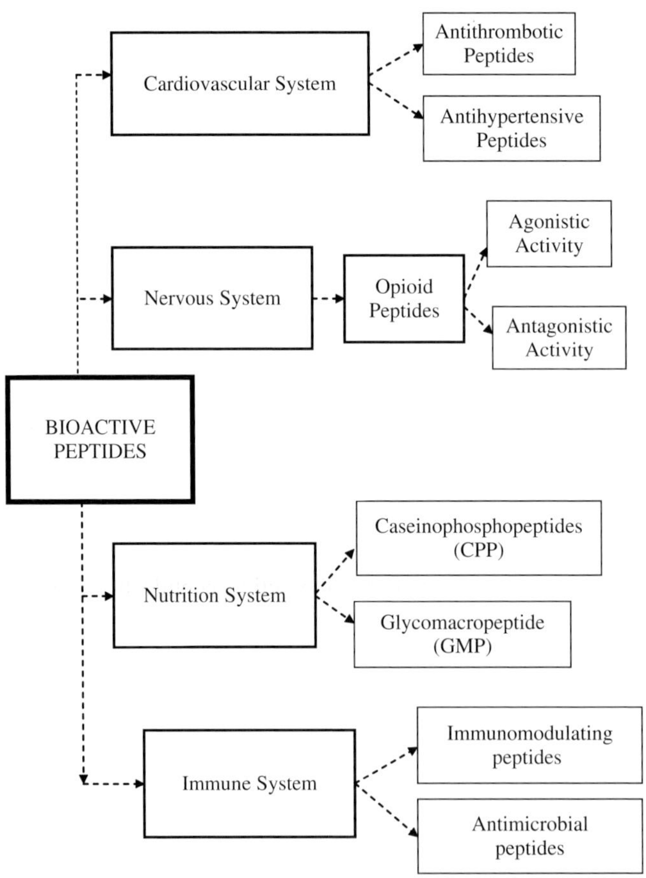

Fig. 4.1 Functions of bioactive peptides from caseins. Silva & Maltaca (2005). Reproduced with permission of Elsevier.

4.2.1 *The cardiovascular system*

Functional peptides derived from casein, present in either milk or dairy products, have been shown to have effects in the cardiovascular system mainly via *antithrombotic* and *antihypertensive* features (Table 4.1).

Antithrombotic peptides

The mechanisms involved in the clotting of milk, defined as the interaction of κ-casein with coagulating enzyme, and in the clotting of blood, defined as the interaction of fibrinogen with thrombin, have been proven to be similar in nature (Jollès 1975; Jollès & Henschen 1982; Fiat *et al.* 1993). In addition, structural homologies between the undecapeptide (residues 106–116) from cow's κ-casein, on the one hand, and the C-terminal dodecapeptide (residues 400–411) of human fibrinogen g-chain, on the other, have been duly reported (Jollès *et al.* 1978). Indeed, three amino acid residues (Ile108, Lys112 and Asp115) of the aforementioned undecapeptide of κ-casein are in a homologous position when compared to the human fibrinogen g-chain (Fiat & Jollès 1989).

Casoplatelins, which are casein-derived peptides (f106–116, f106–112 and f113–116), are inhibitors of both the aggregation of ADP-activated platelets and the binding of human fibrinogen g-chain to a specific receptor region on the platelet surface (Jollès *et al.* 1986; Fiat & Jollès 1989; Schlimme & Meisel 1995). Furthermore, the κ-casein fragment f103–111 can prevent blood clotting through inhibition of platelet aggregation, but is not able to affect fibrinogen binding to ADP-treated platelets (Fiat *et al.* 1993). More recently, it was reported that κ-caseinoglycopeptides from several animal species are a source of antithrombotic peptides. The sequence of amino acids in f106–171 of sheep's κ-casein, known as κ-caseinoglycopeptide, was shown to decrease thrombin- and collagen-induced platelet aggregation in a dose-dependent manner (Qian *et al.* 1995). In the plasma of 5-day-old newborns, after breast feeding or ingestion of cow's milk-based formulae, antithrombotic peptides were detected which had been derived from human and bovine κ-caseinoglycopeptides, respectively (Chabance *et al.* 1995).

Antihypertensive peptides

Blood pressure regulation (and hypertension, when it cannot be ensured) is partially dependent on the renin–angiotensin system; renin acts on angiotensinogen, thus releasing angiotensin-I that is further converted into the active peptide hormone angiotensin-II, a vasoconstrictor, by the angiotensin-converting enzyme (ACE). Angiotensin-II inactivates bradykinin (a vasodilator); moreover, it increases the production of aldosterone which decreases the renal output while increasing water retention (Maruyama *et al.* 1987a, 1987b; Fiat *et al.* 1993; Tirelli *et al.* 1997). Maruyama & Suzuki (1982) reported that tryptic hydrolysates of casein inhibited the *in vitro* activity of ACE. Those peptides derived from casein, known as casokinins, correspond to f(23–24), f(23–27) and f(194–199) of bovine αs1-casein B, as well as to f(177–183) and f(193–202) of bovine β-casein (Maruyama & Suzuki 1982; Maruyama *et al.* 1987a, 1987b; Meisel & Schlimme, 1994).

Pihlanto-Leppälä *et al.* (1998) studied the ACE-inhibitory activity of peptides from milk whey and from casein proteins fermented with lactic acid bacteria and hydrolysed by digestive

Table 4.1 Antithrombotic and antihypertensive peptide sequences in the primary structure of caseins.

Casein type	Peptide sequence	Amino acid segment	References
Antithrombotic peptide sequences			
κ-casein (bovine)	103–111	LSFMAIPPK	Fiat *et al.* (1993)
	106–116	MAIPPKKNQDDK	Jollès *et al.* (1986); Fiat & Jollès (1989); Schlimme & Meisel (1995)
	106–112	MAIPPKK	Jollès *et al.* (1986); Fiat & Jollès (1989); Schlimme & Meisel (1995)
Antihypertensive peptide sequences			
αs1 (bovine)	23–24	FF	Maruyama & Suzuki (1982); Maruyama *et al.* (1987a); Meisel & Schlimme (1994)
	23–27	FFVAP	Maruyama & Suzuki (1982); Maruyama *et al.* (1987a); Meisel & Schlimme (1994)
	102–109	KKYKVPQ	Gómez-Ruiz *et al.* (2002)
	142–147	LAYFYP	Pihlanto-Leppälä *et al.* (1998)
	157–164	DAYPSGAW	Pihlanto-Leppälä *et al.* (1998)
	194–199	TTMPLW	Maruyama & Suzuki (1982); Maruyama *et al.* (1987a); Meisel & Schlimme (1994); Pihlanto-Leppälä *et al.* (1998)
αs2 (bovine)	174–179	FALPQY	Tauzin *et al.* (2002)
	174–181	FALPQYLK	Tauzin *et al.* (2002)
	189–197	AMKPWIQPK	Maeno *et al.* (1996)
	189–193	AMKPW	Maeno *et al.* (1996)
	190–197	MKPWIQPK	Maeno *et al.* (1996)
	198–202	TKVIP	Maeno *et al.* (1996)
β (bovine)	74–76	IPP	Nakamura *et al.* (1995a, b)
	84–86	VPP	Nakamura *et al.* (1995a, b)
	108–113	EMPFPK	Pihlanto-Leppälä *et al.* (1998)
	177–183	AVPYPQR	Maruyama & Suzuki (1982); Maruyama *et al.* (1987a); Meisel & Schlimme (1994)
	193–198	YQEPVL	Pihlanto-Leppälä *et al.* (1998)
	193–202	YQEPVLGPVRGPFPI	Maruyama & Suzuki (1982); Maruyama *et al.* (1987a); Meisel & Schlimme (1994)
	199–204	GPVRGPFPIIV	Nakamura *et al.* (1995)
κ (bovine)	108–110	IPP	Nakamura *et al.* (1995)
γ-casein (bovine)	108–113	EMPFPK	Perpetuo *et al.* (2003)
	114–121	YPVEPFTE	Perpetuo *et al.* (2003)

enzymes. The fermentation of milk proteins by starters did not produce ACE-inhibitory peptides, but subsequent hydrolysis led to ACE-inhibitory activity. The peptides identified were αs1-casein f(142–147), f(157–164) and f(194–199) and β casein f(108–113), f(177–183) and f(193–198), as well as two others from whey proteins. The latter two were shown to be less active than the casein-derived peptides. Among these, the highest ACE-inhibitory activity was accounted for by the peptides derived from αs1-casein.

Other antihypertensive peptides could be pinpointed in the primary sequence of human β- and κ-caseins (Kohmura *et al.* 1989, 1990; Shah, 2000). Few studies exist concerning αs2-casein, which accounts for *c.* 10% w/w of the bovine casein system due to its poor solubility and difficult purification. The few peptides already identified as deriving from bovine αs2-casein exhibit weak ACE inhibition. Those peptides correspond to f(189–193), f(189–197), f(190–197) and f(198–202) (Maeno *et al.* 1996).

Studies by Tauzin *et al.* (2002) on a tryptic digest of bovine αs2-casein allowed the detection of several casokinins. Some exhibited IC50 values close to those that characterise the most potent peptides found in milk protein hydrolysates: two peptides, f(174–181) and f(174–179), had IC50 values very low compared with that of captopril, and were inhibitors of ACE. Such peptides and the whole tryptic hydrolysate which arises *in vivo* are of interest and require further experimental confirmation (Tauzin *et al.* 2002).

Two new peptides generated by tryptic hydrolysis of casein have been found to display bradykinin-potentiating activity on isolated guinea pig ileum. The hexa- and octapeptides, Glu-Met-Pro-Phe-Pro-Lys and Tyr-Pro-Val-Glu-Pro-Phe-Thr-Glu, respectively, originate from the γ-casein sequences f(108–113) and f(114–121); the latter showed an *in vitro* inhibition effect upon ACE (Perpetuo *et al.* 2003). Karaki *et al.* (1990) detected a decrease in blood pressure in spontaneously hypertensive rats following oral administration of tryptic hydrolysates of milk casein.

Two potent ACE-inhibitory peptides from β-casein, f(84–86) which corresponds to Val-Pro-Pro and f(74–76) which corresponds to Ile-Pro-Pro, and one from κ-casein, f(108–110) which corresponds to Ile-Pro-Pro, were purified from 'Calpis', a Japanese soft drink made from bovine skim milk fermented by *Lactobacillus helveticus* and *Saccharomyces cerevisiae* (Nakamura *et al.* 1995a). These two tripeptides have also been found in casein hydrolysates produced by extracellular proteinase of *L. helveticus* CP790 and in milk inoculated with *L. helveticus* (Yamamoto *et al.* 1994a, b).

In a placebo-controlled study, the blood pressure of hypertensive patients decreased significantly after 4–8 weeks of daily ingestion of 95 mL of sour milk containing such two tripeptides (Hata *et al.* 1996). The antihypertensive dipeptide from a yogurt-like product fermented by *L. helveticus* CPN4 containing the sequence Tyr-Pro was identified by Yamamoto *et al.* (1999); it is found within αs1-, β- and κ-casein primary structures. The synthetic peptide yielded significant antihypertensive activity for 2–8 hours after oral administration of 1 mg of peptide per kg of body weight of spontaneously hypertensive rats, and this effect was further proven to be highly dependent on the peptide dosage; however, the peptide possessed a low inhibitory activity against angiotensin-I-converting enzyme (Yamamoto *et al.*1999). The antihypertensive activity of casein-enriched milk, unfermented or fermented by *L. helveticus*, was evaluated by Leclerc *et al.* (2002); a significant reduction of blood pressure occurred in the former situation, whereas the antihypertensive activity measured in the latter case could be explained by the release of ACE-inhibitory peptides from caseins during the digestion process.

Several bioactive peptides with ACE-inhibitory activity have also been found in cheese; their appearance is influenced by proteolysis, but only to a certain degree. A αs1-casein-derived antihypertensive peptide isolated from Parmesan cheese by 6 months of ripening could not be found after 15 months (Addeo *et al.* 1992). Accordingly, the antihypertensive activity found in long-ripened Gouda cheese was half that found in its medium-aged counterpart (Meisel *et al.* 1997). Partially consistent with these findings are those reported by Gómez-Ruiz *et al.* (2002) when studying the ACE-inhibitory peptides in Manchego cheese manufactured with various starter cultures; the antihypertensive activity decreased within the first 4 months, was maximum by 8 months of ripening and decreased again towards 12 months. The protein fragments responsible for such activity were mainly low-molecular-weight peptides; a specific interest arose regarding f(199–204) from β-casein and f(102–109) from αs1-casein. Peptides isolated from Italian cheeses, namely Italico and Gorgonzola, were also found to be inhibitory of ACE activity (Smacchi & Gobbeti, 1998). Following isolation from Gouda cheese, Saito *et al.* (2000) analysed the structure of the antihypertensive features of a few peptides, and concluded that the strongest depressive effect on the systolic blood pressure and the highest ACE-inhibitory capacity is associated with peptides found in 8-month-old cheese. Several new oligopeptides were identified after extraction, semi-quantification and fast on-line identification by high-performance liquid chromatography mass spectrometry (HPLC-MS) of samples of Grana Padano cheese, some of which-contained bioactive sequences showing potential biological activity (Sforza *et al.* 2003).

A new type of ripened, low-fat cheese ('Festivo') was manufactured with classical starter cultures, in combination with *L. acidophilus* and bifidobacteria. Such a novel cheese contains several beneficial components, e.g. bioactive peptides with potential antihypertensive effects, probiotics, conjugated linoleic acid and bioavailable calcium (Ryhanen *et al.* 2001). Conversely, cheeses characterised by low levels of proteolysis, namely Quark, exhibit low ACE-inhibitory capacity (Meisel *et al.* 1997). As already mentioned, some peptides may contain several different bioactivities. Studies encompassing purification and identification of potentially bioactive peptides from enzyme-modified cheese (EMC) (Haileselassie *et al.* 1999) showed that β-casomorphin, as well as other hydrophobic peptides released by Neutrases, might serve as precursors for antihypertensive peptides; this realization suggests that EMC may be an alternative vehicle for production of that class of compounds.

The ACE-inhibitory peptides derived from dairy products are not as potent as the drugs commonly used for hypertension treatment; however, those products that hold a moderate bioactivity behave intrinsically (and naturally) as functional foods, so they can easily be included in the daily diet (Meisel *et al.* 1997).

4.2.2 *The nervous system*

It is a common belief that falling asleep is easier after drinking a glass of milk in the evening and that babies are soothed after breast or bottle feeding. Recent studies have provided evidence that peptides exist in dairy products which play an active role in the nervous system; known as opioid peptides, they can have agonistic or antagonistic activities (Table 4.2).

Table 4.2 Opioid peptide sequences in the primary structure of caseins.

Casein type	Peptide sequence	Amino acid segment	References
Agonistic peptides			
αs1 (bovine)	90–95	RYLGYL	Loukas *et al.* (1983)
	90–96	RYLGYLE	Loukas *et al.* (1983)
	91–95	YLGYL	Loukas *et al.* (1983)
β (human)	41–44	YPSFQ	Fiat *et al.* (1993)
γ (casein)	114–121	YPVEPFTE	Perpetuo *et al.* (2003)
Antagonistic peptides			
κ (bovine)	35–41	YPSYGLN	Xu (1998)
	58–61	YPYY	Xu (1998)
	25–34	YIPIQYVLSR	Xu (1998)
	158–164	EINTVQV	Xu (1998)

Opioid peptides

Located in the nervous endocrine (and also in the immune) system, as well as in the intestinal tract of mammals, opioid receptors (of the m , d- and k-type) interact with endogenous or exogenous ligands. The former 'typical' opioid peptides, e.g. enkephalin, endorphin and dynorphin, are derived from the precursor molecules proopiomelanocortin, proenkephalin and prodynorphin, respectively, and all share the typical N-terminal sequence Tyr-Gly-Gly-Phe. The latter or 'atypical' opioid peptides have agonistic (exorphins or formons: 'food hormones') or antagonistic (casoxins) activities, and are characterised by distinct N-terminal sequences. The common structural feature of endogenous and exogenous opioid peptides (except α-casein opioids) is the presence of a Tyr residue at the N-terminus, coupled with the presence of another aromatic residue, e.g. Phe or Tyr, in the third or fourth position. This is an important structural feature that ensures fitting into the binding site of opioid receptors; furthermore, the negative potential, localised in the vicinity of the phenolic hydroxyl group of tyrosine, seems to be essential for opioid activity. An absence of bioactivity was indeed noticed when no tyrosine residue exists (Meisel 1998).

Several authors have compiled information regarding opioid receptor ligands derived from food proteins, especially milk caseins (Teschemacher *et al.* 1997; Teschemacher 2003). Different opinions have arisen concerning the nature of the ligand–receptor interaction, and the suitability of models for the determination of bioactive conformation of peptide ligands active towards m-opioid receptors have been discussed (Gentilucci & Tolomelli 2004).

The major and the first opioid peptides discovered were the so-called β-casomorphins (Brantl *et al.* 1979) which are fragments of β-casein between the 60th and the 70th residues, mainly f(60–63), f(60–64), f(60–65), f(60–66) and f(60–70) (Smacchi & Gobbettti 2000); these peptides are characterised as m-type ligands (Teschemacher & Brantl 1994). Longobardo *et al.* (2000) reported that replacement of Phe in position 3 of two analogues of bovine β-casomorphin-7 and -5 by a β-homo-Phe enhanced the m-type opioid receptor affinity as compared to the natural peptides, by a factor of 2 and 5, respectively; the most potent is the pentapeptide β-casein f(60–64) (Fiat *et al.* 1993). A similar amino-acid sequence appears in β-casein from sheep milk (Richardson & Mercier, 1979) and water buffalo milk (Petrilli *et al.* 1983). The sequence f(41–44) from human milk β-casein was also claimed to be an opioid peptide, called β-casomorphin, holding agonistic activity (Fiat *et al.* 1993). Three αs-casein-derived exorphins, which typically correspond to bovine αs1-casein f(90–95), f(90–96) and f(91–96), were found to be d-selective receptor ligands (Loukas *et al.* 1983). More rencently, Perpetuo *et al.* (2003) reported a new octapeptide from γ-casein, Tyr-Pro-Val-Glu-Pro-Phe-Thr-Glu, possessing *in vitro* opiate-like activity which was generated by tryptic hydrolysis of casein. Opioid peptides from milk can be obtained *in vitro* by proteolysis of bovine αs-, β- and κ-caseins (Pihlanto-Leppalla *et al.* 1994). Sequential digestion brought about by pepsin and trypsin increases the overall degree of proteolysis, which generates hydrolysates with stronger antimicrobial or antihypertensive activities than a single enzyme treatment (Pihlanto-Leppälä *et al.* 1998, 1999).

Studies *in vivo* have also been performed with opioid agonist peptides. Ermisch *et al.* (1983) have shown that, following intracarotid injection of [3H] β-casomorphin-5 in rats, this compound accumulated to a relatively high degree in the areas free of blood–brain barrier. Higher amounts of β-casomorphin-7 were also encountered in blood from newborn calves, under the form of a precursor (Fiat *et al.* 1993). Meisel & Frister (1989) detected β-casomorphins in the intestinal chyme of minipigs, as well as in the small intestine of humans after ingestion of cow's milk (Svedberg *et al.* 1985). Since their absorption in the gut has not been reported to date, it is generally accepted that physiological influences are restricted to the gastrointestinal tract where they may modulate gastrointestinal function, intestinal transit, amino-acid uptake and water balance. As soon as peptides enter the bloodstream, they are believed to be quickly hydrolysed (Meisel, 1997).

On the other hand, β-casein-derived peptides may pass through the intestinal mucosa in neonates via passive transport, so babies may become calm and sleepy (Sturner & Chang 1988). In pregnant or lactating women, the mammary tissue is also permeable to β-casomorphin (Clare & Swaisgood 2000). Chabance *et al.* (1998) showed that many peptides from αs1-, β-, κ-casein and κ-caseinomacropeptide exist in the stomach of adults following consumption of milk or yogurt, and some fragments of casein were also found in the duodenum. Moreover, the κ-caseinoglycopeptide and the N-terminal peptide of αs1-casein were shown to be absorbed, and were accordingly detected in the plasma.

Exorphins exhibit pharmacological properties similar to opium (morphine) and exert naloxone-inhibitable activities (Meisel & Schlimme 1990); they also induce apnea and irregular breathing, modulate sleep patterns, stimulate pancreatic insulin and gastrointestinal somatostatin release, modulate animal behaviour, stimulate intestinal absorption of electrolytes and water, modulate intestinal motility and stimulate food intake by modifying the endocrine activity of the pancreas, hence causing an increase of insulin output (Xu 1998).

Recently, Sun *et al.* (2003) reviewed novel findings concerning the possible association between β-casomorphin and apnea in sudden infant death syndrome. The β-casomorphin immunoreactive material was first identified by Hamel *et al.* (1985) in cow's milk that had been incubated with various bacterial species including lactic acid bacteria. Matar & Goulet (1996) produced fermented milk enriched with the (opioid) β-casomorphin 1–4 (f(60–63)) using a mutant strain of *L. helveticus* that lacks X-prolyldipeptidyl-aminopeptidase activity. In cheese, β-casomorphins are further hydrolysed in a gradual manner via the proteolytic enzymes of *Lactococcus lactis* subsp. *cremoris* during ripening; this process is strongly dependent on pH and salt concentration prevailing in that dairy product (Muehlenkamp & Warthesen 1996).

Opioid antagonists are those peptides that suppress the agonist activity of enkephalin; known as casoxins, they have been found in both bovine and human κ-casein as well as in αs1-casein (Chiba *et al.* 1989; Yoshikawa *et al.* 1994). Two bovine casoxins are opioid receptor ligands of the m-type, with relatively low antagonistic potency as compared to naloxone (Meisel 1998). Casoxins A and B correspond to amino-acid sequences within bovine κ-casein; Casoxin A corresponds to f(35–41) of κ-casein (i.e. Tyr-Pro-Ser-Tyr-Gly-Leu-Asn), whereas casoxin B is accounted for by f(58–61) of κ-casein (i.e. Tyr-Pro-Tyr-Tyr). Casoxin C is an opioid antagonist obtained from tryptic digests of bovine κ-casein; its sequence corresponds to f(25–34) (i.e. Tyr-Ile-Pro-Ile-Gln-Tyr-Val-Leu-Ser-Arg) and possesses the highest biological potency (Xu, 1998). Casoxin D was generated from αs1-casein (Clare & Swaisgood 2000). Some casoxins were modified by methoxylation during the process of isolation and purification; they were proven to be more active than the corresponding C-terminal, non-methoxylated κ-casein fragment (Schlimme & Meisel 1995).

4.2.3 *The immune system*

The systems involved in the defence of the human body against invaders are rather complex, and diet is known to play an important role. Research concerning the role of functional peptides in this field is relatively new, but already seems very promising. The two main activities are the immunomodulatory activity (i.e. stimulation of the immune system) and the antimicrobial activitity (i.e. inhibition of pathogenic bacteria) (see Table 4.3).

Immunomodulating peptides

Breast feeding facilitates physical transmission of passive immunity via a number of multifunctional factors which have a direct effect on the neonate's resistance to bacterial and viral infections; caseins are included among these factors. In fact, during enzymatic digestion of human and bovine milk caseins, peptides with immunomodulating capacities are released. Jollès *et al.* (1981) were the first to report that extracts of trypsin-hydrolysed human milk possess immunostimulating activity; in particular, the hexapeptide Val-Glu-Pro-Ile-Pro-Tyr corresponding to f(54–59) of human milk β-casein was isolated from its tryptic hydrolysate (Parker *et al.* 1984) and proven to account for such an activity. (It is noteworthy that this peptide represents the C-terminal part of β-casomorphin-11.) Several other peptides were later isolated, namely f(63–68) and f(191–193) from bovine β-casein and f(194–199) from bovine αs1-casein (Parker *et al.* 1984; Migliore-Samour & Jollès 1988). All the peptides mentioned above stimulate phagocytic activity of murine and human macrophages *in vitro*, and exert a protective effect against *Klebsiella pneumoniae* infection in mice *in vivo* (Migliore-Samour *et al.* 1989).

More recently, the pH 4.6-soluble products from the hydrolysis of whole bovine casein by chymosin encompassed peptides possessing immunomodulatory activity, i.e. αs1-casein

Table 4.3 Immune system affecting peptide sequences in the primary structure of caseins.

Casein type	Peptide sequence	Amino acid segment	References
Immunomodulating peptide sequences			
αs1 (bovine)	1–23	RPKHPIKHQGLPQEVLNENLLRF	Minkiewicz *et al.* (2000)
	194–199	TTMPLW	Parker *et al.* (1984); Migliore-Samour & Jollès (1988)
αs2 (bovine)	1–32	KNTMEHVSSSEESIISQETYKQEK-NMAINPSK	Hata *et al.* (1999)
β (bovine)	1–28	RELEELNVPGEIVESLSSSEESITRINK	Hata *et al.* (1999)
	63–68	PGPIPN	Parker *et al.* (1984); Migliore-Samour & Jollès (1988)
	191–193	LLY	Parker *et al.* (1984); Migliore-Samour & Jollès (1988)
	193–209	LLYQEPVLGPVRGPFPIIV	Minkiewicz *et al.* (2000)
β (human)	54–59	VEPIPY	Parker *et al.* (1984)
Antimicrobial peptide sequences			
αs1 (bovine)	1–23	RPKHPIKHQGLPQEVLNENLLRF	Lahov & Regelson (1996)
αs2 (bovine)	164–179	LKKISQRYQKFALPQY	Recio & Visser (1999a)
	165–203	KKISQRYQKFALPQYLKTVYQHQ-KAMKPWIQPKTKVIPY	Zucht *et al.* (1995)
	183–207	VYQHQKAMKPWIQPKTKVIPYVRYL	Recio & Visser (1999a)
β (bovine)	193–209	YQEPVLGPVRGPFPIIV	Sandre *et al.* (2001)
αs1 (bovine)	1–23	RPKHPIKHQGLPQEVLNENLLRF	Lahov & Regelson (1996)
αs2 (bovine)	164–179	LKKISQRYQKFALPQY	Recio & Visser (1999a)

f(1–23) and β-casein f(193–209) (Minkiewicz *et al.* 2000). A tripeptide from κ-casein significantly increased the proliferation of human peripheral blood lymphocytes *in vivo* (Kayser & Meisel 1996), whereas the C-terminal sequence of bovine β-casein (f(193–209)) containing β-casokinin-10 induced similar results in rats (Coste *et al.* 1992). Bioactive peptides present in yogurt actually decreased cell proliferation with IEC-6 or Caco-2 cells which may explain, at least partially, why consumption of yogurt has been associated with a reduced incidence of colon cancer (Ganjam *et al.* 1997). Ru zhen, a product containing bioactive protein and colostral polypeptides, was proved to possess immunomodulatory activity and the potential to improve resistance to disease (Zhang *et al.* 1998). After investigating the modulating activity of peptides corresponding to bioactive sequences of casein and a lyophilised extract of Gouda cheese, Meisel & Gunther (1998) reported that milk proteins are precursors of peptides that modulate the activity of human cells. A commercially available caseinophosphopeptide preparation CPP-III, consisting mainly of f(1–32) of bovine αs2-casein and f(1–28) of β-casein, enhances the proliferative response induced by lipopolysaccharide, phytohaemagglutinin and concavalin A stimulation and immunoglobulin production in mouse spleen cell cultures. This immunostimulating activity was attributed to the o-phospho-l-serine residue, suggesting that such a bioactivity is relatively stable to proteinase action in the intestinal tract (Hata *et al.* 1999).

Studies focusing on the effects of CPP-III on serum and intestinal immunoglobulin G and immunoglobulin A in mice proved that oral use of caseinophosphopeptide is beneficial to mucosal immunity (Otani *et al.* 2000). Matar *et al.* (2001) studied the effects of milks fermented by *L. helveticus* and its non-proteolytic variant on mucosal and tumoral immunity via oral administration of milk to mice (3 mL per mouse per day) for 3, 5 and 7 days. Lactic acid bacteria apparently show important implications upon modulation of the host's immune response, mainly its cellular response, because they are responsible for the release of bioactive peptides. Three fractions from the aforementioned fermented milk were then isolated and fed to mice for various periods (2, 5 and 7 days), and the effect on their tumoral immune system and on fibrosarcoma growth was analysed (LeBlanc *et al.* 2002). The mechanism by which lactic acid bacteria enhance the immune system is not yet clear, but it is accepted that bioactive compounds released in fermented milks contribute to the immunoenhancing and antitumour properties of these dairy products. Supplementation of infant formulae with appropriate milk proteins, namely caseinomacropeptide, may help to promote growth of a host-friendly colonic microflora, mainly bifidobacteria or lactobacilli, and may help to overcome some enteric infections (Bruck *et al.* 2003a, b).

Nowadays, the mechanism, structure and activity by which milk-protein-derived peptides exert their immunomodulatory effect is still questionable. However, the Arg residue at the N- or C-terminal region of a peptide has been suggested to be the leading motif recognisable by specific surface membrane receptors; in this context, it is worth noting that the immunomodulatory ACE inhibitor peptide β-casokinin-10 (among other ACE inhibitory peptides) has Arg in its C-terminus. Opioid peptides are also thought to affect the immunoreactivity of lymphocytes via opiate m-receptors which, for endorphins, are present on T-lymphocytes and human phagocytic leucocytes (Meisel 1998).

Antimicrobial peptides

In vitro inhibition of growth of pathogenic strains occurs in the presence of αs_2-casein or lactoferricin fragments (Tomita *et al.* 1994; Zucht *et al.* 1995). Caseicidin, obtained by chymosin-mediated digestion of casein at neutral pH, was among the first defence peptides actually purified; it exhibited activity against *Staphylococcus* spp., *Sarcina* spp., *Bacillus subtilis*, *Diplococcus pneumoniae* and *Streptococcus pyogenes* (Lahov & Regelson 1996). A cationic fragment from αs2-casein, f(165–203), known as casocidin-I, can inhibit growth of *Escherichia coli* and *S. carnosus* (Zucht *et al.* 1995). Two other peptides were isolated from a peptic hydrolysate of αs2-casein, namely f(183–207) and f(164–179); the former exhibited higher antimicrobial activity than the latter, although both possessed comparable hemolytic effects (Recio & Visser 1999a). Lactoferricin, or f(17–41) from lactoferrin produced by enzymatic cleavage brought about by pepsin, was shown to possess more potent bactericidal properties than the undigested, iron-binding glycoprotein present in most biological fluids of mammals, including milk (Meisel 1998). The other fragment of αs1-casein, namely f(1–23) known as isracidin, has demonstrated antibiotic-type activity *in vivo* versus *S. aureus* and *Candida albicans*; it can protect sheep and cows against mastitis (Lahov & Regelson 1996).

Fragments of human β-casein have also a protective effect against *Klebsiella pneumoniae* in mice (Migliore-Samour *et al.* 1989). The immunomodulatory peptide derived from bovine β-casein, namely f(193–209), was shown to enhance the antimicrobial activity of mouse macrophages, which had been obtained either from germ-free or from human flora-associated mice, without proinflammatory effects (Sandre *et al.* 2001). All these results pertaining to animals have been obtained *in vivo* by injecting them with peptides; at present, no studies are available that provide evidence for their putative effect when ingested.

4.2.4 *The nutrition system*

Some peptides are able to sequester calcium and other minerals, hence acting as biocarriers; they are called phosphopeptides (Table 4.4). Glycomacropeptide (GMP) may also exhibit a number of nutritional features which are discussed in Section 4.3 as these are compounds appearing in it. The term phosphopeptide was pioneered by Mellander (1950), meaning a casein-derived phosphorylated peptide which enhances vitamin-D-independent bone calcification in rachitic infants.

Chabance *et al.* (1998) have proven the occurrence of caseinophosphopeptides(CPPs) in the stomach and duodenum following milk ingestion. CPPs have also been found after *in vitro* and/or *in vivo* digestion of αs1-, αs2- or β-casein (Kitts 1994); f(43–58), f(45–55), f(59–79), f(66–74) and f(106–119) from αs1-casein; f(2–21), f(46–70), f(55–75),

Table 4.4 Caseinophosphopeptide sequences in the primary structure of caseins. Smacchi & Gobbetti (2000). Reproduced with permission of Elsevier.

Casein type	Peptide sequence	Amino acid segment
αs1 (bovine)	43–58	DIGSESTEDQAMEDIK
	45–55	GSESTEDQAME
	59–79	QMEAESISSSEEIVPNSVEQK
	66–74	SSSEEIVPN
	106–119	VPQLEIVPNSAEER
αs2 (bovine)	2–21	NTMEHVSSSEESIISQETYK
	46–70	NANEEEYSIGSSSEESAEVATEEVK
	55–75	GSSSEESAEVATEEVKITVDD
	126–136	EQLSTSEENSK
	138–149	TVDMESTEVFTK
β (bovine)	1–25	RELEELNVPGEIVESLSSSEESITR
	1–28	RELEELNVPGEIVESLSSSEESITRINK
	2–28	ELEELNVPGEIVESLSSSEESITRINK

f(126–136) and f(138–149) from αs2-casein; and f(1–25), f(1–28) and f(2–28) from β-casein have been reported in this particular (Smacchi & Gobbetti 2000). Certain CPPs were detected in ileostomy fluid for the first time by Meisel & FitzGerald (2003), which confirms the ability of such peptides to survive gastrointestinal passage to the human distal ileum. In addition to their presence in whole milk, CPPs have also been claimed to be released by proteolytic enzymes of lactic acid bacteria during ripening of cooked curd cheeses such as Grana Padano (Pellegrino *et al.* 1997) and Comté (Roudot-Algaron *et al.* 1994). Recently, Meisel & FitzGerald (2003) reviewed the structural features and physiological potential of milk-protein-derived caseinophosphopeptides, which were also shown to exert cytomodulatory effects. The phosphate residues, corresponding to *c.* 30% of the phosphorus content in milk, are present as monoesters of serine and occur mainly in clusters; most CPPs share a common feature, i.e. they are constituted by a sequence of three phosphoseryl residues, followed by two glutamic acid residues: SerP-SerP-SerP-Glu-Glu (Meisel *et al.* 1989; Meisel 1997).

The high concentration of negative charges of phosphate peptides makes them resistant to further proteolysis (Meisel & Schlimme 1990; Hirayama *et al.* 1992; Reynolds *et al.* 1994; Kasai *et al.* 1995; FitzGerald 1998; Clare & Swaisgood 2000). Furthermore, the negatively charged side chains (in particular the phosphate groups) of those amino acids represent the binding sites for minerals (Meisel 1998). CPPs have been shown to bind to such macroelements as Ca, Mg and Fe, as well as to such oligoelements as Zn, Ba, Cr, Ni, Co and Se. The Ca-binding activity of the phosphopeptide fractions of casein can be different however, which may be attributed to the influence of further amino acids around the phosphorylated binding sites (Meisel 1998). Iron deficiency, a major worldwide nutritional problem, can be reduced by CPPs; in fact, binding of Fe to phosphopeptides prevents formation of high-molecular-weight ferric hydroxides, which are poorly absorbed.

In vitro studies with rats (Ait-Oukhatar *et al.* 1999) have demonstrated that Fe bound to the phosphoserine residues of low-molecular-weight CPPs, namely β-casein f(1–25), improved their ability to treat anaemia and restore Fe storage tissues as compared to Fe bound to whole casein and inorganic salts. Further studies *in vivo* (Ait-Oukhatar *et al.* 2002) using radiolabelled iron assessed Fe absorption and uptake by tissues involved in Fe metabolism and storage. The metabolism of β-casein f(1–25) [Fe] added to cow's milk was studied in young women. Absorption was not significantly above that of ferrous sulphate, but a significantly higher transient tissue uptake was noticed that vanished by 14 days (hence suggesting that iron was used for metabolic purposes). Zinc absorption can also be enhanced if that metal is bound to CCPs, namely β-casein f(1–25); moreover, its inhibition by other nutrients such as Fe is prevented (Peres *et al.* 1998). An increase in Ca and Zn absorption was noticed after CPP incorporation into rice-based infant food during a human feeding trial (Hansen *et al.* 1997; Meisel & Bockelmann 1999).

Milk and dairy products provide plenty of Ca^{2+} which can form soluble complexes with those phosphopeptides, thus avoiding Ca phosphate precipitation (Sato *et al.* 1986; Berrocal *et al.* 1989) and enhancing intestinal absorption of Ca and retention in the body (Mykkanen & Wasserman 1980; Sato *et al.* 1986). According to Meisel (1997), binding of Ca involves Ser-bound phosphate groups as well as the free carboxyl groups of Glu. The hydrophobic tail protects this complex from further interactions, and hence prevents formation of insoluble calcium phosphate. Ferraretto *et al.* (2001) found that casein phophopeptides influence calcium uptake by cultured human intestinal HT-29 tumour cells. Chemically synthesised casein phosphopeptides β-casein (1–25)4P and αs1-CN(59–79)5P, carrying the characteristic 'acid motif' Ser(P)-Ser(P)-Ser(P)-Glu-Glu, were found to cause an increase of Ca^{2+} concentration due to an influx of extracellular Ca^{2+}. This effect was more pronounced with the β-casein-derived peptide than for the αs1-casein-derived peptide. Moreover, the CPP-promoting effect on calcium concentration was proven to depend on the structural conformation conferred by both the phosphorylated acidic motif and the preceding N-terminal portion (Ferraretto *et al.* 2003).

It has been shown that Ca-binding phosphopeptides have anticariogenic effects via inhibition of caries lesions through recalcification of the dental enamel; their application in the treatment of dental diseases as additives to toothpaste (Reynolds 1994) has therefore been proposed (Reynolds 1987; Tirelli *et al.* 1997; Clare & Swaisgood 2000). The consumption of cheese also has an anticariogenic effect; a significant correlation was found between the high content of casein in cheese and the caries-protective effect (Pause & Lembke 1993). Moreover, anticariogenic activity has also been reported for phosphopeptides from eggs (namely phosvitin and phosphophorin; Reynolds 1994; Tirelli *et al.* 1997). Heating processes affect the bioavailability of CPPs; sterilisation can induce dephosphorylation of phosphoseryl residues that occur in clusters and the formation of dehydroalanine from residues occupying isolated positions in the peptide chain (Meisel *et al.* 1991).

4.3 Whey proteins and their direct and indirect effects on human health

Whey proteins and their important physical properties are listed in Table 4.5.

Table 4.5 The constituents of the total whey proteins. Kinsella & Whitehead 1989. Reproduced with permission of Elsevier.

a/a	Protein	Total whey protein (% w/w)	Molecular weight (Daltons)
1	β-lactoglobulin (β-lg)	55	18 600
2	α-lactalboumin (α-la)	15	14 200
3	Immunoglobulin A (IgA)	Max 15%	420 000
4	Immunoglobulin G1 (IgG1)		163 000
5	Immunoglobulin G2 (IgG2)		150 000
6	Immunoglobulin M (IgM)		1 000 000
7	Bovine serum albumin (BSA)	5	66 000
8	Lactoferrin (Lf)	<1%	76 000
9	Lysozyme (Ly)	<1%	18 000
10	Lactoperoxidase (Lp)	<1%	78 000
11	Casein macropeptides	Less than 10%	7000–33 000
12	Sulphydryl oxidase	<1%	1 000 000
13	Superoxide dismutase	<1%	32 000

4.3.1 *β-lactoglobulin*

β-lactoglobulin (β-lg) is the dominant whey protein with a concentration of 50–60%. The protein exists as a dimmer (molecular weight 36 700 Da; Madureira *et al.* 2007) in solution because of electrostatic interactions between Asp130 and Glu 134 of one monomer with corresponding lysyl residues of another (Xu 1996). Its quaternary structure depends on medium pH. Despite its globular nature, in its isolated form it has low solubility and low ionic strength. The native conformation is sensitive to heat and pH; at temperatures <25°C and pH >7.0, it forms octamers (at pH values 5.2–7.0 and molecular weight 140 000 Da). It is synthesised in the mammary gland and has several genetic variants, the most common being β-lg A. It is composed mainly of β-sheet motifs, and consists of 162 amino acid residues (at pH values 3.5–3.5 and molecular weight 18 277 Da). It can also be found as a monomer with two crysteine residues (at pH 3 or >8).

The biological role of β-lactoglobulin is not completely clear but it is thought that it has a role of transporter, in particular of passive immunity to the newborn and in regulation of phosphorus metabolism at the mammary gland (Farrell *et al.* 1987). It also binds retinol and other small hydrophobic ligands, which are believed to assist in the absorption and availability of vitamin A (Burrington 2002). In milk, and especially in whey, there is a relatively high content of β-lactoglobulin which represents more than half of the present proteins. It acts as an effective emulgator and immunomodulator. Its molecule contains a hydrophobic part which can bind vitamin A, vitamin D, calcium and fatty acids (FA), simplifying their resorption (Brown 1984; Cho *et al.* 1994; Wang *et al.* 1997; Beaulieu *et al.* 2006).

In addition to the bioactivities exerted by the native molecule, β-lactoglobulin can exhibit further physiological functions because of numerous bioactive peptides that are contained within the protein. These peptides are inactive within the sequence of the precursor protein, but they can be released by *in vivo* or *in vitro* enzymic proteolysis. Once released, these peptides play important roles in human health. Their antihypertensive, antithrombotic, opioid, antimicrobial, immunomodulant and hypocholesterolemic properties have all been reported. All β-lactoglobulin-derived peptides also exhibit radical-scavenging activity (Hernández-Ledesma *et al.* 2007). The peptides that are produced through proteolytic degradation of β-lactoglobulin have not yet been sufficiently described. An interesting peptide is β-lactotensin, which stimulates contraction of smooth muscles (Yoshida & Owens 2005).

β-lactoglobulin is responsible for milk allergies, which affects 2–3% of children. In the majority of cases these allergies disappear by the time the child reaches 3 years of age. The peptides originating from β-lactoglobulin may cause oral tolerance and subsequently reduce the production of IgE which is specific for this protein (Pecquet *et al.* 2000), an effect that can be used in the prevention of milk allergies.

Bioactive peptides derived from β-lactoglobulin are currently a subject of intensive research. Various peptides derived from proteolytic digestion of β-lg have been shown to have inhibitory activity against ACE. It has been shown that unhydrolysed β-lg had very poor ACE inhibitory activity (Mullally *et al.* 1997a, b) but the digests of the protein, generated using pepsin, trypsin, chymotrypsin or other commercially available proteases, resulted in high ACE inhibition indices (i.e. 73–90%). Furthermore, these workers showed that the active peptides were usually short (<8 amino acids) and could be enriched from a mixture of protein and other peptides using ultrafiltration with low molecular mass cut-off membranes

(Mullally *et al.* 1997a). A tryptic peptide of β-lg (f(142–148)) was further characterised following reversed-phase chromatographic isolation and shown to have an ACE IC50 value of 42.6 nM (Mullally *et al.* 1997a).

Similarly, several researchers have demonstrated that a number of β-lg-derived peptides have impressive ACE inhibitory activity using a variety of *in vitro* assay techniques (Abubakar *et al.* 1998; Vermeirssen *et al.* 2002a, b). In a study where whey proteins were treated with different lactic acid starters and digestive enzymes it was reported that, following hydrolysis with trypsin or pepsin and characterisation by amino acid and MS-analysis, two peptides from β-lg – f(9–14) and f(15–20) – had ACE inhibitory activity (Pihlanto-Leppälä *et al.* 1998). Four novel ACE inhibitory peptides have been reported from caprine β-lg following hydrolytic treatment with thermolysin and purification (Hernandez-Ledesma *et al.* 2002).

It has been demonstrated that a tetrapeptide isolated from β-lg (f(142–145); Ala-Leu-Pro-Met), termed 'β-lactosin B', had significant antihypertensive activity when administered orally to spontaneously hypertensive rats (SHR) and therefore had potential as a natural antihypertensive agent for inclusion in foods (Murakami *et al.* 2004).

Proteolytic digestion of bovine β-lg by trypsin has been reported to yield four peptide fragments, f(15–20), f(25–40), f(78–83) and f(92–100), with bactericidal activity (Pellegrini *et al.* 2001). These peptides have been isolated and characterised, and found to exert their antimicrobial effects against Gram-positive bacteria only. Modulation of the peptides via targeted amino acid substitution expanded the bactericidal activity of the peptides to include the Gram-negative organisms *Escherichia coli* and *Bordetella bronchiseptica*. The authors concluded that β-lg may exert an antimicrobial function *in vivo* after its partial digestion by endopeptidases of the pancreas, and that small targeted modifications in the sequence of these peptides could be useful in expanding their antimicrobial function (Pellegrini *et al.* 2001).

Peptide fragments of β-lg generated through the action of alcalase, pepsin or trypsin have been shown to be bacteriostatic against *E. coli* and against pathogenic strains of *E. coli*, *Bacillus subtilis* and *Staphylococcus aureus* (Pihlanto-Leppälä *et al.* 1999; El-Zahar *et al.* 2004). For example, the activity of *E. coli* JM103 in the presence of 25 mg mL^{-1} β-lg (or α-la) hydrolysed with pepsin and trypsin was only 21% of the control after incubation for 6 hours (Pihlanto-Leppälä *et al.* 1999). By contrast, the intact β-lg species did not show any antimicrobial activity, even at concentrations as high as 100 mg mL^{-1}. It was also shown that ultrafiltration through 10 and 1 kDa molecular mass cut-off membranes may be used to enrich the bacteriostatic properties of the β-lg-derived peptides (Pihlanto-Leppälä *et al.* 1999). Several studies have reported that β-lg, chemically modified with 3-hydroxyphthalic anhydride to form 3-hydroxyphthaloyl-β-lg, is effective in inhibiting HIV-1, HIV-2, simian immunodeficiency virus, herpes simplex virus Types 1 and 2, and *Chlamydia trachomatis* infection *in vitro*. The authors of these reports conclude that the modified β-lg may be effective as an inhibitor of HIV-1 infection in humans (Berkhout *et al.* 1997; Neurath *et al.* 1997a, b; Wyand *et al.* 1999; Oevermann *et al.* 2003).

β-lg has also been shown to inhibit the replication of rotavirus in a dose-dependent manner (Superti *et al.* 1997). The inhibition of microbial adhesion may prevent colonisation of pathogens at an early stage of infection, and thus prevent or reduce the impact of the infection. The effect of β-lg on adhesion of pathogens to human ileostomy glycoproteins has been the subject of another study (Ouwehand *et al.* 1997). It was found that adhesion of pathogenic strains of *Klebsiella oxytoca* and *E. coli* was inhibited by pre-incubation of immobilised ileostomy glycoproteins with β-lg in a concentration-dependent manner. Further, the disulphide bridges in the β-lg molecule appear to be important to this activity, and the inhibition of pathogen adhesion appears to be mediated by β-lg binding at two distinct sites to the immobilised ileostomy glycoproteins. High-heat-treatment of β-lg appears to adversely affect this anti-adhesion activity of the protein (Ouwehand & Salminen 1998).

β-lg has been implicated in providing protection against development of cancer in animal models when delivered orally. Such activity has been investigated in order to establish the role of these proteins in disease prevention, and to contribute to a basis for their inclusion as ingredients in functional foods. Animal feeding trials have compared the efficacy of dietary whey proteins in retarding chemically induced colon cancer in a rat model of the disease. Dairy proteins, in particular whey protein, were found to be efficacious in retardation of intestinal tumours in young rats compared to other dietary proteins (meat, soy; McIntosh *et al.* 1995). Results also suggested that diets supplemented with β-lg enhanced protection against development of putative tumour precursors (aberrant crypts) in the hind gut wall. The mechanism behind the apparent anticancer activity of dietary whey proteins in these studies may be related to their sulphur amino acid content, for which there is a high requirement in the rat, and hypothesised role in protecting DNA in methylated form.

In a parallel study, a number of potential functional foods containing whey protein (flavoured milk, pasta, ice-cream, desert pudding, muesli and savoury dips) have been developed in preparation for human clinical trials. The foods containing whey protein were generally highly acceptable in sensory trials. These products are expected to be suitable as delivery vehicles for dietary whey protein in studies aimed at substantiating the human health benefits of this protein source, including β-lg (McIntosh *et al.* 1998).

Among other whey proteins, β-lg appears to bind mutagenic heterocyclic amines and thus provide some protection against their carcinogenic properties (Yoshida *et al.* 1991). The effects of whey proteins from bovine milk on melanogenesis in cultured human melanocytes have been studied. Among the major protein components of whey, only β-lg showed a depigmenting effect at a concentration of 1 mg mL^{-1} and also suppressed the activity of tyrosinase in these cells (Nakajima *et al.* 1997).

A tryptic peptide from β-lg (f(71–75); Ile-Ile-Ala-Glu-Lys) has been shown to have hypocholesterolemic activity in animal (rat) trials, and the mechanism of action would appear to relate to inhibition of micellar solubility of the cholesterol which in turn causes suppression of cholesterol absorption by a direct interaction between cholesterol mixed micelles and the tryptic peptide in the jejunal epithelia. The authors claim that their study provides the first direct evidence of a new hypocholesterolemic peptide derived from β-lg that exhibits a greater hypocholesterolemic effect than β-sitosterol in animal trials (Nagaoka *et al.* 2001).

Further, β-lactotensin, a neurotensin agonist derived from β-lg (f(146–149)), shows hypocholesterolemic activity after administration to mice for 2 days at a dose of 30 mg kg^{-1} (by intraperitoneal injection or i.p.) or 100 mg kg^{-1} (*per os* or p.o.) (Yamauchi *et al.* 2003). However, caution is required when extrapolating results from animal studies, particularly using rodent models, to potential effects in humans, as hypocholesterolemic effects can be animal-model specific.

Although β-lg can bind *in vitro* to a variety of hydrophobic substrates including retinol and long-chain fatty acids, its physiological function is still largely unknown and subject to speculation. The retinol and fatty acid binding of β-lg has been widely implicated in the proposed physiological function of β-lg. Fatty acid binding sites have been characterised on β-lg (Perez *et al.* 1992) and it was concluded that β-lg could participate in the digestion of milk lipids during the neonatal period by enhancing the activity of pre-gastric lipase by binding fatty acids that inhibit this enzyme. In addition, it has been shown that β-lg enhanced intestinal uptake of retinol, triglyceride and long-chain fatty acids in pre-ruminant calves (Kushibiki *et al.* 2001), and it was speculated that the protein may play a role in the absorption and subsequent metabolism of fatty acids. One report suggests that bovine β-lg at high concentration (almost $3\,g\,L^{-1}$) exhibits mitogenic activity equal to that of whole whey, as determined by DNA synthesis in hybridoma cultures. This same study indicated that there are variant differences in this mitogenic activity, the B variant of β-lg showing significantly lower activity (Moulti- Mati *et al.* 1991).

Several whey protein fragments have been shown to behave like opioid receptor ligands (Teschemacher *et al.* 1997). Specifically, β-lactorphin, a tetrapeptide (f(102–105); Tyr-Leu-Leu-Phe) derived from β-lg, behaves like an opioid receptor agonist. Recently, β-lactorphin has been shown to improve arterial function in SHR. Notably, β-lactorphin improved vascular relaxation in adult SHR *in vitro*, and additionally enhanced endothelium-independent relaxation (Sipola *et al.* 2002b). While these reports are interesting, only a minority of the opioid activity has been observed upon oral or intra-gastric administration of these peptides or their precursor proteins, and most studies have been performed in animals (Teschemacher 2003). A recent study (Roufik *et al.* 2006) on bioactive peptides derived from bovine β-lg has supported the view that *in vivo* studies are essential to validate the physiological effects of bioactive peptides and that long-chain bioactive peptides require protection from gastrointestinal enzymes when orally administered.

4.3.2 *α-lactalbumin*

Along with β-lactoglobulin and bovine serum albumin, α-lactalbumin (α-la) is the main whey proteins. α-la is between 5 and 25% of total whey protein and its ratio to β-lg in whey is 1:3. It is fully synthesised in the mammary gland. α-la acts as a coenzyme for the biosynthesis of lactose. It contains 123 amino acid residues (including lysine, leucine, threonine, tryptophan and crysteine) and has a molecular weight of 14 175 Da. (Madureira *et al.* 2007). It has a globular structure containing four disulphide bonds, four α helices, two regions of β structure, several regions consisting of a helix containing three turns per ten residues and a Ca^{+2} binding loop. α-lactalbumin is a metalloprotein because it binds metal ions, basically calcium. This structural conformation of α-la is thought to offer protection from thermal denaturation.

In contrast to β-lactoglobulin, α-lactalbumin has low immunogenicity and thus also a low allergy-inducing potential, which makes it a valuable nutrient for children. Its hydrolysis produces peptides with immunomodulatory effects. One of them is the tripeptide Gly-Leu-Phe, which stimulates phagocytosis of macrophages through specific receptors as well as respiratory burst of neutrophiles (Jaziri *et al.* 1992). It is therefore considered to be a substance with an important antimicrobial activity (Pihlanto-Leppälä *et al.* 2000; Pellegrini 2003). Its stress reduction also belongs among the bioactivities of α-lactalbumin (Markus *et al.* 2002). It was shown to raise brain serotonin, reduce cortisol concentration and improve mood under stress. α-lactalbumin may also exert antiulcerative properties; it protects the rat stomach mucosa against ulcerative lesions caused by indomethacin (Mezzaroba *et al.* 2006).

The health effects of α-la for human consumption can be subdivided into three groups of those related to: (1) the intact, whole protein; (2) peptides of the partly hydrolysed protein; and (3) the amino acids of the fully digested protein. Great emphasis has been placed on the latter nutritional aspect, as α-la is a particularly good source of the essential amino acids Trp and Cys which are the precursors of serotonin and glutathione, respectively. Based on the assumption that the nutritional need of a neonate is fully met by human milk, there is a drive to 'humanise' or 'adapt' the formulation to adjust for the different amino acid profile of human and bovine milk (Kelleher *et al.* 2003; Lien 2003). Bovine α-la, with its high homology to human α-la, is an ideal protein to overcome this discrepancy. α-la-enriched whey protein fractions with a reduced β-lg content are therefore of great interest to manufacturers of infant formula. The peptide with the amino acids sequence Tyr-Gly-Leu- Phe (f(50–53)), released from α-la by pepsin treatment, was shown to inhibit ACE with an IC50 value of 733 mM (Mullally *et al.* 1996); this peptide is termed α-lactorphin (Yoshikawa *et al.* 1986). Interestingly, proteolytic fragments of this peptide i.e. the dipeptides Tyr-Gly (f(18–19) and f(50–51)) and Leu-Phe (f(52–53)) were also observed to have an inhibitory effect, having IC50 values of 1523 and 349 mM, respectively (Mullally *et al.* 1996). Other studies detected ACE inhibitory activity in peptides Tyr-Gly-Leu (f(50–52)) at similar IC50 values (409 mM) (Pihlanto-Leppälä *et al.* 2000). In contrast, peptides with higher inhibitory activity were also detected towards the C-terminus of α-la, i.e. Val-Gly-Ile-Asn-Tyr-Trp-Leu-Ala-His-Lys (f(99–108)) exhibited an IC50 of 327 mM. The sequence Trp-Leu-Ala-His-Lys (f(104–108)) exhibited an IC50 value of only 77 mM.

In conscious SHR and in normotensive rats, α-lactorphin lowered blood pressure in a dose-dependent and naloxone inhibitable manner. These effects occurred at 10 mg kg^{-1} dosages. At higher dosages of 100 mg kg^{-1}, maximal reductions in systolic and diastolic blood pressure of 2374 and 1774 mm Hg, respectively, were achieved (Nurminen *et al.* 2000).

The use of α-lactorphin has also been studied for anticancer ability. In particular, it assists in cancer prevention (de Wit 1998). Furthermore, α-la contributes to reducing the risk of incidence of some cancers as it constrains cell division when incubated in distinct mammalian intestinal cell lines (Ganjam *et al.* 1997; Svensson *et al.* 2000). It was also demonstrated that α-la possess antiproliferative effects in colon adenocarcinoma cell lines, delaying initiation of cell apoptosis, after 4 days of growth with low concentrations of 10–25 μg mL^{-1} (Sternhagen & Allen 2001; Madureira *et al.* 2007). A few years ago, a folding variant of human α-la was discovered which selectively enters tumour cells and induces an apoptosis-like mechanism, probably by binding to histones and thereby disrupting the chromatin organisation in the cell nuclei (Düringer *et al.* 2003). This kinetically trapped protein–lipid complex was named HAMLET/BAMLET (human/bovine α-lactalbumin made lethal to tumour cells; Fast *et al.* 2005a). It consists of the calcium depleted-apo form of α-la in the previously mentioned molten globule state, which is stabilised by a fatty acid cofactor.

Note that the α-la/fatty acid interaction is stereo-specific; only unsaturated cis-fatty acids bind to α-la and only the C18:1.9 cis fatty acid (oleic acid) bound to α-la in a compact conformation is active against tumour cells (Svensson *et al.* 2003; Fast *et al.* 2005b). The complex is formed from either co-precipitated α-la in acid-casein or calcium-depleted α-la from whey, on an anionic exchange column that was previously conditioned with either the relevant fatty acid or casein from human milk (which also contains traces of the fatty acid). The active complex is washed off at very high NaCl concentration as it binds tightly to the column matrix. To date, no alternative method of complex formation has been published. *In vitro*, both the human and bovine forms were shown to induce apoptosis in a wide variety of tumour cells (Svensson *et al.* 2003) The specific therapeutic effect of HAMLET *in vivo* has been demonstrated on several examples such as human skin papillomas (Gustafsson *et al.* 2004), human glioblastoma tumour in mice (Fischer *et al.* 2004) and mammary cells of mice (Baltzer *et al.* 2004).

These newly described α-la compounds can be considered potential candidates for therapeutic or prophylactic treatment. However, the health benefits for human digestion (in particular in neonates) of milk or dairy products and whether or not such complexes are formed at any stage during digestion remain highly speculative to date. This requires further scientific and/or clinical investigation.

The α-la complex HAMLET has also been shown to exhibit antimicrobial activity, in particular against *Streptococcus pneumoniae* (both antibiotic-sensitive and -resistant strains) and *Haemophilus influenzae*. It was pointed out that commercially available α-la samples lacked those biological activities (Svanborg & Sabharwal 2004), the most likely reason being the purification method for α-la (size exclusion chromatography) whereby compounds of higher molecular mass are discarded and only monomeric α-la retained. A clinical study using α-la enriched infant formula showed an activity against enteropathogenic *E. coli* O127 and reduced incidences of diarrhoea comparable to that of breast milk (Bruck *et al.* 2003a, b). This action might be related to peptides that are released from α-la during digestion. It is also known that trypsin treatment of α-la has been shown to release two antibacterial peptides: Glu-Gln-Leu-Thr-Lys (f(1–5)) and Gly-Tyr-Gly-Gly-Val-Ser-Leu-Pro-Glu-Trp-Val-Cys-Thr-Thr-Phe (f(17–31)) disulphide-bonded to Ala-Leu-Cys-Ser-Glu-Lys (f(109–114)). Treatment using another intestinal enzyme, chymotrypsin, resulted in one antibacterial peptide, namely Cys-Lys-Asp-Asp-Gln-Asn-Pro-His-Ile-Ser-Cys-Asp-Lys-Phe (f(61–68)) disulphide-bounded to f(75–80). These peptides were mostly active against Gram-positive bacteria; however, weaker effects were observed with Gram-negative bacteria (Pellegrini *et al.* 1999).

Although pepsin did not release any antibacterial peptides in the study by Pellegrini *et al.* (1999), a different study indicated that both pepsin or trypsin released peptides from α-la which inhibited the growth of E. Coli JM103; the peptide concentration was 25 mg mL^{-1}, whereas unhydrolysed α-la did not inhibit the growth at a concentration of 100 mg mL^{-1} (Pihlanto-Leppälä *et al.* 2000). A study has also shown that peptides from hydrolysed α-la have growth-promoting effects on *Bifidobacterium longum* ATCC 15707 (Kee *et al.* 1998). It was further claimed that α-la could act as a prebiotic agent and be used as such in food and food supplements (Maase & Steijns 2002). The sequence of the amino acids Tyr-Gly-Leu-Phe (f(50–53)), released from α-la by pepsin treatment, has structural similarities to the opioid peptide human leu-enkephalin that has the amino acid sequence Tyr-Gly-Gly-Phe, termed α-lactorphin (Horikawa *et al.* 1983).

An opioid-like effect of this synthetic α-la peptide has been reported, having a weak activity both in receptor assay and pharmaco-dynamic measurements in guinea pig ileum and mouse vas deferens preparations *in vitro* (Yoshikawa *et al.* 1986). Tryptophan is a precursor for brain serotonin, which may improve the ability to cope with stress. Studies were carried out to investigate whether α-la might alleviate symptoms of stress in adult subjects. However, large neutral amino acids can compete with transport of tryptophan across the blood brain barrier, preventing uptake of tryptophan. The tryptophan to large neutral amino acid ratio in plasma was observed to be 48% higher after an α-la diet than after a casein diet. Furthermore, in stress-vulnerable subjects higher prolactin concentrations decreased cortisol and a reduction in depressive feelings were observed under stress (Markus *et al.* 2000). In later studies, α-la was observed to improve cognitive performances in stress-vulnerable individuals by increased brain tryptophan and serotonin activity (Markus *et al.* 2002).

Other clinical trials suggested that α-la could be used to improve sleep in adults submitted to nutritional disturbances (Minet-Ringuet *et al.* 2004). Animal studies using rats demonstrated that α-la can protect against ethanol and stress-induced gastric injury (Matsumoto *et al.* 2001) such as stomach ulcers with a dose-dependent effect (optimum 200 mg kg^{-1}). Interestingly, it exhibits a comparable potency to that of the typical antiulcer agent, Selbex. A subsequent study by the same group found that α-la causes an increase in the gastric luminal pH, an increase in gastric fluid and a delay in gastric emptying (Ushida *et al.* 2003).

4.3.3 *Immunoglobulins*

Immunoglobulins, also known as antibodies, represent about 9% of whey protein. They constitute a complex group, the elements of which are produced by B-lymphocytes, and they belong to a family of related structural glycoproteins. Two light and two heavy polypeptide chains combine to form Igs. All chains are joined with disulphide bonds and the whole structure has a molecular weight of about 150 000 Da. Igs have five isotypes – IgG (IgG1 and IgG2), IgA, IgD, IgE and IgM – that differ in structure but have the same size, electrophoretic charge and carbohydrate content.

Immonoglobulins assist in the increase of passive immunity of newborns. Colostrum has been studied in order to take advantage of this property, and the results are encouraging. Cows that had been fed with colostrum had increased levels of antibodies against immunising bacteria and the susceptibility to disease was reduced. There have also been studies on the collaboration of immonoglobulins with human rotavirus serotypes in order to prevent or treat enteric disease caused by viruses. In that direction, there is also a contribution to the treatment of infant gastritis by applying a diet that includes immune milk containing certain anti-helicobacter (Madureira *et al.* 2007).

Immunoglobulins contain antibodies that directly participate in the anti-infectious defence of both the milk gland and the newborn infant. The main immunoglobulin in milk is IgG1, followed by IgM, IgA and IgG2. In addition to the protection against microbial pathogens, they are responsible for activation of complement, stimulation of phagocytosis, preventing adhesion of microbes and neutralisation of viruses and toxins. They also increase the intracellular levels of glutathione, which is the key cell antioxidant (Bounous & Gold 1991). The secretory IgA (sIgA) plays a special role which works in a non-specific way, covering infectious agents and preventing their adhesion to the mucosal epithelium. The concentration of immunoglobulins is higher in colostrum. It can be increased by the immunisation of pregnant mothers, which takes place in practice among farm animals. Using such cow colostrums, food supplements can be prepared not just for veals but also for humans. Their enhanced preventive activity against different microbial infections has been documented in several studies (e.g. Mehra *et al.* 2006).

4.3.4 *Bovine serum albumin*

In comparison to α-lactalbumin and β-lactoglobulin, BSA is not synthesised in the mammary gland but is found in milk because of a passive leakage from the bloodstream. It is a large globular protein with a molecular weight of 66 267 Da. Its single polypeptide chain consists of 582 amino acid residues; it has 17 intrachain disulphide bonds and one free thiol group at residue 34.Due to its large size and its structure, BSA can bind to free fatty acids, other lipids and flavour compounds. BSA has been studied for its ability to inhibit tumour growth; *in vitro* incubation with human breast cancer cell line MCF-7 has provided adequate evidence of this, which is dependent upon the modulation of activities of the autocrine growth regulatory factors (Madureira *et al.* 2007). Recent studies demonstrated the production of the peptide bovine albutensin (BA), which represents f(208–216) of bovine serum albumin and is reported to show strong competitive ACE inhibition. This peptide was produced by tryptic hydrolysis of BSA.

4.3.5 *Lactoferrin*

Lactoferrin (LF) is a monomeric glycoprotein that belongs to the family of the transferrins. It has a molecular weight of *c.* 80 kDa and has the ability to bind two iron ions very tightly with two carbohydrate groups. The presence of iron gives it a characteristic red colour, and that is why it is also known as the red protein. LF has a strong basic character with a pI between 8 and 9. The lactoferrin component of whey consists of *c.* 689 amino acid residues, while human lactoferrin consists of 691 residues.

Lactoferrin can be characterised as a multifunctional protein. It has antimicrobial, anti-inflammatory, anticarcinogenic, immunomodulatory and bone growth factor properties (Brisson *et al.* 2007). Lactoferrin also appears to have antiviral, antifungal and antibacterial properties. The antimicrobial effect is likely more potent on organisms that require iron to replicate, as lactoferrin has the unique ability to chelate iron in a way that deprives micro-organisms of this essential nutrient for growth (Yamaguchi *et al.* 1993). In addition, lactoferrin has the ability to release the outer membrane of Gram-negative bacteria, the lipopolysaccharide component, thus acting as an antibiotic (Appelmelk *et al.* 1994).

Madureira *et al.* (2007) cite several studies based on animals such as mice and rats which provide strong evidence for the therapeutic value of bovine LF (bLF) in treating distinct types of cancer. The iron-binding capacity of bLF has been suggested as the molecular rationale behind such a biological activity. Free iron may act as a mutagenic promoter by inducing oxidative damage to the nucleic acid structure; hence, when BLF binds iron in tissues, it reduces the risk of oxidant-induced carcinomas (Weinberg 1996) and colon adenocarcinomas (Tsuda *et al.* 1998). Studies that pertain to other cancers such as lung, bladder, tongue and oesophagus, and which convey similar results, have also been published (Masuda *et al.* 2000; Tanaka *et al.* 2000).

Human clinical studies using bLF on infants as well as on adults have also been conducted. In infants, increases of Bifidobacterium in the fecal flora and the serum ferritin level were obtained by supplementing infant formula with bLF at 1 mg mL^{-1} (Chierici *et al.* 1992; Roberts *et al.* 1992). After feeding bLF-enriched formula for 2 weeks, the ratio of Bifidobacterium in the fecal flora of low-birth-weight infants was increased, while the ratios of Enterobacteriaceae, Streptococcus and Clostridium showed a tendency to decrease (Kawaguchi *et al.* 1989). These results suggest that feeding of bLF-enriched formula leads to the establishment of a Bifidobacterium-predominant flora in infants.

In adults, it was reported that 7-day bLF treatment increased the eradication rates of the one-week triple therapy with rabeprazole, clarithromycin and tinidazole for *H. pylori* gastric infection (Di Mario *et al.* 2003a). A significant difference was found between LF group (100%) and control group (78.9%). A pilot study showed that the ingestion of bLF for 8 weeks can decrease serum hepatitis C virus (HCV)-RNA levels and serum alanine aminotransaminase (ALT) levels in chronic hepatitis C patients with low viral load (Tanaka *et al.* 1999). In another study, 25 patients with chronic hepatitis C genotype 1b received a 6-month course of bLF (Iwasa *et al.* 2002). The serum level of HCV RNA significantly decreased in patients given bLF during the 6 months of treatment.

The effectiveness of LF was investigated in the treatment of tinea pedis in a placebo-controlled double-blind study (Yamauchi *et al.* 2000). In the analysis limited to subjects with moderate vesicular or interdigital tinea pedis, dermatological symptom scores in the bLF-treated groups were significantly decreased compared with those in the placebo group.

In a study using recombinant human LF (rhLF), a preventive effect on NSAID-induced enteropathy was observed (Troost *et al.* 2003). A study by Troost *et al.* (2001) demonstrated that over 60% of administered bLF survives passage through the adult human stomach and enters the small intestine in an intact form. On the other hand, analysis of the gastric contents revealed that lactoferricin B (LFcin B) was formed at a molar concentration corresponding to 4.5% of ingested bLF (Kuwata *et al.* 1998). When the animals were given free access to milk containing bLF at 40 mg mL^{-1} (482 mmol L^{-1}), the levels of bLF fragments containing the LFcin B region in the contents of the stomach, upper small intestine and lower small intestine were approximately 200, 20 and 1 mmol L^{-1}, respectively (Kuwata *et al.* 1998, 2001). Some parts of ingested LF are not likely to be fully digested and likely to be present in the lower gastrointestinal tract. These intact bLF and partially digested bLF peptides, which retain biological activities, may exert various physiological effects in the digestive tract.

Orally administered LF has been shown to exhibit several beneficial effects at sites other than the digestive tract. In the case of infants or adults with injury to the intestine, it is possible that ingested LF enters the blood circulation and acts systemically. However, neither bLF nor functional bLF fragments (LFcin B-containing peptides and anti-bLF antibody-binding peptides) were at a level higher than the detection limit (1 ng mL^{-1}) detected in the portal blood of normal adult rats orally given bLF at a maximal dose of 5 g kg^{-1} (Wakabayashi *et al.* 2004). LF-related molecules are therefore not likely to be transported from the intestine into the circulation.

Based on the large body of accumulated evidence, it is reasonable to consider that oral bLF or its digested products act initially on the intestinal immune system and then augment the protective immunity systemically (Teraguchi *et al.* 2004). Orally administered bLF enhances interleukin (IL)-18 production in the intestinal epithelial cells and increases the numbers of CD4+ cells, CD8+ cells and natural killer (NK) cells in the intestinal mucosa (Kuhara *et al.* 2000; Wang *et al.* 2000). In the systemic immune system, oral bLF increases the numbers of cells in lymph nodes and the spleen, enhances the activity of peritoneal macrophages and splenic NK cells and enhances the production of Th1 type cytokines (IL-12 and interferon (IFN)-g) (Sekine *et al.* 1997a, b; Takakura *et al.* 2004; Wakabayashi *et al.* 2003, 2004). The circulation of humoral factors such as IL-18 and the recruitment of immune cells may be signals bridging the intestinal immune system and the systemic immune system in the action of oral LF. However, the detailed molecular mechanisms must be elucidated in future studies.

Plasma levels of lactoferrin have been found to be elevated due to its release from neutrophils during infection, inflammation, tumour development and iron overload (Levay & Viljoen 1995). Several studies have revealed that lactoferrin plays a direct role in the body's defence against pathogens, including findings that individuals more susceptible to infection have lower levels of neutrophil lactoferrin (Breton-Gorius *et al.* 1980; Boxer *et al.* 1982).

Several studies have examined the effects of whey protein concentrates on the eradication of *Helicobacter pylori* (Early *et al.* 2001; Di Mario *et al.* 2003a, b) which is is widely accepted as the causative agent in over 90% of duodenal ulcer cases (NIH 1994). To date, the most effective form of eradication of *H. pylori* is by administrating a 1-week course of a triple antibiotic therapy. Due to the expense of treatment and concern for antibiotic resistance, new treatments are being sought. In an open, randomised, single-centre study of 150 individuals with diagnosed *H. Pylori* infection, patients were given antibiotics at varying doses and durations (range 7–10 days) in conjunction with 200 mg encapsulated lactoferrin (Di Mario *et al.* 2003b). Analysis of the study revealed a 100% eradication of *H. pylori* in the group using the 7-day antibiotic course with the addition of lactoferrin. On the other hand, the classic 7-day triple antibiotic group had a success rate of 76.9% while the 10-day treatment group demonstrated a 70.8% success rate.

In a small study, 12 children suffering from chronic pharyngitis were administered a combination of 500 mg erythromycin three times daily and 100 mg bovine lactoferrin in a gargle. All children tested positive for Group A Streptococci prior to the treatment. After 15 days of treatment, fewer intracellular Group A Streptococci were found compared to a group treated with antibiotics alone (Ajello *et al.* 2002).

In a concise review, Shah (2000) discusses the bacteriostatic and bacteriocidal activity of lactoferrin against a number of organisms including *Escherichia coli*, *Salmonella typhimurium*, *Shigella dysenteriae*, *Listeria monocytogenes*, *Bacillus stearothermophilus*, *Bacillus subtilis* and *Micrococcus luteus* (Batish *et al.* 1988; Payne *et al.* 1990; Saito *et al.* 1991; Yamauchi 1992; Shah 2000). Shah further discusses that lactoferrin, when in combination with lysozyme, is a more potent bacteriostatic agent against *Pseudomonas aeruginosa*, *Listeria monocytogenes* and *E. coli* (Suzuki *et al.* 1989; Shah 2000). In addition to the above-mentioned bacteria, lactoferrin has also demonstrated antifungal activity towards *Candida albicans* (Jones *et al.* 1994).

4.3.6 *Lysozyme*

Lysozyme (Ly) is defined as 1,4–b-N-acetylmuramidase. There are two types of lysozymes: those found in hen egg whites, also known as lysozyme c, and those found in goose egg whites which are known as lysozyme g. The lysozyme contained in whey is of both types and the 3D structure of this protein is similar to that of lactalbumin. The molecular weight of this protein is *c.* 18 kDa. The antimicronial activity of lysozyme in milk has long been proven, attributed to a mechanism referred to as 'lysis' which reduces the resistance of the bacterial cell wall and kills it. The lysozyme is active on both Gram-positive and Gram-negative bacteria (Florisa *et al.* 2003; Min *et al.* 2005).

4.3.7 *Lactoperoxidase*

Lactoperoxidase (LP) is a glycoprotein and in whey represents 1% (w/w) of the total protein content. It is also a common enzyme in milk of various species, but not in human milk. Its molecular weight is *c.* 78 kDa with a heme group, and its primary structure is yet unknown. It has an iron content of 0.0689–0.709% while its carbohydrate content is 9.9–10.2%. It is known that at its secondary structure 65% of the molecule exists as β-structures, 23% as α-helix and 12% as uncoordinated structures (Xu 1996).

The activity of lactoperoxidase depends on many factors such as animal species, breed and lactation cycle (Madureira *et al.* 2007). Lactoperoxidase along with lactoferrin and lysozyme comprise the non-immunoglobulin protective system in milk. Many studies have attempted to explain the antimicrobial mechanism of peroxidase, concluding that this is affected by the parameters of the environment where it acts upon. LP as an enzyme is more active at acidic pH (Wever *et al.* 1982); there is a dependence between antimicrobial activity and the ion that plays the role of the electron donor and the LP is completed with the presence of a long thiocyanate ion and hydrogen peroxide (Reiter & Perraudin 1991).

4.3.8 *Casein macropeptides*

Casein macropeptides are casein-derived peptides of whey of a heterogenous nature, derived by the action of proteolysis of caseins by endogenous and exogenous proteinases. This heterogenous group of substance includes members with molecular weight (MW) 7000–33 000 Da.The glycomacropeptide (GMP) is formed during the enzymatic cheese-making process. Rennet or chymosin hydrolyses the peptide bond between residues 105 and 106 of κ-casein and the resulting molecule, GMP, is eluted in the whey. This molecule contains residues f(106–169) of κ-casein. The C-terminal portion of this molecule is more hydrophilic, as it contains the oligosaccharides that are O-linked to threonine and serine. The large GMP molecule cannot be absorbed as such, so it has to be broken down into smaller peptides before an effect on blood components arises. GMP is unique in its amino acid composition: it lacks aromatic amino acids and is rich in branched chain amino acids.

As well as the various bioactivities displayed by GMP (or smaller peptides derived from this), a number of beneficial effects upon the nutrition system can also be described. El Salam *et al.* (1996) indicated that GMP might be useful for diets aimed at controlling several liver diseases and in cases where branched chain amino acids appear to be used as a carbon source. Stan *et al.* (1988) showed that pepsin hydrolysis produced several different peptides with distinct physiological activities. One of these has an opioid effect and another a satiety effect when administered to starving animals; the latter effect was similar to that exhibited following cholecystokinin injection. A well-known disease in infants fed with whey-protein-predominant formulae is hyperthreoninemia. Sweet whey, containing not only whey proteins but also the threonine-rich GMP, is commonly used for the manufacture of those formulae. Rigo *et al.* (2001) tested an experimental formula based on acid whey without GMP against a formula based on whey with GMP upon threonine metabolism; acid whey formulae are accordingly recommended for feeding pre-term infants. Fanaro & Vigi (2002) emphasised that removal of GMP from whey is recommended for the development of infant formulae closer to human milk. Glycomacropeptide is also known to allow absorption of calcium, iron or zinc. Kelleher *et al.* (2003) studied GMP and α-lactalbumin supplementation of infant formulae on the nutritional status in infant rhesus monkeys and found that both increase zinc absorption. This may allow a reduction of formula zinc concentrations and promote a plasma amino acid pattern similar to that of breast-fed infant monkeys, respectively.

4.3.9 *Other enzymes contained in the whey*

The main members of this category are the two enzymes sulphydryl oxidase and superoxide dismutase. According to Meyer *et al.* (2005), the superoxide dismutase (SOD) is an antioxidant protein which has a very low concentration of about 0.15–0.60 mg mL^{-1} of bovine whey.

4.4 The effect of processing on the bioactivity of casein and whey proteins

The effect of several processing methods on the bioactivity of casein and whey proteins has been reviewed by Korhonen *et al.* (1998).

4.4.1 *The effect of heat treatment*

Depending on the intensity of heat treatment, the nutritive value of proteins may be affected either in a positive or negative way. Heat is also used to modify the functional properties of protein ingredients. Heat denatures proteins and may be used, for example, to improve water-binding ability and emulsification. On the other hand, heating usually decreases the solubility of proteins due to their aggregation or coagulation. Heat denaturation temperatures of different dietary proteins vary from 60 to 90°C (Boye *et al.* 1997). Accordingly, the activity of bioactive proteins is reduced by different heat treatments. At a standard pasteurisation temperature of 72°C for 15 s, the bioactive whey proteins retain most of their activity (Fox & Flynn 1992; Finot 1997). Although most thermal denaturation is irreversible in nature, certain proteins may undergo reversible denaturation when the thermal influence is removed. For example, the thermal denaturation of α-lactalbumin is primarily a reversible process with 80–90% renaturation at pH values above 3.3. Below pH 3.3, the ability of the protein to return to the native conformation is reduced. The reversibility of α-lactalbumin is calcium dependent, the regeneration being reduced when a chelator that binds endogenous Ca^{2+} is added (Boye *et al.* 1997). During heat treatment, the lysine residues of proteins can react with reducing carbohydrates of the same food system resulting in the so-called Maillard or non-enzymatic browning reaction (Finot 1997). Depending on the intensity of heat treatment, this reaction affects the sensory properties (aroma, flavour and appearance) of the product and reduces its nutritional value as the bioavailability of lysine is reduced.

4.4.2 *The effect of ultra-high-pressure processing*

Ultra-high-pressure (UHP) processing, defined as exposure to high isostatic pressures of *c.* 100–1000 MPa, can affect protein conformation and lead to protein denaturation, aggregation or gelation depending on the protein system, the applied pressure, the temperature and the duration of the pressure treatment. Low pressures usually induce reversible changes such as dissociation of protein–protein complexes, the binding of ligands and conformational changes (Heremans *et al.* 1997). Pressures higher than *c.* 500 MPa induce irreversible denaturation in most cases. On the other hand, there are considerable differences in protein denaturation and aggregation induced by high pressure compared to heat. β-lactoglobulin appears far more sensitive towards pressure than bovine serum albumin (BSA). The UHP process has been shown to destabilise casein micelles in reconstituted skim milk. The size distribution of the spherical casein micelles changed from *c.* 200 nm to 120 nm after pressurisation. Subsequent heating of skim milk at 30°C and at atmospheric pressure restored the original size distribution. In another study by Nakamura *et al.* (1993) the antigenicity of whey protein hydrolysates treated with high pressure was found to be lower than that of heat-treated hydrolysates.

4.4.3 *The effect of protein hydrolysation*

Proteins can be hydrolysed with acid, alkali or enzymes to yield peptides or, eventually, amino acids. The most commonly used enzymes in the production of whey protein hydrolysates are pepsin, trypsin and chymotrypsin. Plant-originated papain and some bacterial and fungal proteases have also been used in studies reviewed by Lahl & Braun (1994), Panyam & Kilara (1996) and Nielsen (1997). The ability of enzymes to hydrolyse whey proteins is highly variable. Pepsin digests α-lactalbumin and denatured, but not native, β-lactoglobulin (Schmidt & van Markwijk 1993). Trypsin hydrolyses α-lactalbumin slowly but β-lactoglobulin remains almost undegraded (Schmidt & Poll 1991). Chymotrypsin hydrolyses readily α-lactalbumin but β-lactoglobulin is degraded slowly. Lieske & Konrad (1996a, b) demonstrated that BSA and β-lactoglobulin were hydrolysed by papain but α-lactalbumin was resistant. However, α-lactalbumin was hydrolysed completely at acidic pH when calcium binding was absent (Haque 1993). Enzymatic modification of milk proteins by controlled proteolysis can alter their functional properties over a wide pH range and other processing conditions (Lahl & Braun 1994; Schwenek 1997). The hydrolysis of peptide bonds can increase the number of charged groups and hydrophobicity, decrease molecular weight and modify molecular configuration (Haque 1993).

Changes in functional properties are greatly dependent on the degree of hydrolysis. The most common changes in functionality of whey proteins are an increase in solubility and a decrease in viscosity. When the degree of hydrolysis is high, hydrolysates often tolerate strong heating without precipitating and solubility is high even at pH 3.5–4.0. Hydrolysates also have far lower viscosity than intact proteins. The difference is especially striking in solutions with a high protein concentration. Other effects are altered gelation properties, enhanced thermal stability, increased emulsifying and foaming abilities and decreased emulsion and foam stabilities (Nakai & Li-Chan 1989; Lieske & Konrad 1996a, b; Otte *et al.* 1996).

4.4.4 *The role of processing in the production of allergenic or bitter-tasting peptides*

Milk proteins may provoke allergic reactions in sensitised people (Pastorello 1997). Processing may alter the content and/or properties of these allergens, reducing or increasing the allergenicity of the starting material. In earlier studies, conflicting results were obtained with regard to the effect of heat treatment on the allergenicity of whey proteins. Standard pasteurisation of milk does not seem to cause any significant reduction of the antigenicity of milk proteins, while milk sterilisation may even exacerbate allergic reactions (Kilshaw *et al.* 1982). It has been postulated that heat treatment and homogenisation of milk, due to mechanical disintegration of casein micelles and milk fat globules, increase the ability of milk proteins to elicit allergic reactions in sensitised persons (Poulsen *et al.* 1987; Host & Samuelsson 1988). Further studies have not been carried out to substantiate this hypothesis, however. In contrast, many studies have shown that the allergenicity of milk proteins, in particular that of β-lactoglobulin, can be reduced substantially if the proteins are hydrolysed with pepsin or trypsin or using combinations of proteolytic enzymes (Jost *et al.* 1991; Ena *et al.* 1995; Nielsen 1997). Heat treatment or high-pressure treatment prior to hydrolysis increases the DH (degree of hydrolysis) value, further reducing antigenicity of whey proteins (Asselin *et al.* 1989; van Beresteijn *et al.* 1994).

The hydrophilic amino acids, such as lysine, arginine, glutamate and aspartate residues, seem to play important roles in allergenic peptides. The antigenicity of such peptides can probably be lowered further by using proteases specific for these amino acid residues. On the other hand, it has been reported that peptides consisting of only four amino acids can cause an allergic reaction in consumers allergic to milk, even though these small peptides will not be able to sensitise a person (Siemensma *et al.* 1993). Partially or extensively hydrolysed milk proteins have found increasing use in hypoallergenic infant formulas and dietetic products (Wahn *et al.* 1992; Blenford 1994; Cordle 1994; Pastorello 1997).

The problems related to enzymatic hydrolysis of proteins are the production of a high amount of free amino acids, a bitter taste and an increase of the formula osmolarity (Kuwata *et al.* 1997). Current research in this field is focusing on technological reduction in whey of the amount of β-lactoglobulin, which is considered the major allergen because it is absent in human milk. On the other hand, infant formulas enriched with α-lactalbumin are being developed to mimic human milk as it is rich in this protein (Sikorski 1997).

It has long been known that peptides and amino acids can produce many types of taste sensation (Schwenke 1997). A problem in the use of proteolysis for improving functionality and nutritional value has been the formation of bitter peptides which are formed, for example, from β-lactoglobulin (Nielsen 1997). Bitterness is generally related to the hydrophobicity of the amino acids in the peptides. In earlier studies, activated charcoal, various resins, glass fibre and hexyl sepharose were applied for the elimination or reduction of bitter peptides from the hydrolysates. The plastein reaction which can occur when a protein hydrolysate is incubated with a protease is also able to debitter protein hydrolysates (Nielsen 1997). Other methods used for reducing or eliminating bitterness are selective chromatographic separation, masking and enzymatic treatment (Pedersen 1994; Castro *et al.* 1996).

4.5 Bioactive peptides production from milk and whey proteins

Basically, biologically active peptides can be produced from precursor milk proteins (casein and whey proteins) in the following ways: (1) enzymatic hydrolysis by digestive enzymes; (2) fermentation of milk with proteolytic starter cultures; and (3) proteolysis by enzymes derived from micro-organisms or plants. In many studies, a combination of (1) and (2) or (1) and (3) has proven effective in the generation of short functional peptides (Korhonen & Pihlanto 2003b). Examples of bioactive peptides produced by the above treatments are given in the following sections.

4.5.1 *Enzymatic hydrolysis*

The most common method of producing bioactive peptides is through enzymatic hydrolysis of whole protein molecules. Many of the known bioactive peptides have been produced using gastrointestinal enzymes, usually pepsin and trypsin. Angiotensin-converting enzyme (ACE) -inhibitory peptides and calcium-binding phosphopeptides (CPPs), for example, are most commonly produced by trypsin (Meisel & FitzGerald 2003; Yamamoto *et al.* 2003; FitzGerald *et al.* 2004; Gobbetti *et al.* 2004; Vermeirssen *et al.* 2004). Moreover, ACE-inhibitory peptides have recently been identified in the tryptic hydrolysates of bovine αs_2-casein (Tauzin *et al.* 2002) and in bovine, ovine and caprine κ-casein macropeptides (Manso & Lopez-Fandino 2003). Other digestive enzymes and different enzyme combinations of proteinases – including alcalase, chymotrypsin, pancreatin, pepsin and thermolysin as well as enzymes from bacterial and fungal sources – have also been utilised to generate bioactive peptides from various proteins (Kilara & Panyam 2003; Korhonen & Pihlanto 2003a). Apart from conventional production of peptides from natural protein sources by proteolytic enzymes, recombinant DNA techniques have been experimented with for the production of specific peptides or their precursors in micro-organisms. Kim *et al.* (1999) succeeded in expressing recombinant human αs_1-casein in *Escherichia coli* and in purifying it. The trypsin digest of this protein was found to contain several ACE-inhibitory peptides.

4.5.2 *Microbial fermentation*

Many industrially utilised dairy starter cultures are highly proteolytic. Bioactive peptides can therefore be generated by the starter and non-starter bacteria used in the manufacture of fermented dairy products. The proteolytic system of lactic acid bacteria (LAB), e.g. *Lactococcus lactis*, *Lactobacillus helveticus* and *Lb. delbrueckii* ssp. *bulgaricus*, is already well characterised. This system consists of a cell wall-bound proteinase and a number of distinct intracellular peptidases including endopeptidases, aminopeptidases, tripeptidases and dipeptidases (Christensen *et al.* 1999).

Rapid progress has been made in recent years to elucidate the biochemical and genetic characterisation of these enzymes. The fact that the activities of peptidases are affected by growth conditions makes it possible to manipulate the formation of peptides to a certain extent (Williams *et al.* 2002). Table 4.6 provides a list of experimental studies on the release of bioactive peptides upon fermentation of milk using different live proteolytic micro-organisms or proteolytic enzymes derived from such micro-organisms.

Table 4.6 Examples of bioactive peptides released from milk proteins by various micro-organisms and microbial enzymes. Korhonen & Pihlanto (2006). Reproduced with permission of Elsevier.

Micro-organisms used	Precursor protein	Reptide sequence	Bioactivity	References
Lactobacillus helveticus,Saccharomyces cerevisiae	β-casein, κ-casein	Val-Pro-Pro, Ile-Pro-Pro	ACE inhibitory, antihypertensive	Nakamura *et al.* (1995a, b)
Lactobacillus GGenzymes + pepsin & trypsin	β-casein, αs1-casein	Tyr-Pro-Phe-Pro, Ala-Val-Pro-Tyr-Pro-Gln-Arg, Thr-Thr-Met-Pro-Leu-Trp	Opioid, ACE inhibitory, immunostimulatory	Rokka *et al.* (1997)
Lb. helveticus CP90 *proteinase*	β- casein	Lys-Val-Leu-Pro-Val-Pro-(Glu)	ACE inhibitory	Maeno *et al.* (1996)
Lb. helveticus CPN 4	Whey proteins	Tyr-Pro	ACE inhibitory	Yamamoto *et al.* (1999)
Lb. delbrueckii subsp.bulgaricus SS1 *Lactococcuslactis subsp. Cremoris FT4*	β-casein, κ-casein	Many fragments	ACE inhibitory	Gobbetti *et al.* (2000)
Lb. delbrueckii subsp.bulgaricus IFO13953	κ-casein	Ala-Arg-His-Pro-His-Pro-His-Leu-Ser-Phe-Met	Antioxidative	Kudoh *et al.* (2001)
Lb. rhamnosus + digestion withpepsin and Corolase PP	β-casein	Asp-Lys-Ile-His-Pro-Phe, Tyr-Gln-Glu-Pro-Val-Leu,Val-Lys-Glu-Ala-Met-Ala-Pro-Lys	ACE inhibitory, antioxidative	Hernandez-Ledesma *et al.*(2004)
Lb. delbrueckii subsp.Bulgaricus	β-casein	Ser-Lys-Val-Tyr-Pro-Phe-Pro-Gly Pro-Ile	ACE inhibitory	Ashar & Chand (2004)
Streptococcusthermophilus+ Lc. Lactis subsp.Lactis biovar. Diacetylactis	β-casein	Ser-Lys-Val-Tyr-Pro	ACE inhibitory	Ashar & Chand (2004)
Lb. helveticus ICM 1004 cell-free extract	Skim milkhydrolysate	Val-Pro-Pro, Ile-Pro-Pro	ACE inhibitory	Pan *et al.* (2004)

4.5.3 *Proteolysis*

Many recent articles and book chapters have reviewed the release of various bioactive peptides from milk proteins through microbial proteolysis (Korhonen & Pihlanto-Leppälä 2001, 2004; Gobbetti *et al.* 2002, 2004; Matar *et al.* 2003). Most of these studies report the production of ACE-inhibitory or antihypertensive peptides; immunomodulatory, antioxidative and antimicrobial peptides have also been identified. *Lb. helveticus* is widely used as a dairy starter in the manufacture of traditional fermented milk products, such as Emmental cheese. The highly proteolytic *Lb. helveticus* strains capable of releasing ACE-inhibitory peptides have been demonstrated in several studies. The best-known ACE-inhibitory peptides, Val-Pro-Pro (VPP) and Ile-Pro-Pro (IPP), have been identified in milk fermented with *Lb. helveticus* strains (Nakamura *et al.* 1995b; Sipola *et al.* 2002a).

Pihlanto-Leppälä *et al.* (1998) studied the potential formation of ACE-inhibitory peptides from cheese whey and caseins during fermentation with various commercial dairy starters used in the manufacture of yogurt, ropy milk and sour milk. No ACE-inhibitory activity was observed in these hydrolysates. Further digestion of the above samples with pepsin and trypsin resulted in the release of several strong ACE-inhibitory peptides derived primarily from αs1-casein and β-casein.

Gobbetti *et al.* (2000) demonstrated the formation of ACE-inhibitory peptides with two dairy strains, *Lb. delbrueckii* ssp. *bulgaricus* and *Lc. lactis* ssp. *cremoris*, after fermentation of milk separately with each strain for 72 hours. The most inhibitory fractions of the fermented milk mainly contained β-casein-derived peptides with inhibitory concentration (IC50) values ranging from 8.0 to 11.2 mg mL^{-1}.

Yamamoto *et al.* (1999) identified an ACE-inhibitory dipeptide (Tyr-Pro) from a yogurt-like product fermented with *Lb. helveticus* CPN4 strain. This peptide sequence is present in all major casein fractions and its concentration was found to increase during fermentation, reaching a maximum concentration of 8.1 mg mL^{-1} in the product.

Fuglsang *et al.* (2003) tested a total of 26 strains of wild-type LAB, mainly belonging to *Lc. lactis* and *Lb. helveticus*, for their ability to produce a milk fermentate with ACE inhibitory activity. All test strains produced ACE-inhibitory substances in varying amounts, and two of the strains exhibited high ACE inhibition and a high o-phthaldialdehyde (OPA) index, which correlates well with peptide formation. The inhibitory effect of active fermentates on *in vivo* ACE activity was demonstrated in normotensive rats.

A study by Ashar & Chand (2004) identified an ACE-inhibitory peptide from milk fermented with *Lb. delbrueckii* ssp. *bulgaricus*. The peptide showed the sequence Ser-Lys-Val-Tyr-Pro-Phe-Pro-Gly Pro-Ile from β-casein with an IC50 value of 1.7 mg mL^{-1}. In a combination with *Streptococcus thermophilus* and *Lc. lactis* subsp. *lactis* biovar. *diacetylactis*, a peptide structure with a sequence of Ser-Lys-Val-Tyr-Pro was obtained from bcasein with an IC50 value of 1.4 mg mL^{-1}. Both peptides were markedly stable to digestive enzymes, acidic and alkaline pH, as well as during storage at 5 and 10°C for 4 days.

A number of studies have demonstrated that hydrolysis of milk proteins by digestive and/or microbial enzymes may produce peptides with immunomodulatory activities (Gill *et al.* 2000b). Sutas *et al.* (1996) demonstrated that digestion of casein fractions with both pepsin and trypsin produced peptides that provoked immunomodulatory effects on human blood lymphocytes *in vitro*. Peptides derived from total casein and αs_1-casein mainly suppressed the proliferation of lymphocytes, while those derived from β- and κ-casein primarily stimulated the proliferation rate. When the caseins were hydrolysed by enzymes isolated from a probiotic strain of *Lactobacillus* GG var. *casei* prior to pepsin–trypsin treatment, all hydrolysate fractions were immunosuppressive and the highest activity was again found in αs_1-casein. The same hydrolysates also down-regulated *in vitro* the generation of interleukin-4 by lymphocytes (Sutas *et al.* 1996).

Matar *et al.* (2001) fed milk fermented with a *Lb. helveticus* strain to mice for 3 days and detected significantly higher numbers of IgA secreting cells in their intestinal mucosa compared with control mice fed with similar milk incubated with a non-proteolytic variant of the same strain. The immunostimulatory effect of fermented milk was attributed to peptides released from the casein fraction. These results suggest that LAB may modulate the immunological properties of milk proteins prior to or after oral ingestion of the product. Such modulation may be beneficial in the down-regulation of hypersensitivity reactions to ingested proteins in patients with food protein allergies, for example.

In addition to live micro-organisms, proteolytic enzymes isolated from LAB have been successfully employed to release bioactive peptides from milk proteins. Yamamoto *et al.* (1994a) reported that casein hydrolysed by the cell wall-associated proteinase from *Lb. Helveticus* CP790 showed antihypertensive activity in spontaneously hypertensive rats (SHR). Several ACE-inhibitory peptides and one antihypertensive peptide were isolated from the hydrolysate. Maeno *et al.* (1996) hydrolysed casein using the same proteinase and identified a β-casein-derived antihypertensive peptide Lys-Val-Leu-Pro-Val-Pro-Gln, which was shown to be dose-dependent in a rat model at a dosage level from 0.2 to 2 mg of peptide per kg body weight.

In a recent study, Mizuno *et al.* (2004) measured the ACE-inhibitory activity of casein hydrolysates upon treatment with nine different commercially available proteolytic enzymes. Among these enzymes, a protease isolated from *Aspergillus oryzae* showed the highest ACE inhibitory activity *in vitro* per peptide. The *A. oryzae* released peptide had the dose-dependent antihypertensive effect in a rat model with spontaneously hypertensive rats (SHR). Most of the documented ACE-inhibitory peptides are usually short peptides with a proline residue at the carboxyl end terminal.

Proline is also known to be resistant to degradation by digestive enzymes and may pass from the small intestines into the blood circulation in the sequence of short peptides (Yamamoto *et al.* 2003). This hypothesis is supported by a recent study by Pan *et al.* (2004) who hydrolysed skimmed milk with a cell-free extract of *Lb. helveticus* JCM1004 and purified the antihypertensive tripeptides VPP and IPP from the hydrolysate with three runs of HPLC. The IC50 values of the peptides were 9.13 ± 0.21 and 5.15 ± 0.17 mM, respectively. A significant (p 0.01) decrease in systolic blood pressure in SHR was measured after a single gastric intubation of VPP or IPP at 8 or 4 h, respectively. Furthermore, Ueno *et al.* (2004) purified and characterised an endopeptidase from *Lb. helveticus* CM4 and demonstrated that this peptidase can generate the above two antihypertensive peptides using synthetic pro-peptides as a substrate.

4.6 Fractionation and enrichment of bioactive peptides

Commercial production of bioactive peptides from milk proteins has been limited by a lack of suitable large-scale technologies. Until now, membrane separation techniques have provided the best technology available for the enrichment of peptides with a specific molecular weight range (Korhonen & Pihlanto 2003b). Ultrafiltration is routinely employed to enrich bioactive peptides from protein hydrolysates. Enzymatic hydrolysis can be performed through conventional batch hydrolysis or continuous hydrolysis using ultrafiltration membranes. The traditional batch method has several disadvantages, such as the relatively high cost of the enzymes and their inefficiency as compared to a continuous process.

Use of enzymatic membrane reactors for continuous production of specific peptide sequences was introduced during the 1990s. It has already been widely applied for total conversion of food proteins of various origins to hydrolysates with improved nutritional and/or functional properties (Perea & Ugalde 1996; Martin-Orue *et al.* 1999).

Ultrafiltration membrane reactors have been shown to improve the efficiency of enzyme-catalysed bioconversion and to increase product yields, and they can be easily scaled up. Furthermore, ultrafiltration membrane reactors yield a consistently uniform product with desired molecular mass characteristics. The continuous extraction of bioactive peptides in membrane reactors has been mainly applied to milk proteins. Bouhallab & Touze (1995) employed this technique for the recovery of antithrombotic peptides derived from hydrolysed caseinomacropeptide (CMP).

Gauthier & Pouliot (1996) combined enzymatic hydrolysis and ultrafiltration in order to produce emulsifying peptides from β-lactoglobulin. Bordenave *et al.* (1999) demonstrated that α-lactorphin could be successfully generated with continuous hydrolysis of goat whey in an ultrafiltration reactor. On the other hand, they noted severe fouling problems with peptide-membrane interactions, especially in ultrafiltration of casein hydrolysates. Righetti *et al.* (1997) proposed a multicompartment enzyme reactor operating under an electric field for the continuous hydrolysis of milk proteins. This technique allowed for the continuous harvesting of some biologically active peptides, such as phosphopeptides and precursors of casomorphins from the tryptic digest of β-casein.

Stepwise ultrafiltration using cut-off membranes of low molecular mass has been found useful for separating out small peptides from high-molecular-mass residues and remaining enzymes. Turgeon & Gauthier (1990) used a two-step ultrafiltration process and were able to produce a mixture of polypeptides and a fraction rich in small peptides with a molecular mass below 2000 Da. Pihlanto-Leppälä *et al.* (1996) applied selective ultrafiltration membranes (30 and 1 kDa) for the enrichment of opioid peptides (α- and β-lactorphin) from pepsin-hydrolysed α-lactalbumin and from pepsin- and trypsin-hydrolysed β-lg, respectively. The same technique has been successfully used to enrich ACE-inhibitory peptides from purified α-la and β-lg (Pihlanto- Leppälä *et al.* 2000).

Several ion exchange chromatographic methods have been developed for the enrichment of CPPs from casein hydrolysates, but the production costs of this technique were prohibitive for large-scale operation. Ellegard *et al.* (1999) developed a process-scale method for the isolation of high-purity CPPs using acid precipitation, diafiltration and anion exchange chromatography. Recently, ion exchange membrane chromatography has emerged as a promising technique for the enrichment of peptide fractions from protein hydrolysates. Recio & Visser (1999b) described a method where the protein of interest was concentrated within a chromatographic medium and hydrolysed *in situ* by an appropriate enzyme. The resulting active peptides were retained on the ion exchanger, while the other peptides were washed out. The fraction containing the active peptides was eluted from the chromatographic medium. With this method it was possible to isolate and enrich cationic antibacterial peptides from lactoferrin and αs_2-casein, as well as negatively charged phosphopeptides from β-casein. The advantages of this process are that isolation of the precursor protein is unnecessary and the enzyme used in this process can be recovered. This technique provides new possibilities for enriching peptides with a low molecular mass, and can be easily up-scaled to yield gram or even kilogram quantities (Recio *et al.* 2000).

4.7 Commercial applications and future outlook

Bioactive peptides have attracted increasing interest as prominent candidates for various health-promoting functional foods. Such preparations are commercially available and are being introduced into different consumer products, such as dairy and fruit-based drinks, confectionery, chewing gum, pastilles and capsules. Currently marketed products contain peptides with anticariogenic, antihypertensive, mineral-binding, stress-relieving and satiety-inducing properties. Examples of these products are listed in Table 4.7. The potential health benefits of bioactive milk peptides is a subject of growing commercial interest in the context of health-promoting functional foods. So far, antihypertensive, mineral-binding and anticariogenic peptides have been most studied for their physiological effects. A few commercial products have been launched onto the market, and this trend is likely to continue alongside with increasing knowledge about the functionalities of the peptides.

It is anticipated that more peptide-based products will be developed to manage various risk factors of the metabolic syndrome, e.g. hypertension, serum lipid levels, glucose balance and body mass index. In view of the current global trend of increasing obesity and related diseases (Type 2 diabetes in particular), more experimental research should be focused on natural dietary bioactive peptides which can induce satiety and reduce insulin resistance.

Other new areas where more research into bioactive peptides is warranted include impairment of cognitive functions, memory-related diseases and mood control. In this context, antioxidative and opioid properties of many milk peptides may be worth further investigation. The potential to reduce oxidative stress in the body via oral administration of antioxidative peptides may also be of considerable interest in view of the inflammatory events caused by reactive oxygen species (ROS) in the living cells. It is now well-established that persistent low-grade inflammation in the body may lead to the development of various chronic diseases, such as atherosclerosis and Type 2 diabetes (Packard & Libby 2008).

Novel technologies such as chromatographic and membrane separation techniques seem to be appropriate means to enrich active peptide fractions from the hydrolysates of various food proteins. Production of bioactive peptides from protein-rich raw materials may be scaled up to industrial level using controlled fermentation in bioreactors with known LAB. To this end, recombined enzyme technology and specific production strains or peptidases isolated

Table 4.7 Commercial dairy products and ingredients with health or function claims based on bioactive peptides. Silva & Maltaca (2005). Reproduced with permission of Elsevier.

Brand name	Type of product	Claimed functional bioactive peptides	Health/function claims	Manufacturer
Calpis	Sour milk	Val-Pro-Pro, Ile-Pro-Pro, derived from β-casein and κ-casein	Reduction of blood pressure	Calpis Co., Japan
Evolus	Calcium-enriched fermented milk drink	Val-Pro-Pro, Ile-Pro-Pro, derived from β-casein and κ-casein	Reduction of blood pressure	Valio Oy, Finland
BioZate	Hydrolysed whey protein isolate	β-lactoglobulin fragments	Reduction of blood pressure	Davisco, USA
BioPURE-GMP	Whey protein isolate	κ-casein f(106–169) (Gly-comacropeptide)	Prevention of dental caries, influence the clotting of blood, protection against viruses and bacteria	Davisco, USA
PRODIET F200/Lactium	Flavoured milk drink, confectionery, capsules	αs1-casein f(91–100) (Tyr-Leu-Gly Tyr-Leu-Glu-Gln-Leu-Leu-Arg)	Reduction of stress effects	Ingredia, France
Festivo	Fermented low-fat hard cheese	αs1-casein f(1–9), αs1-caseinf(1–7), αs1-casein f(1–6)	No health claim as yet	MTT Agrifood Research Finland
Cysteine Peptide	Ingredient/ hydrolysate	Milk-protein-derived peptide	Aids to raise energy level and improve sleep	DMV International, The Netherlands
C12	Ingredient/ hydrolysate	Casein-derived peptide	Reduction of blood pressure	DMV International, The Netherlands
Capolac	Ingredient	Caseinophospho-peptide	Helps mineral absorption	Arla Foods Ingredients, Sweden
PeptoPro	Ingredient/ hydrolysate	Casein-derived peptide	Improves athletic performance and muscle recovery	DSM Food Specialties, The Netherlands
Vivinal Alpha	Ingredient/ hydrolysate	Whey-derived peptide	Aids relaxation and sleep	Borculo Domo Ingredients(BDI), The Netherlands

from suitable micro-organisms are likely to be employed industrially in the future. In addition to enzymatic hydrolysis, microbial fermentation provides a natural technology applicable for the production of bioactive peptides either from animal or plant proteins. The potential of this approach is already well demonstrated by the presence of bioactive peptides in fermented dairy products. The formation and stability of such peptides is, however, not controlled in traditional dairy products; certain health benefits therefore cannot be guaranteed. New manufacturing technologies, such as supplementation of fermented milk with specific bioactive peptides concentrated from hydrolysates of milk proteins, provide a practical solution for the time being.

Important research topics will continue to be the bioavailability and safety of bioactive peptides. To this end, the stability of chemical structures and biological activity of peptides in different food matrices and subsequent fate during the digestion process will be taken into account in the formulation of products containing bioactive peptides. Again, new emerging technologies such as micro or nanoencapsulation may offer feasible solutions for improving the stability of peptides in various food products and during digestion (Korhonen & Pihlanto 2007). Furthermore, molecular studies are needed to assess the mechanisms by which bioactive peptides exert their activities. For this approach, it is necessary to employ new nutrigenomic techniques, for example, proteomics and metabolomics (O'Donnell *et al.* 2004). By developing such novel facilities it will be possible to study the impact of proteins and peptides on the expression of genes and hence optimise the nutritional and health effects of these compounds. This research area is currently considered highly challenging and will revolutionise protein research in the near future. Also, biosafety aspects of novel peptides generated by new technologies have to be considered before entering into proper product formulation and marketing efforts.

Further challenges with functional food products containing novel bioactive peptides are associated with regulatory issues if any health-related claims will be attached to the products. Currently, health claim regulations seem to vary greatly from one country to another, but regional and continental harmonisation processes are in progress.

4.8 Conclusion

Casein and whey proteins are valuable and useful materials for human health. Their bioactivity is expressed mainly through fragments of their molecular structure, which are called peptides. Various techniques have been developed in order to liberate the bioactive peptides, involving fermentation by certain types of micro-organisms, enzymatic hydrolysis or even processing (e.g. ultra-high-pressure or heat treatment). The biofunctionalities of milk and whey proteins and peptides include: antibacterial, antiviral and antimicrobial action in general; antiflammatory action; antithrombotic action; protection against metabolic syndrome, e.g., hypertension, serum lipid levels, glucose balance and body mass index; and protection against cancer and obesity. Some concerns that the diassociation of milk and whey proteins may produce allergenic substances seem to be without serious grounds, as the allergenicity of main protein involved in allergic reaction β-lactoglobulin is reduced as it is hydrolysed. In conclusion, the utilisation of casein and whey proteins to produce natural pharmaceutical products of high value for human health seems is a highly promising topic for research; many aspects of this subject require to be elucidated.

References

Abubakar, A., Saito, T., Kitazawa, H., Kawai, Y. & Itoh, T. (1998) Structural analysis of new antihypertensive peptides from cheese whey protein by proteinase K digestion. *Journal of Dairy Science*, **81**(12), 3131–3138.

Addeo, F., Chianese, L., Salzano, A., Sacchi, R., Cappuccio, U., Ferranti, P. & Malorni, A. (1992) Characterization of the 12% trichloroacetic acid-insoluble oligopeptides of Parmigiano–Reggiano cheese. *Journal of Dairy Research*, **59**(3), 401–411.

Ait-Oukhatar, N., Bouhallab, S., Arhan, P., Maubois, J.L., Drosdowsky, M. & Bougle, D. (1999) Iron tissue storage and hemoglobin levels of deficient rats repleted with iron bound to the caseinophosphopeptide 1–25 of β-casein. *Journal of Agricultural and Food Chemistry*, **47**(7), 2786–2790.

Ait-Oukhatar, N., Peres, J.M., Bouhallab, S., Neuville, D., Bureau, F., Bouvard, G., Arhan, P. & Bougle, D. (2002) Bioavailability of caseinophosphopeptide-bound iron. *Journal of Laboratory and Clinical Medicine*, **140**(4), 290–292.

Ajello, M., Greco, R., Giansanti, F. et al. (2002) Antiinvasive activity of bovine lactoferrin towards group A Streptococci. *Biochemistry and Cell Biology*, **80**, 119–124.

Appelmelk, B.J., An, Y.-Q., Geerst, M., Thijs, B.G., de Boer, H.A., McClaren, D.M. et al. (1994) Lactoferrin is a lipid A-binding protein. *Infection and Immunity*, **62**, 2628–2632.

Ashar, M.N. & Chand, R. (2004) Fermented milk containing ACEinhibitory peptides reduces blood pressure in middle aged hypertensive subjects. *Milchwissenschaft*, **59**, 363–366.

Asselin, J., Hebert, J. & Amiot, J. (1989) Effects of in vitro proteolysis on the allerginicity of major whey proteins. *Journal of Food Science*, **54**, 1037–1039.

Baltzer, A., Svanborg, C. & Jaggi R (2004) Apoptotic cell death in the lactating mammary gland is enhanced by a folding variant of alpha-lactalbumin. *Cell Molecular Life Science*, **61**(10), 1221–1228.

Batish, V.K, Chander, H., Zumdegeni, K.C. *et al.* (1988) Antibacterial activity of lactoferrin against some common food-borne pathogenic organisms. *Austalian Journal of Dairy Techology*, **5**, 16–18.

Beaulieu, J., Dupont, C. & Lemieux, P (2006) Whey proteins and peptides: beneficial effects on immune health. *Therapy*, **3**, 69–78.

Berkhout, B., Derksen, G.C., Back, N.K., Klaver, B., de Kruif, C.G. & Visser, S. (1997) Structural and functional analysis of negatively charged milk proteins with anti-HIV activity. *AIDS Research and Human Retroviruses*, **13**(13), 1101–1107.

Berrocal, R., Chanton, S., Juillerat, M.A., Pavillard, B., Scherz, J.-C. & Jost, R. (1989) Tryptic phosphopeptides from whole casein. *II. Physicochemical properties related to the solubilization of calcium. Journal of Dairy Research*, **56**(3), 335–341.

Blenford, D.E. (1994) Protein hydrolysates: Functionalities and uses in nutritional products. *International Food Ingredients*, **3**, 45–49.

Bordenave, S., Sannier, F., Ricart, G. & Piot, J.M. (1999) Continuous hydrolysis of goat whey in an ultrafiltration reactor: Generation of alpha-lactorphin. *Preparatory Biochemistry and Biotechnology*, **29**, 189–202.

Bouhallab, S. & Touze, C. (1995) Continuous hydrolysis of caseinomacropeptide in a membrane reactor: Kinetic study and gram-scale production of antithrombotic peptides. *Lait*, **75**, 251–258.

Bounous, G. & Gold, P. (1991) Biologically active whey protein composition, a method for producing it and use of the composition, Patent No. EP0375852, Canada.

Boxer, L.A, Coates, T.D., Haak, R.A. *et al.* (1982) Lactoferrin deficiency associated with altered granulocyte function. *New England Journal of Medicine*, **307**, 404–410.

Boye, J.I., Ma, C.-Y. & Harwalkar, V.R. (1997) Thermal denaturation and coagulation of proteins. In *Food Proteins and their Applications* (eds Damodaran, S. & Paraf, A.) pp. 25–56. Marcel Dekker, Inc., New York.

Brantl, V., Teschemacher, A., Henschen, A. & Lottspeich, F. (1979) Novel opioid peptides derived from casein (β-casomorphins). *Hoppe Seyler's Zeitschrift fur Physiologische Chemie*, **360**(9), 1211–1216.

Breton-Gorius, J., Mason, D.Y., Buriot, D. *et al.* (1980) Lactoferrin deficiency as a consequence of a lack of specific granules in neutrophils from a patient with recurrent infections. *Detection by immunoperoxidase staining for lactoferrin and cytochemical electron microscopy. American Journal of Pathology*, **99**, 413–428.

Brisson, G., Britten, M. & Pouliot, Y. (2007) Electrically-enhanced crossflow microfiltration for seperation of lactoferrin from whey protein mixtures. *Journal of Membrane Science*, **297**, 206–216.

Brown, E.M. (1984) Interaction of β-lactoglobulin and α-lactalbumin with lipids: a review. *Journal of Dairy Science*, **67**, 713–722.

Bruck, W.M., Graverholt, G. & Gibson, G.R. (2003a) A two-stage continuous culture system to study the effect of suplemental a lactalbumin and glycomacropeptide on mixed cultures of human gut bacteria challenged with enteropathogenic Escherichia coli and Salmonella serotype typhimurium. *Journal of Applied Microbiology*, **95**(1), 44–53.

Bruck, W.M., Kelleher, S.L., Gibson, G.R., Nielsen, K.E., Chatterton, D.E. & Lonnerdal, B. (2003b) rRNA probes used toquantify the effects of glycomacropeptide and α-lactalbumin supplementation on the predominant groups of intestinal bacteria of infant rhesus monkeys challenged with enteropathogenic Escherichia coli. *Journal of Pediatric Gastroenterology and Nutrition*, **37**(3), 273–280.

Burrington, K.J. (2002) More than just milk. *Food Production Design*, **12**, 37–64.

Castro, S., Peyronel, D.V. & Cantera, A.M.B. (1996) Proteolysis of whey proteins by a bacillus subtilis enzyme preparation. *International Dairy Journal*, **6**, 285–294.

Chabance, B., Jollès, P., Izquierdo, C., Mazoyer, E., Francoual, C., Drouet, L. & Fiat, A.M. (1995) Characterization of an antithrombotic peptide from κ-casein in new born plasma after milk ingestion. *British Journal of Nutrition*, **73**(4), 583–590.

Chabance, B., Marteau, P., Rambaud, J.C., Migliore-Samour, D., Boynard, M., Perrotin, P., Guillet, R., Jollès, P. & Fiat, A.M. (1998) Casein peptide release and passage to the blood in humans during digestion of milk and yogurt. *Biochimie*, **80**(2), 155–165.

Chiba, H., Tani, F. & Yoshikawa, M. (1989) Opioid antagonist peptides derived from κ-casein. *Journal of Dairy Research*, **56**(3), 363–366.

Chierici, R., Sawatzki, G., Tamisari, L., Volpato, S. & Vigi, V. (1992) Supplementation of an adapted formula with bovine lactoferrin. Effects on serum iron, ferritin and zinc levels. *Acta Paediatrica*, **81**, 475–479.

Cho, Y., Batt, C.A. & Sawyer, L. (1994) Probing the retinol-binding site of bovine β-lactoglobulin. *Journal of Bioogical Chemistry*, **269**, 1102–1107.

Christensen, J.E., Dudley, E.G., Pederson, J.A. & Steele, J.L. (1999) Peptidases and amino acid catabolism in lactic acid bacteria. *Antonie van Leeuwenhoek*, **76**, 217–246.

Clare, D.A. & Swaisgood, H.E. (2000) Bioactive milk peptides: A prospectus. *Journal of Dairy Science*, **83**(6), 1187–1195.

Clare, D.A., Catignani, G.L. & Swaisgood, H.E. (2003) Biodefenseproperties of milk: The role of antimicrobial proteins and peptides. *Current Pharmaceutical Design*, **9**(16), 1239–1255.

Cordle, C.T. (1994) Control of food allergies using protein hydrolysates. *Food Technology*, **10**, 72–76.

Coste, M., Rochet, V., Leonil, J., Molle, D., Bouhallab, S. & Tome, D. (1992) Identification of C-terminal peptides from bovine β-casein that enhance proliferation of rat lymphocytes. *Immunology Letters*, **33**(1), 41–46.

Di Mario, F., Aragona, G., Bo, N.D. *et al.* (2003a) Use of lactoferrin for *Helicobacter pylori* erdication. *Preliminary results. Journal of Clinical Gastroenterology*, **36**, 396–398.

Di Mario, F., Aragona, G., Dal Bo, N. *et al.* (2003b) Use of bovine lactoferrin for *Helicobacter pylori* eradication. *Digest of Liver Disease*, **35**, 706–710.

Düringer, C., Hamiche, A., Gustafsson, L., Kimura, H. & Svanborg, C. (2003) HAMLET interacts with histones and chromatin in tumor cell nuclei. *The Journal of Biological Chemistry*, **278**(43), 42131–42135.

Early, E.M., Hardy, H., Forde, T. & Kane, M. (2001) Bactericidal effect of a whey protein concentrate with anti-*Helicobacter pylori* activity. *Journal of Applied Microbiology*, **90**, 741–748.

Ellegard, K.H., Gammelgard-Larsen, C., Sorensen, E.S. & Fedosov, S. (1999) Process scale chromatographic isolation, characterization and identification of tryptic bioactive casein phosphopeptides. *International Dairy Journal*, **9**, 639–652.

El-Salam, A., El-Shibiny, S. & Buchheim, W. (1996) Characteristics and potential uses of the casein macropeptide. *International Dairy Journal*, **6**(4), 327–341.

El-Zahar, K., Sitohy, M., Choiset, Y., Metro, F., Haertlé, T. & Chobert, J.M. (2004) Antimicrobial activity of ovine whey protein and their peptic hydrolysates. *Milchwissenschaft*, **59**(11–12), 653–656.

Ena, J.M., van Beresteijn, E.C.H., Robben, A.J.P.M. & Schmidt, D.G. (1995) Whey protein antigenicity reduction by fungal proteinases and a pepsin/pancreatin combination. *Journal of Food Science*, **60**, 104–110.

Ermisch, A., Ruhle, H.J., Neubert, K., Hartrodt, B. & Landgraf, R. (1983) On the blood–brain barrier to peptides: 3Hb-casomorphin-5 uptake by eighteen brain regions in vivo. *Journal of Neurochemistry*, **41**(5), 1229–1233.

Fanaro, S. & Vigi, V. (2002) Protein quality and quantity in infant formulas. *A critical look. Minerva Pediatrica*, **54**(3), 203–209.

Farrell, H.M., Bede, M.J. & Enyeart, J.A. (1987) Binding of pnitrophenyl phosphate and other aromatic compounds by β-LG. *Journal of Dairy Science*, **70**, 252–258.

Fast, J., Mossberg, A.K, Svanborg, C. & Linse, S. (2005a) Stability of HAMLET-a kinetically trapped alpha-lactalbumin oleic acid complex. *Protein Science*, **14**(2), 329–340.

Fast, J., Mossberg, A.K., Nilsson, H., Svanborg, C., Akke, M. & Linse, S. (2005b) Compact oleic acid in HAMLET. *FEBS Letters*, **579**(27), 6095–6100.

Ferraretto, A., Signorile, A., Gravaghi, C. & Fiorilli, A. (2001) Casein phosphopeptides influence calcium uptake by cultured human intestinal HT-29 tumor cells. *Journal of Nutrition*, **131**(6), 1655–1661.

Ferraretto, A., Gravaghi, C., Fiorilli, A. & Tettamanti, G. (2003) Casein-derived bioactive phosphopeptides: Role of phosphorylation and primary structure in promoting calcium uptake by Ht-29 tumor cells. *FEBS Letters*, **551**(1–3), 92–98.

Fiat, A.M. & Jollès, P. (1989) Caseins of various origins and biologically active casein peptides and oligosaccharides: Structural and physiological aspects. *Molecular and Cellular Biochemistry*, **87**(1), 5–30.

Fiat, A.-M., Migliore, D. & Jollès, P. (1993) Biologically active peptides from milk proteins with emphasis on two examples concerning antithrombotic and immunostimulating activities. *Journal of Dairy Science*, **76**(1), 301–310.

Finot, P.A. (1997) Effects of processing and storage on the nutritional value of food proteins. In *Food Proteins and their Applications* (eds Damodaran, S. & Paraf, A.) pp. 551–577. Marcel Dekker, Inc., New York.

Fischer, W., Gustafsson, L., Mossberg, A.K. *et al.* (2004) Human alpha-lactalbumin made lethal to tumor cells (HAMLET) kills human glioblastoma cells in brain xenografts by an apoptosis-like mechanism and prolongs survival. *Cancer Research*, **64**, 2105–2112.

FitzGerald, R.J. (1998) Potential uses of caseinophosphopeptides. *International Dairy Journal*, **8**(5/6), 451–457.

FitzGerald, R.J., Murray, B.A. & Walsh, D.J. (2004) Hypotensive peptides from milk proteins. *Journal of Nutrition*, **134**, 980S–988S.

Florisa, R., Recio, I., Berkhout, B. & Visser, S. (2003) Antibacterial and antivirial effects of milk proteins and derivatives thereof. *Current Pharmaceutical Design*, **9**(16), 1257–1275.

Fox, P.F. & Flynn, A. (1992) Biological properties of milk proteins. In *Advanced Dairy Chemistry*, Vol. 1: Proteins (ed. Fox, P.F.) pp. 255–284. Elsevier, London.

Fuglsang, A., Rattray, F.P., Nilsson, D. & Nyborg, N.C.B. (2003) Lactic acid bacteria: Inhibition of angiotensin converting enzyme *in vitro* and *in vivo*. *Antonie van Leeuwenhoek*, **83**, 27–34.

Ganjam, L.S., Thornton, W.H., Marshall, R.T. & MacDonald, R.S. (1997) Antiproliferative effects of yogurt fractions obtained by membrane dialysis on cultured mammalian intestinal cells. *Journal of Dairy Science*, **80**(10), 2325–2329.

Gauthier, S.F. & Pouliot, M. (1996) Use of ultrafiltration for the preparation of enzymatic hydrolysates from milk proteins. *IDF Bulletin*, **311**, 31–32.

Gentilucci, L. & Tolomelli, A. (2004) Recent advances in the investigation of the bioactive conformation of peptides active at the m-opioid receptor. *Conformational analysis of endomorphins. Current Topics in Medicinal Chemistry*, **4**(1), 105–121.

Gill, H.S., Rutherford, K.J. & Cross, M.L. (2000a) Bovine milk: a unique source of immunomodulatory ingredients for functional foods. In *Functional Foods II –Claims and Evidence* (eds Buttriss, J. &Saltmarsh, M.), pp. 82–90. Royal Society of Chemistry Press, Cambridge, England.

Gill, H.S., Doull, F., Rutherfurd, K.J. & Cross, M.L. (2000b) Immunoregulatory peptides in bovine milk. *British Journal of Nutrition*, **84**(Suppl. 1), S111–S117.

Gobbetti, M., Ferranti, P., Smacchi, E., Goffredi, F. & Addeo, F. (2000) Production of angiotensin-I-converting-enzyme-inhibitory peptides in fermented milks started by Lactobacillus delbrueckii subsp. bulgaricusSS1 and Lactococcus lactis subsp. cremoris FT4. *Applied and Environmental Microbiology*, **66**, 3898–3904.

Gobbetti, M., Stepaniak, L., De Angelis, M., Corsetti, A. & Di Cagno, R. (2002) Latent bioactive peptides in milk proteins: Proteolytic activation and significance in dairy processing. *Critical Reviews in Food Science and Nutrition*, **42**, 223–239.

Gobbetti, M., Minervini, F. & Rizzello, C.G. (2004) Angiotensin I converting- enzyme-inhibitory and antimicrobial bioactive peptides. *International Journal of Dairy Technology*, **57**, 172–188.

Gómez-Ruiz, J.A., Ramos, M. & Recio, I. (2002) Angiotensinconverting enzyme-inhibitory peptides in Manchego cheeses manufactured with different starter cultures. *International Dairy Journal*, **12**(8), 697–706.

Gustafsson, L., Leijonhufvud, I., Aronsson, A., Mossberg, A.K. & Svanborg C (2004) Treatment of skin papillomas with topical alpha-lactalbumin-oleic acid. *New England Journal of Medicine*, **350**(26), 2663–2672.

Haileselassie, S.S., Lee, B.H. & Gibbs, B.F. (1999) Purification and identification of potentially bioactive peptides from enzymemodified cheese. *Journal of Dairy Science*, **82**(8), 1612–1617.

Hamel, U., Kielwein, G. & Teschmacher, H. (1985) β-Casomorphin immunoreactive materials in cow's milk incubated with various bacterial species. *Journal of Dairy Research*, **52**(1), 139–148.

Hansen, M., Sandstrom, B., Jensen, M. & Sorensen, S.S. (1997) Casein phosphopeptides improve zinc and calcium absorption from rice-based but not from whole grain infant cereal. *Journal of Pediatric Gastroenterology and Nutrition*, **24**(1), 56–62.

Haque, Z.U. (1993) Influence of milk peptides in determining the functionality of milk proteins: a review. *Journal of Dairy Science*, **76**, 311–320.

Hata, H., Yamamoto, M., Ohni, M., Nakajima, K. & Nakamura, Y. (1996) A placebo-controlled study of the effect of sour milk on blood pressure in hypertensive subjects. *American Journal of Clinical Nutrition*, **64**(5), 767–771.

Hata, I., Ueda, J. & Otani, H. (1999) Immunostimulatory action of a commercially available casein phosphopeptide preparation, CPPIII, in cell cultures. *Milchwissenschaft*, **54**(1), 3–7.

Heremans, K., Van Camp, J. & Huyghebaert, A. (1997) High-pressure effects on proteins. In *Food Proteins and their Applications* (eds Damodaran, S. & Paraf, A.), pp. 473–502. Marcel Dekker, Inc., New York.

Hernandez-Ledesma, B., Recio, I., Ramos, M. & Amigo, L. (2002) Preparation of ovine and caprine β-lactoglobulin hydrolysates withACE-inhibitory activity. Identification of active peptides from caprine β-lactoglobulin hydrolysed with thermolysin. *International Dairy Journal*, **12**(10), 805–812.

Hernandez-Ledesma, B., Amigo, L., Ramos, M. & Regio, I. (2004) Angiotensin converting nzymeinhibitory activity in commercial fermented prosucts. *Formation of peptides under simulated gastrointestinal digestion. Journal of Agricultural and Food Chemistry*, **52**, 1504–1510.

Hernandez-Ledesma, B., Amigo, L., Recio I. & Bartolome, B. (2007) ACE-inhibitory and radical-scavenging activity of peptides derived from β-lactoglobulin *f* (19-25). Interactions with ascorbic acid. *Journal of Agricultural and Food Chemistry*, **55**, 3392–3397.
Hirayama, M., Toyota, K., Yamaguchi, G., Hidaka, H. & Naito, H. (1992) HPLC analysis of commercial casein phosphopeptides (CPP). *Bioscience Biotechnology and Biochemistry*, **56**(7), 1126–1127.
Holt, C. (1997) The milk salts and their interaction with casein. In *Advanced Dairy Chemistry* (ed. Fox, P.F.), pp. 233–256. Chapman & Hall, London.
Horikawa, S., Takai, T., Toyosato, M., Takahashi, H., Noda, M., Kakidani, H. et al. (1983) Isolation and structural organization of the human preproenkephalin B gene. *Nature*, **306**(5943), 611–614.
Host, A. & Samuelsson, E.G. (1988) Allergic reactions to raw, pasteurized and homogenized, pateurized cow milk: a comparison. *Allergy*, **43**, 113–118.
Iwasa, M., Kaito, M., Ikoma, J., Takeo, M., Imoto, I., Adachi, Y. et al. (2002) Lactoferrin inhibits hepatitis C virus viremia in chronic hepatitis C patients with high viral loads and HGV genotype 1b. *American Journal of Gastroenterology*, **97**, 766–767.
Janecka, A., Fichna, J. & Janecki, T. (2004) Opioid receptors and their ligands. *Current Topics in Medicinal Chemistry*, **4**(1), 1–17.
Jazirim, M., Migliore, S.D., Casabianca, M.R.P., Keddad, K., Morgat, J.L. & Jolles, P. (1992) Specific binding sites on human phagocytic blood cells for Gly-Leu-Phe and Val-Glu-Pro-Ile-Pro-Tyr, immunostimulating peptides from human milk proteins. *Biochimica et Biophysica Acta*, **1160**, 251–261.
Jensen, R.G. (1995) *Handbook of Milk Composition*. Academic Press, San Diego, CA.
Jollès, P. (1975) Structural aspects of the milk clotting process. *Comparative features with the blood clotting process. Molecular and Cellular Biochemistry*, **7**(2), 73–85.
Jollès, P. & Henschen, A. (1982) Comparison between the clotting of blood and milk. *Trends in Biochemical Science*, **7**, 325–328.
Jollès, P., Loucheux-Lefebvre, M.H. & Henschen, A. (1978) Structural relatedness of κ-casein and fibrinogen g-chain. *Journal of Molecular Evolution*, **11**(4), 271–277.
Jollès, P., Parker, F., Floch, F., Migliore, D., Alliel, P., Zerial, A. & Werner, G.H. (1981) Immunostimulating substances from human casein. *Journal of Immunopharmacology*, **3**(3–4), 363–369.
Jollès, P., Levy-Toledano, S., Fiat, A.M., Soria, C., Gillesen, D., Thomaidis, A., Dunn, F.W. & Caen, J. (1986) Analogy between fibrinogen and casein: Effect of an undecapeptide isolated from k casein on platelet function. *European Journal of Biochemistry*, **158**(2), 379–382.
Jones, E.M., Smart, A., Bloomberg, G. *et al.* (1994) Lactoferricin, a new antimicrobial peptide. *Journal of Applied Bacteriology*, **77**, 208–214.
Jost, R., Fritsche, R. & Pahud, J.J. (1991) Reduction of milk protein allerginicity through processing in food allergy and food intolerance. In *Food Allergy and Food Intolerance* (eds Somogyi, J.C., Muller, H.R. & Ockhuizen,Th.), pp. 127–137. Karger, Basel.
Karaki, H., Dot, K., Sugano, S., Uchiwa, H., Sugai, R., Murakami, U. & Takemoto, S. (1990) Antihypertensive effect of tryptic hydrolysate of milk casein in spontaneously hypertensive rats. *Comparative Biochemistry and Physiology*, **96**(2), 367–371.
Kasai, T., Iwasaki, R., Tanaka, M. & Kiriyama, S. (1995) Caseinophosphopeptides (CPP) in faeces and contents in digestive tract of rats fed casein and CPP preparations. *Bioscience Biotechnology and Biochemistry*, **59**(1), 26–30.
Kawaguchi, S., Hayashi, T., Masano, J., Okuyama, K., Suzuki, T. & Kawase, K. (1989) A study concerning the effect of lactoferrin enriched infant formula on low birth weight infants, with special reference to changes in intestinal flora. *Perinatal Medicine*, **19**, 557–562 (in Japanese).
Kayser, H. & Meisel, H. (1996) Stimulation of human peripheral blood lymphocytes by bioactive peptides derived from bovine milk proteins. *FEBS Letters*, **383**(1–2), 18–20.
Kee, H.J., Kim, E.R., Jung, H.K., Yun, S.S., Juhn, S.L. & Hong, Y.H. (1998) Effect of enzymatically hydrolyzed α-LA fractions with pepsin on growth-promoting of Bifidobacterium longum ATCC 15707. *Korean Journal of Dairy Science*, **20**(1), 61–68.

Kelleher, S.L., Chatterton, D., Nielsen, K. & Lonnerdal B. (2003) Glycomacropeptide and α-lactalbumin supplementation of infant formula affects growth and nutritional status in infant rhesus monkeys. *American Journal of Clinical Nutrition*, **77**, 1261–1268.

Kilara, A. & Panyam, D. (2003) Peptides from milk proteins and their properties. *Critical Reviews in Food Science and Nutrition*, **43**(6), 607–633.

Kilshaw, P.J., Heppel, L.M. & Ford, J.E. (1982) Effects of heat treatment of cow's milk and whey on the nutritional quality and antigenic properties. *Archives of Dissabled Child*, **57**, 842–847.

Kim, Y.E., Yoon, S., Yu, D.Y., Lonnerdal, B. & Chung, B.H. (1999) Novel angiotensin-I-converting enzyme inhibitory peptides derived from recombinant human αs1-casein expressed in Escherichia coli. *Journal of Dairy Research*, **66**, 431–439.

Kitts, D. (1994) Bioactive peptides in food: identification and potential uses. *Canadian Journal of Physiology and Pharmacology*, **72**(4), 423–434.

Kitts, D.D. & Weiler, K. (2003) Bioactive proteins and peptides from food sources. *Applications of bioprocesses used in isolation and recovery. Current Pharmaceutical Design*, **9**(16), 1309–1323.

Kohmura, M., Nio, N., Kubo, K., Minoshima, Y., Munekata, E. & Ariyoshi, Y. (1989) Inhibition of angiotensin-converting enzyme by synthetic peptide fragments of human β-casein. *Agricultural and Biological Chemistry*, **53**(8), 2107–2114.

Kohmura, M., Nio, N. & Ariyoshi, Y. (1990) Inhibition of angiotensin-converting enzyme by synthetic peptide fragments of human κ-casein. *Agricultural and Biological Chemistry*, **54**(3), 835–836.

Korhonen, H. & Pihlanto-Leppälä, A. (2001) Milk protein-derived bioactive peptides: novel opportunities for health promotion. *IDF Bulletin*, **363**, 17–26.

Korhonen, H. & Pihlanto, A. (2003a) Food-derived bioactive peptides: opportunites for designing future foods. *Current Pharmaceutical Design*, **9**(16), 1297–1308.

Korhonen, H. & Pihlanto, A. (2003b) Bioactive peptides: Novel applications for milk proteins. *Applied Biotechnology: Food Science and Policy*, **1**, 133–144.

Korhonen, H. & Pihlanto-Leppälä, A. (2004) Milk-derived bioactive peptides: Formation and prospects for health promotion. In *Handbook of Functional Dairy Products* (eds Shortt, C. & O'Brien, J.), pp. 109–124. CRC Press, Functional Foods and Nutraceuticals Series 6.0, Florida, US.

Korhonen, H. & Pihlanto, A. (2006) Bioactive peptides: production and functionality. *International Dairy Journal*, **16**, 945–960.

Korhonen, H. & Pihlanto, A. (2007) Technological options for the production of health-promoting proteins and peptides derived from milk and colostrums. *Current Pharmaceutical Design*, **13**, 829–843.

Korhonen, H., Pihlanto-Leppälä, A., Rantamaki, R. & Tupasela, T. (1998) Impact of processing onbioactive proteins and peptides. *Trends in Food Science & Technology*, **9**, 307–319.

Kudoh, Y., Matsuda, S., Igoshi, K. & Oki T. (2001) Antioxidative peptide from milk fermented with *Lactobacillus delbrueckii* subsp. Bulgaricus IFO13953. *Journal of the Japanese Society of Food Science*, **48**, 44–50.

Kuhara, T., Iigo, M., Itoh, T., Ushida, Y., Sekine, K., Terada, N. *et al.* (2000) Orally administered lactoferrin exerts an antimetastatic effect and enhances production of IL-18 in the intestinal epithelium. *Nutrition and Cancer*, **38**, 192–199.

Kushibiki, S., Hodate, K., Kurisaki, J., Shingu, H., Ueda, Y., Watanabe, A. *et al.* (2001) Effect of β-lactoglobulin on plasma retinol and triglyceride concentrations, and fatty acid composition in calves. *Journal of Dairy Research*, **68**(4), 579–586.

Kuwata, T., Yajima, T. & Kaneko, T. (1997) Recent and future improvements of protein fraction in cow's milk-based infant formula in milk composition, production and biotechnology. In *Milk Composition, Production and Biotechnology* (eds Welch, R.A.S., Burns, D.J.W., Davis, S.R., Popay, A.I. & Prosser, C.G.), pp. 215–230. Cab International, Wallingford, UK.

Kuwata, H., Yip, T.T., Tomita, M. & Hutchens, T.W. (1998) Direct evidence of the generation in human stomach of an antimicrobial peptide domain (lactoferricin) from ingested lactoferrin. *Biochimica et Biophysica Acta*, **1429**, 129–141.

Kuwata, H., Yamauchi, K., Teraguchi, S., Ushida, Y., Shimokawa, Y., Toida, T. et al. (2001) Functional fragments of ingested lactoferrin are resistant to proteolytic degradation in the gastrointestinal tract of adult rats. *Journal of Nutrition*, **131**, 2121–2127.

Lahl, W. J. & Braun, S.D. (1994) Enzymatic production of protein hydrolysates for food use. *Food Technology*, **48**, 68–71.

Lahov, E. & Regelson, W. (1996) Antibacterial and immunostimulating casein-derived substances from milk: Caseicidin, isracidin peptides. *Food and Chemical Toxicology*, **34**(1), 131–145.

LeBlanc, J.G., Matar, C., Valdez, J.C., LeBlanc, J. & Perdigon, G. (2002) Immunomodulating effects of peptidic fractions issued from milk fermented with Lactobacillus helveticus. *Journal of Dairy Science*, **85**(11), 2733–2742.

Leclerc, P.L., Gauthier, S.F., Bachelard, H., Sature, M. & Roy, D. (2002) Antihypertensive activity of casein-enriched milk fermented by Lactobacillus helveticus. *International Dairy Journal*, **12**(12), 995–1004.

Levay, P.F. & Viljoen, M. (1995) Lactoferrin: a general review. *Haematologica*, **80**, 252–267.

Lien, E.L. (2003) Infant formulas with increased concentrations of alpha-lactalbumin. *American Journal of Clinical Nutrition*, **77**, 1555S–1558S.

Lieske, B. & Konrad, G. (1996a) Physico-chemical and functional properties of whey proteins as affected by limited papain proteolysis and selective ultrafltration. *International Dairy Journal*, **6**, 13–31.

Lieske, B. & Konrad, G. (1996b) Interrelation between pH and availability of alpha-lactalbumin and beta-lactoglobulin for proteolysis by papain. *International Dairy Journal*, **6**, 359–370.

Longobardo, L., Melck, D., Siciliano, R., Santini, A., di Marzo, V. & Cammarota, G. (2000) β-Casomorphin: Substitution of the phenylalanine with β-homo-phenylalanine increases the m-type opioid receptor activity. *Bioorganic and Medicinal Chemistry Letters*, **10**(11), 1185–1188.

Lonnerdal, B. (2003) Nutritional and physiologic significance of human milk proteins. *American Journal of Clinical Nutrition*, **77**(6), 1537–1543.

Loukas, S., Varoucha, D., Zioudrou, C., Streaty, R.A. & Klee, W.A. (1983) Opioid activities and structures of α-casein-derived exorphins. *Biochemistry*, **22**(19), 4567–4573.

Maase, K. & Steijns, J.M.J.M. (2002) Use of α-lactalbumin as prebiotic agent. Campina Melkunie BV (The Netherlands) Patent number: WO 02/060279.

Madureira, A.R., Pereira, C.I., Gomes, A.M.P., Pintado, M.E. & Malcata, F.X. (2007) Bovine whey proteins: overview of their main biological properties. *Food Research International*, **40**, 1197–1211.

Maeno, M., Yamamoto, N. & Takano, T. (1996) Identification of an antihypertensive peptide from casein hydrolysate produced by a proteinase from Lactobacillus helveticus Cp790. *Journal of Dairy Science*, **79**(8), 1316–1321.

Manso, M.A. & Lopez-Fandino, R. (2003) κ-Casein macropeptides from cheese whey: physicochemical, biological, nutritional, and technological features for possible uses. *Food Reviews International*, **20**, 329–355.

Markus, C.R., Olivier, B., Panhuysen, G.E., Van Der Gugten, J., Alles, M.S., Tuiten, A. *et al.* (2000) The bovine protein α-lactalbumin increases the plasma ratio of tryptophan to the other large neutral amino acids, and in vulnerable subjects raises brain serotonin activity, reduces cortisol concentration, and improves mood under stress. *American Journal of Clinical Nutrition*, **71**(6), 1536–1544.

Markus, C.R., Olivier, B. & de Haan, E.H. (2002) Whey protein rich in α-lactalbumin increases the ratio of plasma tryptophan to the sum of the other large neutral amino acids and improves cognitive performance in stress-vulnerable subjects. *American Journal of Clinical Nutrition*, **75**(6), 1051–1056.

Martin-Orue, C., Henry, G. & Bouhallab, S. (1999) Tryptic hydrolysis of κ-caseinomacropeptide: Control of the enzymatic reaction in a continuous membrane reactor. *Enzyme Microbiology and Technology*, **24**, 173–180.

Maruyama, S. & Suzuki, H. (1982) A peptide inhibitor of angiotensinI-converting enzyme in the tryptic hydrolysate of casein. *Agricultural and Biological Chemistry*, **46**, 1393–1394.

Maruyama, S.H., Mitachi, H., Awaja, J., Kurono, M., Tomizaka, N. & Suzuki, H. (1987a) Angiotensin I-converting enzyme inhibitor activity of the C-terminal hexapeptide of αs1-casein. *Agricultural and Biological Chemistry*, **51**, 2557–2561.

Maruyama, S.H., Mitachi, H., Tanaka, H., Tomizuka, N. & Suzuki, H. (1987b) Studies on the active site and antihypertensive activity of angiotensin I-converting enzyme inhibitors derived from casein. *Agricultural and Biological Chemistry*, **51**, 1581–1586.

Masuda, C., Wanibushi, H., Sekine, K., Yano, Y., Otani, S., Kishimoto,T. *et al.* (2000) Chemopreventive effect of bovine lactoferrin N-butyl-N-(4-hydroxybutyl)-nitrosamine-induced bladder carcinogenesis. *Japan Journal of Cancer Research*, **91**, 582–588.

Matar, C. & Goulet, J. (1996) β-Casomorphin 4 milk fermented by a mutant of Lactobacillus helveticus. *International Dairy Journal*, **6**(4), 383–397.

Matar, C., Valdez, J.C., Medina, M., Rachid, M. & Perdigon, G. (2001) Immunomodulating effects of milks fermented by Lactobacillus helveticus and its non-proteolytic variant. *Journal of Dairy Research*, **68**(4), 601–609.

Matar, C., LeBlanc, J.G., Martin, L. & Perdigon, G. (2003) Biologically active peptides released in fermented milk: role and functions. In *Handbook of Fermented Functional Foods* (ed. *Farnworth, E.R.)*, pp. 177–201. CRC Press, Functional Foods and Nutraceuticals Series, Florida, US.

Matsumoto, H., Shimokawa, Y., Ushida, Y., Toida, T. & Hayasawa, H. (2001) New biological function of bovine α-lactalbumin: Protective effect against ethanol- and stress-induced gastric mucosal injury in rats. *Bioscience Biotechnology and Biochemistry*, **65**(5), 1104–1111.

McIntosh, G.H., Regester, G.O., Le Leu, R.K., Royle, P.J. & Smithers, G.W. (1995) Dairy proteins protect against dimethylhydrazineinduced intestinal cancers in rats. *Journal of Nutritional Biochemistry*, **125**(4), 809–816.

McIntosh, G.H., Royle, P.J., Le Leu, R.K., Regester, G.O., Johnson, M.A., Grinsted, R.L. *et al.* (1998) Whey proteins as functional food ingredients? *International Dairy Journal*, **8**(5–6), 425–434.

Mehra, R., Marnila, P. & Korhonen, H. (2006) Milk immunoglobulins for health promotion. *International Dairy Journal*, **16**(11), 1262–1271.

Meisel, H. (1997) Biochemical properties of regulatory peptides derived from milk proteins. *Biopolymers*, **43**(2), 119–128.

Meisel, H. (1998) Overview on milk protein-derived peptides. *International Dairy Journal*, **8**(5/6), 363–373.

Meisel, H. & Frister, H. (1989) Chemical characterization of bioactive peptides from in-vivo digestion of casein. *Journal of Dairy Research*, **56**(3), 343–349.

Meisel, H. & Schlimme, E. (1990) Milk proteins: precursors of bioactive peptides. *Trends in Food Science and Technology*, **1**(2), 41–43.

Meisel, H. & Schlimme, E. (1994) Inhibitors of angiotensin converting enzyme derived from bovine casein (casokinins). In: β-*Casomorphins and related peptides: Recent developments* (eds Brantl, V. & Teschemacher, H.), pp. 27–33. VCH, Weinheim, Germany.

Meisel, H. & Gunther, S. (1998) Food proteins as precursor for peptides modulating human cell activity. *Nahrung/Food*, **42**,175–176.

Meisel, H. & Bockelmann, W. (1999) Bioactive peptides encrypted in milk proteins: Proteolytic activation and thropho-functional properties. *Antoine van Leeuwenhoek*, **76**(1–4), 207–215.

Meisel, H. & FitzGerald, R.J. (2003) Biofunctional peptides from milk proteins: Mineral binding and cytomodulatory effects. *Current Pharmaceutical Design*, **9**, 1289–1295.

Meisel, H., Frister, H. & Schlimme, E. (1989) Biologically active peptides in milk proteins. *Zeitschrift fur Ernaehrungswissenschaft*, **28**(4), 267–278.

Meisel, H., Andersson, H.B., Buhl, K., Erbersdobler, H.F. & Schlimme, E. (1991) Heat-induced changes in casein-derived phosphopeptides. *Zeitschrift fur Ernaehrungswissenschaft*, **30**(3), 227–232.

Meisel, H., Goepfert, A. & Gunther, S. (1997) ACE-inhibitory activities in milk products. *Milchwissenschaft*, **52**(6), 307–311.

Mellander, O. (1950) The physiological importance of the casein phosphopeptide calcium salts. *Peroral calcium dosage of infants. Acta of the Society of Medicine of Uppsala*, **55**, 247–255.

Meyer, A., Hansen, D.B., Gomes, C.S.G., Hobley, T.J., Thomas, O.R.T. & Franzreb, M. (2005) Demonstration of a strategy for product purification by high-gradient magnetic fishing: recovery of superoxide dismutase from unconditioned whey. *Biotechnology Progress*, **21**, 244–254.

Mezzaroba, L.F.H., Carvalho, J.E., Ponezi, A.N., Antonio, M.A., Monteiro, K.M. *et al.* (2006) Antiulcerative properties of bovine α-lactalbumin. *International Dairy Journal*, **16**, 1005–1112.

Migloire-Samour, D. & Jollès, P. (1988) Casein, a pro-hormone with immunomodulating role to the newborn? *Experientia*, **44**(3), 188–193.

Migloire-Samour, D., Floch, F. & Jollès, P. (1989) Biologically active casein peptides implicated in immunomodulation. *Journal of Dairy Research*, **56**(3), 357–362.

Min, S., Harris, L.J., Han, J.H. & Krochta, J.M. (2005) Listeria monocytogenes inhibition by whey protein films and coatings incorporating lysozyme. *Journal of Food Protection*, **68**(11), 2317–2325.

Minet-Ringuet, J., Le Ruyet, P.M., Tome, D. & Even, P.C. (2004) A tryptophan-rich protein diet efficiently restores sleep after food deprivation in the rat. *Behavioural Brain Research*, **152**(2), 335–340.

Minkiewicz, P., Slangen, C.J., Dziuba, J., Visser, S. & Mioduszewska, H. (2000) Identification of peptides obtained via hydrolysis of bovine casein by chymosin using HPLC and mass spectrometer. *Milchwissenschaft*, **55**(1), 14–17.

Mizuno, S., Nishimura, S., Matsuura, K., Gotou, T. & Yamamoto, N. (2004) Release of short and proline-rich antihypertensive peptides from casein hydrolysate with an Aspergillus oryzae protease. *Journal of Dairy Science*, **87**, 3183–3188.

Moulti-Mati, F., Mati, A., Capiaumont, J., Belleville, F., Linden, G. & Nabet, P. (1991) β-Lactoglobulin part in mitogenic activity of bovine whey. *Le Lait*, **71**(5), 543–553.

Muehlenkamp, M. R. & Warthesen, J. J. (1996) β-Casomorphins: Analysis in cheese and susceptibility to proteolytic enzymes from Lactococcus lactis ssp. cremoris. *Journal of Dairy Science*, **79**(1), 20 26.

Mullally, M.M., Meisel, H. & FitzGerald, R.J. (1996) Synthetic peptides corresponding to α-lactalbumin and β-lactoglobulin sequences with angiotensin-I-converting enzyme inhibitory activity. *Biological Chemistry Hoppe-Seyler*, **377**, 259–260.

Mullally, M., Meisel, H. & Fitzgerald, R. (1997a) Angiotensin-I-enzyme inhibitory activities of gastric and pancreatic proteinase digests of whey protein. *International Dairy Journal*, **7**, 299–303.

Mullally, M.M., Meisel, H. & Fitzgerald, R.J. (1997b) Identification of a novel angiotensin-I-converting enzyme inhibitory peptide corresponding to a tryptic fragment of bovine β-lactoglobulin. *FEBS Letters*, **402**(2–3), 99–101.

Murakami, M., Tonouchi, H., Takahashi, R., Kitazawa, H., Kawai, Y., Negishi, H. *et al.* (2004) Structural analysis of a new anti-hypertensive peptide (β-lactosin B) isolated from a commercial whey product. *Journal of Dairy Science*, **87**(7), 1967–1974.

Mykkanen, H.M. & Wasserman, R.H. (1980) Enhanced absorption of calcium by casein phosphopeptides in rachitis and normal chicks. *Journal of Nutrition*, **110**(11), 2141–2148.

Nagaoka, S., Futamura, Y., Miwa, K., Awano, T., Yamauchi, K., Kanamaru, Y. *et al.* (2001) Identification of novel hypocholesterolemic peptides derived from bovine milk β-lactoglobulin. *Biochemical and Biophysical Research Communications*, **281**(1), 11–17.

Nakai, S. & Li-Chan, E. (1989) Chemical and enzymatic modification of milk proteins. In *Developments in Dairy Chemistry* (ed. Fox, P.F.), pp. 347–376. Elsevier Science Publishers, London.

Nakajima, M., Shinoda, I., Mikogami, T., Iwamoto, H., Hashimoto, S., Miyauchi, H. *et al.* (1997) β-Lactoglobulin suppresses melanogenesis in cultured human melanocytes. *Pigment Cell Research*, **10**(6), 410–413.

Nakamura, T., Syukunobe, Y., Sakurai, T. & Idota, T. (1993) Enzymatic production of hypoallergenic peptides from casein. *Milchwissenschaft*, **48**, 11–14.

Nakamura, Y., Yamamoto, N., Sakai, K. & Takano, T. (1995a) Antihypertensive effect of sour milk and peptides isolated from it that are inhibitors to angiotensin I-converting enzyme. *Journal of Dairy Science*, **78**, 1253–1257.

Nakamura, Y., Yamamoto, N., Sakai, K., Okubo, A., Yamazaki, S. & Takano, T. (1995b) Purification and characterization of angiotensin-I-converting enzyme inhibitors from sour milk. *Journal of Dairy Science*, **78**(4), 777–783.

Neurath, A.R., Debnath, A.K., Strick, N., Li, Y.Y., Lin, K. & Jiang, S. (1997a) 3-Hydroxyphthaloyl-β-lactoglobulin. I. Optimization of production and comparison with other compounds considered for chemoprophylaxis of mucosally transmitted human immunodeficiency virus type 1. *Antiviral Chemistry and Chemotherapy*, **8**(2), 131–139.

Neurath, A.R., Debnath, A.K., Strick, N., Li, Y.Y., Lin, K. & Jiang, S. (1997b) 3-Hydroxyphthaloyl-β-lactoglobulin. II. anti-human immunodeficiency virus type 1 activity in in vitro environments relevant to prevention of sexual transmission of the virus. *Antiviral Chemistry and Chemotherapy*, **8**(2), 141–148.

Nielsen, P.M. (1997) Functionality of protein hydrolysates in food proteins and their applications. In *Food Proteins and their Applications* (eds Damodaran, S. & Paraf, A.) pp. 443–472. Marcel Dekker, Inc., New York.

NIH (1994) *Helicobacter pylori* in peptic ulcer disease. NIH Consensus Development Panel on *Helicobacter pylori*. Peptic Ulcer Disease. *Journal of American Medical Association*, **272**, 65–69.

Nurminen, M.L, Sipola, M., Kaarto, H., Pihlanto-Leppälä, A., Piilola, K., Korpela, R. *et al.* (2000) α-Lactorphin lowers blood pressure measured by radiotelementry in normotensive and spontaneously hypertensive rats. *Life Sciences*, **66**, 1535–1543.

O'Donnell, R., Holland, J.W., Deeth, H.C. & Alewood, P. (2004) Milk proteomics. *International Dairy Journal*, **14**, 1013–1023.

Oevermann, A., Engels, M., Thomas, U. & Pellegrini, A. (2003) The antiviral activity of naturally occurring proteins and their peptide fragments after chemical modification. *Antiviral Research*, **59**(1), 23–33.

Otani, H., Kihara, Y. & Minkyu, P. (2000) The immunoenhancing property of a dietary casein phosphopeptide preparation in mice. *Food and Agricultural Immunology*, **12**(2), 165–173.

Otte, J., Ju, Z.Y., Skriver, A. and Qvist, K.Q. (1996) Effects of limited proteolysis on the microstructure of heat-induced whey protein gels at varying pH. *Journal of Dairy Science*, **79**, 782–790.

Ouwehand, A.C. & Salminen, S.J. (1998) Adhesion inhibitory activity of β-lactoglobulin isolated from infant formulae. *Acta Paediatrica*, **87**(5), 491–493.

Ouwehand, A.C., Salminen, S.J., Skurnik, M. & Conway, P.L. (1997) Inhibition of pathogen adhesion by β-lactoglobulin. *International Dairy Journal*, **7**(11), 685–692.

Packard, R.R.S. & Libby, P. (2008) Inflammation in atherosclerosis: From vascular biology to biomarker discovery and risk prediction. *Clinical Chemistry*, **54**, 24–38.

Pan, D., Luo, Y. & Tanokura, M. (2004) Antihypertensive peptides from skimmed milk hydrolysate digested by cell-free extract of Lactobacillus helveticus JCM1004. *Food Chemistry*, **91**, 123–129.

Panyam, D. & Kilara, A. (1996) Enhancing the functionality of food proteins by enzymatic modification. *Trends in Food Science and Technology*, **7**, 120–125.

Parker, F., Migliore-Samour, D., Floch, F., Zerial, A., Werner, G.H., Jollès, J., Casaretto, M., Zahn, H. & Jollès, P. (1984) Immunostimulating hexapeptide from human casein: Amino acid sequence, synthesis and biological properties. *European Journal of Biochemistry*, **145**(3), 677–682.

Pastorello, E.A. (1997) *Study of nutritional factors in food allergies and food intolerances*. European Commission Directorate General, Brussels, Luxembourg.

Pause, B. & Lembke, J. (1993) Comparative examination of anticaries effect of cheese. II. Results and conclusions. *Milchwissenschaft*, **48**(3), 137–141.

Payne, K.D., Davidson, P.M. & Oliver, S.P. (1990) Influence of bovine lactoferrin on the growth of *Listeria monocytogenes*. *Journal of Food Protection*, **53**, 468–472.

Pecquet, S., Bovetto, L. & Maynard, F. (2000) Peptides obtained by tryptic hydrolysis of bovine β-lactoglobulin induce specific oral tolerance in mice. *Journal of Allergy and Clinical Immunology*, **105**, 514–521.

Pedersen, P. (1994) Removing bitterness from protein hydrolysates. *Food Technology*, **48**, 96–98.

Pellegrini, A. (2003) Antimicrobial peptides from food proteins. *Current Pharmaceutical Design*, **9**(16), 1225–1238.

Pellegrini, A., Thomas, U., Bramaz, N., Hunziker, P. & von Fellenberg, R. (1999) Isolation and identification of three bactericidal domains in the bovine α-lactalbumin molecule. *Biochimica et Biophysica Acta*, **1426**, 439–448.

Pellegrini, A., Dettling, C., Thomas, U. & Hunziker, P. (2001) Isolation and characterization of four bactericidal domains in the bovine β-lactoglobulin. *Biochimica et Biophysica Acta*, **1526**(2), 131–140.

Pellegrino, L., Batteli, G., Resmini, P., Ferranti, P., Barone, F. & Adeo, F. (1997) Alkaline phosphate inactivation during Grana Padano cheese-making and related effects on cheese characterization and ripening. *Le Lait*, **77**, 217–220.

Perea, A. & Ugalde, U. (1996) Continuous hydrolysis of whey proteins in a membrane recycle reactor. *Enzyme Microbiology and Technology*, **18**, 29–34.

Peres, J.M., Bouhallab, S., Petit, C., Bureau, F., Maubois, J.L., Arhan, P. & Bougle, D. (1998) Improvement of zinc intestinal absorption and reduction of zinc/iron interaction using metal bound to the caseinophosphopeptide 1–25 of β-casein. *Reproduction, Nutrition & Development*, **38**(4), 465–472.

Perez, M.D., Sanchez, L., Aranda, P., Ena, J.M., Oria, R. & Calvo, M. (1992) Effect of β-lactoglobulin on the activity of pregastric lipase. Apossible role for this protein in ruminant milk. *Biochimica et Biophysica Acta*, **1123**(2), 151–155.

Perpetuo, E.A., Juliano, L. & Lebrun, I. (2003) Biochemical and pharmacological aspects of two bradykinin-potentiating peptides obtained from tryptic hydrolysis of casein. *Journal of Protein Chemistry*, **22**(7/8), 601–606.

Petrilli, P.F., Addeo, F. & Chianese, L. (1983) Primary structure of water buffalo β-casein tryptic and CNBr peptides. *Italian Journal of Biochemistry*, **32**(5), 336–344.

Pihlanto, A. & Korhonen, H. (2003) Bioactive peptides and proteins. *Advances in Food and Nutrition Research*, **47**, 175–276.

Pihlanto-Leppälä, A., Antila, P., Mantsala, P. & Hellman, J. (1994) Opioid peptides produced by in vitro proteolysis of bovine caseins. *International Dairy Journal*, **41**(4), 291–301.

Pihlanto-Leppälä, A., Koskinen, P., Paakkari, I., Tupasela, T. & Korhonen, H. (1996) Opioid whey protein peptides obtained by membrane filtration. *IDF Bulletin*, **311**, 36–38.

Pihlanto-Leppälä, A., Rokka, T. & Korhonen, H. (1998) Angiotensin I-converting enzyme inhibitory peptides derived from bovine milk proteins. *International Dairy Journal*, **8**(4), 325–331.

Pihlanto-Leppälä, A., Marnila, P., Hubert, L., Rokka, T. & Korhonen, H.J.T. (1999) The effect of α-lactalbumin and blactoglobulin hydrolysates on the metabolic activity of Escherichia coli JM103. *Journal of Applied Microbiology*, **87**(4), 540–545.

Pihlanto-Leppälä, A., Koskinen, P., Piilola, K., Tupasela, T. & Korhonen, H. (2000) Angiotensin I-converting enzyme inhibitory properties of whey protein digests: Concentration and characterization of active peptides. *Journal of Dairy Research*, **67**(1), 53–64.

Poulsen, O.M., Hau, J. & Kollerup, J. (1987) Effect of homogenization and pasteurization on the allerginicity of bovine milk analysed by a murine anaphylactic shock model. *Clinical Allergy*, **17**, 449–458.

Qian, Z.Y., Jollès, P., Migliore-Samour, D., Schoentgen, F. & Fiat, A.M. (1995) Sheep κ- casein peptides inhibit platelet aggregation. *Biochimica et Biophysica Acta*, **1244**(2–3), 411–417.

Recio, I. & Visser, S. (1999a) Identification of two distinct antibacterial domains within the sequence of bovine αs2-casein. *Biochimica et Biophysica Acta*, **1428**(2/3), 314–326.

Recio, I. & Visser, S. (1999b) Two ion-exchange methods for the isolation of antibacterial peptides from lactoferrin: in situ enzymatic hydrolysis on an ion-exchange membrane. *Journal of Chromatography*, **831**,191–201.

Recio, I., Floris, R. & Visser, S. (2000) Bioactive peptides from food proteins: A new isolation method. *Agro-Food-Industry Hi-Tech*, **2**, 9–11.

Reiter, B. & Perraudin, J.P. (1991) Lactoperoxidase, biological functions. In *Peroxidases in Chemistry and Biology*, Volume 1 (eds Everse, J., Everse, K.E. & Grisham, M.B.) pp. 144–180. CRC Press, Boca Raton, Florida, US.

Reynolds, E.C. (1987) The prevention of surface demineralisation of bovine enamel and change in plaque composition by casein in intra-oral model. *Journal of Dental Research*, **66**(6), 1120–1127.

Reynolds, E.C. (1994) Anticariogenic casein phosphopeptide. In *Proceedings of 24th International Dairy Congress*, Melbourne, Australia, 18–22 September.

Reynolds, E.C., Riley, P.F. & Adamson, N.J. (1994) A selective purification procedure for multiple phosphoseryl-containing peptides and methods for their identification. *Analytical Biochemistry*, **217**(2), 277–284.

Richardson, B.C. & Mercier, J.C. (1979) The primary structure of the bovine β-caseins. *European Journal of Biochemistry*, **99**(2), 285–297.

Righetti, P.G., Nembri, F., Bossi, A. & Mortarino, M. (1997) Continuous enzymatic hydrolysis of beta-casein and isoelectric collection of some of the biologically active peptides in an electric field. *Biotechnology Progress*, **13**, 258–264.

Rigo, J., Boehm, G., Georgi, G., Jelinek, J., Nyambugabo, K., Sawatzki, G. & Studzinski, F. (2001) An infant formula free of glycomacropeptide prevents hypothreoninemia in infant formulafed preterm infants. *Journal of Pediatric Gastroenterology and Nutrition*, **32**(2), 127–130.

Roberts, A.K., Chierici, R., Sawatzki, G., Hill, M.J., Volpato, S. & Vigi, V. (1992) Supplementation of an adapted formula with bovine lactoferrin: 1. Effect on the infant faecal flora. *Acta Paediatrica*, **81**, 119–124.

Rokka, T., Syvaoja, E.-L., Tuominen, J. & Korhonen, H. (1997) Release of bioactive peptides by enzymatic proteolysis of *Lactobacillus* GG fermented UHT-milk. *Milchwissenschaft*, **52**, 675–678.

Roudot-Algaron, F., le Bars, D., Kerhoas, L., Einhorn, J. & Gripon, J.C. (1994) Phosphopeptides from Comté cheese: nature and origin. *Journal of Food Science*, **59**(3), 544–547.

Roufik, S., Gauthier, S.F. & Turgeon, S.L. (2006) In vitro digestibility of bioactive peptides derived from bovine β-lactoglobulin. *International Dairy Journal*, **16**(4), 294–302.

Ryhanen, E.L., Pihlanto-Leppälä, A. & Pahkala, E. (2001) A new type of ripened, low-fat cheese with bioactive properties. *International Dairy Journal*, **11**(4–7), 441–447.

Saito, H., Miyakawa, H., Tamura, Y. et al. (1991) Potent bactericidal activity of bovine lactoferrin hydrolysate produced by heat treatment at acidic pH. *Journal of Dairy Science*, **74**, 3724–3730.

Saito, T., Nakamura, T., Kitazawa, H., Kawai, Y. & Itoh, T. (2000) Isolation and structural analysis of antihypertensive peptides that exist naturally in Gouda cheese. *Journal of Dairy Science*, **83**(7), 1434–1440.

Sandre, C., Gleizes, A., Forestier, F., Gorges-Kergot, R., Chilmonczyk, S., Leonil, J., Moreau, M.C. & Labarre, C. (2001) A peptide derived from bovine β-casein modulates functional properties of bone marrow-derived macrophages from germfree and human flora-associated mice. *Journal of Nutrition*, **131**(11), 2936–2942.

Sato, R., Naguchi, T. & Naito, H. (1986) Casein phosphopeptide (CPP) enhances calcium absorption from the ligated segment of rat small intestine. *Journal of Nutritional Science and Vitaminology*, **32**(1), 67–76.

Schlimme, E. & Meisel, H. (1995) Bioactive peptides derived from milk proteins. Structural, physiological and analytical aspects. *Die Nahrung*, **39**(1), 1–20.

Schmidt, D.G. & Poll, J.K. (1991) Enzymatic hydrolysis of whey proteins: Hydrolysis of alpha-lactalbumin and beta-lactoglobulin in buffer solutions by proteolytic enzymes. *Netherlands Milk & Dairy Journal*, **45**, 225–240.

Schmidt, D.G. & van Markwijk, B.W. (1993) Enzymatic hydrolysis of whey proteins: Influence of heat treatment of α-lactalbumin and β-lactoglobulin on their proteolysis by pepsin and papain. *Netherlands Milk & Dairy Journal*, **47**, 15–22.

Schwenke, K.D. (1997) Enzyme and chemical modification of proteins. In *Food Proteins and their Applications* (eds Damodaran, S. & Paraf, A.) pp. 393–423. Marcel Dekker, Inc., New York.

Sekine, K., Ushida, Y., Kuhara, T., Iigo, M., Baba-Toriyama, H., Moore, M.A. *et al.* (1997a) Inhibition of initiation and early stage development of aberrant crypt foci and enhanced natural killer activity in male rats administered bovine lactoferrin concomitantly with azoxymethane. *Cancer Letters*, **121**, 211–216.

Sekine, K., Watanabe, E., Nakamura, J., Takasuka, N., Kim, D.J., Asamoto, M. *et al.* (1997b) Inhibition of azoxymethane-initiated colon tumor by bovine lactoferrin administration in F344 rats. *Japanese Journal of Cancer Research*, **88**, 523–526.

Sforza, S., Ferroni, L., Galaverna, G., Dossena, A. & Marchelli, R. (2003) Extraction, semi-quantification, and fast on-line identification of oligopeptides in Grana Padano cheese by HPLC-MS. *Journal of Agricultural and Food Chemistry*, **51**(8), 2130–2135.

Shah, N.P. (2000) Effects of milk-derived bioactives: an overview. *British Journal of Nutrition*, **84**, S3-S10.

Siemensma, A., Weijer, W.J. & Bak, H.J. (1993) The importance of peptide lengths in hypoallergenic infant formula. *Trends in Food Science and Technology*, **4**, 16–21.

Sikorski, Z.E. (1997) Food components and their role in food quality in chemical and functional properties of food components. In *Chemical and Functional Properties of Food Proteins* (ed. Sikorski, Z. E.) pp. 1–8. Technomic Publishing Co. Inc., Lancaster.

Silva, S.V. & Maltaca, F.X. (2005) Caseins as source of bioactive peptides. *International Dairy Journal*, **15**(1), 1–15.

Sipola, M., Finckenberg, P., Korpela, R., Vapaatalo, H. & Nurminen, M.L. (2002a) Effect of long-term intake of milk products on blood pressure in hypertensive rats. *Journal of Dairy Research*, **69**, 103–111.

Sipola, M., Finckenberg, P., Vapaatalo, H., Pihlanto-Leppälä, A., Korhonen, H., Korpela, R. *et al.* (2002b) α-Lactorphin and blactorphin improve arterial function in spontaneously hypertensive rats. *Life Sciences*, **71**(11), 1245–1253.

Smacchi, E. & Gobbetti, M. (2000) Bioactive peptides in dairy products: Synthesis and interaction with proteolytic enzymes. *Food Microbiology*, **17**(2), 129–141.

Stan, E.I., Ekimovskii, A.P., Aleinik, S.I. & Zhuravlev, B.V. (1988) Heterogeneity and physiological activity of bovine κ-casein proteolysis products. *Voprosy Pitaniia*, **1**, 39–43.

Sternhagen, L.G. & Allen, J.C. (2001) Growth rates of a human colon adenocarcinoma cell line are regulated by the milk protein alpha-lactalbumin. *Advances in Experimental Medicine and Biology*, **501**, 115–120.

Sturner, R.A. & Chang, K.J. (1988) Opioid peptide content in infant formulas. *Pediatric Research*, **23**, 4–10.

Sun, Z., Zhang, Z., Wang, X., Cade, R., Elmir, Z. & Fregly, M. (2003) Relation of β-casomorphin to apnea in sudden infant death syndrome. *Peptides*, **24**(6), 937–943.

Superti, F., Ammendolia, M.G., Valenti, P. & Seganti, L. (1997) Antirotaviral activity of milk proteins: Lactoferrin prevents rotavirus infection in the enterocyte-like cell line HT-29. *Medical Microbiology and Immunology*, **186**(2–3), 83–91.

Sutas, Y., Hurme, M. & Isolauri, E. (1996) Down-regulation of anti-CD3 antibody-induced IL-4 production by bovine caseins hydrolysed with Lactobacillus GG-derived enzymes. *Scandinavian Journal of Immunology*, **43**, 687–689.

Suzuki, T., Yamauchi, K., Kawase, K. *et al.* (1989) Collaborative bacteriostatic activity of bovine lactoferrin with lysozyme against E. coli. *Agricultural Biological Chemistry*, **53**, 1705–1706.

Svanborg, C. & Sabharwal, H. (2004) Protein complex from ion exchange chromatography for treatment of bacterial infections. Pub. No. US 2004/0106541 Patent Application Publications.

Svedberg, J., de Haas, J., Leimenstoll, G., Paul, F. & Teschemacher, H. (1985) Demonstration of a β-casomorphin immunoreactive material in in vivo digest of bovine milk and in small intestine contents after bovine milk ingestion in adult humans. *Peptides*, **6**(5), 825–830.

Svensson, M., Hakansson, A., Mossberg, A.K., Linse, S. & Svanborg, C. (2000) Conversion of α-lactalbumin folding to a protein inducing apoptosis. *Proceedings of National Academy of Science, USA*, **97**, 4221–4226.

Svensson, M., Mossberg, A.K., Pettersson, J., Linse, S. & Svanborg, C. (2003) Lipids as cofactors in protein folding: stereo-specific lipid-protein interactions arerequired to form HAMLET (human alpha-lactalbumin made lethal to tumor cells). *Protein Science*, **12**(12), 2805–2814.

Takakura, N., Wakabayashi, H., Ishibashi, H., Teraguchi, S., Tamura, Y., Yamaguchi, H. *et al.* (2003) Oral lactoferrin treatment of experimental oral candidiasis in mice. *Antimicrobial Agents and Chemotherapy*, **47**, 2619–2623.

Tanaka, K., Ikeda, M., Nozaki, A., Kato, N., Tsuda, H., Saito, S. *et al.* (1999) Lactoferrin inhibits hepatitis C virus viremia in patients with chronic hepatitis C: A pilot study. *Japanese Journal of Cancer Research*, **90**, 367–371.

Tanaka, T., Kawabata, K., Kohno, H., Honjo, S., Murakami, M., Ota, T. *et al.* (2000) Chemopreventive effect of bovine lactoferrin on 4-nitroquinoline 1-oxide induced tongue carcinogenesis in male F344 rats. *Japanese Journal of Cancer Research*, **91**, 25–33.

Tauzin, J., Miclo, L. & Gaillard, J. (2002) Angiotensin I-converting enzyme inhibitory peptides from tryptic hydrolysate of bovine αs2-casein. *FEBS Letters*, **531**, 369–374.

Teraguchi, S., Wakabayashi, H., Kuwata, H., Yamauchi, K. & Tamura, Y. (2004) Protection against infections by oral lactoferrin: Evaluation in animal models. *Biometals*, **17**, 231–234.

Teschemacher, H. (2003) Opioid receptor ligands derived from food proteins. *Current Pharmaceutical Design*, **9**(16), 1331–1344.

Teschemacher, H. & Brantl, V. (1994) Milk protein derived atypical opioid peptides and related compounds with opioid antagonist activity. In *β-Casomorphin and Related Peptides: Recent Developments* (eds Brantl, V. & Teschemacher, H.), pp. 3–17. VCM, Weinheim, Germany.

Teschemacher, H., Koch, G. & Brantl, V. (1997) Milk protein-derived opioid receptor ligands. *Biopolymers*, **43**(2), 99–117.

Tirelli, A., de Noni, I. & Resmini, P. (1997) Bioactive peptides in milk products. *Italian Journal of Food Science*, **9**(2), 91–98.

Tomita, M., Takase, M., Bellami, W. & Shimamura, S. (1994) A review: The active peptide of lactoferrin. *Acta Paediatrica Japonica*, **36**(5), 585–591.

Troost, F.J., Steijns, J., Saris, W.H.M. & Brummer, R.-J.M. (2001) Gastric digestion of bovine lactoferrin in vivo in adults. *Journal of Nutrition*, **131**, 2101–2104.

Troost, F.J., Saris, W.H.M. & Brummer, R.J.M. (2003) Recombinant human lactoferrin ingestion attenuates indomethacin-induced enteropathy in vivo in healthy volunteers. *European Journal of Clinical Nutrition*, **57**, 1579–1585.

Tsuda, H., Sekine, K., Nakamura, J., Ushida, Y., Kuhara, T., Takasuda, N. *et al.* (1998) Inhibition of azoxymethane-initiated colon tumour and aberrant crypt foci development by bovine lactoferrin administration in F344 rats. *Advances in Experimental Medicine and Biology*, **443**, 273–284.

Turgeon, S.L. & Gauthier, S.F. (1990) Whey peptide fractions obtained with a two-step ultrafiltration process: Production and characterization. *Journal of Food Science*, **55**, 106–110.

Ueno, K., Mizuno, K. & Yamamoto, N. (2004) Purification and characterization of an endopeptidase that has an important role in the carboxyl terminal processing of antihypertensive peptides in Lactobacillus helveticus CM4. *Letters in Applied Microbiology*, **39**, 313–318.

Ushida, Y., Shimokawa, Y., Matsumoto, H., Toida, T. & Hayasawa, H. (2003) Effects of bovine α-lactalbumin on gastric defense mechanisms in naive rats. *Bioscience, Biotechnology and Biochemistry*, **67**(3), 577–583.

van Beresteijn, E.C.H., Peeters, R.A., Kaper, J., Meijer, R.J.G.M., Robben, A.J.P.M. and Schmidt, D.G. (1994) Molecular mass distribution, immunological properties and nutritive value of whey protein hydrolysates. *Journal of Food Protection*, **57**, 619–625.

Vermeirssen, V., Deplancke, B., Tappenden, K.A., Van Camp, J., Gaskins, H.R. & Verstraete, W. (2002a) Intestinal transport of the lactokinin Ala-Leu-Pro-Met-His-Ile-Arg through a Caco-2 Bbe monolayer. *Journal of Peptide Science*, **8**(3), 95–100.

Vermeirssen, V., Van Camp, J. & Verstraete, W. (2002b) Optimisation and validation of an angiotensin-converting enzyme inhibition assay for the screening of bioactive peptides. *Journal of Biochemical and Biophysical Methods*, **51**(1), 75–87.

Vermeirssen, V., van Camp, J. & Verstraete, W. (2004) Bioavailability of angiotensin I-converting enzyme inhibitory peptides. *British Journal of Nutrition*, **92**, 357–366.

Wahn, U., Wahl, R. & Rugo, R. (1992) Comparison of the residual allergenic activity of six different hydrolyzed protein formulas. *Journal of Pediatrics*, **121**, 80–84.

Wakabayashi, H., Takakura, N., Teraguchi, S. & Tamura, Y. (2003) Lactoferrin feeding augments peritoneal macrophage activities in miceintraperitoneally injected with inactivated Candida albicans. *Microbiology and Immunology*, **47**, 37–43.

Wakabayashi, H., Kuwata, H., Yamauchi, K., Teraguchi, S. & Tamura, Y. (2004) No detectable transfer of dietary lactoferrin or its functional fragments to portal blood in healthy adult rats. *Bioscience, Biotechnology and Biochemistry*, **68**, 853–860.

Wal, J.M. (2002) Cow's milk proteins/allergens. *Annals of Allergy, Asthma & Immunology*, **89**(6 Suppl. 1), 3–10.

Walker-Smith, J. (2003) Hypoallergenic formulas: Are they really hypoallergenic? *Annals of Allergy, Asthma & Immunology*, **90**(6 Suppl. 3), 112–114.

Wang, Q., Allen, J.C., Swaisgood, H.E. (1997) Binding of vitamin D and cholesterol to β-lactoglobulin. *Journal of Dairy Science*, **80**, 1054–1059.

Wang, W.P., Iigo, M., Sato, J., Sekine, K., Adachi, I. & Tsuda, H. (2000) Activation of intestinal mucosal immunity in tumor-bearing mice by lactoferrin. *Japanese Journal of Cancer Research*, **91**, 1022–1027.

Weinberg, E.D. (1996) The role of iron in cancer. *European Journal of Cancer Prevention*, **5**, 19–36.

Wever, R., Kast, W.M., Kasinoedin, J.H. & Boelend, R. (1982) The peroxidation of thiocyanate catalysed by myeloperoxidase and lactoperoxidase. *Biochimica et Biophysica Acta*, **709**, 212–219.

Williams, A.G., Noble, J., Tammam, J., Lloyd, D. & Banks, J.M. (2002) Factors affecting the activity of enzymes involved in peptide and amino acid catabolism in non-starter lactic acid bacteria isolated from Cheddar cheese. *International Dairy Journal*, **12**, 841–852.

Wyand, M.S., Manson, K.H., Miller, C.J. & Neurath, A.R. (1999) Effect of 3-hydroxyphthaloyl-β-lactoglobulin on vaginal transmission of simian immunodeficiency virus in rhesus monkeys. *Antimicrobial Agents and Chemotherapy*, **43**(4), 978–980.

Xu, R.J. (1998) Bioactive peptides in milk and their biological and health implications. *Food Reviews International*, **14**(1), 1–16.

Xu, Y. (1996) Isolation and characterization of components from whey. Ph.D. thesis, University of Western Sydney, Hawkesbury, Volume I.

Yamaguchi, K., Tomita, M., Giehl, T.J. & Ellison, R.T. (1993) Antibacterial activity of lactoferrin and a pepsin-derived lactoferrin peptide fragment. *Infection and Immunity*, **61**, 719–728.

Yamamoto, N., Akino, A. & Takano, T. (1994a) Antihypertensive effects of the peptides derived from casein by an extracellular proteinase from Lactobacillus helveticus CP790. *Journal of Dairy Science*, **77**(4), 917–922.

Yamamoto, N., Akino, A. & Takano, T. (1994b) Antihypertensive effects of different kinds of fermented milk in spontaneously hypertensive rats. *Bioscience, Biotechnology and Biochemistry*, **58**, 776–778.

Yamamoto, N., Maeno, M. & Takano, T. (1999) Purification and characterization of an antihypertensive peptide from a yogurt-like product fermented by Lactobacillus helveticus CPN4. *Journal of Dairy Science*, **82**(7), 1388–1393.

Yamamoto, N., Ejiri, M. & Mizuno, S. (2003) Biogenic peptides and their potential use. *Current Pharmaceutical Design*, **9**, 1345–1355.

Yamauchi, K. (1992) Biologically functional proteins in milk and peptides derived from milk proteins. *Bulletin of International Dairy Federation*, **272**, 51.

Yamauchi, K., Hiruma, M., Yamazaki, N., Wakabayashi, H., Kuwata, H., Teraguchi, S. *et al.* (2000) Oral administration of bovine lactoferrin for treatment of tinea pedis. A placebo-controlled, double-blind study. *Mycoses*, **43**, 197–202.

Yamauchi, R., Ohinata, K. & Yoshikawa, M. (2003) β-Lactotensin and neurotensin rapidly reduce serum cholesterol via NT2 receptor. *Peptides*, **24**(12), 1955–1961.

Yoshida, S., Ye, X. & Nishiumi, T. (1991) The binding ability of α-lactalbumin and β-lactoglobulin to mutagenic heterocyclic amines. *Journal of Dairy Science*, **74**(11), 3741–3745.

Yoshida, T. & Owens, G.K. (2005) Molecular determinant of vascular smooth muscle diversity. *Circular Research*, **96**, 280–291.

Yoshikawa, M., Tani, F., Yoshimura, T. & Chiba, H. (1986) Opioid peptides from milk proteins. *Agricultural Biological Chemistry*, **50**, 2419–2421.

Yoshikawa, M., Tani, F., Shiota, H., Usui, H., Kurahashi, K. & Chiba, H. (1994) Casoxin D, an opioid antagonist/ileum-contracting/ vasorelaxing peptide derived from human casein. In: β-*Casomorphins and Related Peptides: Recent Developments* (eds Brantl, V. & Teschemacher, H.) pp. 43–48. VCH, Weinheim, Germany.

Zhang, C., Hou, Y. & Su, Y. (1998) Study on immunologic regulation action of Ru Zhen. *China Dairy Industry*, **26**(2), 26–27.

Zucht, H.D., Raida, M., Adermann, K., Magert, H.J. & Forssman, W.G. (1995) Casocidin-I: A casein αs2-derived peptide exhibits antibacterial activity. *FEBS Letters*, **372**(2–3), 185–188.

5 Sports Nutrition and Dairy Products

G. Davison

School of Sport and Exercise Sciences, University of Kent,
United Kingdom

5.1 Introduction

Sport and exercise nutrition refers to the nutritional and dietary guidelines and recommendations for individuals undertaking physical activity. This may range from individuals taking part in physical activity for health, to recreational athletes, to those involved in regular strenuous and structured training programmes (such as the professional athlete). These guidelines are based on available scientific knowledge, derived from the scientific literature. They should be kept distinct from anecdotal information and the unfounded and unsubstantiated marketing claims that are sometimes made (often based on flawed theory, poor 'science' and/or misinterpreted or exaggerated evidence) about nutritional products or supplements. It is important for athletes (of all levels) to pay due attention to their nutritional practices as foods provide the energy for mechanical work, but also essential nutrients for the synthesis and maintenance of tissues (in order for recovery and physiological adaptations to occur: the 'training responses'). Sport and exercise nutrition is not about providing secret potions or supplements that can enhance athletic performance (although this is often the perception from some coaches and athletes), but rather about providing advice and guidelines that can minimise the likelihood of nutritional deficiencies or inadequacies occurring, and ensuring that nutritional intake is optimal to enhance performance, recovery and adaptations and to keep the athletes healthy. Inadequate nutrition can have detrimental effects on performance and training and this is more likely in some athletes or specific groups (e.g. those with extremely high requirements; those with very low needs). Correcting a deficiency will usually reverse the performance detriments but consuming more than the required amount does not generally provide any further benefits.

The general consensus is therefore that athletes at all levels can meet their requirements through an appropriate and well-balanced diet (the use of supplements is not usually needed). In other words, athletes can obtain all of the benefits (or avoid detriments) by consuming 'foods' rather than expensive supplements. Many dairy products contain a good balance of macronutrients and some important micronutrients. As part of a well-balanced diet, they can help to ensure that dietary requirements are achieved (giving the athletes the best chance of attaining optimal performance). A number of dairy products are also considered as functional foods, meaning they may have properties that confer health (or other) benefits beyond simply supplying nutrients (see Chapter 7), and some of these may potentially be of benefit to the athlete.

Milk and Dairy Products as Functional Foods, First Edition. Edited by A. Kanekanian.

In this chapter, a brief outline of some important dietary recommendations derived from the sport and exercise nutrition literature is presented before discussing the role of dairy products as suitable foods (or, in exceptional circumstances, supplements) that may assist in the attainment of these recommendations. The role of some functional dairy foods for athletes is also discussed.

Another reason why diet and nutrition are of paramount importance to athletes is because of the effect it can have on athletes' health and resistance to infection (which have indirect effects on performance, as discussed in Section 5.3.2). Before discussing the role of nutrition in maintaining athletes' immune function and reducing infection risk (and hence loss of training days), it is first necessary to briefly introduce the area of *exercise immunology*. It is generally accepted that moderate amounts of exercise improve immune system functions and hence reduce the risk of infection. However, there is now strong scientific support for the notion that athletes engaged in regular prolonged and/or strenuous exercise (e.g. endurance athletes such as cyclists, runners, swimmers, triathletes) have a higher than normal incidence of minor infections, especially of the upper respiratory tract (URT) (Nieman 1994; Pedersen *et al.* 1994; Gleeson 2007). Overtraining or training with insufficient recovery can further exacerbate these effects. In general, *exercise immunology* is the study of how exercise affects immunity and infection risk.

Exercise immunology research has clearly and consistently demonstrated that many components of the immune system are temporarily reduced (immunodepression) after strenuous and/or prolonged bouts of exercise (Gleeson 2007). This exercise-induced immunodepression may persist for as little as a few hours or as long as a few days. This depends largely on the nature of the exercise (in terms of intensity and duration). However, another key component is the frequency of exercise and/or recovery between bouts. For example, if subsequent bouts are commenced too soon, before the immune system has fully recovered, then a progressive accumulation of immunodepression may ensue, resulting in chronically low immunity and increased infection risk.

Periods of depressed immunity, whether small periods after individual sessions or more chronic periods, are referred to as 'open windows' during which athletes are more susceptible to infection. If an athlete should pick up an infection then this will undoubtedly have an effect on performance either by reducing performance, preventing an athlete from competing altogether and/or interfering with training. Obviously, this is detrimental and may cause considerable set-backs for athletes wishing to maintain a high level of performance and/or peak for a specific competition. Although an appropriately structured and periodised training programme can go some way to minimising the risks, endurance athletes must train intensively if they are to be successful and are consequently at an increased risk (compared to non-athletes) for considerable periods of time. In addition to the direct effects of exercise, other factors such as psychological stress of competition, personal or life stresses, inadequate diet, lack of sleep, travel for competition, etc. may have further negative effects on the immune system and thus resistance to infection. Taking measures to maximise the immune system (and minimise risk and exposure to infection) is particularly beneficial to such athletes.

Sections 5.2–5.5 discuss the various factors involved in nutrition and excerise performance, while Sections 5.6 and 5.7 cover the role of functional foods in sports.

Key Point 1: Athlete infection risk

Depressed immune function may increase the chances of picking up an infection. This is also dependent on exposure to pathogens however, which may be beyond their control. Factors such as contact with lots of other people (e.g. crowded places, work, travel); increased ventilation during exercise; skin abrasions; environmental factors, etc. may also increase the exposure to or entry of pathogens into the body.

5.2 Energy balance

Energy is required in order to perform work and metabolic activities, for maintenance or recovery of homeostasis and for tissue remodelling processes, all of which are essential for optimal exercise performance or training adaptation. Obtaining enough energy from the diet is therefore an essential requirement for most athletes. Some athletes may wish to increase or decrease body mass to obtain an optimal physique or weight for performance in their sport. In addition to undertaking the correct type of activity and training, they must achieve this through the correct intake of food (in the correct proportions). For example, a period of negative energy balance is necessary in order to achieve a loss of body mass and positive energy balance is needed to achieve an increase.

Athletes are often tempted to follow extreme (or fad) dietary regimes to achieve these changes in body composition quickly but this, more often than not, leads to negative effects on health and performance. It would be prudent for athletes to seek guidance from a suitably qualified professional, but general guidelines usually dictate that such changes require time and should be achieved gradually to avoid negative health or performance effects. For most athletes, energy balance is the goal and this should be achieved by consuming the correct amounts of the energy containing nutrients, as discussed in Sections 5.3–5.4.

5.3 Carbohydrates

When considering carbohydrate intake within *sport and exercise nutrition* there are often a number of recommendations relating to: (1) the total quantity of carbohydrates needed; (2) the timing of intake throughout the day and relative to the timing of exercise; and (3) the composition and type of carbohydrate consumed.

5.3.1 *Total dietary carbohydrate requirement*

Carbohydrate is often considered the most important nutrient for sporting performance, perhaps due to its important role in providing energy. Carbohydrate is the predominant fuel for many activities and types of exercise performance. However, carbohydrate availability is particularly important in prolonged (>1.5 hours) exercise (e.g. endurance and ultra-endurance events). It is accepted that depletion of the relatively small body stores of carbohydrate (mainly liver and muscle glycogen in the active/exercising muscles) contributes to the development of fatigue during such activities. Since the seminal studies of Bergström & Hultman (1966) and Bergström *et al.* (1967), many studies have shown that endurance exercise performance and capacity (see Key Point 2) are associated with carbohydrate stores and that these can be manipulated by dietary modifications.

Key Point 2: Exercise performance and exercise capacity tests

Many sport and exercise nutrition studies assess either exercise performance (subjects must cover a fixed distance or amount of work as fast as possible, for example a time trial) or exercise capacity (subjects must exercise at a fixed intensity and time until volitional exhaustion is recorded). Performance tests are considered to be more realistic, that is, in most sports, performance requires a distance to be covered as quickly as possible rather than tolerating exercise at a fixed intensity for as long as possible. Exercise capacity test results do correlate well with performance measures, but it is important to be aware of the different types of test when considering research on exercise performance and its applicability to athletes.

Carbohydrate intake is not often associated with sprint-type or power performance, as total dietary carbohydrate intake (and body carbohydrate stores, such as muscle glycogen) may have little direct effect on performance in short-term high-intensity activities such as sprinting or power lifting. However, it will undoubtedly have an effect on training (in which repeated sprints or lifts and/or exhaustive exercise may be performed), recovery and adaptation, and therefore a potential indirect influence on subsequent performance. For example, Gaitanos *et al.* (1993) showed that a session of 10 repeated 'maximal' sprints, each lasting 6 seconds with short recoveries (30 seconds), significantly reduced muscle glycogen content. In this study, muscle biopsy samples were obtained from the vastus lateralis before and after the first and last sprints in a group of 8 subjects. The authors demonstrated that the first sprint caused a 14% reduction of muscle glycogen concentration and, by the end of the last sprint, muscle glycogen had decreased by 36%.

In team sports such as football and rugby in which repeated high-intensity activities (sprints, jumps, tackling, etc.) are interspersed with periods of lower-intensity activities, significant glycogen depletion can also occur during match play and training. For example, Nicholas *et al.* (1999) showed that a 90 minute protocol, which simulates the physiological demands of soccer, resulted in a *c.* 60% reduction of muscle glycogen. Therefore, dietary carbohydrate is important for optimal performance in this type of athlete.

This is highlighted by the classical work of Saltin & Karlsson (1973) who studied the activity patterns of soccer players during 2 matches separated by 3 days. Half of the players consumed a normal mixed diet and half a high-carbohydrate diet between the two games. The authors observed that the players in the high-carbohydrate group performed more high-intensity activities such as sprinting and less walking, suggesting they were less fatigued and able to perform better in the second game than the players on a normal diet. Since then there have been many studies confirming this assertion, both in field settings and in well-controlled laboratory simulations in which many parameters of performance (such as exercise capacity, repeated sprint performance, skills performance and cognitive functions or performance) have been monitored (e.g. Balsom *et al.* 1999; Welsh *et al.* 2002; Ali *et al.* 2007).

Taken together, these findings highlight the notion that physically active individuals have a high requirement for dietary carbohydrate, not only to optimise performance (e.g. endurance events) but also to support a high training load. However, the specific requirements of any individual will vary depending on the type of sport or activity that they participate in and individual features of the athlete (e.g. age, sex, body mass). As a starting point, however, some generally agreed-upon guidelines do exist: the majority of daily energy intake should come from carbohydrate. Basic recommendations suggest that athletes should obtain around 60% of daily energy intake from carbohydrates. However, for athletes with 'extreme' requirements, following such guidelines based on a percentage of total energy intake may be inappropriate (e.g. the very small gymnast with a low requirement for energy may not consume enough carbohydrate if following such guidelines, whereas the heavyweight rower or endurance cyclist may find difficulty in consuming such quantities of bulky carbohydrate-containing foods). Indeed, more appropriate guidelines are usually based on an absolute requirement, relative to body mass, to ensure that they meet their demands. It is now accepted that, in order to meet the demands of strenuous training, athletes should aim to consume between 7 and 10 g of carbohydrate per kilogram body mass per day (g kg^{-1} day^{-1}) (Maughan 2002; Tarnopolsky *et al.* 2005). The small female gymnast weighing *c.* 40 kg may require *c.* 300–400 g day^{-1}, whereas larger athletes may require 500–800 g day^{-1} (or more).

5.3.2 *Dietary carbohydrate, exercise, immune function and infection risk in athletes*

As discussed in the introduction, exercise-induced immunodepression may be a concern during periods of strenuous or intensive training, with the potential for detrimental effects on training and/or performance. Dietary deficiencies can exacerbate and prolong the duration of exercise-induced immune perturbations (i.e. a larger 'open window'), increasing opportunistic infection risk to the athlete. Of all the macronutrients, carbohydrate availability appears to have the greatest impact on exercise-induced immune changes. It has been demonstrated that athletes who consume less than the recommended amount of carbohydrate will be at greater risk of immune disruption during exercise (Gleeson *et al.* 2004). It is therefore important that athletes consume enough total energy (energy balance) and sufficient carbohydrate in order to avoid increasing their risk further.

5.3.3 *Dairy products as a source of dietary carbohydrate*

In a typical 'western' diet, the major sources of dietary carbohydrate are cereals, grains and staples (e.g. bread, pasta, rice, potatoes). Although dairy products may contribute to total carbohydrate intake, it would not be prudent to advise athletes to modify a well-balanced diet or include more dairy carbohydrate sources for the sole reason of achieving dietary carbohydrate requirements. This is much better achieved from the other sources mentioned above. Likewise, the omission of dairy from the diet is unlikely to result in a dietary deficiency of carbohydrate (although it may cause other issues) as it is more likely that the additional energy intake will be made up by additional carbohydrate-rich foods. In terms of obtaining adequate amounts of total carbohydrate in the diet, the role of dairy is relatively small. However, the timing of carbohydrate intake in relation to exercise performance and the composition of the carbohydrates are of key importance; it is in this area where there is potential for dairy sources to be of benefit to athletes and sporting performance, as discussed in the following sections.

5.3.4 *Timing and composition of carbohydrate intake*

With regard to the timing of carbohydrate intake, we can broadly consider intake before, during and after exercise. When considering intake before exercise, this includes the dietary intake in the days before (see Section 5.3.1), the hours before and immediately before exercise. The major concerns are providing adequate carbohydrate to ensure that glycogen stores are 'full', without having a negative impact upon performance (e.g. negative effects that may occur as a result of consuming food too close to exercise). Intake during exercise is usually aimed at providing an additional, exogenous, energy substrate supply. Intake after exercise is usually directed towards recovery (e.g. replenishment of muscle glycogen, repair of damaged tissues) and adaptation prior to subsequent performance.

5.3.5 *Carbohydrate before and during exercise*

Strictly speaking, total dietary carbohydrate intake (meeting the daily requirements) could be classified as carbohydrate intake before exercise and can impact upon performance as discussed above. However, another important consideration is the acute consumption of carbohydrate (i.e. between a number of hours before and immediately before exercise) and the consumption of carbohydrate during exercise. Indeed, the consumption of carbohydrate during exercise has been shown to delay fatigue and enhance performance by a number of mechanisms; these include the provision of exogenous carbohydrate/energy (Coggan & Coyle 1987) as well as central mechanisms, such as activation of reward and pleasure centres within the central nervous system (CNS)/brain, which can reduce central and psychological fatigue and enhance performance (Chambers *et al.* 2009).

When carbohydrate is consumed during exercise, there is some evidence that carbohydrate mixtures that contain different types of carbohydrates that are transported across the intestine via different transporters (e.g. glucose and fructose) allow a greater intestinal absorption of total carbohydrate, and consequently higher rates of exogenous carbohydrate oxidation (Jeukendrup 2008). This may enhance endurance performance, thus demonstrating that the composition of the ingested carbohydrate is also of key importance.

When the carbohydrates are consumed in the short-term period before prolonged, endurance or intermittent activities then the composition and type of the carbohydrate can impact upon the physiological responses and sometimes performance (if carbohydrate is not to be consumed during the subsequent exercise). For example, a series of studies by Stevenson and colleagues (Stevenson *et al.* 2005a–c, 2006) have demonstrated that the glycaemic index (GI) of meals consumed in the hours before exercise and/or the evening before (i.e. 1–12 hours before) can significantly affect metabolic responses, substrate utilisation and sometimes performance during subsequent exercise. In one study, Stevenson *et al.* (2005b) recruited subjects to exercise on two consecutive days. The exercise on day 1 required subjects to run on a treadmill for 90 minutes (to reduce glycogen stores) and on day 2 they were required to run to exhaustion (exercise capacity test) at 70% of maximal oxygen uptake. After the exercise on day 1, they consumed a high-carbohydrate diet (8 g kg^{-1} day^{-1}) for the rest of the day. Total diet for that day was prescribed as either low GI (35) or high GI (70) in a cross-over design, separated by 1 week. Time to exhaustion on the second day was over 10 minutes longer when the subjects had consumed the low-GI diet compared to the high-GI diet, which could be used to the advantage of athletes needing to train or compete on consecutive days.

Li *et al.* (2004) also showed that pre-exercise low-GI meals are associated with less-severe alterations in immune parameters. In this crossover study, subjects ran on a treadmill for 90 minutes, 3 hours after consuming a high-carbohydrate-content meal with either a high (77) or low (37) GI. The results showed that leukocyte numbers were increased post-exercise (part of the stress response) but to a smaller extent in the low GI trial. There was also a smaller post-exercise change in the neutrophil: lymphocyte ratio (a marker of immunological stress), which may be indicative of less stress to the immune system and less exercise-induced immunodepression.

5.3.6 *Dairy products and carbohydrate intake before and during exercise*

When considering the above studies on GI, the importance of the type of carbohydrate and the associated glycaemic responses are clearly evident. It is possible that dairy carbohydrate is beneficial as lactose has a low GI (lactose GI = 46, whole milk GI = 30). In all of these studies the low-GI meals were achieved by the selection of foods with a high abundance of low-GI carbohydrates (mainly fructose and, to a lesser extent, lactose). All of the low-GI test meals contained dairy products, for example, about half a pint of skimmed milk and one portion of yogurt with breakfast, one portion of cheese with lunch and/or dinner and a large portion of yogurt for snacks, depending on the study and whether these meals were included in the design. However, it must be noted that these studies were not designed to compare dairy-based carbohydrate *per se* but simply to determine the effects of different GI meals. While it is possible that the dairy sources contributed to the beneficial effects, it is also possible that the effects were due to other factors (for example the abundance of fructose).

Before clear conclusions can be drawn about dairy carbohydrate, further research is required to specifically address these questions. It is also important to be mindful of the fact that the consumption of too much lactose may be associated with negative symptoms and gastrointestinal complaints, especially in susceptible individuals. If dairy carbohydrate is to be beneficial for performance it may be more appropriate to achieve this by combination with other carbohydrate sources rather than as a sole source.

Lee *et al.* (2008) investigated the potential performance benefits of milk-based drinks compared to commercial carbohydrate-sports drinks, when consumed before and during exercise. Generally, sports drinks enhance performance via energy provision mechanisms (due to their carbohydrate content) but also via fluid provision mechanisms (due to their water and electrolyte content; see Section 5.5). Lee *et al.* (2008) observed that prolonged exercise capacity tended to be greater with the milk and sports drinks compared to a placebo, but there were no differences between the milk and sports drinks. One interpretation is that the milk-based drink is as beneficial as a commercial sports drink, although there did tend to be higher feelings of gut fullness with the milk-based drinks which could be negative under some circumstances. However, it is not possible to determine whether this benefit was due to the carbohydrate content, the fluid and electrolytes or a combination, although evidence from carbohydrate and sports drinks studies would suggest the effects are combined. At present, although there is sound theory as to why adding dairy sources of carbohydrate may be beneficial to performance (e.g. low GI), there are only a few studies that have investigated the direct benefits of dairy, meaning that further research is required.

5.3.7 *Carbohydrate intake after exercise*

As discussed previously the depletion of glycogen stores within the body (liver and active muscles) is associated with fatigue. An important part of the recovery process after glycogen-depleting exercise is therefore the restoration of the depleted stores (glycogen resynthesis). If 24 hours or more pass before the athlete exercises again, then full glycogen restoration can be achieved as long as an adequate diet with sufficient total carbohydrate intake is consumed (although from other perspectives, such as minimising immune disruption, it is important to refuel and replenish stores as soon as possible). However, when exercise must be repeated again on the same day, optimal restoration becomes more difficult. For optimal short-term recovery of carbohydrate stores it is important to consume carbohydrate in the first 2 hours post-exercise (Ivy *et al.* 1988). Common strategies are to consume 50 g or 1 g kg^{-1} in the first 30 minutes and a further 1–2 g kg^{-1} every subsequent 30 minutes (or consume a high-carbohydrate meal within 2 hours), and these strategies have been shown to enhance subsequent performance after limited recovery time (e.g. between 2 and 6 hours).

Much research has been conducted with carbohydrate-based sports drinks, which offer a convenient and practical means to attain these recommendations for the athlete. There is also some evidence that the addition of protein to the carbohydrate beverage can further augment the recovery of glycogen restoration, which may be useful when recovery time is limited (e.g. 4 hours; Betts & Williams 2010). The literature has mixed results on carbohydrate-protein mixtures but some generally agreed-upon notions are emerging. The addition of protein may be beneficial if carbohydrate intake is sub-optimal, but if the maximal absorbable amount of carbohydrate is ingested (e.g. 60 g $hour^{-1}$ or 1.1 g kg^{-1} $hour^{-1}$) then the addition of protein provides no further benefit. Note however that there may be other benefits from additional protein, such as providing protein to maintain positive protein balance and support protein synthesis, important for the long-term/training adaptation responses to occur as discussed in Section 5.4.

5.3.8 *Dairy products and carbohydrate intake after exercise*

Dairy products such as milk naturally have a carbohydrate–protein balance comparable to the commercially available carbohydrate–protein sports and recovery drinks (i.e. 5–6% carbohydrate and 1–4% protein). It is therefore possible that they may be effective as recovery foods/drinks to optimise the restoration of glycogen stores following exercise-induced depletion (and/or enhance other aspects of recovery and adaptation, as discussed in Sections 5.4 and 5.5 on protein and fluids, respectively).

Indeed, there is an emerging body of research that has investigated dairy-based products as a potential nutritional strategy to promote recovery after exercise and enhance subsequent performance. In one study, Karp *et al.* (2006) found that chocolate milk better enhanced performance (exercise capacity) in the second of two exercise bouts with a 4 hour recovery interval, compared to a sports drink (containing similar amounts of carbohydrate and protein). However, it is not possible to determine from this study whether the beneficial effect that was seen with the chocolate milk was mediated by the carbohydrate or protein content, or by some other mechanisms related to the chocolate or cocoa content.

A few other studies have demonstrated that milk-based drinks are similar to sports drinks in promoting recovery of performance after demanding exercise, but few have been able to determine the mechanisms. Further research is required on recovery, specifically of carbohydrate stores and performance. Other potential beneficial effects due to the protein content and the fluid and electrolytes in dairy-based products are discussed in Sections 5.4 and 5.5.

5.4 Protein

Protein is often perceived as the most important nutrient for strength and power-type athletes, perhaps due to the important role of protein in tissue (e.g. muscle) structure and building (protein synthesis). While it is generally accepted that some athletes may have higher-than-normal requirements, this can almost always be obtained from a normal balanced diet (even in vegetarians). Indeed, protein availability is of key importance as proteins (or amino acids) are required for the physiological training adaptations to occur (Hawley *et al.* 2007). This is true for all training responses, not just in the strength or power athlete. While the requirements may be more obvious to the non-expert (e.g. for muscle hypertrophy and increasing strength), the adaptations that take place in response to other types of training are also dependent on protein. For example, tissue remodelling and repair after strenuous intermittent or endurance exercise and the up-regulation of metabolic enzymes within the muscles of endurance athletes, or hypertrophy of the cardiac muscle, will also require an increase in protein synthesis and need protein/amino acids to take place. As for carbohydrate intake, it is important to consider protein intake in the context of: (1) the total quantity of protein needed; (2) the timing of intake throughout the day and relative to exercise/training; and (3) the composition/type of protein consumed.

5.4.1 *Total dietary protein intake*

It is generally accepted that athletes require a higher amount of protein compared to the recommended amount for a normal individual (which is *c.* $0.8\,g\,kg^{-1}\,day^{-1}$). There is still some controversy as to how much is optimal, and this will depend on the individual and their training and performance requirements. However, the general consensus is that the optimal requirement for the vast majority of athletes (from endurance to strength/power athletes) is 1.4–1.8 g kg^{-1} day^{-1} (Lemon 1995, 2000; Bilsborough & Mann, 2006). However, some athletes (e.g. strength athletes) are known to consume well in excess of this. This is often due to the consumption of protein supplements and/or imbalanced diets. This may be unnecessary, however, as a well-balanced diet will easily provide this amount. Also, if excessively high protein intakes occur at the expense of other nutrients, this can lead to deficiencies in other areas of the diet which may result in negative effects on health and performance.

5.4.2 *Timing of protein intake and protein composition*

Feeding must take place (ideally with carbohydrate to provide energy and high 'biological value' protein, meaning a good balance of essential amino acids) to promote the positive protein balance (synthesis > degradation) that is required for the physiological adaptations (training adaptations) to occur. This can be seen, for example, in the study of Phillips *et al.* (1997) in which post-exercise protein balance (the net balance between degradation and synthesis) remains negative post-exercise (i.e. degradation > synthesis) if the subjects remain fasted. It is therefore important for athletes to consume food as soon as possible following training, and the inclusion of small amounts of high-quality protein may be particularly important in maximising recovery, adaptation and training responses.

5.4.3 *Dairy products and dietary protein intake*

Dairy foods in the diet help to contribute to a good balance of essential (indispensible) amino acids. While it is theoretically possible that the exclusion of dairy from the diet could increase the risk of protein (or certain amino acids) deficiency, even vegetarian and vegan athletes can achieve adequate protein intake through an appropriate and varied diet, without the need for supplementation (Fogelholm 2003). Some athletes do choose to supplement or follow an imbalanced (very high protein) diet, however. At the most extreme, these practices can have negative consequences; for example, intakes greater than 3–3.5 g kg^{-1} day^{-1} may exceed the body's capacity to dispose of the excess nitrogen, leading to toxicity (Bilsborough & Mann 2006). The body will adapt to a high-protein intake by up-regulating clearance and excretion mechanisms. When this has occurred, it is possible to experience negative protein balance if a 'normal' intake is suddenly reverted to. In fact, this is often the reason why athletes with excessive protein intakes continue with such practices. They may notice the effects of a slight negative protein balance (maybe small losses of lean tissue) if they go back to a normal intake. If they returned more gradually however, and maintained the 'normal' intake, the body would have time to adjust the clearance and excretion mechanisms, meaning they could achieve protein balance with the lower (normal) intake.

Despite the fact that ample protein can be obtained from a balanced diet, supplementation can have a place for some athletes as long as it is done sensibly (i.e. within the balanced diet). For example, the timing of protein intake is important and can be used to promote optimal training responses and adaptation (Phillips *et al.* 1997; Hawley *et al.* 2007). Supplemental proteins can be convenient and practical for some athletes (e.g. powdered proteins, protein drinks) as they can be easily taken to training for consumption shortly before or after exercise.

The most common source of protein for use in this way, casein and whey proteins, are derived from dairy sources. These are potentially beneficial to athletes as they have a good amino acid profile (a high-biological-value source of protein) and are readily absorbed.

One study with dairy-based products (Wilkinson *et al.* 2007) was conducted to compare the effects of different drinks on post-exercise (resistance training) protein balance. They compared a fat-free milk drink with a soy-protein drink. Drinks were matched on energy, nitrogen and macronutrient content and contained 745 kJ energy, 18.2 g protein, 1.5 g fat and 23 g carbohydrate. The post-exercise increase in protein balance and muscle protein synthesis was greater when the milk drink was consumed. It was demonstrated that exercise induced similar increases in degradation, but synthesis was up-regulated to a greater extent with the milk-based drink, resulting in a greater net positive protein balance. It is therefore possible that the milk drink could result in faster and/or larger training responses or physiological adaptation and contribute to more beneficial chronic adaptations over the course of a training programme.

Indeed, one recent study attempted to study this directly by employing a 10-week training programme (resistance training 3 times per week) to compare two drinks consumed post-exercise. The drinks in this study were either low-fat chocolate milk or a commercially available carbohydrate sports drink with equivalent amounts of energy (Rankin *et al.* 2004). Similar improvements in strength and body composition were observed in both groups, but there was a trend for the milk group to have greater increases in lean tissue (muscle) which could be beneficial in the long-term.

Taking it a step further, Hartman *et al.* (2007) employed a more intensive training programme (resistance training 5 times per week) over a slightly longer period (12 weeks). They compared the training responses and adaptations that occurred with different post-exercise drinks. They compared three drinks – fat-free milk, a soy drink matched with the milk drink for energy and macronutrient content and a carbohydrate (maltodextrin) drink containing the same amount of total energy, but no protein – and observed greater muscle hypertrophy in the milk group. They also observed a greater loss of body fat in the milk group. The findings for body fat loss could be an indirect effect of increased lean tissue (and hence increased resting metabolic rate) but could also be related to the increased calcium intake in this group.

There is some evidence to suggest that calcium intake and/or dairy products can influence adipocyte metabolism and aid the loss of body fat (Zemel 2004). Taken together, these studies provide some support for the notion that consumption of dairy-based products, within the limits of a well-balanced and healthy diet, may contribute to promoting a more anabolic environment within the body, conducive to promoting physiological training adaptations.

5.5 Fluid and hydration

It is accepted that fluid balance is important to exercise performance. Considerable amounts of fluid can be lost during exercise as the body produces large amounts of heat. This must be dissipated in order to maintain homeostasis and prevent dangerous increases in core body temperature. Sweating is a major mechanism by which the body can effectively remove heat. The environment obviously has a large effect on fluid balance, with hotter and more humid environments resulting in greater sweat losses. However, with prolonged and/or strenuous exercise, there may be considerable energy expenditure which results in a large amount of heat being generated by the body. Even in cooler environments, the body can lose significant amounts of fluid predominantly via sweating (Gleeson 1998; Shirreffs 2007). It is well known that changes in hydration status (i.e. loss of body water) can contribute to fatigue and negatively affect physical and mental performance in some specific circumstances.

In order to minimise such negative effects it is important to optimise hydration status, including being optimally hydrated (a state of euhydration) before exercise, minimising the negative effects of water loss during exercise and recovering/replacing that which is lost after exercise to ensure the next session (especially if recovery time is limited) is not compromised. Like many other aspects of nutrition, fluid intake is therefore important before, during and after exercise.

5.5.1 *Fluid intake before exercise*

The loss of body water (dehydration) leading to a state of hypohydration, if severe enough, can have a negative effect on performance and contribute to fatigue (Armstrong *et al.* 1985). If exercise is commenced in an already hypohydrated state, performance may suffer. Even if the level of hypohydration is not sufficient to limit performance immediately, it is obviously more likely that a detrimental level will be reached during exercise than if the exercise was commenced in an euhydrated state. Fluid consumption before exercise is therefore important. The environment and activity levels of an individual will also have a large effect on fluid losses and hence requirements.

The total daily requirement for water is often quoted as 2 L per day for normal individuals (i.e. non-athletes) or 1 mL per kcal of daily energy expenditure for active individuals (including athletes). While these figures are generally accepted and supported by the scientific literature, they are often misinterpreted. For example, this does not necessarily mean that individuals must consume 2 L of plain water each day. It is important to acknowledge that other drinks (other than plain water) and foods (especially fruit and vegetables) contribute, to some extent, to daily fluid intake. In fact, up to 50% of the daily requirement is met by water contained in food and by metabolic water production within the body. Depending on diet, individuals may only need to consume about half of their daily requirement in the form of drinks.

5.5.2 *Dairy products and total fluid intake*

The absorption and retention of ingested fluid can be influenced by the composition of the fluid (e.g. energy density and electrolyte concentration). Dairy-based fluids such as milk naturally have an abundance of electrolytes. They may therefore have a beneficial role in obtaining normal hydration status before exercise, although products with a higher energy density can contribute to a slower rate of gastric emptying. For athletes undertaking training once per day or less, the body's normal regulatory mechanisms (along with a normal balanced diet) are usually adequate to obtain this over the course of a day, so special attention is not normally required. However, there may be important roles for dairy products during and after exercise. Furthermore, given the potential benefits of milk in restoring hydration status after hypohydration has occurred, it is possible that the acute consumption of such drinks in the short-term period before exercise (i.e. within 2–4 hours before) may be a suitable strategy to ensure optimal hydration is achieved before exercise onset, although this has not yet been studied directly.

5.5.3 *Fluid intake during exercise*

Since hypohydration can negatively affect performance, it seems obvious that offsetting fluid loss by the consumption of fluids during exercise can be beneficial. The amount of fluid required is dependent upon the amount and rate of loss. This is also an individually specific physiological response, but is affected by factors such as the intensity and duration of exercise, environment, clothing, etc. Prolonged/endurance exercise (e.g. over 1.5 hours duration) and/or exercise in a warm or hot environment are most affected by slight to moderate hypohydration, i.e. a loss of 2–7% body mass, which is equivalent to losing *c.* 1.5–5 L of sweat (i.e. a considerable amount) for an average-sized male athlete (Cheuvront *et al.* 2003; Shirreffs 2007). Fluid intake may therefore be particularly important when exercise is prolonged and/or in hot (or unfamiliar) environments. However, it is not always possible, practical, tolerable, or necessary, to consume as much fluid as is lost, especially if sweat rates are high. Nevertheless, the consumption of some fluid will reduce the rate of decline in body water/hydration status to detrimental levels.

The composition of the ingested fluid is a key consideration because this has a large effect on the rate of gastric emptying and the absorption of fluid across the intestine. For example, drinks with a high energy density, especially high-carbohydrate and high-osmolality fluids, generally slow the gastric emptying rate. If carbohydrate provision is more important than fluid, i.e. carbohydrate depletion is more likely to limit performance, then it may be necessary to formulate the drink to optimise carbohydrate delivery. However, for prolonged exercise in a warm or hot environment, hyperthermia and loss of fluid may cause more marked detriments to exercise performance meaning fluid delivery is more important. If this is the case, then a drink with lower energy and carbohydrate content may be more beneficial. In these situations, the electrolyte (especially sodium) content of the drinks may also be crucial. For example, if an athlete consumes a large volume of fluid with low osmolality and sodium content, it can lead to an acute decrease in plasma osmolality and sodium content, which stimulates diuresis. When the drink sodium content is higher (e.g. 30–50 mmol L^{-1}) then this helps to better maintain plasma sodium concentration and osmolality, and prevent or reduce such diuretic effects (Maughan & Leiper 1995; Merson *et al.* 2008). Sodium also stimulates the uptake of glucose and water in the small intestine, which can assist in the delivery of fluids and also carbohydrate into the body (Murray 1987; Shirreffs 2007).

5.5.4 *Dairy products and fluid intake acutely before and during exercise*

To date, there are few studies that have investigated the effects on performance of consuming dairy products immediately before and during exercise. In one study, already discussed in Section 5.3.6, beneficial effects to exercise capacity when a milk-based drink was consumed immediately before and during exercise were suggested (Lee *et al.* 2008). This is likely caused by the macronutrient composition but the electrolyte and fluid content, which will contribute to fluid delivery and balance during exercise, may also contribute to any beneficial effects observed.

5.5.5 *Fluid intake after exercise*

The main aim of fluid intake post-exercise is to promote recovery of fluid balance (rehydration) in order to achieve euhydration. This is particularly important if exercise is to be performed again within a relatively short period of time. As is the case with fluid ingestion during exercise, the composition (and volume) of the ingested fluid also has a significant effect on the recovery of fluid balance. For example, gastric emptying rate is higher when the stomach is full, so larger volumes (or regular 'topping up') promote greater fluid delivery into the body. However, drinks with a high energy density, carbohydrate content and total osmolality can slow the rate of gastric emptying. Electrolytes (especially sodium) content is also important to prevent an acute increases in diuresis and to promote fluid and carbohydrate uptake in the small intestine (Murray 1987; Maughan & Leiper 1995; Shirreffs 2007), although the concurrent ingestion of food containing sodium and other electrolytes can have a similar effect.

In order to achieve complete rehydration, it is recommended that the athlete consumes a volume of fluid greater than the volume lost. Even in the hypohydrated state, there is a continued loss of fluid and obligatory urine production in order to remove waste products and maintain renal functions (Shirreffs *et al.* 1996; Shirreffs 2007). It is therefore often recommended to consume 150% of the estimated fluid loss. For example, if the athlete has lost 2 L of sweat (which can be estimated by acute changes in body mass) they need to drink 3 L of fluid over the ensuing recovery period to achieve optimal rehydration. It is important to note that this should be distributed over a number of hours of recovery, and athletes should not attempt to consume this volume in a single dose. A good general and practical (but non-specific) recommendation is to consume 200–400 mL every 15–20 minutes post-exercise for the first hour and then undertake normal dietary practices (if the next session is at least 24 hours away), or continue drinking 200–400 mL until the target 150% is achieved if subsequent exercise is required in the short-term (i.e. within 4–6 hours).

5.5.6 *Dairy products and post-exercise rehydration*

Within the field of sport and exercise nutrition, it is accepted that the consumption of plain water (or other drinks with low electrolytes content) is not optimal for promoting rehydration post-exercise. Such drinks can stimulate diuresis and do not promote optimal recovery of fluid balance. The consumption of some energy (e.g. a drink containing carbohydrate, and maybe a small amount of protein) is often recommended to promote the physiological recovery and adaptation processes. This is one reason why athletes are often recommended to consume sports drinks (containing carbohydrate and electrolytes) or recovery drinks (e.g. carbohydrate and protein mixtures) post-exercise. Such drinks may also contribute to the recovery of fluid balance, especially if they contain sodium (or are consumed with sodium-containing foods).

There has been a great deal of interest recently in the potential benefit of dairy-based drinks (mainly milk-based drinks) as effective rehydration and recovery drinks. The carbohydrate and protein content of such drinks has been covered in Sections 5.3 and 5.4. Milk and dairy products naturally have a high concentration of electrolytes, which may aid in fluid absorption and retention. A number of recent studies have tested this hypothesis and compared the effectiveness of milk-based drinks and commercial sports drinks as part of a post-exercise rehydration strategy.

Shirreffs *et al.* (2007) compared skimmed-milk, water and sports drinks. Subjects first exercised in a controlled hot environment (in an environmental chamber) until they lost 1.8% of their body mass (i.e. equivalent to *c.* 1.3 L of sweat for a 70 kg subject). They then consumed a volume of drink equivalent to 150% of fluid loss (i.e. *c.* 1.9 L) over a 60-minute period while urine output and hydration status/fluid balance were monitored over the 5 hour period post-exercise. They compared four drinks in total: (1) water as control; (2) a commercial sports drink; (3) skimmed milk; and (4) skimmed milk with additional sodium (20 mmol L^{-1} sodium chloride) added. Urine output was lower in the milk trials (suggesting less diuresis). At the end of the 5-hour recovery period, positive fluid balance was only observed in the milk trials, suggesting that the milk-based drinks may be more beneficial than water or sports drinks in optimising post-exercise rehydration. There were no significant differences between the milk drink and the milk drink with added sodium, so it is possible that the milk drink contained the optimal amount of sodium naturally and additional sodium provides no further benefit. The additional sodium was a relatively small dose, however, and more trials with wider ranges of added sodium are required to confirm this.

These findings were confirmed in a subsequent study from the same group in which a milk-based drink was compared to a commercial sports drink (Watson *et al.* 2008). In the latter study however, subjects were also required to perform an exercise capacity test after 4 hours of recovery. Performance was not different between the milk-based drink and sports drink trials, suggesting that the milk drink was as effective as the sports drink in promoting recovery for subsequent performance. Together these studies provide strong support for the potential beneficial use of milk-based drinks as part of a post-exercise rehydration and recovery strategy.

5.6 Micronutrients

Micronutrients are important for optimal health in general; athletes with high-energy requirements may have an increased need for many micronutrients (Manore 2000; Woolf & Manore 2007). This increased requirement is usually achieved with an appropriate increase in energy intake to maintain energy balance, and deficiencies are unlikely if the diet is well balanced and varied. Under certain circumstances however, athletes (or certain groups) may be at an increased risk of dietary deficiency (e.g. those adopting an imbalanced diet or excluding certain foods and those experiencing increased losses or gastrointestinal and malabsorption problems). A deficiency of certain micronutrients can have direct negative effects on performance. It can also have negative effects on general health, immune function and resistance to infection (and consequently sporting performance). This often leads some athletes and coaches to the common misconception that additional intakes of these substances (i.e. vitamins and minerals) will boost performance and/or health; the mentality that 'if *some* is good for me, then *more* must be better' is common among athletes. However, if a short-term deficiency is corrected, the detrimental performance and health effects are usually restored to 'normal', although more prolonged and serious deficiencies may need medical treatment. For most micronutrients, additional intake above the normal (individual-specific) requirement does not provide any additional benefits. For those athletes with a very high requirement, or those most at risk of deficiency, the most likely nutrients to be deficient in are some B vitamins, calcium, iron, magnesium and zinc (Woolf & Manore 2007), all of which can have a negative effect upon performance.

The B vitamins are generally involved in energy metabolism and/or the synthesis of important blood components (such as red blood cells and haemoglobin), all of which are important for the physically active individual and may impact upon performance. Thiamnin, riboflavin, viatmin b_6, viatmin b_{12} and folate are some B vitamins which are more likely to be deficient in some athletes/groups. For example, Manore (2000, 2002) has demonstrated that female athletes are much more likely to have a dietary deficiency of a number of B vitamins, which may be related to the generally lower total energy requirement and intake in this group. Females are also more likely to be deficient in dietary minerals such as calcium, iron, magnesium and zinc.

Deficiencies of iron and magnesium can lead to fatigue and have negative effects on performance. For example, iron deficiency can lead to anaemia and magnesium deficiency can lead to fatigue and muscle weakness, which can have severe detrimental effects on performance. Zinc deficiency can have negative effects on immune function and resistance to infection, which can be negative to athlete health and performance. Calcium deficiency can often go unnoticed in athletes as, unless the deficiency is very severe, the major consequences are usually not noticed until many years later when poor bone health (e.g. low bone mineral content, osteopenia or osteoporosis) may become apparent. However, calcium intake may also have a role in the maintenance of optimal fat metabolism and body composition, so the avoidance of calcium deficiency may have an indirect effect on performance via this mechanism.

Dairy is a major source of dietary calcium and dairy foods may play an important role in meeting daily requirements for this nutrient. Indeed, individuals most at risk of dietary deficiency are those who exclude dairy from their diet. Within a healthy and balanced diet, dairy products may also contribute to meeting daily requirements for a number of other important micronutrients, such as some B vitamins and minerals.

Table 5.1 shows the extent to which a single portion of selected dairy products may go towards achieving the recommended daily intake of some of the key micronutrients listed above (e.g. those for which certain groups of athletes, such as females, are more likely to be deficient in). For example, for females, one portion of low-fat yogurt provides 23% of the reference nutrient intake (RNI; see COMA 1991) of thaimin, 30% of riboflavin, 14% of folate, 33% of vitamin B_{12}, 35% of calcium and 13% of zinc (although dairy products are generally poor sources of iron and magnesium). Table 5.1 shows similar information for a selection of other dairy products that athletes may include in a normal diet. It is clear to see that, as part of a well-balanced and healthy diet, dairy products can help athletes to meet their daily requirements for a number of important micronutrients.

Table 5.1 Contribution to RNI per portion of selected dairy products (%) for both male (M) and female (F) subjects aged 19–50 years. Green (yellow) shaded cells indicate that a single portion of this food will provide at least 20% (10%) of the RNI. Source: COMA (1991) and FSA (2002). (See colour section.)

Example Food	Thiamin		Riboflavin		Vitamin B_6		Folate		Vitamin B_{12}		Calcium		Iron		Magnesium		Zinc	
	M	F	M	F	M	F	M	F	M	F	M	F	M	F	M	F	M	F
Milk (semi-skimmed)	6%	8%	38%	45%	9%	10%	10%	10%	127%	127%	35%	35%	0%	0%	-	-	8%	11%
Low fat yoghurt	18%	23%	25%	30%	1%	2%	14%	14%	33%	33%	35%	35%	2%	1%	-	-	9%	13%
Cheese (cheddar)	1%	1%	9%	11%	3%	3%	5%	5%	47%	47%	32%	32%	1%	1%	-	-	13%	17%
Rice pudding	2%	3%	20%	24%	1%	2%	0%	0%	-	-	25%	25%	2%	1%	0%	0%	11%	14%

KEY : M: Males, F: Females. Green shaded cells indicate that a single portion of this food will provide at least 20% of the RNI; Yellow shaded cells indicate that a single portion of this food will provide at least 10% of the RNI.

Information on food composition taken from Food Standards Agency (2002) McCance and Widdowson's The Composition of Foods, Sixthsummary edition. Cambridge: Royal Society of Chemistry. The percentage of RNI values are based on UK values for adults (aged 19–50 years) from COMA (1991) Dietary reference values for food and energy nutrients for the U.K. Department of health report on health and social subjects #41 HMSO: London.

5.7 Bovine colostrum

Bovine colostrum and its potential health benefits is specifically covered in Chapter 3. Briefly, bovine colostrum ('early milk') is the initial milk produced by cows, usually obtained within the first 48 hours post-partum. Like normal milk, it contains a rich source of nutrition both in terms of macronutrients and micronutrients but is also abundant in bioactive components including immune, growth and antimicrobial factors (Shing *et al.* 2009). As a nutritional supplement, bovine colostrum has regained popularity among athletes in recent years and has been suggested to confer performance, immune, health and recovery benefits. Following this trend, a considerable amount of research has been conducted to investigate these claims. It is important to note here that all of the studies on (and reference to) bovine colostrum in this section refer to normal bovine colostrum from normal dairy cows (most suppliers use only organic dairy sources) and *not* hyperimmune colostrum. The latter is obtained from cows that have been inoculated or vaccinated with specific pathogens, meaning that the colostrum contains high concentrations of antibodies to a specific pathogen(s), usually for medical use. Hyperimmune colostrum is not considered here in relation to healthy athletes and active individuals.

A number of studies have sought to determine whether daily supplementation with bovine colostrum, normally for periods of between 1 and 12 weeks, has a beneficial effect on exercise performance either via direct mechanisms (i.e. directly ergogenic to performance) or indirect mechanisms (e.g. by improving health, recovery or reducing illness), as discussed in the following sections.

5.7.1 *Body composition and strength*

Some early studies suggested that colostrum may enhance body composition by promoting muscle gain and fat loss. In conjunction with resistance training, there is some evidence that bovine colostrum does promote greater gains in muscle (and strength) and losses of body fat when compared to a placebo (Kerksick *et al.* 2001), although numerous studies have shown no differences (Buckley *et al.* 2003; Brinkworth *et al.* 2004).

The limited evidence for benefits of this nature does not prove that this is caused by colostrum *per se*. For example, as discussed in Section 5.6 on protein, the timing and composition of post-exercise protein can have a large influence on the physiological training adaptations that follow. In the colostrum studies mentioned above, however, the placebo is whey protein, which is absorbed much more rapidly than other dairy proteins (i.e. casein). It is therefore likely that the colostrum in these studies (which will contain more casein) resulted in a more sustained, gradual and prolonged release of amino acids into the circulation, which could be the reason for any beneficial effects (or may contribute to any additional effects actually caused by colostrum *per se*).

Furthermore, the colostrum supplements likely contained more calcium than purified whey protein. As mentioned in Section 5.6, calcium has an important role in lipid metabolism and supplemental doses may affect fat metabolism and body composition in athletes. It has been suggested that the intake of additional calcium, in the region of 600–1000 mg day^{-1}, can increase fat loss or prevent fat gain during periods of positive energy balance (Zemel 2004). Depending on the specific study design, subjects may have been consuming an extra 250–750 mg of calcium per day in the colostrum group.

That is not to say that colostrum does not have beneficial effects from colostrum-specific bioactive components within it, just that the studies to date cannot prove with any certainty which components of colostrum are responsible for any benefits that have been observed. More research is required, with better-matched placebo and control conditions, in order for this to be determined. However, both casein and whey proteins are derived from dairy sources and dairy is a good source of calcium, demonstrating the potential benefit of dairy products in general in certain situations for athletes.

5.7.2 *Endurance performance*

There is an emerging body of evidence suggesting that bovine colostrum supplementation may enhance recovery from demanding exercise (and subsequent performance), endurance performance and prevent performance detriments induced by acute periods of overtraining or overreaching. In one study (Buckley *et al.* 2002), subjects were provided with daily supplements of powdered bovine colostrum (60 g day^{-1}) or placebo (whey protein) during an 8 week intervention period. During this time they trained three times per week (45 minutes of running in each session). Performance (exercise capacity) was assessed, pre-intervention (baseline) and again at week 4 and week 8. This involved two repeated incremental treadmill running tests with a 20 minute recovery interval between each. There was no difference in running performance between the colostrum and placebo groups at baseline or week 4, but by week 8 the colostrum group covered more distance in the second treadmill test. The study was not designed to determine the mechanisms responsible for this beneficial effect, so it is only possible to speculate at present. Indeed, the mechanisms could include changes in immune function and illness (meaning less missed training and sick days), improved buffer capacity or enhanced physiological training responses.

More recently, Shing *et al.* (2006) conducted a study with cyclists in which they demonstrated that daily supplementation with 10 g day^{-1} of concentrated bovine colostrum powder for 6 weeks prevented the decrease in time-trial performance that was caused by a 5-day period of very intensive training in week 6 (although there was no benefit to time-trial performance during normal training).

Coombes *et al.* (2002) assessed endurance performance in cyclists before and after 8 weeks of supplementation with placebo, 20 g day^{-1} or 60 g day^{-1} of bovine colostrum. They observed that performance improved more in the colostrum groups compared to the placebo group. There was no difference between the two colostrum groups however, suggesting that there may be a plateau in the dose-response effects. Further research is required to determine the optimal doses, which is likely to be specific and different for each individual athlete.

In summary, there is some evidence from well-conducted placebo-controlled experimental trials suggesting that bovine colostrum supplementation may be beneficial to endurance performance, especially during periods of intensified training (a strategy sometimes employed by endurance athletes).

5.7.3 *High-intensity exercise (i.e. sprint and repeated sprint) performance*

Fatigue in high-intensity activities such as sprinting or repeated sprints with short recovery intervals is caused by multiple factors. However, it is generally accepted that the decline in muscle energy pools (and thus adenosine-triphosphate or ATP) and the accumulation of metabolic by-products play significant roles. The accumulation of hydrogen ions (H^+) and consequent lowering of pH within muscles has been implicated in fatigue during high-intensity exercise, especially for repeated sprints and sustained efforts lasting between 2 and 7 minutes (e.g. middle-distance events in athletics). A number of studies have demonstrated that this type of performance can be enhanced by increasing the buffer capacity through nutritional manipulation, either intracellular buffer capacity (i.e. within muscles) or extracellular buffer capacity (allowing a greater efflux of H^+ out of the muscle cells).

A few studies have been conducted to examine whether bovine colostrum supplementation has any effect on exercise performance via changes in buffer capacity. In one study, Brinkworth et al. (2002) showed that supplementation with 60 g day^{-1} of bovine colostrum powder, during a 9-week training programme in female rowers, was associated with a significant increase in blood buffer capacity compared to the placebo (60 g day^{-1} whey protein). However, training is a potent stimulus for increasing the body's buffering capacity, so it is not clear whether this was a direct effect of bovine colostrum or an indirect effect via promoting a more favourable training response. Furthermore, there was no difference in performance in two repeated incremental rowing tests, separated by 15 minutes of recovery. However, as just noted, the type of exercise most likely to be affected by an increase in buffer capacity are activities that cause exhaustion in 2–7 minutes, meaning that the incremental test was not ideal for the purposes of assessing performance in this case. There is convincing evidence for performance benefits in this type of activity (2–7 minutes) with other nutritional interventions that increase buffer capacity; if bovine colostrum can increase buffer capacity, then it could theoretically improve performance. It is however clear that further research, specifically with bovine colostrum, is required before this can be determined.

5.7.4 *Gastrointestinal integrity/health*

Many stressors, such as heat stress, certain types of medication and oxidative stress are known to cause disturbance to gut function and integrity. A number of these stressors may be induced by strenuous exercise (e.g. heat stress, oxidative stress), which has been shown to increase intestinal permeability (colloquially referred to as 'leaky gut' syndrome) and injury. This increases permeability to luminal toxins and bacteria, allowing them into the systemic circulation (endotoxaemia). This may place additional stress on the immune system, which then has to deal with these bacteria and/or toxins, and this may be one factor contributing to exercise-induced immunodepression.

Increased intestinal permeability may also contribute to increased incidences of some acute gastrointestinal complaints and symptoms in athletes, such as stomach cramps, nausea and dizziness and diarrhoea (Jeukendrup *et al.* 2000). A high prevalence of such symptoms have been noted in some groups of athletes after strenuous competitive events, for example up to 50% incidence after marathon and ultra-endurance events or even >90% incidence during and after a very strenuous ultra-endurance event (an Ironman Triathlon consisting of a 2.4 mile open water swim, 115 mile cycle then a full marathon of 26.2 miles) (Jeukendrup *et al.* 2000). Such symptoms have potentially negative effects on performance if incurred during an event or subsequent recovery, dietary intake, training or performance if they develop afterwards. Under extreme circumstances (e.g. exhaustive exercise in a hot and humid environment) this can cause severe endotoxaemia, leading to acute inflammation, sepsis, shock and organ failure, which can be fatal in extreme situations. Indeed, it has been suggested that damage to the gut can be a contributing factor in exertional heat stroke (Leon & Helwig 2010).

Supplementation with bovine colostrum has been shown to be beneficial in maintaining intestinal integrity and permeability. For example, Marchbank *et al.* (2011) conducted a placebo-controlled counterbalanced crossover study to determine the effects of bovine colostrum supplementation on exercise-induced changes in gut permeability. Subjects were required to run on a treadmill for 20 minutes (which caused core body temperature to increase by between 1 and 2°C) after 2 weeks of daily supplementation with 20 g day^{-1} of bovine colostrum or a protein and energy-matched placebo (milk protein concentrate). Running caused a significant increase in gut permeability in the placebo trial, but this was almost completely blunted in the colostrum trial. Similar beneficial effects of colostrum have been observed in human, animal and *in vitro* studies with cultured epithelial and gut cell lines subjected to physical, chemical or thermal stressors.

In summary, bovine colostrum supplementation may be beneficial in preventing exercise-induced increases in gut permeability. There is some preliminary evidence that this may be beneficial to athletes (e.g. by indirectly impacting upon training and performance), especially in those required to exercise or compete in hot (or unfamiliar) environments.

5.7.5 *Immunity*

Bovine colostrum is of obvious importance for protection against infection in newborn calves and there is growing interest in this dairy product as a potential dietary supplement to counteract immunodepression in athletes. It has been demonstrated that bovine colostrum supplementation can enhance a number of human immune functions, from the functioning of peripheral blood leukocytes assessed *in vitro* (Sugisawa *et al.* 2002, 2003; Biswas *et al.* 2007) to *in vivo* immune functions such as the humoral immune response to an orally administered vaccine (He *et al.* 2001). Athletes may be at increased risk of infection during periods of strenuous training and stress, and this may be detrimental to training and performance. A number of recent studies have therefore investigated whether bovine colostrum is able to enhance immunity and/or reduce infections in athletes.

Crooks *et al.* (2006) have shown that a 12-week bovine colostrum supplementation intervention in a group of distance runners resulted in a 79% increase in resting salivary secretary IgA (s-IgA). A 33% increase in salivary s-IgA has also been found after only 2 weeks of supplementation (Mero *et al.* 2002). The concentration and/or secretion of salivary s-IgA has been shown to be a good predictor of upper respiratory tract illness (or infection) risk in athletes, so it has been proposed that the increases observed with bovine colostrum may be beneficial and confer protection in athletes. Indeed, Brinkworth & Buckley (2003) demonstrated a lower incidence of self-reported (illness logs) upper respiratory tract (URT) illness symptoms in a group of men taking daily bovine colostrum compared to a placebo group.

Other studies have shown that bovine colostrum supplementation improves other markers of immune function and/or prevents the magnitude of decrease following strenuous exercise (i.e. exercise-induced immunodepression), which may contribute to the protective effects in athletes. For example, Davison & Diment (2010) observed that 4 weeks of daily bovine colostrum supplementation (20 g day^{-1}) prevented the prolonged-exercise-induced decrease of salivary lysozyme concentration and secretion and speeded the recovery of neutrophil functions post-exercise. These mechanisms may also contribute to the protective effects against illness in athletes.

Not all studies have observed beneficial effects, however. For example, Shing *et al.* (2007) investigated the effects of bovine colostrum supplementation on salivary s-IgA in addition to some components of innate immunity. They observed some beneficial effects on some immune measures but not salivary s-IgA. Likewise, in the study of Davison & Diment (2010), salivary s-IgA was not affected. Such findings highlight the fact that variations in research design, dosage (and perhaps quality of supplements) and duration of supplementation, as well as variations between different individuals (e.g. non-responders) may have a large effect on the nature of any benefits provided by bovine colostrum for athletes. Furthermore, the underlying mechanisms responsible for such immune-enhancing effects of colostrum are not completely clear at present, although preliminary evidence suggests that some bioactive metabolites become available after ingestion, which directly stimulate and prime key aspects of immunity (Sugisawa *et al.* 2002, 2003; Davison & Diment 2010).

Further research is required (e.g. dose-response studies in broad populations) before we can fully understand the potential benefits to athletes. The existing evidence does however support the notion that bovine colostrum is beneficial for athletes involved in strenuous training (e.g. endurance athletes) in terms of immunity and resistance to infection, which may ultimately be beneficial to performance.

5.8 Probiotics

Probiotics are reviewed in Chapter 8. Although probiotics may have numerous health benefits (such as gastrointestinal health, immune and blood health), a recent systematic review (Nichols 2007) concluded that no studies could be identified that have investigated potential direct effects on performance. This author is not aware of any subsequent studies (in humans) that have changed the current state of knowledge (although there is one in horses; Art *et al.* 1994), meaning that it is only possible to speculate about potential indirect benefits that may arise as a result of the other benefits (e.g. immune and gut health).

Art *et al.* (1994) conducted a study with thoroughbred horses during a 12-week training programme. The horses were randomly assigned to either a control group or a probiotics group, receiving a daily commercial probiotics supplement. The training-induced improvements in physiological measures (e.g. maximal oxygen uptake) occurred earlier during the training programme and/or were greater in magnitude in the probiotics group. This suggests that probiotics may be beneficial, but it is not possible to determine whether this was due to a direct or indirect mechanism (although an indirect mechanism seems more plausible). Whether this finding in horses is applicable to humans also remains to be determined, and would be an interesting area for further study.

Although there is limited research on probiotics and performance, there are a few relevant studies with athletes (e.g. gut health, immunity and infection incidence). Supplementation with probiotics is able to modulate the intestinal microbial flora and enhance gut immune function. As the gut microbial flora is able to interact with intestinal epithelial cells and a number of specific immune cells, this can exert beneficial effects on the skin and upper respiratory tract which could have important implications for athletes in heavy training (West *et al.* 2009).

In a recent study, Cox *et al.* (2010) compared self-reported URT illness symptoms in 20 male athletes during the winter months who were taking either a daily probiotics supplement or placebo. They reported a lower number of days with symptoms (30 days compared to 72 days) and lower severity ratings in the probiotics group compared to the placebo group. However, Kekkonen *et al.* (2007) reported no significant differences in URT illness symptoms or gastrointestinal symptoms in 141 runners randomised into either placebo or probiotics groups taking the supplements daily for 3 months prior to a marathon race, and for the 2 week period post-race. However, in those who were affected they did observe a trend for the duration of symptoms to be shorter in the probiotics group. Theoretically, beneficial effects on the gut may be beneficial to host immune defences in athletes, and the studies of Cox *et al.* (2010) and, to a lesser extent, Kekkonen *et al.* (2007) provide preliminary evidence for this. Further research is required to explore the usefulness of probiotics within a sport and exercise context.

5.9 Summary

Dairy products can have both direct and indirect effects on health and performance and may confer benefits to athletes and active individuals if consumed as part of a healthy and balanced diet. Dairy products in the diet may contribute to attainment of the daily requirements for a number of nutrients that are important for optimal training, performance, recovery and adaptation in athletes. In some situations, they may also be useful as supplements to help optimise training and performance. Dairy products (as foods) may also be suitable alternatives to supplements or commercial products such as sports drinks for use by athletes. However, a number of the beneficial effects have only been investigated by a relatively small number of researchers and in relatively small-scale studies. It is therefore important for research into the potential performance benefits of dairy products to be continued in the future.

Recommended further reading

Fogelholm, M. (2003) Dairy products, meat and sports performance. *Sports Medicine*, **33**(8), 615–631.
Gleeson, M. (2006) *Immune Function in Sport and Exercise*. Edinburgh, New York: Churchill Livingstone Elsevier.
MacLaren, D. (2007) *Nutrition and Sport*. Edinburgh, London, New York: Churchill Livingstone Elsevier.
Roy, B.D. (2008) Milk: the new sports drink? A Review. *Journal of the International Society of Sports Nutrition*, **2**, 5–15.

References

Ali, A., Williams, C., Nicholas, C.W. & Foskett, A. (2007) The influence of carbohydrate-electrolyte ingestion on soccer skill performance. *Medicine and Science in Sports and Exercise*, **39**(11), 1969–1976.

Armstrong, L.E., Costill, D.L. & Fink, W.J. (1985) Influence of diuretic-induced dehydration on competitive running performance. *Medicine and Science in Sports and Exercise*, **17**(4), 456–461.

Art, T., Votion, D., McEntee, K., Amory, H., Linden, A., Close, R. & Lekeux, P. (1994) Cardio-respiratory, haematological and biochemical parameter adjustments to exercise: effect of a probiotic in horses during training. *Veterinary Research*, **25**(4), 361–370.

Balsom, P.D., Wood, K., Olsson, P. & Ekblom, B. (1999) Carbohydrate intake and multiple sprint sports: with special reference to football (soccer). *International Journal of Sports Medicine*, **20**(1), 48–52.

Bergström, J. & Hultman, E. (1966) Muscle glycogen synthesis after exercise: an enhancing factor localized to the muscle cells in man. *Nature*, **210**(5033), 309–310.

Bergström, J., Hermansen, L., Hultman, E. & Saltin, B. (1967) Diet, muscle glycogen and physical performance. *Acta Physiologica Scandinavica*, **71**(2), 140–150.

Betts, J.A. & Williams, C. (2010) Short-term recovery from prolonged exercise: exploring the potential for protein ingestion to accentuate the benefits of carbohydrate supplements. *Sports Medicine*, **40**(11), 941–959.

Bilsborough, S. & Mann, N. (2006) A review of issues of dietary protein intake in humans. *International Journal of Sports Nutrition and Exercise Metabolism*, **16**(2), 129–152.

Biswas, P., Vecchi, A., Mantegani, P., Mantelli, B., Fortis, C. & Lazzarin, A. (2007) Immunomodulatory effects of bovine colostrum in human peripheral blood mononuclear cells. *New Microbiology*, **30**(4), 447–454.

Brinkworth, G.D. & Buckley, J.D. (2003) Concentrated bovine colostrum protein supplementation reduces the incidence of self-reported symptoms of upper respiratory tract infection in adult males. *European Journal of Nutrition*, **42**(4), 228–232.

Brinkworth, G.D., Buckley, J.D., Bourdon, P.C., Gulbin, J.P. & David, A. (2002) Oral bovine colostrum supplementation enhances buffer capacity but not rowing performance in elite female rowers. *International Journal of Sports Nutrition and Exercise Metabolism*, **12**(3), 349–365.

Brinkworth, G.D., Buckley, J.D., Slavotinek, J.P. & Kurmis, A.P. (2004) Effect of bovine colostrum supplementation on the composition of resistance trained and untrained limbs in healthy young men. *European Journal of Applied Physiology*, **91**(1), 53–60.

Buckley, J.D., Abbott, M.J., Brinkworth, G.D. & Whyte, P.B. (2002) Bovine colostrum supplementation during endurance running training improves recovery, but not performance. *Journal of Science and Medicine in Sport*, **5**(2), 65–79.

Buckley, J.D., Brinkworth, G.D. & Abbott, M.J. (2003) Effect of bovine colostrum on anaerobic exercise performance and plasma insulin-like growth factor I. *Journal of Sports Science*, **21**(7), 577–588.

Chambers, E.S., Bridge, M.W. & Jones, D.A. (2009) Carbohydrate sensing in the human mouth: effects on exercise performance and brain activity. *Journal of Physiology*, **15**(Pt 8), 1779–1794.
Cheuvront, S.N., Carter, R. 3rd, & Sawka, M.N. (2003) Fluid balance and endurance exercise performance. *Current Sports Medicine Reports*, **2**(4), 202–208.
Coggan, A.R. & Coyle, E.F. (1987) Reversal of fatigue during prolonged exercise by carbohydrate infusion or ingestion. *Journal of Applied Physiology*, **63**(6), 2388–2395.
COMA (1991) *Dietary reference values for food and energy nutrients for the UK Department of Health Report on Health and Social Subjects #41*. London: HMSO.
Coombes, J.S., Conacher, M., Austen, S.K. & Marshall, P.A. (2002) Dose effects of oral bovine colostrum on physical work capacity in cyclists. *Medicine and Science in Sports and Exercise*, **34**(7), 1184–1188.
Cox, A.J., Pyne, D.B., Saunders, P.U. & Fricker, P.A. (2010) Oral administration of the probiotic Lactobacillus fermentum VRI-003 and mucosal immunity in endurance athletes. *British Journal of Sports Medicine*, **44**(4), 222–226.
Crooks, C.V., Wall, C.R., Cross, M.L. & Rutherfurd-Markwick, K.J. (2006) The effect of bovine colostrum supplementation on salivary IgA in distance runners. *International Journal of Sports Nutrition and Exercise Metabolism*, **16**(1), 47–64.
Davison, G. & Diment, B.C. (2010) Bovine colostrum supplementation attenuates the decrease of salivary lysozyme and enhances the recovery of neutrophil function after prolonged exercise. *British Journal of Nutrition*, **103**(10), 1425–1432.
Fogelholm, M. (2003) Dairy products, meat and sports performance. *Sports Medicine*, **33**(8), 615–631.
Food Standards Agency (2002) *McCance and Widdowson's Composition of Foods*, 6th edition. Cambridge: Royal Society of Chemistry.
Gaitanos, G.C., Williams, C., Boobis, L.H. & Brooks, S. (1993) Human muscle metabolism during intermittent maximal exercise. *Journal of Applied Physiology*, **75**(2), 712–719.
Gleeson, M. (1998) Temperature regulation during exercise. *International Journal of Sports Medicine*, **19**(Suppl 2), S96–99.
Gleeson, M. (2007) Immune function in sport and exercise. *Journal of Applied Physiology*, **103**(2), 693–699.
Gleeson, M., Nieman, D.C. & Pedersen, B.K. (2004) Exercise, nutrition and immune function. *Journal of Sports Science*, **22**(1), 115–125.
Hartman, J.W., Tang, J.E., Wilkinson, S.B., Tarnopolsky, M.A., Lawrence, R.L., Fullerton, A.V. & Phillips, S.M. (2007) Consumption of fat-free fluid milk after resistance exercise promotes greater lean mass accretion than does consumption of soy or carbohydrate in young, novice, male weightlifters. *American Journal of Clinical Nutrition*, **86**(2), 373–381.
Hawley, J.A., Gibala, M.J. & Bermon, S. (2007) Innovations in athletic preparation: role of substrate availability to modify training adaptation and performance. *Journal of Sports Science*, **25**(Suppl 1), S115–124.
He, F., Tuomola, E., Arvilommi, H. & Salminen, S. (2001) Modulation of human humoral immune response through orally administered bovine colostrum. *FEMS Immunology and Medical Microbiology*, **31**, 93–96.
Ivy, J.L., Katz, A.L., Cutler, C.L., Sherman, W.M. & Coyle, E.F. (1988) Muscle glycogen synthesis after exercise: effect of time of carbohydrate ingestion. *Journal of Applied Physiology*, **64**(4), 1480–1485.
Jeukendrup, A.E. (2008) Carbohydrate feeding during exercise. *European Journal of Sport Science*, **8**(2), 77–86.
Jeukendrup, A.E., Vet-Joop, K., Sturk, A., Stegen, J.H., Senden, J., Saris, W.H. & Wagenmakers, A.J. (2000) Relationship between gastro-intestinal complaints and endotoxaemia, cytokine release and the acute-phase reaction during and after a long-distance triathlon in highly trained men. *Clinical Science (London)*, **98**(1), 47–55.

Karp, J.R., Johnston, J.D., Tecklenburg, S., Mickleborough, T.D., Fly, A.D. & Stager, J.M. (2006) Chocolate milk as a post-exercise recovery aid. *International Journal of Sport Nutrition and Exercise Metabolism*, **16**(1), 78–91.

Kekkonen, R.A., Vasankari, T.J., Vuorimaa, T., Haahtela, T., Julkunen, I. & Korpela, R. (2007) The effect of probiotics on respiratory infections and gastrointestinal symptoms during training in marathon runners. *International Journal of Sports Nutrition and Exercise Metabolism*, **17**(4), 352–363.

Kerksick, C., Kreider, R., Rasmussen, C. *et al.* (2001) Effects of bovine colostrum supplementation on training adaptations II: performance. *FASEB Journal*, **15**, LB315 (abstract).

Lee, J.K., Maughan, R.J., Shirreffs, S.M. & Watson, P. (2008) Effects of milk ingestion on prolonged exercise capacity in young, healthy men. *Nutrition*, **24**(4), 340–347.

Lemon, P.W. (1995) Do athletes need more dietary protein and amino acids? *International Journal of Sport Nutrition*, **5**, S39–61.

Lemon, P.W. (2000) Beyond the zone: protein needs of active individuals. *Journal of American College of Nutrition*, **19**(5 Suppl), 513S–521S.

Leon, L.R. & Helwig, B.G. (2010) Heat stroke: role of the systemic inflammatory response. *Journal of Applied Physiology*, **109**, 1980–1988.

Li, T.L., Wu, C.L., Gleeson, M. & Williams, C. (2004) The effects of pre-exercise high carbohydrate meals with different glycemic indices on blood leukocyte redistribution, IL-6, and hormonal responses during a subsequent prolonged exercise. *International Journal of Sports Nutrition and Exercise Metabolism*, **14**(6), 647–656.

Manore, M.M. (2000) Effect of physical activity on thiamine, riboflavin, and vitamin B-6 requirements. *American Journal of Clinical Nutrition*, **72**(2 Suppl), 598S–606S.

Manore, M.M. (2002) Dietary recommendations and athletic menstrual dysfunction. *Sports Medicine*, **32**(14), 887–901.

Marchbank, T., Davison, G., Oakes, J.R., Ghatei, M.A., Patterson, M., Moyer, M.P. & Playford, R.J. (2011) The nutriceutical, bovine colostrum, truncates the increase in gut permeability caused by heavy exercise in athletes. *American Journal of Physiology: Gastrointestinal and Liver Physiology*, **300**(3), G477–487.

Maughan, R. (2002) The athlete's diet: nutritional goals and dietary strategies. *Proceedings of the Nutrition Society*, **61**(1), 87–96.

Maughan, R.J. & Leiper, J.B. (1995) Sodium intake and post-exercise rehydration in man. *European Journal of Applied Physiology and Occupational Physiology*, **71**(4), 311–319.

Mero, A., Kähkönen, J., Nykänen, T., Parviainen, T., Jokinen, I., Takala, T., Nikula, T., Rasi, S. & Leppäluoto, J. (2002) IGF-I, IgA, and IgG responses to bovine colostrum supplementation during training. *Journal of Applied Physiology*, **93**(2), 732–739.

Merson, S.J., Maughan, R.J. & Shirreffs, S.M. (2008) Rehydration with drinks differing in sodium concentration and recovery from moderate exercise-induced hypohydration in man. *European Journal of Applied Physiology*, **103**(5), 585–594.

Murray, R. (1987) The effects of consuming carbohydrate-electrolyte beverages on gastric emptying and fluid absorption during and following exercise. *Sports Medicine*, **14**(5), 322–351.

Nicholas, C.W., Tsintzas, K., Boobis, L. & Williams, C. (1999) Carbohydrate-electrolyte ingestion during intermittent high-intensity running. *Medicine and Science in Sports and Exercise*, **31**(9), 1280–1286.

Nichols, A.W. (2007) Probiotics and athletic performance: a systematic review. *Current Sports Medicine Reports*, **6**(4), 269–273.

Nieman, D.C. (1994) Exercise, upper respiratory tract infection, and the immune system. *Medicine and Science in Sports and Exercise*, **26**(2), 128–139.

Pedersen, B.K., Kappel, M., Klokker, M., Nielsen, H.B. & Secher, N.H. (1994) The immune system during exposure to extreme physiologic conditions. *International Journal of Sports Medicine*, **15**(Suppl 3), S116–121.

Phillips, S.M., Tipton, K.D., Aarsland, A., Wolf, S.E. & Wolfe, R.R. (1997) Mixed muscle protein synthesis and breakdown after resistance exercise in humans. *American Journal of Physiology*, **273**(1 Pt 1), E99–107.

Rankin, J.W., Goldman, L.P., Puglisi, M.J., Nickols-Richardson, S.M., Earthman, C.P. & Gwazdauskas, F.C. (2004) Effect of post-exercise supplement consumption on adaptations to resistance training. *Journal of American College of Nutrition*, **23**(4), 322–330.

Saltin, B. & Karlsson, J. (1973) Die ernahrung des sportlers. In: *Zentrale Themen de Sportmedizin* (ed. W Hollman), pp. 245–260. Berlin: Springer.

Shing, C.M., Jenkins, D.G., Stevenson, L. & Coombes, J.S. (2006) The influence of bovine colostrum supplementation on exercise performance in highly trained cyclists. *British Journal of Sports Medicine*, **40**(9), 797–801.

Shing, C.M., Peake, J., Suzuki, K., Okutsu, M., Pereira, R., Stevenson, L., Jenkins, D.G. & Coombes, J.S. (2007) Effects of bovine colostrum supplementation on immune variables in highly trained cyclists. *Journal of Applied Physiology*, **102**(3), 1113–1122.

Shing, C.M., Hunter, D.C. & Stevenson, L.M. (2009) Bovine colostrum supplementation and exercise performance: potential mechanisms. *Sports Medicine*, **39**(12), 1033–1054.

Shirreffs, S. (2007) Maintenance of fluid balance in sport and exercise. In *Nutrition and Sport* (ed. D. MacLaren), pp. 137–151. Edinburgh, London, New York: Churchill Livingstone Elsevier.

Shirreffs, S.M., Taylor, A.J., Leiper, J.B. & Maughan, R.J. (1996) Post-exercise rehydration in man: effects of volume consumed and drink sodium content. *Medicine and Science in Sports and Exercise*, **28**(10), 1260–1271.

Shirreffs, S.M., Watson, P. & Maughan, R.J. (2007) Milk as an effective post-exercise rehydration drink. *British Journal of Nutrition*, **98**(1), 173–180.

Stevenson, E., Williams, C. & Biscoe, H. (2005a) The metabolic responses to high carbohydrate meals with different glycemic indices consumed during recovery from prolonged strenuous exercise. *International Journal of Sports Nutrition and Exercise Metabolism*, **15**(3), 291–307.

Stevenson, E., Williams, C., McComb, G. & Oram, C. (2005b) Improved recovery from prolonged exercise following the consumption of low glycemic index carbohydrate meals. *International Journal of Sports Nutrition and Exercise Metabolism*, **15**(4), 333–349.

Stevenson, E., Williams, C. & Nute, M. (2005c) The influence of the glycaemic index of breakfast and lunch on substrate utilisation during the postprandial periods and subsequent exercise. *British Journal of Nutrition*, **93**(6), 885–893.

Stevenson, E.J., Williams, C., Mash, L.E., Phillips, B. & Nute, M.L. (2006) Influence of high-carbohydrate mixed meals with different glycemic indexes on substrate utilization during subsequent exercise in women. *American Journal of Clinical Nutrition*, **84**(2), 354–360.

Sugisawa, H., Itou, T., Ichimura, Y. & Sakai, T. (2002) Bovine milk enhances the oxidative burst activity of polymorphonuclear leukocytes in low concentrations. *Journal of Veterinary Medical Science*, **64**(12), 1113–1116.

Sugisawa, H., Itou, T., Saito, M., Moritomo, T., Miura, Y. & Sakai, T. (2003) A low-molecular-weight fraction of bovine colostrum and milk enhances the oxidative burst activity of polymorphonuclear leukocytes. *Veterinary Research Communications*, **27**(6), 453–461.

Tarnopolsky, M.A., Gibala, M., Jeukendrup, A.E., Phillips, S.M. (2005) Nutritional needs of elite endurance athletes. *Part I: Carbohydrate and fluid requirements. European Journal of Sports Science*, **5**(1), 3–14.

Watson, P., Love, T.D., Maughan, R.J. & Shirreffs, S.M. (2008) A comparison of the effects of milk and a carbohydrate-electrolyte drink on the restoration of fluid balance and exercise capacity in a hot, humid environment. *European Journal of Applied Physiology*, **104**(4), 633–642.

Welsh, R.S., Davis, J.M., Burke, J.R. & Williams, H.G. (2002) Carbohydrates and physical/mental performance during intermittent exercise to fatigue. *Medicine and Science in Sports and Exercise*, **34**(4), 723–731.

West, N.P., Pyne, D.B., Peake, J.M. & Cripps, A.W. (2009) Probiotics, immunity and exercise: a review. *Exercise Immunology Review*, **15**, 107–126.

Wilkinson, S.B., Tarnopolsky, M.A., Macdonald, M.J., Macdonald, J.R., Armstrong, D. & Phillips, S.M. (2007) Consumption of fluid skim milk promotes greater muscle protein accretion after resistance exercise than does consumption of an isonitrogenous and isoenergetic soy-protein beverage. *American Journal of Clinical Nutrition*, **85**(4), 1031–1040.

Woolf, K. & Manore, M.M. (2007) Micronutrients important for exercise. In *Nutrition and Sport* (ed. D. MacLaren), pp. 119–136. Edinburgh, London, New York: Churchill Livingstone Elsevier.

Zemel, M.B. (2004) Role of calcium and dairy products in energy partitioning and weight management. *American Journal of Clinical Nutrition*, **79**(5), 907S–912S.

6 Dairy Fat: Perceptions and Realities

A.L. Lock[1], D.I. Givens[2] and D.E. Bauman[3]

[1]Department of Animal Science, Michigan State University, East Lansing, United States of America
[2]Animal Science Research Group, School of Agriculture, Policy and Development, University of Reading, Reading, United Kingdom
[3]Department of Animal Science, Cornell University, Ithaca, United States of America

6.1 Introduction

The importance of animal-derived foods in meeting the food security needs of the global population is well recognised (Demment & Allen 2003; Randolph *et al.* 2007). Dairy products are an important source for many key nutrients including high-quality protein, energy and many essential minerals and vitamins, although the nutritional importance of dairy fats is often less well understood. The dairy sector has made continuous advancement over the years and today there is a wide variety of milks and dairy products readily available to the consumer. In response to considerable scientific research on the nutritional value of milk, dietary guidelines around the world have recommended daily consumption of dairy products for the overall health of the population; for example, *Dietary Guidelines for Americans* has emphasised that dairy products provide critical amounts of three of the five 'nutrients of concern' for children and four of the seven nutrients of concern for adults (USDA/USDHHS 2010). With the projected growth in world population and the increased demand for animal-derived food products as living standards improve, dairy products will undoubtedly continue to be an important dietary source of nutrients.

Fat is the most variable component of milk and accounts for many of the physical properties, manufacturing characteristics and organoleptic qualities of dairy products. Milk fat content and fatty acid (FA) composition can be significantly altered through nutrition of the dairy cow, and this has been extensively reviewed elsewhere (e.g. Chilliard *et al.* 2000; Lock & Shingfield 2004). This research has often involved studies designed to achieve shifts in the ratio of saturated FA (SFA) to polyunsaturated FA (PUFA). While modest changes have been achieved, this can often negatively affect cow performance and lead to challenges related to the quality and stability of dairy products.

For over half a century the concept of eating healthy has become synonymous with avoiding dietary fat and cholesterol, especially saturated fat; on a population basis, a diet low in saturated fat remains at the heart of nutritional advice in many countries for lowering plasma cholesterol and reducing cardiovascular disease (CVD) risk. In the case of dairy products, there has been a general perception that a food containing saturated fat is unlikely

Milk and Dairy Products as Functional Foods, First Edition. Edited by A. Kanekanian.

Table 6.1 Partial list of bioactive components in milk that have human health implications. Adapted from Bauman *et al.* (2006b). Reproduced with permission of Elsevier.

Milk fat components	Milk protein components	Other
Conjugated linoleic acids	Whey proteins	Calcium
Omega-3 fatty acids	Casein	Lactose
Oleic acid	Lactoferrin	Vitamins A, D and K
Vaccenic acid	α-lactalbumin	Oligosaccharides
Sphingolipids	Peptides	Nucleosides
Butyric acid	Milk fat-globule membrane proteins	Probiotics
13-methyltetradecanoic acid		Potassium
Ether lipids		Phosphorus

to be beneficial to health. Research however continues to unravel the complexities associated with individual FA and fats from different sources and it is becoming increasingly apparent that not all FA, or SFA, have the same biological effects. As will be highlighted in this review, from this research it is clear that broad generalisations about fats can be misleading and often inaccurate; rather, biological effects and nutritional value must be considered on the basis of individual FA and within a whole diet context (Lock *et al.* 2008; Parodi 2009).

Nutritional quality is increasingly an important consideration in food choices by consumers because of their growing awareness of the link between diet and health. This has led to an increased recognition that foods can contain specific 'bioactive' components that can affect health, and scientists are being asked to clarify the role of specific foods and food components in health maintenance and disease prevention. This has led to the phrase 'functional foods', often being used as a collective term for any food or food component that provides additional health benefits beyond that associated with traditional nutrients. Consequently, the bioactive properties of a number of components in milk have been examined with regard to a range of health-related variables. Of special interest are the components associated with the prevention of chronic human diseases; results have demonstrated that milk contains specific proteins, peptides and FA that are bioactive, while the production of fermented milk products has also been shown to have the potential to elicit beneficial effects on health-related variables. A partial list of bioactive components in dairy products is highlighted in Table 6.1.

In the following sections we will first provide background on rumen lipid metabolism and milk fat synthesis and then address the state of knowledge regarding milk FA and human health implications.

6.2 The dairy cow dimension

6.2.1 *Rumen lipid metabolism*

Dietary lipids are extensively metabolised in the rumen and this has a major impact on the profile of FA available for absorption and subsequent use by body tissues. FA in dietary lipids are mainly esterified and two major transformations occur when they enter the rumen: the first is hydrolysis of the ester linkages to yield free FA and the second is the biohydrogenation of unsaturated FA (Palmquist *et al.* 2005). Biohydrogenation involves only a few of the rumen bacteria species, and they carry out these reactions as a protective mechanism against the toxic effects of PUFA. Linoleic acid (18:2 n-6) and α-linolenic acid (18:3 n-3) are the main PUFA in ruminant diets and their biohydrogenation results in stearic acid (18:0) being the major FA in outflow from the rumen and absorption from the small intestine (Jenkins *et al.* 2008). During the biohydrogenation process, transitory FA intermediates are also produced and some escape via rumen outflow and are subsequently absorbed (Palmquist *et al.* 2005). Two such FA that were initially identified are *trans*-11 18:1 (vaccenic acid; VA) and *cis*-9, *trans*-11 conjugated linoleic acid (rumenic acid; RA). Improvements in analytical techniques have established that rumen biohydrogenation processes are considerably more complex than first thought, and a remarkable range of *trans* 18:1 and conjugated linoleic acid (CLA) isomers have been identified in rumen outflow (Palmquist *et al.* 2005) and subsequently incorporated into milk fat (Table 6.2). Thus, in addition to the pathway involving the formation of VA and RA as intermediates, there must be many minor pathways of biohydrogenation. Furthermore, modifications in diet and rumen environment impact the relative importance of these biohydrogenation pathways, resulting in substantial changes in FA intermediates produced in the rumen and ultimately incorporated into milk fat (Palmquist *et al.* 2005).

Table 6.2 Range of positional and geometric isomers (percentager of total isomers) of *trans*-18:1 and conjugated 18:2 fatty acids in milk and dairy products. Adapted from Lock & Bauman (2004). Reproduced with permission of Springer Science + Business Media.

Trans 18:1		Conjugated 18:2	
Isomer	Total *trans* 18:1 isomers (%)	Isomer	Total CLA isomers (%)
trans-4	0.30–2.30	*trans*-7, *cis*-9	1.20–8.89
trans-5	<0.01–1.40	*trans*-7, *trans*-9	0.02–2.39
trans-6-8	0.50–11.30	*trans*-8, *cis*-10	0.06–1.47
trans-9	3.00–18.20	*trans*-8, *trans*-10	0.19–0.37
trans-10	3.40–29.80	*cis*-9, *trans*-11	72.56–91.16
trans-11	24.50–74.90	*trans*-9, *trans*-11	0.77–2.87
trans-12	1.90–17.60	*trans*-10, *cis*-12	0.03–1.51
trans-13 + 14	<0.01–23.10	*trans*-10, *trans*-12	0.28–1.31
trans-15	3.30–11.10	*cis*-11, *trans*-13	0.18–4.70
trans-16	1.70–12.50	*trans*-11, *cis*-13	0.07–8.00
		trans-11, *trans*-13	0.28–4.24
		cis-12, *trans*-14	0.04–0.80
		trans-12, *trans*-14	0.33–2.76
		cis-*cis* isomers	0.06–4.80

6.2.2 *Milk fat synthesis*

Cows' milk contains approximately 3–5% fat depending on breed, nutrition and stage of lactation. This fat is present as complex globules with structural properties distinct from other biological sources of fats. Milk fat predominantly contains triglycerides (95–98%), which are composed of FA of different carbon chain length (4–24 atoms), degree of saturation and positional specificity on the glycerol backbone; these triglycerides form the core of the milk fat droplet, with the remainder of the lipids composed of small amounts of phospholipids, cholesterol and cholesterol esters found in the globule membrane surrounding the milk fat droplet (Lock & Bauman 2004). It is one of the most complex naturally occurring fats with more then 400 different FA (Jensen 2002), most originating as intermediates formed during lipid metabolism by rumen bacteria (e.g. Table 6.2). These FA are mainly present in trace amounts with only about a dozen FA making up the majority of milk fat, however. The typical FA composition of bovine milk fat is depicted in Figure 6.1; the specific FA profile will, however, depend on diet, rumen fermentation and physiological variables (Chilliard *et al.* 2000; Lock & Shingfield 2004). A large diversity of dairy foods of widely differing composition is manufactured from this unique raw material.

Milk FA originate from two sources: *de novo* synthesis and the uptake of preformed FA (Bauman & Griinari, 2003). Substrates for *de novo* synthesis are acetate and β-hydroxybutyrate produced during rumen fermentation. They are used by mammary epithelial cells for the synthesis of short- and medium-chain FA (4:0 to 14:0) plus a portion of the 16 carbon FA in milk fat. The remainder of the 16 carbon and all of the longer-chain FA are derived from mammary uptake of circulating long-chain FA that arise from digestive tract absorption and the mobilisation of body fat reserves. In the dairy cow *de novo* synthesis

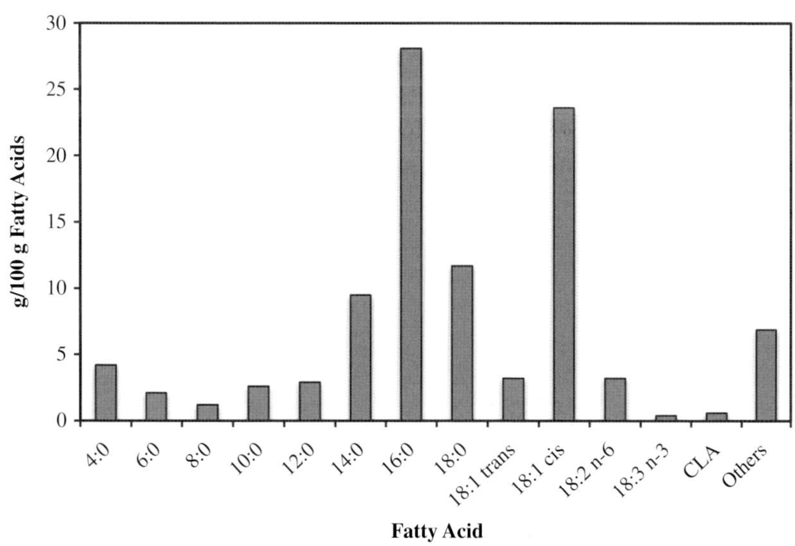

Fig. 6.1 Regional and seasonal average fatty acid composition of retail fluid milk in the United States. Data from O'Donnell-Megaro *et al.* (2011).

and the uptake of preformed FA contribute approximately equally to milk FA (molar basis), although this can vary according to physiological state. For example, the contribution of FA from body fat reserves can vary from about 5% of milk FA in a well-fed animal to over 20% in early lactation when cows are in negative energy balance (Bauman & Griinari 2003).

6.3 Saturated fatty acids

SFA typically comprise 60–70% of milk FA (Fig. 6.1) and thus milk fat is often considered a saturated fat. For over half a century the concept of healthy eating has become synonymous with avoiding dietary fat, especially saturated fat, and on a population basis a low saturated fat diet remains at the heart of nutritional advice for lowering plasma cholesterol and reducing the risk of CVD. The 2009 American Heart Association Paediatrics and Adult Nutrition Guidelines re-affirmed a target of reducing saturated fat intake to <7% of total energy intake (Gidding *et al.* 2009). One method of achieving this is by selecting fat-free (skim) and low-fat milk and dairy products, and this has been a central public health recommendation in many countries. Recent estimates indicate that *c.* 30% of the US dietary intake of saturated fat comes from dairy products, with cheese being the major source (Fig. 6.2).

Initial research indicated that SFA intake was the major determinant of circulating cholesterol. However, the relationship of fats, cholesterol and health is far more complex than initially thought and today many dietary, genetic and secondary factors causing hypercholesterolemia have been identified (Grundy & Vega 1990; Parodi 2009). In reviewing the history and politics behind the diet–heart hypothesis Taubes (2001) concluded that, after 50 years of research, there was little evidence that a diet low in saturated fat prolongs life. This conclusion is reinforced by recent results from several large-scale investigations. For example, the Women's Health Initiative, an 8-year randomised dietary modification trial involving *c.* 50,000 women, represents the largest dietary intervention study ever undertaken; results demonstrated no differences in risk of stroke or CVD for the group in which the dietary intervention reduced total fat intake and increased intakes of vegetables, fruits and grains

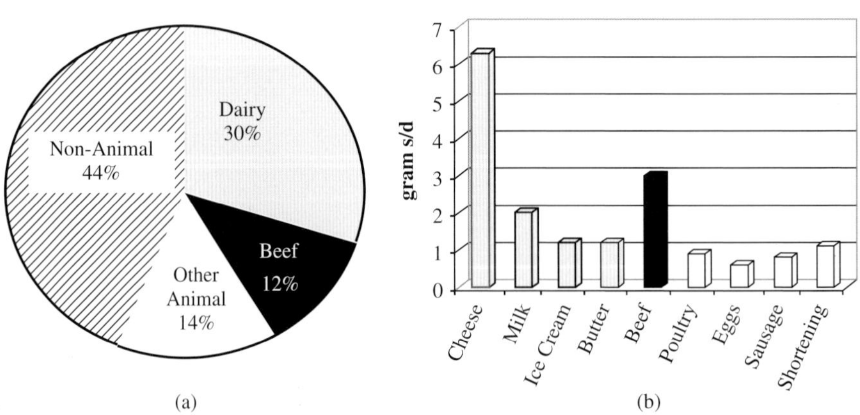

Fig. 6.2 Saturated fatty acid intake in the US population presented as (a) source and (b) quantity from different animal-derived foods. Data from Ervin *et al.* (2004).

(Howard *et al.* 2006). Likewise, a recent meta-analysis of 21 prospective epidemiologic studies reported that 'there is no significant evidence for concluding that dietary saturated fat is associated with an increased risk of CVD' (Siri-Tarino *et al.* 2010). Clearly, the relationship between fats including saturated fats, cholesterol and CVD is more complex than initially thought and the risk of CVD is multifaceted (Parodi 2009).

It is important to appreciate that individual SFA in milk vary in their structure and many have no effect on plasma cholesterol. This was highlighted by The Nutrition Committee of the American Heart Association who emphasised the diversity of the biological effects of individual FA and the need to evaluate specific FA with respect to a range of variables related to the risk of CVD (Kris-Etherton *et al.* 2001). Of the SFA in milk, most have no effect on circulating cholesterol; only lauric (12:0), myristic (14:0) and palmitic (16:0) acids have been shown to increase blood levels of total cholesterol (TC) and low-density lipoprotein (LDL) -cholesterol when added as dietary supplements, but these represent only one-quarter to one-third of total milk FA (see Fig. 6.1). Further advances in this area have established that lauric and myristic acids also increase circulating levels of high-density lipoprotein (HDL) -cholesterol and this change is associated with a reduction in the risk of CVD (Mensink *et al.* 2003). The pattern of changes of circulating cholesterol in different lipid fractions is therefore an important consideration, and several investigations suggest that comparisons of the ratio of TC to HDL-cholesterol is among the best indicator of atherogenic risk (German *et al.* 2009).

For example, a meta-analysis by Mensink *et al.* (2003) provides convincing evidence of differences in the effects of individual FA on risk factors for CVD, but also challenges conventional wisdom about the atherogenic risk from specific SFA. Using data from 60 clinical studies, they evaluated CVD risk on the basis of the plasma ratio of TC to HDL-cholesterol. It has been shown that ratios for lauric, myristic and palmitic acids provide little or no evidence for an atherogenic effect when compared to an isoenergetic substitution with carbohydrate; in fact the ratio for lauric acid was significantly decreased. Furthermore, when compared by FA group, the meta-analysis also revealed no effect of SFA when compared to carbohydrate substitution on an isoenergetic basis (Mensink *et al.* 2003). However, changes in this ratio were indicative of the well-established anti-atherogenic effect of *cis* monounsaturated FA (MUFA) and PUFA and the increased atherogenic risk of *trans* fatty acids (TFA). Although epidemiological associations provide little support for SFA being causative for CVD, the meta-analysis of Mensink *et al.* (2003) does suggest that replacement of SFA with MUFA and PUFA fats will reduce CVD risk based on changes in plasma cholesterol pools. The case for replacing SFA with PUFA fats was supported by a recent systematic review and meta-analysis of randomised control trials with CVD events (myocardial infarction, heart disease death and/or sudden death) as outcomes (Mozaffarian *et al.* 2010). Interestingly, they found no suitable randomised control trials where SFA were replaced with MUFA fats. This is an issue that clearly requires further attention.

Despite the large number of studies on the effects of changing dietary lipid type it is worth noting that only a very limited number of intervention studies have examined the benefits of reduced-SFA dairy products on CVD risk factors in humans (Givens & Minihane 2009). Those that have been done were of fairly short duration and relied almost entirely on effects on plasma cholesterol (e.g. Noakes *et al.* 1996; Tholstrup *et al.* 1998, 2006; Poppitt *et al.* 2002; Seidel *et al.* 2005;). In most cases, replacing some SFA in milk fat led to improvements in cholesterol profile. For example, Noakes *et al.* (1996) reported a 4.3% ($P < 0.001$)

and 5.3% ($P < 0.001$) reduction in TC and LDL-cholesterol, respectively, in healthy adults following the inclusion of modified dairy products (51% of SFA, 39% of *cis*-MUFA) versus normal dairy (70% of SFA, 28% of *cis*-MUFA) into a low-fat diet for three weeks. While these changes would be expected to reduce CVD risk somewhat (Givens 2008), given the considerable interest in SFA-reduced dairy products there is a great need for longer-term comparative studies with more functional outcome measures than have so far been used.

National and international bodies continue to point to SFA as a major cause of CVD, and have recommend restricting dietary intake of SFA to less than 10% of calories (e.g. EFSA 2010; USDA/USDHHS 2010). However, peer-reviewed publications have recently challenged these recommendations, concluding that they were based on an inaccurate representation and summarisation of the scientific literature (Hite *et al.* 2010; Hoenselaar 2012). Indeed, recent large-scale studies and literature reviews indicate that the majority of studies have failed to support an adverse association between the intake of dairy products and CVD regardless of milk fat levels and, in many cases, a long-term beneficial effect was observed (e.g. Elwood *et al.* 2010; Soedamah-Muthu *et al.* 2011; de Oliveira Otto *et al.* 2012; Kratz *et al.* 2013). Overall, a growing body of scientific knowledge supports a dramatic paradigm shift in conclusions regarding the relationship between saturated fat, especially milk fat, and cardiovascular health. This is discussed in further detail in Section 6.6.

6.4 *Trans* fatty acids

Milk fat contains *c.* 2–4% TFA, mainly *trans*-18:1 with minor amounts of *trans*-PUFA such as CLA (Table 6.2). The double bond in unsaturated FA present in foods is typically of a *cis* configuration, but TFA have been under considerable scrutiny in recent years due to their association with increased risk of CVD and other chronic diseases (Institute of Medicine Food and Nutrition Board 2002). *Trans* double bonds are introduced into FA by one of two means: (1) as chemical processes during the formation of partially hydrogenated vegetable oils (PHVO; industrial sources) or (2) as FA intermediates formed during rumen biohydrogenation (ruminant sources). PHVO have been used extensively in many prepared foods such as bakery products, cooking fats, margarine and fried products.

A number of countries have established policies aimed at reducing TFA intake in the human diet. These efforts have primarily involved nutritional labelling of the content of TFA in food products and/or legislation limiting the use of PHVO in industrially prepared foods. The United States, as well as other public health agencies in North America and Europe, has enacted labelling requirements for the TFA content of food products, regardless of their TFA source (Lock *et al.* 2005b). The specific labelling requirements (i.e. labelling thresholds) as well as definitions for what constitutes a 'TFA' differ substantially between countries. The Danish Government chose a different approach to reduce dietary intake of TFA; they decided to use a legislative approach to limit the amount of TFA in food products with any oils and fats used as ingredients in foods currently not allowed in Denmark if they contain >2% TFA. In effect, this legislation banned the use of PHVO in industrially produced foods. It is important to note that this law only applies to the use of industrially produced oils. Based on their extensive review of the scientific literature, Denmark has exempted ruminant-derived foods from this ruling (Lock *et al.* 2005b).

Estimates for worldwide consumption of TFA have recently been reviewed (Craig-Schmidt & Rong 2009). Over the last decade, available data from the US indicates that mean or median adult intake of *trans* fat is *c*. 2–3 g day^{-1} or 1–2% of daily calories as estimated from semi-quantitative food frequency questionnaires and 4–8 g day^{-1} or 2–3% of daily calories based on diet recalls and diet records; these estimates are lower than those published in the mid 1990s (Craig-Schmidt & Rong 2009). In response to the various labelling requirements for TFA, the food industry has been transitioning to alternative practices that allow for a marked reduction in the use of PHVO in processed food products. As a consequence the TFA intake from industrial sources is declining, whereas the intake of TFA from ruminant sources has remained more or less constant; the net result is a reduction in the dietary intake of TFA but the proportion of total TFA intake from ruminant sources is gradually increasing. As Craig-Schmidt & Rong (2009) recently highlighted, within a few years the TFA in the food supply will be mostly limited to the 'natural' supply present in ruminant-derived fats in meat and dairy products. Thus, understanding the biological effects of TFA found in dairy products and differences between TFA isomers are of great importance.

PHVO generally contain about 40–60% TFA and the isomer profile is a Gaussian distribution that centres on *trans*-9, *trans*-10, *trans*-11 and *trans*-12 18:1 (Fig. 6.3). In contrast, the

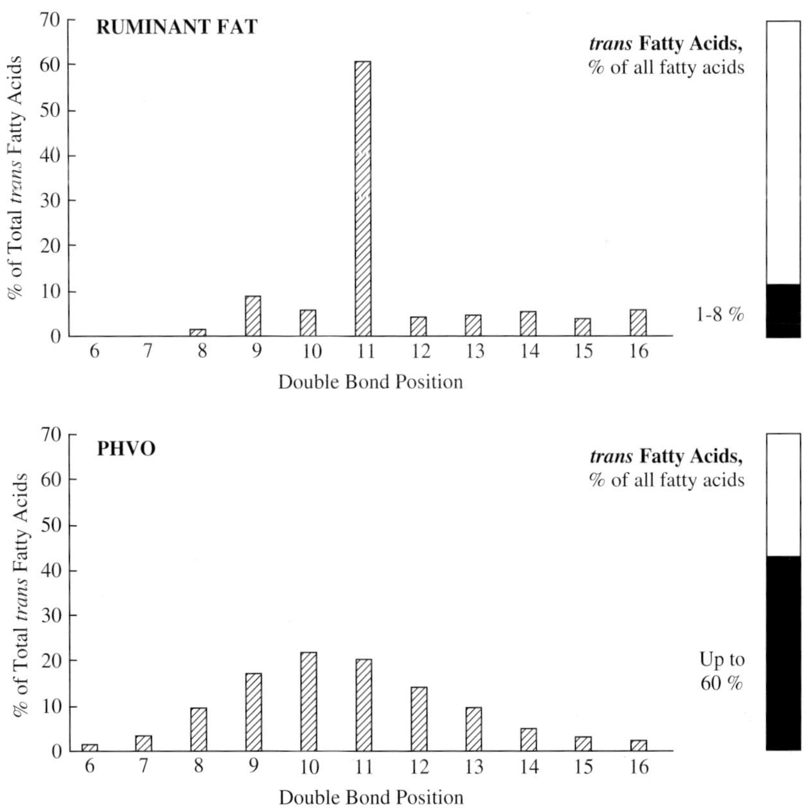

Fig. 6.3 Typical distribution of *trans* 18:1 fatty acids in ruminant fat and partially hydrogenated vegetable oils (PHVO). Adapted from Lock *et al*. (2005b). Reproduced with permission of Dairy Industry Association of Australia.

major TFA isomer in ruminant fat (milk and meat) is VA. Differences in isomer profile are of importance because the position of the *trans*-double bond can influence both physiological properties and the rate of biochemical reactions (Parodi 2006; Lock *et al.* 2005b; Gebauer *et al.* 2011).

Over the past 50 years the relationship between dietary TFA intake and plasma lipid levels and human health, particularly CVD, has been extensively investigated. Data from controlled human intervention studies have consistently demonstrated that diets containing TFA result in increased serum TC and LDL-cholesterol and decreased HDL-cholesterol (Mensink *et al.* 2003). Prospective epidemiological studies have consistently supported findings from intervention studies further indicating that higher intakes of TFA are associated with increased risk of CVD. These results have been broadly extrapolated to imply that high consumption of all sources of TFA is associated with an increased risk of CVD. Further examination of the epidemiological investigations reveals that the positive association with risk of CVD can be explained entirely by the intake of industrial TFA, however (Fig. 6.4). In contrast, the absolute intake of TFA from ruminant sources is but a fraction of that from industrial sources; furthermore, the relationship between intake of ruminant-derived TFA and risk of CVD observed for these studies is a significant negative association, an inverse non-significant association or no association (Fig. 6.4).

The difference between TFA sources and the risk of CVD probably relates to differences in the TFA isomer profile. For example, a recent study in hamsters revealed that PHVO derived from different plant oils (high-linoleic acid safflower oil versus high-oleic sunflower oil) resulted in divergent plasma lipoprotein and subclass profiles despite both PHVO containing a similar concentration of total TFA (Kraft *et al.* 2010). Furthermore (again using the hamster), it was shown that while PHVO consumption resulted in changes in plasma lipoprotein cholesterol concentrations indicative of increased risk of atherosclerosis, consumption of a pure source of *trans*-9 18:1, typically the most abundant *trans*-18:1 isomer in PHVO, did not elicit the same negative effects. VA did not cause the same negative effect observed with the PHVO treatment either (Tyburczy *et al.* 2009). In addition, an important related aspect is that the VA in milk fat can be converted to RA via Δ9-desaturase (also known as stearoyl-CoA desaturase), discussed in Section 6.5.3.

Several studies have established that humans are capable of this conversion with *c.* 20% of VA converted to RA in humans, thereby doubling RA supply (Palmquist *et al.* 2005). This enzyme system may therefore be a key factor in differentiating VA from other *trans* 18:1 fatty acid isomers. Consistent with this, biomedical studies with the hamster model have demonstrated that a TFA-enriched milk fat had positive effects on plasma lipoprotein biomarkers (Lock *et al.* 2005a).

Finally, the impact of milk fat naturally enriched in TFA has been recently examined in human clinical studies (e.g. Chardigny *et al.* 2008; Motard-Bélanger *et al.* 2008; Lacroix *et al.* 2012). It is important to note that the *trans* fats found in dairy products are consumed in low amounts and results from these clinical studies indicate that current levels of intake have little or no significant impact on CVD risk factors. Recent reviews by Jakobson & Overvad (2009), Malpuech-Brugère *et al.* (2009), Gebauer *et al.* (2011), Brouwer *et al.* (2013), and Wang & Proctor (2013) provide additional information on the effects of TFA on CVD risk factors based on epidemiological and clinical evidence.

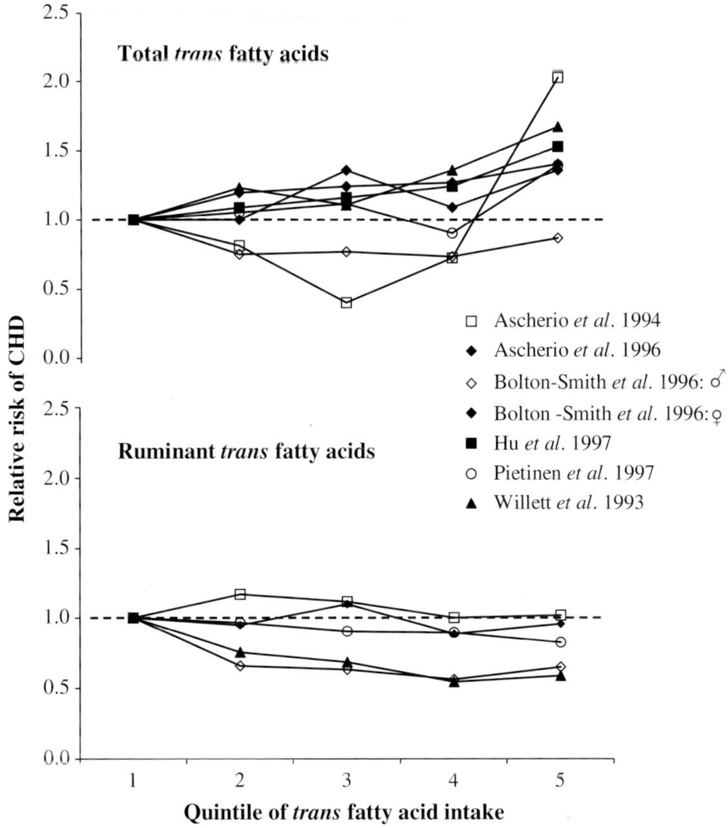

Fig. 6.4 Relative risk of coronary heart disease with increasing relative intake (quintiles) of total and ruminant-derived *trans* fatty acids. Risks are relative to the risk in the lowest quintile of *trans* fatty acid intake; the fully adjusted model is presented for each study. Adapted from Lock *et al.* (2005b). Reproduced with permission of Dairy Industry Association of Australia.

6.5 Unsaturated fatty acids

Several unsaturated FA are recognised as functional food components having beneficial effects on health maintenance and disease prevention. For milk fat these include oleic acid, very long-chain omega-3 FA and RA (Table 6.1). Enhancing their content in milk fat requires an understanding of the interrelationship between dietary supply of lipid, rumen fermentation and mammary synthesis (Lock & Bauman 2004). Dietary unsaturated FA typically undergo extensive biohydrogenation in the rumen and, as a consequence, minimal amounts of unsaturated FA are absorbed. Even though linoleic and α-linolenic acids are prevalent in the dairy cattle feedstuffs, the milk fat content of PUFA is therefore only *c.* 4% of total milk FA (Fig. 6.1). An alternative method to minimise their rumen metabolism is to feed a lipid supplement formulated to protect against rumen biohydrogenation. The most common method for rumen protection of lipid supplements is the use of calcium salts; while this method minimises adverse effects of unsaturated FA on fibre digestion, recent studies have demonstrated it offers little protection against ruminal biohydrogenation (Castaneda-Gutierrez *et al.* 2007).

As proof of concept several studies have abomasally infused plant oils as a convenient means to bypass rumen biohydrogenation; results have uniformly shown that increasing the supply of unsaturated FA results in an increase in the unsaturated FA content of milk fat. However, undesirable effects are often noted when plant oils are supplied in the diet because of adverse effects on rumen fermentation, manifest by a reduction in dry matter intake and milk production (Castaneda-Gutierrez *et al.* 2007). Thus, increasing milk fat content of unsaturated FA by increasing the supply of unsaturated FA has limitations in devising unsaturated FA formulations that will bypass rumen biohydrogenation while avoiding any adverse effects on performance.

6.5.1 *Oleic acid*

Milk fat is an excellent source of oleic acid (*cis*-9 18:1), with this MUFA representing *c.* 25% of total milk FA (Fig. 6.1). As a consequence of rumen biohydrogenation, dairy cows absorb very little oleic acid. Rather, the oleic acid in milk fat originates predominately from endogenous synthesis involving mammary uptake of stearic acid and its conversion to oleic acid by the enzyme Δ9-desaturase. In fact, about two-thirds of the stearic acid taken up by the mammary gland is converted to oleic acid, and this has major affects on the fluidity properties of milk fat. Additional milk FA synthesised endogenously by mammary Δ9-desaturase include myristoleic acid (*cis*-9 14:1, derived from 14:0), palmitoleic acid (*cis*-9 16:1, derived from 16:0) and RA (derived from VA); these desaturase products are present at much lower concentrations than oleic acid, however (Bauman *et al.* 2006a). As a proxy for Δ9-desaturase activity, researchers often calculate a desaturase index that is based on the substrate: product ratio for the Δ9-desaturase FA pairs (Kelsey *et al.* 2003). Investigations have demonstrated that the desaturase index in milk fat varies considerably among individual cows and the heritability of this index is *c.* 0.2–0.3 (Soyeyrt *et al.* 2008; Garnsworthy *et al.* 2010). The potential therefore exists to select cows for high mammary expression of Δ9-desaturase or utilise nutritional management practices that increase mammary activity of this enzyme.

6.5.2 *Omega-3 fatty acids*

The milk fat content of omega-3 FA are of special interest because these FA are essential for growth and development and are beneficial in the maintenance of human health and prevention of chronic diseases including CVD, inflammatory diseases, and neurological disorders (Yashodhara *et al.* 2009; Deckelbaum & Torrejon 2012). Milk fat content of omega-3 FA is generally very low (<0.5% of total FA) and this is mainly α-linolenic acid (Jensen 2002). α-Linolenic acid is functional for some processes, but its conversion to very long chain omega-3 FA (eicosapentaenoic acid, EPA and docosahexaenoic acid, DHA) is essential for many physiological effects including the prevention of chronic diseases. This conversion requires Δ6-desaturase (Fig. 6.5), an enzyme that is rate limiting in humans (James *et al.*

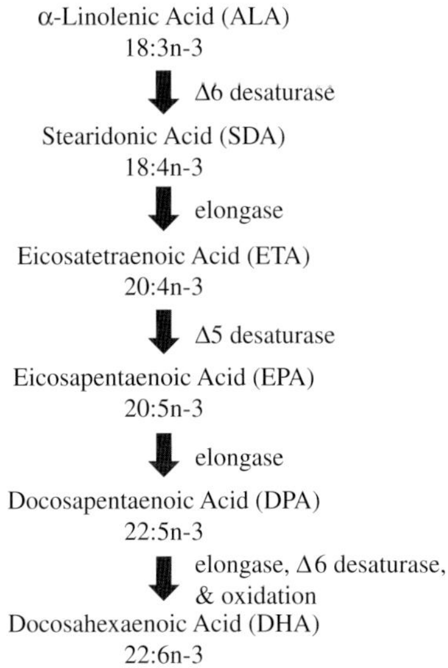

Fig. 6.5 Biosynthesis of omega-3 fatty acids. Bernal-Santos *et al.* (2010). Reproduced with permission of Elsevier.

2003; Whelan & Rust 2006). Efforts to increase the omega-3 FA content of milk fat have often involved feeding rations that have a high content of α-linolenic acid (e.g. flaxseed, linseed oil, pasture; Lock & Bauman 2004; Palmquist 2009). While this dietary approach marginally increases the milk fat content of α-linolenic acid, it has little or no effect on milk fat levels of EPA or DHA because of limited Δ6-desaturase activity in the mammary gland of the dairy cow (Hagemeister *et al.* 1991); omega-3 FA-enhanced dairy products produced by this approach therefore lack the specific omega-3 FA isomers that are of apparent greatest importance for health maintenance and disease prevention.

Feeding fish oil to dairy cattle leads to modest increases of EPA and DHA in milk and, in some markets, speciality milk and dairy products are currently available which have been enriched with EPA and DHA through adding fish oil or fish by-products to the cow's diet. Another approach is the use of plant sources of stearidonic acid (18:4 n-3; SDA) that bypasses the Δ6-desaturase reaction (Fig. 6.5). Advances in genetic modification of oil seeds allows for the production of SDA-enhanced oils, and a recent study demonstrated that dairy cows supplied with SDA-enhanced soybean oil had markedly increased omega-3 FA in their milk fat including SDA and EPA (Bernal-Santos *et al.* 2010). However, feeding large amounts of unsaturated FA from oils, plant seeds or other by-products often has undesirable effects on rumen function and the biohydrogenation process; when this occurs there is often an increased rumen outflow of *trans*-18:1 and CLA biohydrogenation intermediates, some of which have negative effects on mammary fat synthesis resulting in low milk fat tests (Bauman *et al.* 2011). The application of such a technology therefore requires that the PUFA be of a formulation that offers rumen protection.

6.5.3 *Conjugated linoleic acids*

The interest in health benefits of CLA has its genesis in the research by Pariza and associates who first demonstrated that CLA isomers are bioactive food components (see review by Pariza 1999). Because of this discovery, research on CLA isomers has increased exponentially over the last 20 years and a number of potential health benefits of CLA isomers have been reported. The presence of CLA isomers in ruminant fat is related to the biohydrogenation of PUFA. Over 20 different isomers of CLA have been identified in milk fat (Kraft *et al.* 2003; Lock & Bauman 2004). Particularly noteworthy is the fact that RA, the major CLA isomer in milk fat, is bioactive when supplied as a natural food component in the form of RA-enriched milk fat. For a broad overview of the biological effects of RA, particularly its biological effects when supplied as a natural component of the diet, see Lock *et al.* (2009).

The major source of CLA in human diets is ruminant-derived foods: mainly fat from milk and dairy products. As shown in Table 6.2, RA is the predominant CLA in milk fat (*c.* 90% of total CLA). RA was originally thought to be derived from rumen outflow of bioyhdrogenation intermediates, but studies over the last decade have unequivocally established that it originates predominantly via endogenous synthesis; this involves Δ9-desaturase with the substrate being VA, another intermediate in rumen biohydrogenation (Fig. 6.6; Bauman & Lock 2006). Strategies to increase the milk fat content of RA are therefore based on enhancing rumen output of VA and increasing tissue activity of Δ9-desaturase (Fig. 6.6). Diet has the most influence on RA content of milk fat and this has been extensively reviewed elsewhere (e.g. Chilliard *et al.* 2000; Palmquist *et al.* 2005). By choosing combinations of feed

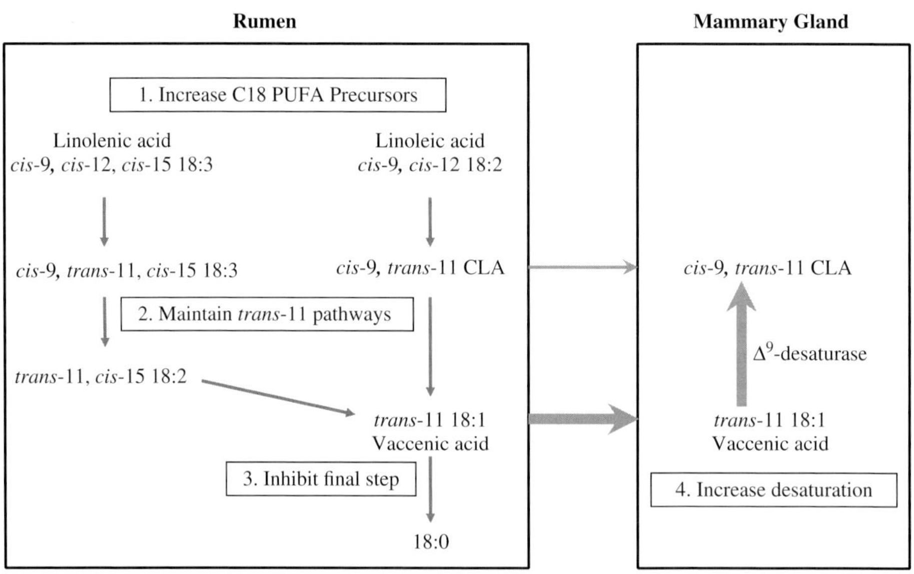

Fig. 6.6 Pathways for ruminal and endogenous synthesis of *cis*-9, *trans*-10 CLA in dairy cows and strategies to increase milk fat content of CLA. Adapted from Bauman & Lock (2006). Reproduced with permission of Springer Science + Business Media.

ingredients that affect the dietary supply of PUFA and/or affect the rumen environment to alter the rate and completeness of biohydrogenation, milk fat content of RA can be markedly affected. The RA content of milk fat is also influenced by physiological factors and individual variation among cows is *c.* three-fold (Bauman & Lock 2006). In addition, a genetic component for Δ9-desaturase activity has been identified using the desaturase index to establish heritability estimates for Δ9-desaturase (e.g. Soyeyrt *et al.* 2008; Garnsworthy *et al.* 2010).

Interest in RA as a functional food component relates to its potential anticancer and anti-atherogenic effects. Most RA research has utilised mixed isomer preparations from synthetic sources, typically synthesised from PUFA vegetable oils. More recently, a number of research studies have utilised ruminant-derived sources of RA, manufactured using modified cow nutrition programs designed to increase the milk fat content of RA and VA (Lock *et al.* 2009). It is important to note that such feeding practices result in an overall increase in the TFA content of milk fat; research from these studies is therefore pertinent to the general discussion regarding ruminant-derived TFA and human health.

Although other CLA isomers are present in milk fat, they appear to be present at concentrations much too low to have a significant biological effect as bioactive food components. RA appears to play no role in the anti-obesity and antidiabetogenic effects that have been observed with mixed CLA isomers. Rather, it is the *trans*-10, *cis*-12 CLA isomer that is responsible for these biological effects. Furthermore, recent investigations have reported that, under some situations, CLA supplements may cause undesirable effects related to hyperinsulinemia and insulin resistance (Wang & Jones 2004); these effects have also been specifically attributed to *trans*-10, *cis*-12 CLA and have not been observed with RA (Lock *et al.* 2009). A detailed discussion of specific biological effects of CLA isomers can be found elsewhere (e.g. Belury 2002; Bauman *et al.* 2006a; Churruca *et al.* 2009; Lock *et al.* 2009).

There are biomedical models for most types of cancer and many of these have been used to investigate the role of RA as an anticarcinogen. These include the use of human cancer cell lines, transplanted cell lines and *in situ* organ site carcinogenesis models. The protective effect of RA against cancer has been consistently observed in animal and cell culture models, being especially effective in models for breast cancer (Bauman *et al.* 2006a). The majority (but not all) of these studies indicate a protective effect of RA at all stages of the tumourigenic process from preventing the occurrence to modifying the spread of tumours (Lock *et al.* 2009).

Since RA is found predominately in dairy fats of human diets, a series of studies have used the rat prepubertal mammary cancer model to investigate the anticarcinogenic potential of RA when supplied as naturally enriched milk fat. The initial investigation established that RA was an effective anticarcinogen when it was supplied as a dietary food in a natural form (esterified in triglycerides; Ip *et al.* 1999). Importantly, tissue concentrations of RA were greater in rats fed the VA/RA-enriched butter than in rats fed a comparable amount of chemically synthesised RA, suggesting the possibility of endogenous synthesis from VA. Subsequent investigations established that dietary VA derived from VA/RA-enriched butter also resulted in a dose-dependent increase in the accumulation of RA in the mammary fat pad, which was accompanied by a parallel decrease in tumour incidence and tumour number (Corl *et al.* 2003), and that the anticarcinogenic effects of VA were predominately (perhaps exclusively) mediated through its conversion to RA via Δ9-desaturase (Lock *et al.* 2004).

VA and RA derived from milk fat are therefore both anticarcinogenic and this series of pre-clinical investigations clearly demonstrate the feasibility of a functional food approach using dairy products enriched in VA and RA in the prevention of mammary cancer. It is noteworthy that RA may be more effective in preventing the development of tumours, a role consistent with its potential value as a functional food component, rather than having an impact as a cancer therapy (Bauman *et al.* 2006a). Taken together, *in vitro* and *in vivo* evidence of the role of RA in cancer prevention is persuasive; this role remains to be proven directly in humans however because data obtained from epidemiological studies are not conclusive and cannot be taken as definitive (Lock *et al.* 2009). Assessing the role of dietary RA (particularly as a natural component of our diet) for the prevention of chronic diseases such as cancer presents some unusual difficulties, and many of the traditional approaches to evaluate human health effects have substantial limitations (Bauman *et al.* 2006a).

While the anticancer effects of RA have been particularly impressive in biomedical models of breast cancer, by comparison, investigations of RA effects on atherosclerosis are more limited. To date, the anti-atherogenic effects of RA have been established in animal models and humans, with effects due to changes in both lipoprotein and cholesterol metabolism and anti-inflammatory effects (Lock *et al.* 2009). Of special importance, butter naturally enriched in RA has been shown to be effective in biomedical studies with atherogenic animal models. The ability of RA to alter blood lipid profiles has been tested in various animal models, such as the apoE knock-out mouse, the Golden Syrian hamster and the New Zealand white rabbit; results from these studies indicate that RA may improve cholesterol levels, and certainly not adversely affect them (Lock *et al.* 2009). When butter enriched with RA (*c.* 1%) and VA was fed to Golden Syrian hamsters, TC, LDL-cholesterol, and the non-HDL-cholesterol to HDL-cholesterol ratio were lower than those fed a control diet (Valeille *et al.* 2006). This is similar to the results of Lock *et al.* (2005a) who used the same animal model to examine the potential of RA when fed as a component of a functional food (VA/RA-enriched butter) as part of a diet that was high in cholesterol (0.2%) and fat (20%). Compared with the control animals, those fed the VA/RA-enriched butter showed a number of beneficial effects, including reduced plasma TC, VLDL-cholesterol and LDL-cholesterol, suggesting that RA may modify the production of atherogenic lipoproteins by the liver. In addition, the VA/RA-enriched butter produced a less atherogenic profile than an equivalent diet in which the VA/RA-enriched butter was replaced with TFA from PHVO (Lock *et al.* 2005a). A similar study in New Zealand white rabbits determined that RA-enriched butter had a neutral effect on plasma lipoprotein cholesterol (Bauchart *et al.* 2007).

Although RA has been widely regarded for its ability to improve plasma cholesterol and lipoprotein profiles, perhaps more impressive is evidence demonstrating its ability to reduce the size of atherosclerotic lesions and halt the progression of such lesions in animal models. For example, when butter was naturally enriched with either VA/RA or with *trans*-10 18:1, the *trans*-10 18:1 enriched butter was found to have detrimental effects, as evidenced by increases in plasma concentrations of TC, LDL-cholesterol, VLDL-cholesterol and the non-HDL-cholesterol to HDL-cholesterol ratio. In contrast, the VA/RA-enriched butter had neutral effects on blood lipid profiles and tended to reduce aortic lipid deposition (Roy *et al.* 2007).

The use of surrogate biomarkers for disease risk is more readily achievable for atherosclerosis than for cancer in humans. However, epidemiological studies that have examined the intake of RA derived from foods with the risk of atherosclerosis are lacking. Several human intervention studies involving dietary supplements of RA in the form of capsules have reported plasma lipid variables as secondary observations, but most utilised mixed isomers of CLA and found inconsistent results (see Bauman *et al.* 2006a).

Tricon *et al.* (2006) provided foods naturally enriched with RA to healthy men; at the conclusion of 6 weeks of RA-enriched milk, butter and cheese alongside a normal diet, there were no significant changes in TC, LDL-cholesterol, or HDL-cholesterol. It is important to note that during the consumption of these RA-enriched dairy products, subjects consumed twice the level of dairy fat compared to baseline values. The enriched dairy foods changed LDL-cholesterol FA composition, but had no significant effect on LDL particle size or susceptibility to oxidation (Tricon *et al.* 2006). Overall, results indicate that RA has beneficial effects in reducing risk factors for atherosclerosis. In particular, the available data indicate that consumption of dairy foods rich in RA, although naturally high in SFA, may be protective against the development of atherosclerosis (Lock *et al.* 2009).

Extending anticarcinogenic and anti-atherogenic research from biomedical models to implications of RA (and VA) as bioactive food components in humans is limited and problematic. These evaluations are challenging because the development of cancer and atherosclerosis has long latency periods and there is a lack of consensus biomarkers, especially for cancer. Further, there have been significant advances in the analysis of RA and specific TFA, and these techniques have not yet been applied to updating values in food databases. These challenges are discussed further by Bauman *et al.* (2006a).

6.6 Dairy-derived fats in foods and human health

The concept of eating healthy has become synonymous with avoiding dietary fat and cholesterol, especially saturated fat. On a population basis, a diet low in saturated fat remains at the heart of nutritional advice in many countries for lowering plasma cholesterol and reducing CVD risk. In the case of dairy products, there has been a general perception that a food containing saturated fat is unlikely to be beneficial to health. As indicated earlier, a growing body of scientific literature challenges this perception. Indeed, over the last decade evidence has accumulated that the composition and quantities of dietary fat is very important in determining the relative risk to diseases such as CVD and cancer, and that milk-derived fat may offer significant health benefits compared to some common sources of dietary fats. It is particularly important to recognise that individuals consuming dairy fats do not just consume SFA and the other groups of FA discussed in this review, but rather as fats in the whole dairy food which is highly complex and may contain beneficial ingredients.

The appropriateness of recommendations regarding the intake of dairy products (particularly in relation to reducing saturated fat intake by reducing dairy consumption) has been challenged by conclusions from a number of recent meta-analysis and data summaries including those by Elwood *et al.* (2008, 2010), German *et al.* (2009), Kliem & Givens (2011),

Kratz *et al.* (2013), and Rice *et al.* (2013). The long-term effects of milk and dairy product consumption on health would ideally be tested in adequately powered randomised control intervention studies with disease/death events as the key outcomes. So far no adequate studies of this type have been reported. As a result, Elwood *et al.* (2008, 2010) concluded that the most valuable evidence on associations between milk and dairy products and health and survival would be provided by long-term prospective cohort studies. These avoid the weaknesses of case-control studies and of using markers of risk such as plasma cholesterol. The cholesterol weakness is important since while it may demonstrate some increased risk associated with the fats present in milk, it would not account for any counterbalancing properties in milk as a whole.

Elwood *et al.* (2008) reported on meta-analyses that examined the associations between milk and dairy products and health and survival. This showed convincing evidence that a high intake of milk can provide long-term reductions in the risk of CVD. The relative risk (RR) of stroke and ischemic heart disease in subjects with high milk or dairy consumption was shown to be 0.79 (95% confidence interval or CI 0.75, 0.82) and 0.84 (95% CI 0.76, 0.93) respectively, relative to the risk in those with low consumption. Figure 6.7 shows the numbers of deaths in England and Wales in 2008 from various causes, as well as the risks for these causes in subjects with the highest milk/dairy consumption relative to subjects with the lowest milk/dairy consumption. This work has been extended to examine the evidence for differential effects of milk, cheese and butter on incidence of vascular disease (Elwood *et al.* 2010). This essentially found that there were very few prospective cohort studies available for cheese (five) and butter (six). For butter, only three studies were suitable for meta-analysis yielding a non-significant RR for high vs. low consumption (0.93, 95% CI 0.84–1.02). For cheese, only two studies were suitable for meta-analysis. While there is some additional evidence related to butter and cheese from retrospective case-control studies, this is weaker

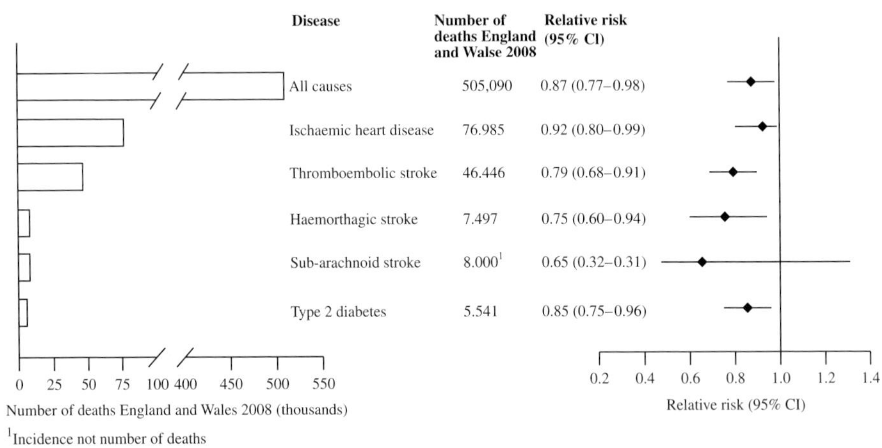

Fig. 6.7 The number of deaths in England and Wales in 2008 from various causes and the risks for these causes in subjects with the highest milk/dairy consumption relative to subjects with the lowest milk/dairy consumption. Adapted from Elwood *et al.* (2010). Reproduced with permission of Springer Science + Business Media.

evidence than from cohort studies and overall this highlights a large gap in knowledge. Similarly, there are few studies that have adequately compared the effects of full-fat milk with fat-reduced milk.

These data provide the best evidence available that those who consume large quantities of milk and dairy products are at no greater risk of CVD than those who consume little; indeed, there appears to be a small but valuable reduction in risk of CVD from increased milk and dairy product consumption. As noted above, the evidence for cheese and butter is inconclusive. These findings are in broad agreement with the recently reported outcome of a remarkable 61-year follow-up of the Boyd–Orr cohort. This study involved the recruitment of 4999 children in England and Scotland in 1937–1939 with causes of death recorded from 1948 (van der Pols *et al.* 2009). This study showed that a family diet in childhood which was high in dairy products did not give rise to a greater risk of CVD or stroke mortality. Indeed, all-cause mortality was lowest in those with the highest dairy product and milk intake (basic hazard ratio for both, 0.69; 95% CI 0.57, 0.84; P for trend <0.002).

These findings are therefore suggestive that, despite milk fat being rich in SFA, milk at least has other properties that are beneficial in respect of CVD risk. A key effect is likely to be that on blood pressure. For example, in a six-year study with 2245 older subjects in the Rotterdam Cohort Study, a substantial reduction in incident hypertension (hazard ratio 0.84; 95% CI 0.70, 1.01) was attributed to consumption of low-fat dairy products (Engberink *et al.* 2009). The hypotensive effect of dairy foods may be because, in addition to the effect of calcium, milk proteins release certain peptides during digestion which may have a beneficial effect on blood pressure by inhibiting the angiotension-converting enzyme-I, thus modulating endothelial function and leading to vasodilation (Clare & Swainsgood 2000). Interestingly, a recent study has shown that while both casein and whey proteins were able to reduce blood pressure in overweight and obese subjects, only whey proteins provided improved vascular function (Pal *et al.* 2010) which is a more valuable long-term indicator of risk.

Overall, the available evidence does not support the concept that consumption of dairy products adversely affects the risk of CVD. Indeed, linking the benefits of milk consumption with deaths from key chronic diseases led Elwood *et al.* (2010) to conclude that high milk consumers probably have an 'overall survival advantage' (Fig. 6.7). It is unfortunate that, due to a focus on the small rise in blood cholesterol with milk drinking, the debate on milk has never achieved a reasonable balance in the evaluation of risks and benefits. As previously discussed, several bioactive FA found in milk fat, milk proteins and other components have potential benefits for health maintenance and the reduction of chronic disease risk, and this reinforces the need for the dietetic community to reconsider current recommendations on dairy products and human health.

For additional information on the scientific evidence related to the impact of dairy product consumption and milk fat in human diets on overall health, the reader is directed to the recent symposium papers published in the *Journal of the American College of Nutrition* (Lock *et al.* 2008) and reviews by Kliem & Givens (2011) and Rice *et al.* (2013). The review by German *et al.* (2009) also provides suggestions for future milk-fat–human-health research.

References

Ascherio, A., Hennekens, C.H., Buring, J.E., Master, C., Stampfer, M.J. & Willet, W.C. (1994) *Trans*-fatty acids intake and risk of myocardial infarction. *Circulation*, **89**, 94–101.

Ascherio, A., Rimm, E.B., Giovannucci, E.L., Spiegelinan, D., Stampfer, M.J. & Willet, W.C. (1996) Dietary fat and risk of coronary heart disease in men, cohort follow up study in the United States. *British Medical Journal*, **313**, 84–90.

Bauchart, D., Roy, A., Lorenz, S., Chardigny, J.M., Ferlay, A., Gruffat, D., Sebedio, J.L., Chilliard, Y. & Durand, D. (2007) Butters varying in *trans* 18:1 and *cis*-9, *trans*-11 conjugated linoleic acid modify plasma lipoproteins in the hypercholesterolemic rabbit. *Lipids*, **42**, 123–133.

Bauman, D.E. & Griinari, J.M. (2003) Nutritional regulation of milk fat synthesis. *Annual Review of Nutrition*, **23**, 203–227.

Bauman, D.E. & Lock, A.L. (2006) Conjugated linoleic acid: Biosynthesis and nutritional significant. In: *Advanced Dairy Chemistry, Volume 2: Lipids* (eds Fox, P.F. & Sweeney, P.L.H.), pp. 93–136. NY, USA: Springer.

Bauman, D.E., Corl, B.A. & Peterson, D.G. (2003) The biology of conjugated linoleic acids in ruminants. In: *Advances in Conjugated Linoleic Acid Research*, Vol. II (eds Sebedio, J.L., Christie, W.W. & Adlof, R.O.), pp. 146–173. Champaign, IL: AOCS Press.

Bauman, D.E., Lock, A.L., Corl, B.A., Ip, C., Salter, A.M. & Parodi, P.W. (2006a) Milk fatty acids and human health: potential role of conjugated linoleic acid and *trans* fatty acids. In: *Ruminant Physiology: Digestion, Metabolism and Impact of Nutrition on Gene Expression, Immunology and Stress* (eds Sejrsen, K., Hvelplund, T. & Nielsen, M.O.), pp. 529–561. Wageningen, The Netherlands: Wageningen Academic.

Bauman, D.E., Mather, I.H., Wall, R.J. & Lock, A.L. (2006b) Major advances associated with the biosynthesis of milk. *Journal of Dairy Science*, **89**, 1235–1243.

Bauman, D.E., Harvatine, K.J. & Lock, A.L. (2011) Nutrigenomics, rumen-derived bioactive fatty acids, and the regulation of milk fat synthesis. *Annual Review of Nutrition*, **31**, 299–319.

Belury, M.A. (2002) Dietary conjugated linoleic acid in health: physiological effects and mechanisms of action. *Annual Review of Nutrition*, **22**, 505–531.

Bernal-Santos, G., O'Donnell, A.M., Vicini, J.L., Hartnell, G.F. & Bauman, D.E. (2010) Enhancing omega-3 fatty acids in milk fat of dairy cows by using stearidonic acid-enriched soybean oil from genetically modified soybeans. *Journal of Dairy Science*, **93**, 32–37.

Bolton-Smith, C., Woodward, M., Fenton, S. & Brown, C.A. (1996) Does dietary *trans* fatty acid intake relate to the prevalence of coronary heart disease in Scotland? *European Heart Journal*, **17**, 837–845.

Brouwer, I.A., Wanders, A.J. & Katan, M.B. (2013) *Trans* fatty acids and cardiovascular health: research completed? *European Journal of Clinical Nutrition*, **67**, 541–547.

Castaneda-Gutierrez, E., de Veth, M.J., Lock, A.L., Dwyer, D.A., Murphy, K.D. & Bauman, D.E. (2007) Effect of supplementation with calcium salts of fish oil on n-3 fatty acids in milk fat. *Journal of Dairy Science*, **90**, 4149–4156.

Chardigny, J.-M., Destaillats, F., Malpuech-Brugère, C., Moulin, J., Bauman, D.E. *et al.* (2008) Do *trans* fatty acids from industrially produced sources and from natural sources have the same effect on cardiovascular diseases risk factors in healthy subjects? Results of the *trans* fatty acids collaboration (TRANSFACT) study. *American Journal of Clinical Nutrition*, **87**, 558–66.

Chilliard, Y., Ferlay, A., Mansbridge, R.M. & Doreau, M. (2000) Ruminant milk fat plasticity: Nutritional control of saturated, polyunsaturated, *trans* and conjugated fatty acids. *Annals of Zootechnology*, **49**, 181–205.

Churruca, I., Fernandez-Quintela, A. & Portillo, M.P. (2009) Conjugated linoleic acid isomers: differences in metabolism and biological effects. *International Union of Biochemistry and Molecular Biology*, **35**, 105–111.

Clare, D.A. & Swainsgood, H.E. (2000) Bioactive milk peptides: A prospectus. *Journal of Dairy Science*, **83**, 1187–1195.

Corl, B.A., Barbano, D.M., Bauman, D.E. & Ip, C. (2003) *cis*-9, *trans*-11 CLA derived endogenously from *trans*-11 18:1 reduces cancer risk in rats. *Journal of Nutrition*, **133**, 2893–2900.

Craig-Schmidt, M.C. & Rong, Y. (2009) Evolution of worldwide consumption of *trans* fatty acids. In: *Trans Fatty Acid in Human Nutrition*, 2nd edition (eds Destaillats, F., Sébédio, J.-L., Dionisi, F. & Chardigny, J.-M.), pp. 329–380. Bridgewater, UK: The Oily Press.

Deckelbaum, R.J. & Torrejon, C. (2012) The omega-3 fatty acid nutritional landscape: health benefits and sources. *Journal of Nutrition*, **142**(Suppl), 587S–591S.

Demment, M.W. & Allen, L.H. (2003) Animal source foods to improve micronutrient nutrition and human function in developing countries. *Journal of Nutrition*, **133**, 3875S–4062S.

de Oliveira Otto, M., Mozaffarian, D., Kromhout, D., Bertoni, A.G., Sibley, C.T., Jacobs Jr., D.R. & Nettleton, J.R. (2012) Dietary intake of saturated fat by food source and incident cardiovascular disease: the multi-ethnic study of atherosclerosis. *American Journal of Clinical Science*, **96**, 397–404.

EFSA (Panel on Dietetic Products, Nutrition, and Allergies) (2010) Scientific opinion on dietary reference values for fats, including saturated fatty acids, polyunsaturated fatty acids, monounsaturated fatty acids, trans fatty acids, and cholesterol. *EFSA Journal*, **8**, 1461. Available at http://www.efsa.europa.eu/en/efsajournal/pub/1461.htm (accessed October 2013).

Elwood, P.C., Givens, D.I., Beswick, A.D., Fehily, A.M., Pickering, J.E. & Gallacher, J. (2008) The survival advantage of milk and dairy consumption: an overview of evidence from cohort studies of vascular diseases, diabetes and cancer. *Journal of American College of Nutrition*, **27**, 723S–734S.

Elwood, P.C., Pickering, J.E., Givens, D.I. & Gallacher, J.E. (2010) The consumption of milk and dairy foods and the incidence of vascular disease and diabetes: An overview of the evidence. *Lipids*, **45**, 925–939.

Engberink, M.F., Hendriksen, M.A., Schouten, E.G., van Rooij, F.J.A., Hofman, A., Witteman, J.C.M. & Geleijnse, J.M. (2009) Inverse association between dairy intake and hypertension: The Rotterdam Study. *American Journal of Clinical Nutrition*, **89**, 1877–1883.

Ervin, R.B., Wright, J.D., Wang, C.-Y. & Kennedy-Stephenson, J. (2004) Dietary intake of fats and fatty acids for the United States population: 1999-2000. US Department of Health and Human Services, Advanced Data Number 348, November 8, 2004.

Garnsworthy, P.C., Feng, S., Lock, A.L. & Royal, M.D. (2010) Heritability of milk fatty acid composition and stearoyl-CoA desaturase indices in dairy cows. *Journal of Dairy Science*, **93**, 1743–1748.

Gebauer, S.K., Chardigny, J.-M., Jakobsen, M.U., Lamarche, B., Lock, A.L., Proctor, S.D. & Baer, D.J. (2011) Effects of ruminant *trans* fatty acids on cardiovascular disease and cancer: a comprehensive review of epidemiological, clinical, and mechanistic studies. *Advances in Nutrition*, **2**, 332–354.

German, J.B., Gibson, R.G., Krauss, R.M., Nestel, P., Lamarche, B., van Staveren, W.A., Steijns, J.M., de Groot, L.C., Lock, A.L. & Destaillats, F. (2009) A reappraisal of the impact of dairy foods and milk fat on cardiovascular disease risk. *European Journal of Nutrition*, **48**, 191–203.

Gidding, S.S., Lichtenstein, A.H., Faith, M.S., Karpyn, A.K., Mennella, J.A., Popkin, B., Rowe, J., Van Horn, L. & Whitsel, L. (2009) Implementing American Heart Association pediatric and adult nutrition guidelines: A scientific statement from the American Heart Association Nutrition Committee. *Circulation*, **119**, 1161–1175.

Givens, D.I. (2008) Impact on CVD risk of modifying milk fat to decrease intake of saturated fatty acids and increase intake of *cis*-monounsaturates. *Proceedings of the Nutrition Society*, **67**, 419–427.

Givens, D.I. & Minihane, A.-M. (2009) Dairy products: their role in the diet and effects on cardiovascular disease. In: *Fatty Acids in Health Promotion and Disease Causation* (ed. Watson, R.R.), pp. 163–180. Urbana, IL, USA: AOCS Publications.

Grundy, S.M. & Vega, G.L. (1990) Causes of high blood cholesterol. *Circulation*, **81**, 412–427.

Hagemeister, H.D., Precht, D., Franzen, M. & Barth, C.A. (1991) α-Linolenic acid transfer into milk fat and its elongation by cows. *Fat Science Technology*, **93**, 387–391.

Hite, A.H., Feinman, R.D., Guzman, G.E., Satin, M., Schoenfeld, P.A. & Wood, R.J. (2010) In the face of contradictory evidence: report of the Dietary Guidelines for Americans Committee. *Nutrition*, **26**, 915–924.

Hoenselaar, R. (2012) Saturated fat and cardiovascular disease: the discrepancy between the scientific literature and dietary advice. *Nutrition*, **28**, 118–123.

Howard, B.V., Van Horn, L., Hsia, J., Manson, J.E., Stefanick, M.L. *et al.* (2006) Low-fat dietary pattern and risk of cardiovascular disease: The Women's Health Initiative Randomized Controlled Dietary Modification Trial. *Journal of American Medical Association*, **295**, 655–666.

Hu, F.B., Stampfer, M.J., Manson, J.E., Rimm, E., Colditz, G.A., Rosner, B.A., Hennekens, C.H. & Willett, W.C. (1997) Dietary fat intake and the risk of coronary heart disease in women. *New England Journal of Medicine*, **337**, 1491–1499.

Institute of Medicine Food and Nutrition Board (2002). Letter report on dietary reference intakes for *trans* fatty acids. *National Academy of Science Press*, Washington, DC.

Ip, C., Banni, S., Angioni, E., Carta, G., McGinley, J., Thompson, H.J., Barbano, D. & Bauman, D.E. (1999) Conjugated linoleic acid-enriched butter fat alters mammary gland morphogenesis and reduces cancer risk in rats. *Journal of Nutrition*, **129**, 2135–2142.

Jakobsen, M.U. & Overvad, K. (2009) Observational epidemiological studies on intake of *trans* fatty acids and risk of ischaemic heart disease. In: *Trans Fatty Acid in Human Nutrition*, 2nd Edition (eds Destaillats, F., Sébédio, J.-L., Dionisi, F. & Chardigny, J.-M.), pp. 255–306. Bridgewater, UK: The Oily Press.

James, M.J., Ursin, V.M. & Cleland, L.G. (2003) Metabolism of stearidonic acid in human subjects: comparison with the metabolism of other n-3 fatty acids. *American Journal of Clinical Nutrition*, **77**, 1140–1145.

Jenkins, T.C., Wallace, R.J., Moate, P.J. & Mosley, E.E. (2008) Board-invited review: Recent advances in biohydrogenation of unsaturated fatty acids within the rumen microbial ecosystem. *Journal of Animal Science*, **86**, 397–412.

Jensen, R.G. (2002) Invited review: The composition of bovine milk lipids: January 1995 to December 2000. *Journal of Dairy Science*, **85**, 295–350.

Kelsey, J.A., Corl, B.A., Collier, R.J. & Bauman, D.E. (2003) The effect of breed, parity, and stage of lactation on conjugated linoleic acid (CLA) in milk fat from dairy cows. *Journal of Dairy Science*, **86**, 2588–2597.

Kliem, K.E. & Givens, D.I. (2011) Dairy products in the food chain: their impact on health. *Annual Review of Food Science*, **2**, 21–36.

Kraft, J., Collomb, M., Mockel, P., Sieber, R. & Jahreis, G. (2003) Differences in CLA isomer distribution of cow's milk lipids. *Lipids*, **38**, 657–664.

Kraft, J., Preseault, C., Rice, B.H. & Lock, A.L. (2010) Partially hydrogenated vegetable oils from different sources have divergent effects on lipoprotein metabolism in male F1B hamsters *FASEB Journal*, **24**, 336.6 (abstract).

Kratz, M., Baars, T. & Guyenet, S. (2013) The relationship between high-fat dairy consumption and obesity, cardiovascular, and metabolic disease. *European Journal of Nutrition*, **52**, 1–24.

Kris-Etherton, P., Daniels, S.R., Eckel, R.H., Engler, M., Howard, B.V. *et al.* (2001) AHA scientific statement: Summary of the scientific conference on dietary fatty acids and cardiovascular health. Conference summary from the nutritional committee of the American Heart Association. *Journal of Nutrition*, **131**, 1322–1326.

Lacroix, E., Charest, A., Cyr, A., Baril-Gravel, L., Lebeuf, Y., Paquin, P., Chouinard, P.Y., Couture, P. & Lamarche, B. (2012) Randomized controlled study of the effect of a butter naturally enriched in *trans* fatty acids on blood lipids in healthy women. *American Journal of Clinical Nutrition*, **95**, 318–325.

Lock, A.L. & Bauman, D.E. (2004) Modifying milk fat composition of dairy cows to enhance fatty acids beneficial to human health. *Lipids*, **39**, 1197–1206.

Lock, A.L. & Shingfield, K.J. (2004) Optimising milk composition. In: *UK Dairying: Using Science to Meet Consumers' Needs* (eds Kebreab, E., Mills, J. & Beever, D.E.), pp. 107–188. Nottingham, UK: Nottingham University Press.

Lock, A.L., Corl, B.A., Bauman, D.E., Barbano, D.M. & Ip, C. (2004) The anticarcinogenic effect of *trans*-11 18:1 is dependent on its conversion to *cis*-9, *trans*-11 CLA by Δ9-desaturase in rats. *Journal of Nutrition*, **134**, 2698–2704.

Lock, A.L., Horne, C.A.M., Bauman, D.E. & Salter, A.M. (2005a) Butter naturally enriched in conjugated linoleic acid and vaccenic acid alters tissue fatty acids and improves the plasma lipoprotein profile in cholesterol-fed hamster. *Journal of Nutrition*, **135**, 1934–1939.

Lock, A.L., Parodi, P.W. & Bauman, D.E. (2005b) The biology of *trans* fatty acids: Implications for human health and the dairy industry. *Australian Journal of Dairy Technology*, **60**, 134–142.

Lock, A.L., Destaillats, F., Kraft, J. & German, J.B. (2008) Introduction to the proceedings of the symposium 'Scientific update on dairy fats and cardiovascular disease'. *Journal of American College of Nutrition*, **27**, 720S–722S.

Lock, A.L., Kraft, J., Rice, B.H. & Bauman, D.E. (2009) Biosynthesis and biological activity of rumenic acid: a natural CLA isomer. In: *Trans Fatty Acid in Human Nutrition*, 2nd Edition (eds Destaillats, F., Sébédio, J.-L., Dionisi, F. & Chardigny, J.-M.), pp. 195–230. Bridgewater, UK: The Oily Press.

Malpuech-Brugère, C., Morio, B. & Mensink, R.P. (2009) Dietary *trans* fatty acids and cardiovascular disease risk. In: *Trans Fatty Acid in Human Nutrition*, 2nd Edition (eds Destaillats, F., Sébédio, J.-L., Dionisi, F. & Chardigny, J.-M.), pp. 317–388. Bridgewater, UK: The Oily Press.

Mensink, R.P., Zock, P.L., Kester, A.D.M. & Katan, M.B. (2003) Effects of dietary fatty acids and carbohydrates on the ratio of serum total to HDL cholesterol and on serum lipids and apolipoproteins: A meta-analysis of 60 controlled trials. *American Journal of Clinical Nutrition*, **77**, 1146–1155.

Motard-Bélanger, A., Charest, A., Grenier, G., Paquin, P., Chouinard, Y. *et al*. 2008. Study of the effect of *trans* fatty acids from ruminants on blood lipids and other risk factors for cardiovascular disease. *American Journal of Clinical Nutrition*, **87**, 593–599.

Mozaffarian, D., Micha, R. & Wallace, S. (2010) Effects on coronary heart disease of increasing polyunsaturated fat in place of saturated fat: A systematic review and meta-analysis of randomized controlled trials. *PLoS Medicine*, **7**, 1–10.

Noakes, M., Nestel, P.J. & Clifton, P.M. (1996) Modifying the fatty acid profile of dairy products through feedlot technology lowers plasma cholesterol of humans consuming the products. *American Journal of Clinical Nutrition*, **63**, 42–46.

O'Donnell-Megaro, A.M., Barbano, D.M. & Bauman, D.E. (2011) Survey of fatty acid composition of retail milk in the United States including regional and seasonal variations. *Journal of Dairy Science*, **94**, 59–65.

Pal, S., Ellis, V. & Dhaliwal, S. (2010) Effects of whey protein isolate on body composition, lipids, insulin and glucose in overweight and obese individuals. *British Journal of Nutrition*, **104**, 716–723.

Palmquist, D.L. (2009) Omega-3 fatty acids in metabolism, health, and nutrition and for modified animal product foods. *Professional Animal Scientist*, **25**, 207–249.

Palmquist, D.L., Lock, A.L., Shingfield, K.J. & Bauman, D.E. (2005) Biosynthesis of conjugated linoleic acid in ruminants and humans. *Adances in Food and Nutrition Research*, **50**, 179–218.

Pariza, M.W. (1999) The biological activities of conjugated linoleic acid. In: *Advances in Conjugated Linoleic Acid Research*, Vol. 1 (eds Yurawecz, M.P., Mossoba, M.M., Kramer, J.K.G., Pariza, M.W. & Nelson, G.), pp. 12–20. Champaign, IL: AOCS Press.

Parodi, P.W. (2006) Nutritional significance of milk lipids. In: *Advanced Dairy Chemistry, Volume 2: Lipids* (eds Fox, P.F. & Sweeney, P.L.H.), pp. 601–639. NY, USA: Springer.

Parodi, P.W. (2009) Has the association between saturated fatty acids, serum cholesterol and coronary heart disease been over emphasized? *International Dairy Journal*, **19**, 345–361.

Pietinen, P., Ascherio, A., Korhonen, P., Hartman, A.M., Willett, W.C., Albanes, D. & Virtamo, J. (1997) Intake of fatty acids and risk of coronary heart disease in a cohort of Finnish men. The alpha-tocopherol, beta-carotene cancer prevention study. *American Journal of Epidemiology*, **145**, 876–887.

Poppitt, S.D., Keogh, G.F., Mulvey, T.B., McArdle, B.H., MacGibbon, A.K. & Cooper, G.J. (2002) Lipid-lowering effects of a modified butter-fat: a controlled intervention trial in healthy men. *European Journal of Clinical Nutrition*, **56**, 64–71.

Randolph, T.F., Schelling, E., Grace, D., Nicholson, C.F., Leroy, J.L., Cole, D.C., Demment, M.W., Omore, A., Zinsstag, J. & Ruel, M. (2007) Invited review: Role of livestock in human nutrition and health for poverty reduction in developing countries. *Journal of Animal Science*, **85**, 2788–2800.

Rice, B.H., Quann, E.E. & Miller, G.D. (2013) Meeting and exceeding dairy recommendations: effects of dairy consumption on nutrient intakes and risk of chronic disease. *Nutrition Reviews*, **71**, 209–223.

Roy, A., Chardigny, J.M., Bauchart, D., Ferlay, A., Lorenz, S., Durand, D., Gruffat, D., Faulconnier, Y., Sebedio, J.L. & Chilliard, Y. (2007) Butters rich either in *trans*-10 C18:1 or in *trans*-11 C18:1 plus *cis*-9, *trans*-11 CLA differentially affect plasma lipids and aortic fatty streak in experimental atherosclerosis in rabbits. *Animal*, **1**, 467–476.

Seidel, C., Deufel, T. & Jahreis, G. (2005) Effects of fat-modified dairy products on blood lipids in humans in comparison with other fats. *Annals of Nutrition and Metabolism*, **49**, 42–48.

Siri-Tarino, P.W., Sun, Q., Hu, F.B. & Krauss, R.M. (2010) Meta-analysis of prospective cohort studies evaluating the association of saturated fat with cardiovascular disease. *American Journal of Clinical Nutrition*, **91**, 535–546.

Soedamah-Muthu, S.S., Ding, E.L., Al-Delaimy, W.K., Hu, F.B., Engberink, M.F., Willett, W.C. & Geleijnse, J.M. (2011) Milk and dairy consumption and incidence of cardiovascular diseases and all-cause mortality: dose-response meta-analysis of prospective cohort studies. *American Journal of Clinical Nutrition*, **93**, 158–171.

Soyeyrt, H., Dehareng, F., Mayeres, P., Bertozzi, C. & Gengler, N. (2008) Variation of Δ^9-desaturase activity in dairy cattle. *Journal of Dairy Science*, **91**, 3211–3224.

Taubes, G. (2001) The soft science of dietary fat. *Science*, **291**, 2535–2541.

Tholstrup, T., Sandstrom, B., Hermansen, J.E. & Holmer, G. (1998) Effect of modified dairy fat on postprandial and fasting plasma lipids and lipoproteins in healthy young men. *Lipids*, **33**, 11–21.

Tholstrup, T., Raff, M., Basu, S., Nonboe, P., Sejrsen, K. & Straarup, E.M. (2006) Effects of butter high in ruminant *trans* and monounsaturated fatty acids on lipoproteins, incorporation of fatty acids into lipid classes, plasma C-reactive protein, oxidative stress, hemostatic variables, and insulin in healthy young men. *American Journal of Clinical Nutrition*, **83**, 237–243.

Tricon, S., Burdge, G.C., Jones, E.L., Russell, J.J., El-Khazen, S., Moretti, E., Hall, W.L., Gerry, A.B., Leake, D.S., Grimble, R.F., Williams, C.M., Calder, P.C. & Yaqoob, P. (2006) Effects of dairy products naturally enriched with *cis*-9, *trans*-11 conjugated linoleic acid on the blood lipid profile in healthy middle-aged men. *American Journal of Clinical Nutrition*, **83**, 744–753.

Tyburczy, C., Major, C., Lock, A.L., Destaillats, F., Lawrence, P., Brenna, J.T., Salter, A.M., Bauman, D.E. (2009) Individual *trans* octadecenoic acids and partially hydrogenated vegetable oil differentially affect hepatic lipid and lipoprotein metabolism in golden Syrian hamsters. *Journal of Nutrition*, **139**, 257–263.

USDA/USDHHS (2010) Dietary Guidelines for Americans 2010. Report of the Dietary Guidelines Advisory Committee. USDA and USDHHS, Washington DC. Available at http://www.cnpp.usda.gov/DGAs2010-DGACReport.htm (accessed October 2013).

Valeille, K., Ferezou, J., Parquet, M., Amsler, G., Gripois, D., Quignard-Boulange, A. & Martin, J.C. (2006) The natural concentration of the conjugated linoleic acid, *cis*-9, *trans*-11 in milk fat has antiatherogenic effects in hyperlipidemic hamsters. *Journal of Nutrition*, **136**, 1305–1310.

van der Pols, J.C., Gunnell, D., Williams, G.M., Holly, J.M.P., Bain, C. & Martin, R.W. (2009) Childhood dairy and calcium intake and cardiovascular mortality in adulthood: 65-year follow-up of the Boyd-Orr cohort. *Heart*, **19**, 1600–1606.

Wang, Y.W. & Jones, P.J. (2004) Conjugated linoleic acid and obesity control: efficacy and mechanisms. *International Journal of Obesity and Related Metabolic Disorders*, **28**, 941–955.

Wang, Y. & Proctor, S.D. (2013) Current issues surrounding the definition of *trans*-fatty acids: implications for health, industry and food labels. *British Journal of Nutrition*, **110**, 1369–1383.

Whelan, J. & Rust, C. (2006) Innovative dietary sources of n-3 fatty acids. *Annual Review of Nutrition*, **26**, 75–103.

Willett, W.C., Stampfer, M.J., Manson, J.E., Colditz, G.A., Speizer, F.E., Rosner, B.A., Sampson, L.A. & Hennekens, C.H. (1993) Intake of *trans* fatty acids and risk of coronary heart disease among women. *Lancet*, **341**, 581–585.

Yashodhara, B.M., Umakanth, S., Pappachan, J.M., Bhat, S.K., Kamath, R. & Choo, B.H. (2009) Omega-3 fatty acids: a comprehensive review of their role in health and disease. *Postgraduate Medical Journal*, **85**, 84–90.

7 Milk Ingredients as Functional Foods

A.S. Kvistgaard[1], J.B. Schroder[1], E. Jensen[1],
A. Setarehnejad[2] and A. Kanekanian[2]

[1]Arla Foods Ingredients Group P/S, Aarhus, Denmark
[2]Centre for Nutrition, Dietetics and Food Science, Cardiff Metropolitan University, Cardiff, Wales, United Kingdom

7.1 Infant nutrition

7.1.1 *Milk ingredients and intestinal protection and maturation*

Specific proteins and components found in bovine whey fractions have been shown to protect the infant gut. The ingredients protect the infant by promoting the growth of beneficial bacteria as well as inhibiting colonisation and infection by pathogens. Whey components can also support the physiological development of the gut on different levels.

Breast-fed infants seem to have an intestinal microbiota dominated by bifidobacteria, while formula-fed infants have a more adult-like complex microbiota. A predominance of bifidobacteria is therefore seen as an indicator of intestinal health (Brück *et al.* 2002). Within the first week of life, strictly anaerobic genera such as bifidobacteria and bacteroides become established. In breast-fed infants, bifidobacteria outnumber other bacteria by a factor of at least 100. This has been explained by the presence of glycoproteins, lactoferrin, a low protein concentration and a low buffering capacity of human milk (Brück *et al.* 2003a), among other factors.

A higher proportion and variety of bifidobacteria may explain why breast-fed infants experience fewer infections than formula-fed infants. One way to mimic the functionality of human milk in infant formulas more closely is to supplement infant formulas with milk components that promote the growth of beneficial bacteria and/or inhibit the proliferation of pathogens.

Between birth and weaning, the physiology of the infant gut undergoes dramatic changes in order to adapt to oral nutrition and to prepare for the change from milk to solid food. Among other factors, intestinal maturation is associated with increased cell proliferation that results in a thickening of the mucosa due to the growth of villi and crypts. The mucosal brush border enzyme activity also adapts throughout maturation as lactase activity decreases and maltase as well as sucrase activities increase.

Bovine whey contains components with similar functional properties as human milk for facilitating intestinal protection and maturation. Alpha-lactalbumin (α-la), caseinoglycomacropeptide (CGMP) and milk fat globular membrane (MFGM) have been studied for their growth-promoting effects on beneficial intestinal bacteria as well as antipathogenic

Milk and Dairy Products as Functional Foods, First Edition. Edited by A. Kanekanian.

effects. Alpha-lactalbumin is a major protein in human milk; it is present at concentrations of 2–3 g L^{-1} and comprises 25–35% of the total protein content (Lönnerdal 2003). CGMP are multiglycosylated peptides from κ-casein. The fat globules in milk consist of a triglyceride core surrounded by a thin membrane called MFGM. MFGM contains polar lipids and proteins with different functionalities. Up to one-third of the MFGM polar lipid fraction comprises sphingolipids, while as much as 25–70% is made out of mainly glycoproteins. MFGM also has an effect on intestinal maturation.

These milk components show beneficial effects on supporting a healthy gut flora, protection against pathogens (decoy effect, antimicrobial effect and diarrhoea prevention/reduction) and intestinal growth and maturation.

Supporting a healthy gut flora

The sugar portion – but also the polypeptide portion of CGMP – has shown a strong promotion of the growth of beneficial intestinal bacteria such as *Bifidobacterium breve*, *B. bifidum*, *B. infantis in vitro* (Azuma *et al.* 1984; Bouhallab *et al.* 1993; Idota *et al.* 1994). These bifidobacteria are, along with *B. longum*, the most abundant bifidobacteria in infants receiving human milk. α-la and some proteolytic fragments of α-la have also shown bifidogenic functions *in vitro* by stimulating bifidobacteria such as *B. breve* and *B. infantis* (Petschow & Talbott 1991; Lönnerdal 2010).

So far there are only two human intervention studies describing prebiotic effects of α-la and CGMP. A prebiotic effect of α-la was observed in a study including infants 5–14 days old. A significant increase in bifidobacteria from baseline to week 8 was observed both in the α-la supplement and the breast-fed group. It was not possible to establish a clear prebiotic effect in the other study. Breast-fed infants were enrolled at 6 weeks of age, and it turned out that all groups had similar total bifidobacteria counts at baseline. This could be explained by the fact that all the infants were breast-fed initially, and included in the trial when they were too old (Brück *et al.* 2006; Bettler & Kullen 2007).

Prebiotic studies are difficult to perform as there is a large intra-individual difference in the intestinal microbiota. Furthermore, the baseline level of bacteria and early inclusion seem to be important for the possibility of observing changes. Several *in vitro* studies show a bifidogenic effect of α-la and CGMP. Nevertheless, further carefully designed trials need to be performed in order to establish the prebiotic effect of α-la and CGMP in infants.

Protection against pathogens

CGMP has antipathogenic activity that is connected to its carbohydrate part containing N-acetylneuraminic acid (sialic acid) as well as to its polypeptide part. Sialic acid on the surface on intestinal cells is often required for infection of a variety of pathogens. Oligosaccharides can act as releasable decoy ligands for bacterial adhesions and limit attachment of pathogens to cell surface molecules. A milk ingredient containing sialic acid may therefore work as a decoy, preventing bacterial attachment, which is the essential first step prior to infection. CGMP has been studied *in vitro* and was shown to inhibit the adhesion of bacteria, viruses and toxins to the cell membrane through binding to the receptor sites of the pathogen. An *in vitro* study was conducted in human HT29 tissue cell cultures to investigate the ability of CGMP to inhibit adhesion of three strains of verotoxigenic *E. coli* and three strains of enteropathogenic *E. coli*. It was found that CGMP exhibited effective anti-adhesive properties at a dose of 2.5 mg mL^{-1}. CGMP reduced adhesion of the verotoxigenic *E. coli* by up to 50% of the control group and the enteropathogenic *E. coli* to between 10–80% of the control group (Rhoades *et al.* 2005).

The efficacy of α-la and CGMP in inhibiting association with infective bacteria has also been studied in human intestinal caco-2 cells. Undigested pepsin, pepsin/prancreatin-digested α-la and CGMP all inhibited attachment of *E. coli* (EPEC), *Salmonella typhimurium* and *Shigella flexneri* (Brück *et al.* 2006).

Bactericidal activity has been observed *in vitro* for α-la. An α-la-enriched infant formula significantly reduced potentially pathogenic bacteria including clostridia, bacteroides and *E. coli* similar to human milk in a two-stage continuous culture system. Vessels containing α-la and CGMP also showed stable bifidobacteria counts similar to vessels with breast milk (Brück *et al.* 2002; Lönnerdal 2010).

The anti-infectious effect of α-la and CGMP on intestinal bacteria was studied *in vivo* in infant rhesus monkeys that were challenged with enteropathogenic *E. coli*. After being challenged, breast-fed infants and infants fed with α-la-supplemented formula had no diarrhoea. Infants fed with CGMP-supplemented formula had intermittent diarrhoea. In infants fed with control formula the diarrhoea was acute. Alpha-lactalbumin also shows a gastrointestinal tolerance profile similar to human breast-fed infants. However, in this study *Bifidobacterium spp* and *Lactobacillus spp* numbers were not significantly higher in breast-fed infants or infants fed with formula supplemented with α-la than in infants fed with CGMP or control formula. Antimicrobial activities of α-la or peptides formed from α-la in the gastrointestinal tract may therefore explain the observed effect. It is hypothesised that the peptides affect the membrane integrity of the pathogen. Since α-la was present in both the α-la- and the CGMP-supplemented formula, the protective effect of α-la may be dependent upon dose (Brück *et al.* 2003b; Davies *et al.* 2007).

Mucin 1 (MUC1), immunoglobulin G (IgG) and sphingomyelin are three of the many interesting components of MFGM. Studies have shown that breast-fed infants are better protected against infections such as rotavirus infection that occurs in the small intestine. Both bovine MUC1 and IgG have shown to be effective against rotavirus infection *in vitro* and in mice studies, in both cases probably through binding to the virus. In the case of MUC1, the binding is probably through its heavy glycosylation (Kvistgaard *et al.* 2004). MUC1 also functions as part of an early epithelial response to bacterial pathogens such as *Campylobacter jejuni* and *E. coli* (McAuley *et al.* 2007; Sando *et al.* 2009).

Many viruses and bacteria as well as their toxins use glycosphingolipids to bind to the intestinal epithelium, although they can also bind to similar receptors on MFGM (Dewettinck *et al.* 2007). The digestion product lysosphingomyelin and other hydrolysates have shown high bactericidal effect as well as against several pathogens such as *C. jejuni*, *Listeria monocytogenes*, *Clostridium perfringens*, *E. coli* and *Salmonella typhimurium* (Rueda *et al.* 1998; Sprong *et al.* 2001; Dewettinck *et al.* 2007; Clare *et al.* 2008).

Intestinal growth and maturation

Sphingomyelin, which comprises 38% of human milk phospholipids, plays an important role in neonatal gut maturation. Sphingomyelin participates in the regulation of cell growth, induction of cell differentiation and apoptosis and it has been seen to promote intestinal maturation in rats by reducing the lactase activity (Motouri *et al.* 2003). The change in lactase is normally seen in breast-fed infants during postnatal development and before weaning. Sphingomyelin also reduces the number of vacuolated cells and increases crypt depth, suggesting that sphingomyelin accelerates enterocyte turnover and promotes crypt cell proliferation when replacing vacuolated cells with mature cells. Vacuoles appear when immunoglobulin absorption is high in the small intestine, thus a decrease in immunoglobulin absorption is involved in gut closure. An increase in the area of Auerbach's plexus was also observed. The area of Auerbach's plexus is part of the enteric nervous system that controls the intestine, and is located between the intestinal muscles. Cells in Auerbach's plexus are connected to the sympathetic ganglia and the central nervous system. It plays an important role in intestinal peristalsis, secretion and absorption. There are therefore several indications that sphingomyelin may influence mucosal growth and development in the gut, but further studies are essential.

Several bovine milk ingredients have shown to be candidates for supporting important functions in early infant intestinal development such as intestinal protection and maturation in a way that can be compared to the effects obtained with human milk. These effects have primarily been found in *in vitro* and *in vivo* animal studies. To further substantiate the effects of bovine milk ingredients in infant intestinal development, more human clinical trials with different ingredients and mixtures of ingredients are needed.

7.1.2 *Brain development and cognitive function in infant nutrition*

The link between infant nutrition and cognitive function was first recognised in the late 1920s. Since then, more than 50 clinical studies have been published on the subject of of infant feeding regime (breast- or formula-fed) and cognitive function. These studies have consistently demonstrated that breast-feeding has a small but significant positive influence on cognitive function compared to formula feeding. However, the magnitude of influence becomes smaller when control for a range of relevant factors is carried out, including ratio and length of breast-feeding, parental intelligence and socioeconomic status.

In order to evaluate the connection between breast-feeding and cognitive development, a meta-analysis by Anderson *et al.* (1999) identified three inclusion criteria. First of all it compared predominantly breast-fed with predominantly formula-fed subjects. Secondly, primary outcome measurement was a widely applied test of cognitive development or ability, yielding a single score for the purpose of comparison. Thirdly, subjects were between infancy and adolescence. A total of 20 studies met the inclusion criteria; 11 of these presented both unadjusted and adjusted results when controlled for a minimum of 5 covariants.

The unadjusted results demonstrated a benefit of 5.32 points for breast-fed infants compared to formula-fed infants. Adjusting after covariant, the benefit of breast-feeding was 3.16 points compared to formula feeding. Furthermore, the difference was more significant in low-birth-weight infants, indicating that breast-feeding has a stimulatory effect on cognitive development in premature infants.

While there is a general scientific acceptance that breast-feeding promotes and stimulates infant cognitive function, there is still a request for high-quality studies controlling all relevant confounding factors. Further, residual confounding factors have to be identified in order to make a true comparison of the influence of breast-feeding and formula-feeding on cognitive development.

Breast-feeding and cognitive function

There are several potential mechanisms and theories proposed regarding how breast-feeding influences cognitive function. The mechanisms include several confounding factors related to the choice of breast-feeding, physical stimuli during breast-feeding to components in human milk influencing cognitive development. One plausible theory is that long-chain polyunsaturated fatty acids (LCPUFA), such as docosahexaenoic acid (DHA) and arachidonic acid (AA or ARA), are present in human milk compared to infant formulas. DHA and AA/ARA are found in exceptionally high concentrations in the cell membranes of the central nervous system (CNS). They accumulate during the growth spurt between the last trimester of pregnancy and the first year of life (Gale *et al.* 2010). However, while some clinical studies on LCPUFA supplementation have demonstrated a beneficial effect on cognitive function, other studies have failed to do so. There is therefore still a need to investigate in depth the influence of LCPUFA on cognitive development and performance in infancy.

LCPUFA in human milk may not be the only explanation for the stimulating effect seen from breast-feeding in infant cognitive development. Human milk contains a range of bioactive components that have other purposes than nutrition. Sialic acid and gangliosides are two components gaining more focus in connection with infant cognitive development.

Sialic acid and infant cognitive development

While sialic acid is widely distributed in human tissue, the concentration is 20-fold higher in neuronal cell membranes compared to other cellular membranes. The major part of sialic acid in the CNS is lipid bound and incorporated into gangliosides constituting 65% of the sialic acid pool while 32% is protein bound.

The incorporation takes place from early gestation and continues until approximately the age of 5 (Wang 2009). A study on piglets feed supplemented with sialic acid at different concentrations investigated the effect on learning performance and memory by using different levels of cues of feed and its accessibility in a maze. The study demonstrated that high dietary levels of sialic acid led to faster learning and a higher amount of sialic acid in the frontal cortex. It also showed increased mRNA levels of two genes that are associated with learning in the hippocampus. It may up-regulate the sialic acid synthesis in the developing brain (Wang *et al.* 2007).

The highest concentration of sialic acid is found in human colostrum and declines gradually as the lactation stage matures. The sialic acid concentration in bovine milk, and thus in bovine-milk-based infant formulas, is significantly lower (<25%; Wang *et al.* 2001). Sialic acid can be found as free sialic acid or as protein-, carbohydrate- or lipid-bound. The composition of human and bovine milk differs significantly. This is not only reflected in the total sialic acid concentration but also in the form it is found in. Whereas human milk primarily contains oligosaccharide-bound sialic acid (>70%) and *c.* 23% bound to glycoproteins, the ratio is the opposite for bovine milk and thus infant formulas. It is still to be investigated if the source of sialic acid has any influence on its effect.

The different exogenous concentration of sialic acid content in human milk and commercial infant formulas may be reflected in the sialic acid levels of the infant. In the early 1990s a study demonstrated higher levels of sialic acid in saliva in breast-fed infants compared to formula-fed infants. In a similar manner a study of brain material of sudden infant death syndrome (SIDS) infants has demonstrated that breast-fed infants have a 32% higher lipid-bound sialic acid content and a 23% higher protein-bound sialic acid compared to formula-fed infants (Wang & Brand-Miller 2003). While the mechanism of the absorption of dietary sialic acid remains to be fully understood, these studies suggest that dietary sialic acid may/can influence endogenous levels.

There are several indications that sialic acid can be a valuable component in supporting the infant cognitive development process, such as: the different levels of sialic acid in human milk and in tissue of breast-fed compared to formula-fed infants; and the capabilities of sialic acid influencing learning and the memory ability in piglets. While infants are capable of synthesising sialic acid (and the CNS experiences a very rapid development in the first year of life), dietary sialic acid could support this process and thus secure the optimal cognitive health.

Gangliosides and infant cognitive development

Much of the complex lipid structure in the brain has to do with the high concentration of gangliosides, given that these lipids are significantly concentrated in the brain. Between weeks 16 and 22 of human gestation, tissue ganglioside concentration increases significantly in the frontal cortex and hippocampus. The physiological role of gangliosides in the brain is not fully understood, although the lipids have both structural and functional roles including modulators of neuronal transmission. Besides, gangliosides are the major source of sialic acid in the brain material, whose properties are described above. Several studies have investigated the effect of gangliosides on brain function, including rodent studies that demonstrated the enhancement of learning and memory abilities of neonatal, adult and aged rats when injecting gangliosides into the brain (McJarrow *et al.* 2009). A rat study demonstrated that dietary gangliosides increase ganglioside concentration in brain, retina and plasma. Dietary ganglioside can therefore influence endogen ganglioside levels (Park *et al.* 2005).

The highest ganglioside concentration is found in human colostrum and declines during lactation stage. The ganglioside concentration in bovine milk is, similarly to sialic acid, significantly lower compared to human milk (McJarrow *et al.* 2009). Breast-fed infants therefore receive higher ganglioside levels than formula-fed infants. This dietary difference could be part of the explanation that brain material from breast-fed SIDS-infants had significantly higher ganglioside concentration than formula-fed SIDS-infants.

Research is still needed to expand the understanding of ganglioside function in infant cognitive development. However, rodent studies on its functionality and the evidence that dietary gangliosides are capable of increasing brain ganglioside concentration support the theory that these lipids are relevant components in securing optimal cognitive health of infants, and thus important components in very early nutrition.

7.2 Clinical nutrition

7.2.1 *Immobility-caused muscle loss*

Injury or illness often results in a period of immobilisation or bed rest. Prolonged bed rest may result in a deteriorating nitrogen balance, muscle loss and reduction in functional capacity (Stehno-Bittel *et al.* 1998; Ingle *et al.* 1999; Stevens *et al.* 2006). Clearly physical activity would be effective to prevent muscle loss. However, it is not always possible or practical for bed rest patients to be physically active. The negative consequences of bed rest are caused by an imbalance between skeletal muscle protein synthesis and breakdown (Paddon-Jones *et al.* 2006). Several studies suggest that the muscle protein synthesis is reduced, whereas the muscle breakdown is unchanged. This imbalance will consequently cause a loss of muscle mass (Stuart *et al.* 1990; Paddon-Jones *et al.* 2004, 2006). Obviously the main issue should be supporting an increase in protein synthesis in order to reduce or avoid immobility-caused muscle loss during bed rest. Increasing dietary protein can better the protein synthesis.

In a study where protein intake was increased from 0.6 g protein kg^{-1} day^{-1} to more than 1.0 g protein kg^{-1} day^{-1}, it was shown that the nitrogen balance was maintained during 7 days of bed rest (Stuart *et al.* 1990). Test subjects on the low-protein diet had an unchanged protein breakdown and a decreased protein synthesis. Some authors have suggested that the beneficial effects of increased protein intake may be caused by the branched chain amino acids (BCAA) part of the protein (May & Buse 1989; Stein *et al.* 1999).

A large dose of essential amino acids (EAA) has also been seen to produce a rapid increase in plasma amino acid levels and stimulate net protein synthesis better than a mixed meal or a solution containing non-essential amino acids (Tipton *et al.* 1999; Borsheim *et al.* 2002). A 10 day bed rest study (Ferrando *et al.* 2010) with older test subjects (68±5 years) demonstrated that it was possible to avoid loss in muscle protein synthesis when the diet was supplemented with EAAs, whereas the control group lost 30% of their muscle protein synthesis. Additionally, it has in fact been demonstrated that EAAs alone are primarily responsible for the stimulation of net protein synthesis in skeletal muscle (Tipton *et al.* 1999; Volpi *et al.* 2003).

Whey proteins are an excellent choice to help prevent immobility-caused muscle loss since they have an amino acid profile that meets or exceeds all the EAA requirements set by the Food and Agriculture Organisation/World Health Organisation (FAO/WHO). In addition, whey proteins have the highest concentration of EAA combined with the best biological value compared to other protein sources (see Table 7.1).

Bed rest is also associated with developing resistance to the hypoglycaemic action of insulin. This resistance is most prominent in skeletal muscle (Ferrando *et al.* 1996). Some studies have shown that insulin promotes muscle anabolism primarily through an increased uptake of amino acids in muscle cells (Tipton *et al.* 1995). Several studies have shown a strong insulinotrophic effect of whey proteins, especially of whey protein hydrolysates (Belobrajdic *et al.* 2004; Frid *et al.* 2005; Power *et al.* 2009). It may therefore be possible to increase protein synthesis during bed rest by intake of whey proteins or whey protein hydrolysates, and by doing so prevent muscle loss.

Table 7.1 Concentration of EAA and biological value of various proteins. Data from Arla Foods Ingredients Group P/S (unpublished data), Danish Food Composition Databank, version 7.0, 2008 and Hoffman & Falvo (2004).

	Whey	Casein	Milk	Eggs	Beef
Concentration of EAA (% of total protein)	50	35	45	45	37
Biological value	104	77	91	100	80

Table 5.1 Contribution to RNI per portion of selected dairy products (%) for both male (M) and female (F) subjects aged 19–50 years. Green (yellow) shaded cells indicate that a single portion of this food will provide at least 20% (10%) of the RNI. Source: COMA (1991) and FSA (2002).

Example Food	Thiamin		Riboflavin		Vitamin B_6		Folate		Vitamin B_{12}		Calcium		Iron		Magnesium		Zinc	
	M	F	M	F	M	F	M	F	M	F	M	F	M	F	M	F	M	F
Milk (semi-skimmed)	6%	8%	38%	45%	9%	10%	10%	10%	127%	127%	35%	35%	0%	0%	-	-	8%	11%
Low fat yoghurt	18%	23%	25%	30%	1%	2%	14%	14%	33%	33%	35%	35%	2%	1%	-	-	9%	13%
Cheese (cheddar)	1%	1%	9%	11%	3%	3%	5%	5%	47%	47%	32%	32%	1%	1%	-	-	13%	17%
Rice pudding	2%	3%	20%	24%	1%	2%	0%	0%	-	-	25%	25%	2%	1%	0%	0%	11%	14%

KEY : M: Males, F: Females. Green shaded cells indicate that a single portion of this food will provide at least 20% of the RNI; Yellow shaded cells indicate that a single portion of this food will provide at least 10% of the RNI.

Information on food composition taken from Food Standards Agency (2002) McCance and Widdowson's The Composition of Foods, Sixthsummary edition. Cambridge: Royal Society of Chemistry. The percentage of RNI values are based on UK values for adults (aged 19–50 years) from COMA (1991) Dietary reference values for food and energy nutrients for the U.K. Department of health report on health and social subjects #41 HMSO: London.

Milk and Dairy Products as Functional Foods, First Edition. Edited by A. Kanekanian.

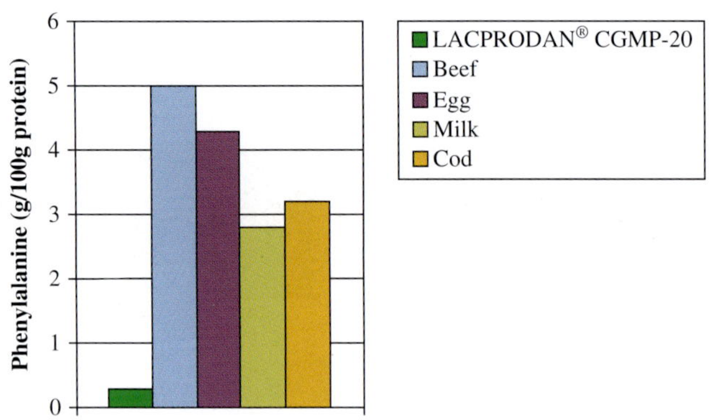

Fig. 7.2 Comparison of phenylalanine levels in various protein sources. By Danish Food Composition Databank, version 7.0, 2008 and Arla Foods Ingredients Group P/S (unpublished data).

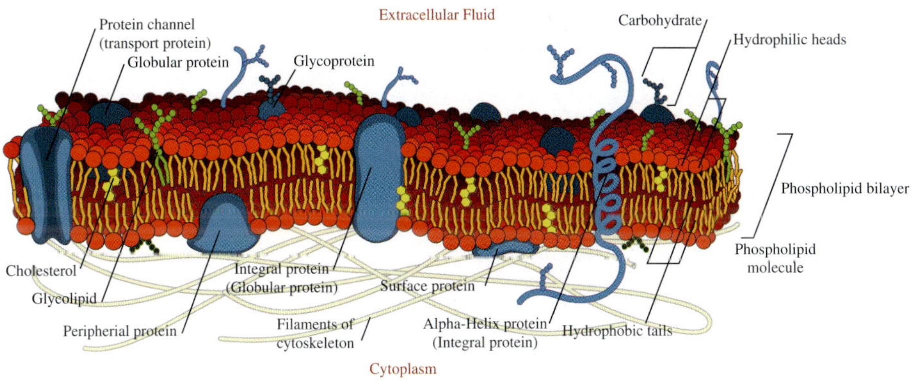

Fig. 7.3 Biological membrane, bilayer. By LadyofHats Mariana Ruiz (Public domain), via Wikimedia Commons.

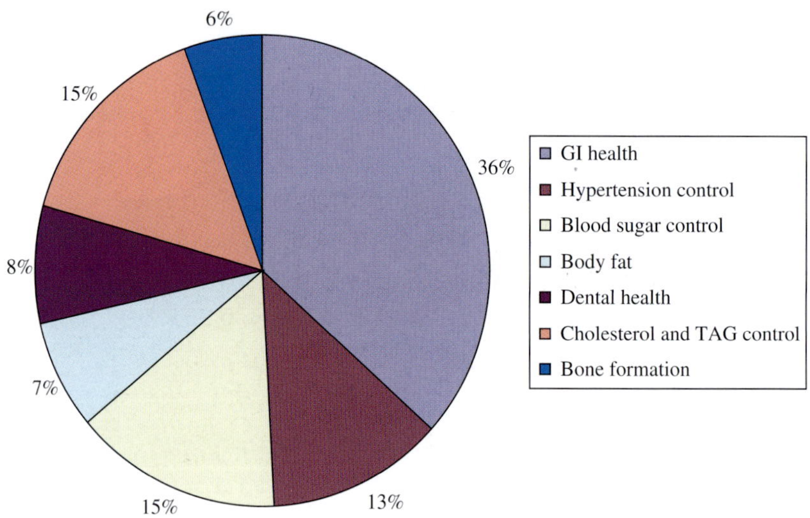

Fig. 10.5 FOSHU by health effect as a percentage of total approvals. Based on Consumer Affairs Agency (Japan) (2011).

7.2.2 *Diabetes*

In recent years the awareness of the insulinotrophic effects of milk has grown (Ostman *et al.* 2001). It seems that milk proteins – the whey fraction in particular – have a stimulating effect on insulin secretion in both healthy subjects and subjects with Type 2 diabetes (Nilsson *et al.* 2004; Frid *et al.* 2005). It is hypothesised that the insulinotrophic effect of whey could be a valuable tool in managing Type 2 diabetes given that it might postpone medical treatment (Frid *et al.* 2005).

The following sections focus on whey proteins and their potential effect in Type 2 diabetes. Special attention is paid to the aetiology and its possible role as a tool in managing Type 2 diabetes, including the suggested mechanism of action. As an introduction, the prevalence and characteristics of Type 2 diabetes are summarised.

Prevalence and characteristics of Type 2 diabetes

The major forms of diabetes are Type 1 and Type 2 diabetes, the former amounting to 10% of all incidences and the latter about 90% (Schrezenmeir & Jagla 2000). In 2010 it was estimated that about 285 million people (4% of the world's adult population) suffer from diabetes worldwide. This number is expected to rise to 438 million people in 2030. At that time, India, China and the US are estimated to be in the top three (Wild *et al.* 2004). The world market for diabetes is estimated to be worth US$376 billion in 2010 and is expected to grow to US$490 billion in 2030 (Novo Nordisk 2009).

Obesity is an important risk factor for developing Type 2 diabetes. Nevertheless, it is estimated that even if the prevalence of obesity remains stable until 2030, the number of people with diabetes will more than double as a consequence of population aging and changing lifestyle caused by urbanisation (Schrezenmeir & Jagla 2000; Wild *et al.* 2004).

The aetiology of Type 2 diabetes remains unclear. Type 2 diabetes is characterised by insulin resistance and hyperinsulinemia. In long-term Type 2 diabetic subjects there is an impaired ability to make and secrete insulin in response to a rise in glucose concentration. Type 2 diabetes may be treated with diet and later with oral antidiabetics. The antidiabetics are given either to improve insulin sensitivity or to stimulate insulin production before the so-called secondary failure of the β-cells may induce insulin requirements (Schrezenmeir & Jagla 2000).

Glycemic and insulinemic response

As early as 1936 it was suggested that high-protein diet might be useful for diabetic subjects (Conn & Newburgh 1936). Since then a number of studies have explored this hypothesis, and in Nuttall *et al.* found in 1984 that protein given with glucose increased insulin secretion and reduced plasma glucose rise in Type 2 diabetic subjects (Nuttall *et al.* 1984).

Two years later Gannon *et al.* found skim milk to be a particularly potent insulin secretagogue in Type 2 diabetic subjects (Gannon *et al.* 1986). While plasma glucose response to ingestion of fruits and milk products could be predicted from the amount of carbohydrate present, the serum insulin response could not. The predicted and observed insulin responses were significantly different for milk ($p<0.05$).

The observed value for skim milk was approximately 5-fold greater than predicted; this finding has since been confirmed by a number of studies. Consequently, it is now well accepted that milk, and in particular the whey fraction, causes a discrepancy between glycemic index (GI) and insulinemic index (II) in both Type 2 diabetic subjects and healthy subjects (Nilsson *et al.* 2004; Frid *et al.* 2005).

In long-term Type 2 diabetic subjects, hyperglycemia is no longer accompanied by compensatory hyperinsulinemia. Consequently, it could be assumed that the insulin secretory capacity of the β-cells is reduced. However, a study in long-term Type 2 diabetic subjects found that insulin response following carbohydrate ingestion could be substantially increased by the co-ingestion of a free amino acid/protein mixture (50% wheat protein, 25% free leucine and 25% free phenylalanine).

It has been proposed that increasing insulin secretion may lead to early β-cell failure, the so-called 'β-cell exhaustion hypothesis' (Frid *et al.* 2005). However, epidemiologic evidence suggests that consumption of milk and insulin resistance are not correlated. An animal study has even indicated that a high protein diet (32%) containing whey protein concentrate can increase insulin sensitivity (Belobrajdic *et al.* 2004). This was also confirmed in a recent human study with overweight and obese subjects, where whey protein isolate significantly increased insulin sensitivity compared to casein (Pal *et al.* 2010).

Mechanism of action

The key mechanism of the insulinotrophic effects of milk proteins is not known, but elevated concentrations of specific insulinogenic amino acids as well as bioactive peptides, either originally present in whey or formed during digestion, may play a role (Frid *et al.* 2005). The effect might be mediated through the release of glucose-dependent insulinotrophic polypeptide (GIP) and glucagons-like peptide 1 (GLP-1). Several metabolic pathways are activated in the β-cell, depending on the amino acid (Nilsson *et al.* 2007). Recently leucine, phenylalanine and tyrosine in healthy subjects were found to be some of the amino acids with the maximum insulinotrophic effect when co-ingested with carbohydrate (Van Loon *et al.* 2000). The insulinotrophic effect was also seen when the mixture was given to long-term Type 2 diabetic subjects (Van Loon *et al.* 2003).

The physiological role of individual amino acids may vary with different conditions, e.g. leucine that may be a better secretagogue of insulin in children compared to adults. Dietary composition may also modulate the ability of amino acids to elicit hormone secretion and metabolism, which makes it important to look at mixed meal studies (Nair & Short 2005). Furthermore, differences in the physical form of various food proteins may play a role. Liquid whey exits the stomach faster and is digested and absorbed more rapidly than a solid protein; this results in a short pronounced increase in postprandial plasma amino acid (Boirie *et al.* 1997).

Whey proteins as a tool in managing Type 2 diabetes

There are a few studies on the impact of milk as a part of a mixed meal on postprandial metabolic responses. Liljeberg & Bjorck (2001) were the first to study this, and investigated the impact of milk at two levels (200 mL or 400 mL) added to a high-GI white bread meal, compared to a low-GI spaghetti meal, on postprandial blood glucose and insulin response in

10 healthy subjects. They found that even an ordinary amount of milk (200 mL) added to the low-GI meal increased the insulin response to the same level as seen with the high GI.

In 2005 Frid *et al.* studied the insulinotrophic effect of whey when used as a supplement to a breakfast and a subsequent lunch. A total of 14 diet-treated subjects with Type 2 diabetes were included in the study. On two separate test days the subjects were served breakfast and lunch after four hours, with or without the addition of whey. Blood samples were drawn before and after eating both meals. After breakfast no significant difference was observed between the whey meal and the reference meal when examining blood glucose. However, the glycemia was significantly decreased after lunch, most likely related to the higher insulin response when whey was included in the meal.

Frid and co-workers suggest that the insulinotrophic effect of milk proteins (and in particular the whey proteins) are a valuable tool in managing Type 2 diabetes. Today sulfonylurea agents are commonly used to stimulate insulin secretion and to attenuate postprandial blood glucose. This treatment may occasionally cause hypoglycemia, which does not seem to be the case with whey. Future studies are required to determine whether these results are reproducible and, if so, how long the benefit of a whey supplement will last.

To some extent it seems that the powerful insulinotrophic stimulants in milk products, and in particular the whey fraction, can be stimulated by specific amino acid mixtures (Nilsson *et al.* 2007). There appears to be an additional effect when whey protein is ingested, resulting in a larger insulin response. In a study by Nilsson *et al.* it was discovered that a mixture of leucine, isoleucine, valine, lysine and threonine resulted in glycemic and insulinemic responses closely mimicking those seen after whey ingestion. Even so, the insulin response was still higher when whey was ingested. This might be explained by the additional response in gastric inhibitory polypeptide (GIP) seen after ingestion of whey (Nilsson *et al.* 2007).

It has also been demonstrated that whey protein hydrolysate produces a higher-GIP response compared to whole whey protein (Calbet & Holst 2004). The higher GIP response will induce insulin secretion and lower blood glucose levels. This may be interesting in clinical situations with insulin resistance and glucose intolerance.

Whey proteins have a stimulating effect on insulin secretion in both healthy subjects and subjects with Type 2 diabetes. The key mechanism of the insulinotrophic effects of whey proteins is not known. However, elevated concentrations of specific insulinogenic amino acids as well as bioactive peptides, either originally present in whey or formed during digestion, may play a role.

The use of whey as a food supplement may help to reduce postprandial blood glucose levels and increase insulin sensitivity in Type 2 diabetic subjects. Additionally, whey protein hydrolysates may provide a higher GIP and insulin response than intact whey proteins do when consumed.

7.2.3 *Inflammatory bowel disease*

Inflammatory bowel disease (IBD) can be divided into two forms called ulcerative colitis and Crohn's disease. Ulcerative colitis is limited to the colon; although Crohn's disease can involve any part of the gastrointestinal tract from the mouth to the anus, it most commonly affects the small intestine and/or the colon. IBD is a lifelong disease with periods of active disease alternating with periods of remission. The aetiologies of the diseases are unknown, but current opinion suggests that genetic and environmental components and an overreactive immune system are involved. It is estimated that up to two million people worldwide are affected by IBD, mainly from industrialised countries (Calder *et al.* 2009).

Mucin formation

IBD causes intestinal barrier dysfunction where the mucus layer and the production of mucin are impaired. The mucus layer is a viscoelastic gel that functions as a protectant of the gastrointestinal tract and lungs. The mucus provides protection from acids, bile, digestive enzymes and enteric bacteria and their toxins.

Mucus mainly consists of mucins, which are glycoproteins, and they are continuously secreted from the surfaces of the gastrointestinal tract and lungs. Mucins are large macromolecules composed of a peptide core and oligosaccharide side chains. The oligosaccharide side chains are linked via O-glycosidic bonds to serine and threonine in the peptide core. Threonine constitutes about 22% of the peptide core. An administration of a therapeutically effective dose of threonine could therefore potentially have a beneficial effect on gut health by ensuring enough threonine to the synthesis of mucin.

Commercial CGMP products (Arla Foods Ingredients Group) contain about 18% threonine. This level is about three times higher than the level in a standard whey protein concentrate and provides a highly concentrated dairy source of threonine.

The effect of threonine supplementation on the mucin synthesis rate was analysed *in vitro* in HT29-MTX cells (human colonic carcinoma cells). It was demonstrated that the mucoprotein fractional synthesis rate was increased from 0.2 mM to 0.8 mM, generally associated with an increased level of threonine in the culture medium (Ballevre *et al.* 2003).

The effect of threonine content in the diet on the mucin synthesis rate was analysed *in vivo* in growing rats. Diets contained 30–150% of the threonine requirement for rats. The results showed that if the threonine requirement is not covered by diet, then growth and mucoprotein synthesis rate will decrease (Ballevre *et al.* 2003).

Stimulation of the immune system

Two *in vivo* rat studies investigated the effect of CGMP on inflammatory bowel disease (Daddaoua *et al.* 2005; Requena *et al.* 2008). The anti-inflammatory effect was investigated by chemical induction of colitis with trinitrobenzenesulfonic acid (TNBS). The rats were given CGMP daily starting from two days prior to colitis induction. It was found that CGMP had a dose-dependent anti-inflammatory effect that could be observed in e.g. lower body weight loss, decreased anorexia, lower colonic damage and lower colonic weight to length ratio. The effect was comparable to that of established drugs, such as sulphasalazine and 5-aminosalicylic acid. The mechanism of action may involve Th17 and regulatory T cells and perhaps macrophages, but probably not Th1 lymphocytes. It was concluded that CGMP might play a role in managing patients with inflammatory bowel disease.

CGMP is a versatile ingredient which works on two aspects to help ameliorate IBD. Firstly, it helps support the integrity of the mucus layer in the gut. This is due to a high level of threonine in CGMP that is a key component in the mucus layer. Secondly, CGMP stimulates the immune system resulting in better protection against IBD.

Phenylketonuria (PKU)

PKU is an inherited metabolic disease characterised by a deficiency in the hepatic enzyme phenylalanine hydroxylase. This enzyme is necessary to metabolise the amino acid phenylalanine (Phe) to the amino acid tyrosine. Consequently, toxic levels of phenylalanine may build up in the body and can cause irreversible mental retardation if untreated. Fortunately this can be prevented by lifelong consumption of a low phenylalanine diet. Since most proteins contain around 5% phenylalanine, such a diet must be extremely low in protein and include a protein substitute consisting of amino acids instead. The drawback to the amino acids is that they have an unpleasant taste and smell that the PKU patients may reject, which is why some patients will have difficulties in complying with the PKU diet.

CGMP is a neutral-tasting peptide derived from casein that does not contain any phenylalanine and, for that reason, may be a suitable supplement to an amino acid diet. CGMP is produced during cheese processing when bovine κ-casein is cleaved by chymosin into para-κ-casein and CGMP. CGMP is a 64-amino acid glycophosphopeptide, depicted by Figure 7.1.

A study (Ney *et al.* 2008) of PKU mice deficient in phenylalanine hydroxylase aimed to assess how the ingestion of diets containing CGMP supports growth and affects concentrations of amino acids in plasma and brains. The PKU mice were fed diets containing 20% protein from amino acids or CGMP supplemented with limiting indispensable amino acids (IAA). Those on the CGMP diet showed body weight gains, feed efficiency and a protein efficiency ratio equivalent to that of the amino acid diet. This demonstrated that CGMP supplemented with limiting IAA provides a nutritionally adequate source of dietary protein for growing mice with PKU. Additionally, it was found that the concentration of phenylalanine in five sections of the brain and in plasma was significantly reduced in mice fed with CGMP supplemented with limiting IAA compared to the amino acid diet.

The high content of threonine in CGMP has also been shown to reduce plasma phenylalanine levels in patients with hyperphenylalaninemia (Sanjurjo *et al.* 2003). One hypothesis for this mechanism is that threonine decreases the intestinal absorption of phenylalanine

106 110 120
1 10
H.Met – Ala – Ile – Pro –Pro –Lys – Lys - Asn – Gln– Asp – Lys- Thr – glu – Ile – Pro – Thr – Ile

20 130 30
– Asn- Thr – Ile- Ala – Ser – Gly – Glu – Pro – Thr – Ser – Thr – Pro – Thr– Thr – Glu – Ala – Val
140 150 **Ile**

35 40 45 50
– Glu – Ser – Thr – Val – Ala – Thr – Leu – Glu – Asp – Ser – Pro – Glu - Val – Ile – Glu – Ser –
160 **Ala** P

55 60 64
Pro – Pro – Glu – Ile – Asn – Thr – Val – Gln – Val – Thr – Ser – Thr – Ala – Val.OH

Fig. 7.1 Primary structure of bovine glycomacropeptide including variant A, containing THr in position 136 and Asp in position 148, and variant B with Ile and Ala replacing these two positions. By Brody 2000 (reproduced with permission of Cambridge University Press) and Thomä-Worringer *et al.* 2006 (reproduced with permission of Elsevier).

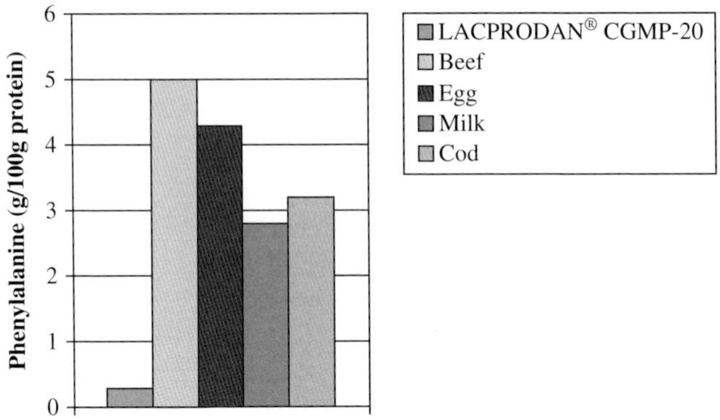

Fig. 7.2 Comparison of phenylalanine levels in various protein sources. By Danish Food Composition Databank, version 7.0, 2008 and Arla Foods Ingredients Group P/S (unpublished data). (See plate section for the color version of this figure.)

due to competition in a common intestinal transport system. However, current knowledge is not sufficient to determine which mechanism is behind the phenylalanine-lowering effect of threonine.

Van Calcar *et al.* conducted a human study in 2009 with 11 subjects. They were fed a CGMP diet supplemented with limiting amino acids (AA), which was compared to an amino acid diet. The study demonstrated that CGMP is a safe protein source and also that the subjects preferred the CGMP diet.

Another study has demonstrated that it is possible to make palatable, low-phenylalanine foods and beverages using CGMP as the protein source in a PKU diet (Lim 2007). Puddings, fruit leather, crackers, beverages with chocolate, strawberry and caramel and sports drinks with orange and mango peach are among the tested food and beverage products.

A commercial CGMP product that contains less than 0.28% of total product of phenylalanine is available from Arla Foods Ingredients Group. It is possible to incorporate high dosages in a PKU diet. For example, a consumption of 18 g Lacprodan CGMP-20 corresponds to an intake of 50 mg phenylalanine, equivalent to the phenylalanine content of 1 g protein from meat. A comparison of phenylalanine levels in various protein sources compared to Lacprodan CGMP-20 is shown in Figure 7.2.

7.3 Milk ingredients in sports nutrition

Whey protein provides all the necessary amino acids for muscle synthesis in approximate proportion to their ratios in the skeletal muscle. All the essential amino acids (EAA: phenylalanine, tryptophan, lysine, methionine (+cysteine) and threonine) including branched chain amino acids (BCAAs: leucine, isoleucine and valine) are present in a high level compared to other natural food protein sources (Table 7.2). The oxidation of the BCAAs in skeletal muscles is increased by up to 25% during exercise (McKenzie *et al.* 2000). BCAAs are directly metabolised by muscle tissue and this makes them highly important to athletes. Other amino acids are metabolised by the liver.

Table 7.2 Branched chain amino acids (BCAA) content of key protein. Data from Arla Foods Ingredients Group P/S (unpublished data) and Danish Food Composition Databank, version 7.0, 2008.

Protein source	BCAA (g/100 g protein)
Whey protein isolate	26
Egg white powder	22
Milk protein isolate	20
Soy protein isolate	17

The nutritive value of a protein depends on its capacity to provide nitrogen and amino acids in adequate amounts to meet the requirements of an organism. As shown in Table 7.3, the nutritive value of whey proteins is high compared to some other protein sources.

Athletes have to cope with several physiological challenges; dehydration, depletion of muscle fuels and deficient recovery are just some of them. A large volume of scientific research demonstrates that whey proteins provide the athlete with a number of benefits for coping with some of these challenges. As described in the following sections, whey proteins may have an effect on: muscle synthesis; muscle strength; endurance; recovery; body composition; and the immune system.

7.3.1 *Muscle synthesis*

Whey protein offers EAA and also the muscle building BCAA. Leucine and the other EAA play important roles in the protein synthetic pathway, both as signalling molecules for initiation and as a substrate for the synthesis of new proteins (Layman 2002). Leucine is a unique amino acid that has independently been proven to stimulate muscle protein synthesis. A study by Borsheim *et al.* (2002) found that there is a dose-dependent effect of EAA ingestion on muscle protein synthesis.

Table 7.3 Nutritive value of key proteins. Data from Hoffman & Falvo (2004), Schaafsma (2000) and Arla Foods Ingredients Group P/S (unpublished data).

Protein source	Biological value	Protein efficiency ratio	Net protein utilisation	PDCAAS	PDCAAS (Acc. FAO/WHO)
Whey	104	3.6	92	1.15	1.00
Cow's milk	91	3.1	87	1.21	1.00
Casein	77	2.9	76	1.23	1.00
Soy protein	74	2.1	70	0.91	0.91
Whole egg	100	3.8	98	1.18	1.00
Beef	80	2.9	78	0.92	0.92

7.3.2 *Muscle strength*

In several trials involving strength training, whey protein supplementation provides significant enhancement in muscle strength compared to carbohydrate and/or protein control groups (Cribb *et al.* 2002, 2006; Hayes & Cribb 2008). In one of these trials they found that whey protein isolates gave a significant increased strength in exercises such as the barbell bench press and squat (10–20% better gains than matched control groups). Strength increase in these exercises is thought to enhance the athlete's capacity to improve performance in many power- and strength-based events.

7.3.3 *Endurance*

The role of the glutathione (GSH) antioxidant system is to protect cells against oxidative damage. In addition to this, a direct link between GSH and exercise performance has been established. During exercise there is an increase in oxidative damage and this is thought to be a major contributor to muscle fatigue and poor athletic performance (Sen *et al.* 1994). Maintaining GSH status is shown to minimise oxidative stress and boost athletic performance (Sen *et al.* 1994). Whey protein is a rich source of cysteine. Furthermore studies have shown that compared to other sources, whey proteins are exclusive in their ability to boost GSH production (Lands *et al.* 1999; Cribb 2004; Mariotti *et al.* 2004; Middleton *et al.* 2004). It has also been observed that GSH increases with the amount of whey protein ingested (Zavorsky *et al.* 2007).

In a study by Ivy *et al.* (2003) into endurance performance of trained cyclists, a combination of carbohydrate and protein supplementation significantly increased the time to exhaustion compared to the test group supplemented with only carbohydrates. The combination of carbohydrates and proteins nearly doubled the time to fatigue. A study by MacDermid & Stannard (2006) also noted an increased endurance when using whey compared to a carbohydrate diet in cycling performance.

7.3.4 *Recovery*

It is important to athletes that they have adequate stores of glycogen in their tissues, as glycogen is the body's storage form of energy that fuels exercise. Low tissue glycogen level is associated with fatigue and poor performance. Blood insulin level plays an important role in the body's uptake and storage of nutrients as well as carbohydrates. Ingestion of protein combined with carbohydrate can induce an up to 100% higher plasma insulin concentration compared to carbohydrate alone (Calbet & MacLean 2002) and a greater muscle protein net balance (Ivy *et al.* 2003). Whey protein hydrolysate in particular provides a great plasma insulin response and amino acid concentration compared to glucose and other protein sources (Calbet & MacLean 2002; Morifuji *et al.* 2010). Recovery may be influenced by the effect of whey protein to stimulate insulin secretion. A study using whey protein supplementation has shown that, compared to carbohydrate supplementation, it accelerates recovery after intense resistant exercise (Borsheim *et al.* 2004) with respect to maximum strength and probably also muscle damage. It has been found that whey protein isolate can result in a more rapid recovery of muscle force generation capacity compared to carbohydrates (Cooke *et al.* 2004). Whey protein hydrolysate also gives a more rapid recovery than non-hydrolysed whey (Buckley *et al.* 2010).

7.3.5 *Body composition*

Scientific evidence indicates that whey proteins are tailored to promote a better body composition, especially when combined with resistance exercise. Several studies have shown that incorporating whey in the diet results in less fat storage, more lean tissue and higher insulin sensitivity in muscles compared to other protein sources (Boza 2000; Bouthegourd *et al.* 2002; Belobrajdic *et al.* 2004). In 2002 Bouthegourd *et al.* examined the effect of various supplements during exercise. They revealed that whey provided the best results in body composition - especially the α-la-enriched whey decreased adiposity in the rats. A study by Lands *et al.* (1999) found that whey supplementation (20 g day^{-1} for 12 weeks) provided a significant decrease in body fat percentage in healthy young adults.

7.3.6 *Immunity*

The immune system is highly influenced by exercise. During moderate exercise the immune system may be boosted just as it may be suppressed after high-intensity or long-duration exercise. As mentioned in Section 7.3.3, the GSH antioxidant system is also important for the immune system (Wu *et al.* 2004). Whey protein, especially α-la-enriched whey, is a rich source of cysteine and other EAAs. A rat study performed by Mariotti *et al.* (2004) showed that a single cysteine intake as part of a protein intake can have a large effect on liver GSH during exercise. They also stated that the effect was important with an α-la-rich meal because of its high cysteine content.

7.4 Milk ingredients in functional foods

Whey proteins have been used as functional foods for some time. There are several claimed health effects of whey, and the following sections describe the topics of weight management, cognitive performance and natural defence.

7.4.1 *Whey proteins and milk calcium for weight management*

The prevalence of obesity is increasing in the world. As early as in 1998 the World Health Organisation (WHO) reported a global increasing prevalence of obesity. The consequences of obesity are detrimental to the health status of a population. Obese persons are three times more likely to develop Type 2 diabetes, insulin resistance and cardiovascular disease (CVD).

Milk calcium

Recent studies (Davies *et al.* 2000; Carruth & Skinne 2001; Zemel 2003) show that intake of milk calcium from dairy can reduce body weight. In a PhD thesis, Lorenzen (2009) concludes that dietary calcium intake may have the potential to increase fat excretion to an extent that could be relevant to prevent weight regain (Lorenzen 2009). Firstly, the proposed mechanisms provide a regulation of human adipocyte metabolism (Zemel 2003). Secondly, the high intake of calcium can increase faecal energy excretion (Jacobsen *et al.* 2005). A calcium appetite concept (that high calcium intake may lead to better appetite control during weight loss program) has also been proposed (Astrup *et al.* 2010).

Whey proteins

It is well known that increased dietary protein may influence weight management in several ways. The influence is observed in a number of areas: through increased postprandial and post-absorptive satiety; increased thermogenesis; preservation of fat-free body mass; and lower energy efficiency (Weigle *et al.* 2005; Westerterp-Plantenga *et al.* 2009), but the number of studies comparing different protein types or fractions of proteins is low. A study by Veldhorst (2009) compared the different kinds of proteins and the effect on short-term satiety. When focusing on short-term satiety effects, that is, effects on satiety and subsequent food intake induced by a single meal, they found that different proteins (α-la, gelatine, gelatine+tryptophan (Trp)) were 40% more satiating than other proteins (casein, soy, whey, whey CGMP) and that they could induce a related 20% reduction of subsequent energy intake. Another study in rats supported these results (Pilvi *et al.* 2009). A proposed increased effect on satiety obtained by α-la is a well-known effect of proteins. Alpha-lactalbumin is also known to have a high content of tryptophan, a precursor of serotonin in the brain, which influences a person's appetite.

Baer *et al.* (2011) investigated the effect of consumption of a daily supplement with 56 g whey protein, soy protein or an isoenergetic amount of carbohydrate for 23 weeks. The aim of the study was to examine body weight and composition performed in free-living overweight and obese persons. They found that a supplement with whey protein could assist long-term maintenance of body weight without energy restriction in overweight and obese individuals.

For a while it was postulated that CGMP found in whey protein concentrate has an effect on weight management if influenced by the secretion of cholecystokinin (CCK), a hormone involved in the regulation of satiety (Beucher *et al.* 1994). However, there are several controversial results. Some of the most promising findings (Burton-Freeman 2008) are that CGMP alone is not critical in pre-meal whey-induced satiety and that CGMP might have a unique role in managing the regulation of daily energy intake.

7.4.2 *Milk phospholipids for cognitive performance in adults*

Many of the richest dietary sources of phospholipids are also very high in cholesterol and fat. Egg yolk, meat and offal are good sources of phospholipids. Modern diets are depleted of phospholipids, and changes in diets have lead to a decrease in daily consumption of phospholipids. The different kinds of phospholipids and the content of phospholipids in milk are listed in Table 7.4.

Table 7.4 Percentage of total phospholipid. Adapted from Burling & Graverholt (2008). Reproduced with permission of John Wiley & Sons.

	Soya	Egg	Milk
Phosphatidylcholine	34	75	27
Phosphatidylethanolamine	21	15	25
Phosphatidylinositol	18	0.4	8
Sphingomyelin	0	1.5	24
Phosphatidylserine	0.5	0	12
Phosphatidic acid	9	0	0
Others	17.5	8.1	4

Today's food contains only approximately one-third of the phospholipid level compared to that of the beginning of the last century. Phospholipids are diglycerides that are covalently bonded to a phosphate group by an ester linkage; they are a major component of biological membranes of animals, higher plants and micro-organisms (Fig. 7.3).

The specific phospholipid (phosphatidylserine or PS) is a naturally occurring phospholipid present in all biological membranes. PS is found most concentrated in the brain in humans where it comprises up to 15% of the total phospholipid pool. Neurons are cells transmitting signals from one cell to the next, and phosphatidylserine is thought to be especially vital to the neuron membrane. This membrane is particularly important for the communication between neurons, since networks of communicating neurons store memories. Phosphatidylserine is particularly important in neuronal functions as maintaining the state of the neuron membrane, increasing numbers of receptors for neurotransmitters and also stimulating release of neurotransmitters from neurons. PS is therefore characterised as a brain-specific nutrient

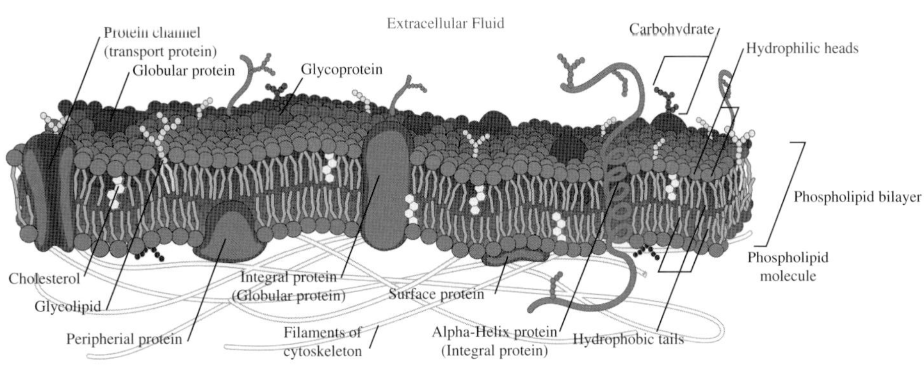

Fig. 7.3 Biological membrane, bilayer. By LadyofHats Mariana Ruiz, via Wikimedia Commons. (See Plate section for the color version of this figure.)

(McDaniel *et al.* 2003). It is also the reason we have chosen to focus specifically on the effect of PS in stress and memory in the following section.

Benefits of phosphatidylserine in relation to stress and memory

Prolonged stress can result in predictable systemic effects. The systemic effects of stress include change in levels of stress hormones such as cortisol, a decline in certain aspects of immune system function and a disruption of the gastrointestinal microflora balance. These systematic changes might be a substantial contributor to many of the stress-related declines in health.

Some researchers have suggested that chronic, oral administration of PS might counteract stress-induced activation of hypothalamic-pituitary-adrenal axis (HPA). PS in a dosage of 50 or 75 mg has been seen to influence stress in a positive manor by blunting the adrenocorticotropic hormone (ACTH) and cortisol responses (Monteleone *et al.* 1990).

Helhammer *et al.* (2004) conducted a study on healthy persons to investigate the effect of PS on endocrine, autonomic and psychological measures evoked by the Trier social stress test (TSST). Their results show that 400 mg PS day^{-1} significantly decreased stress compared to a placebo, measurable by a significantly decreased ACTH response. The persons getting PS also reported a reduced feeling of stress and positive feeling of well-being.

Milk is a natural source of phospholipids and especially of PS. Some studies have investigated the cognitive effects of phospholipids from milk during acute stress (Hellhammer *et al.* 2004; Hero *et al.* 2009; Schubert *et al.* 2011). These studies used a fat-reduced cream powder, rich in phospholipids (Lacprodan® PL-20, Arla Foods Ingredients Group). It was found that milk phospholipids could improve cognitive performance and memory during acute stress in highly stressed persons.

7.4.3 *Milk ingredients for natural defence*

'Natural defence' is defined as the defence provided by natural barriers against disease. In humans this includes natural barriers to handle infection, such as the skin and mucus layer in the intestinal tract, and blood defence systems such as production of antibodies. Several bioactive milk ingredients (e.g. MFGM, α-la, CGMP, lactoferrin, sialic acids, IgG, phospholipids and gangliosides) have shown effects on natural defence through different mechanisms. Some of these mechanisms and effects are described in Section 7.1.1 and 7.2.3 on infant nutrition and clinical nutrition; in this section the focus is on adults.

The effect of CGMP, lactoferrin and α-la on natural defence could be through a direct effect on viruses and bacteria. Studies have shown that sialic acid in CGMP can be involved in the binding of virus and bacteria, and prevent viral and bacterial infections (Brück *et al.* 2003b; Nakajima *et al.* 2005; Zimeckl & Kruzel 2007). On the other hand, the effect of lactoferrin is to prevent entry of the viruses into the host cell (Van der Strate *et al.* 2001).

Milk calcium is another milk component which has an effect on natural defence. Results from Bovee-Oudenhoven *et al.* (1997) have shown that milk calcium can inhibit intestinal colonisation and translocation of invasive salmonella. The dietary calcium can also decrease the severity of salmonella and *E. coli* infections (Bovee-Oudenhoven *et al.* 1999, 2003).

Initially, the proposed mechanisms stimulate the microflora and in particular endogenous lactobacilli. Secondly, it binds pathogens directly to amorphous calcium phosphate. By this binding an insoluble complex is formed in the human small intestine, preventing a mucosal attachment of these pathogens. Several bioactive components in the milk can influence and increase natural defence in adults.

7.5 Milk protein and dental health

Dental health is generally associated with firm healthy gums and smooth enamel surfaces of the teeth, while dental deterioration is usually linked with caries and dental erosion with increased roughness of the enamel surface. Such roughness facilitates further microbial growth on the tooth enamel and the formation of more plaque which harbour oral bacteria. The main sign of dental disease is bleeding gums due to gingivitis and periodontitis which eventually cause the destruction of the tooth structure. In the case of dental erosion and caries, the erosion of the enamel occurs directly or indirectly in the presence of acids in the vicinity of the tooth enamel.

In dental caries, the acid is generated *in situ* during the metabolism of dietary sugars by oral bacteria such as *Streptococcus mutans* present in the plaque, which utilise the remaining sugars in the oral cavity causing caries (Meurman *et al.* 1987; Aimutis 2004; Cheng *et al.* 2009). While some have suggested vaccination or manipulation of oral bacteria to combat dental caries (Loimarante *et al.* 1999; Russel *et al.* 2004), this would not be expected to have any influence on the more general non-microbial and purely chemical processes at low pH giving rise to dental erosion and demineralisation of the enamel, especially the calcium.

Calcium remineralisation occurs naturally from the saliva to keep the calcium balance, but with frequent consumption of acidic food and drink this balance is compromised and cannot be easily restored, giving rise to dissolution and erosion of the enamel (Lendenmann *et al.* 2000). It has also been established that erosion by acid softens the enamel and abrasion by brushing further exacerbates the problem within a short time (Rios *et al.* 2008).

It is also worth mentioning here that enamel erosion could also occur under alkaline condition. Obviously, such alkalinity is not usually found in food or beverages but it has been reported that enamel erosion can occur through some unusual unprotected industrial environment. Taubee *et al.* (2010) reported that enamel loss, distinct pores on the surface of the enamel and the destruction of the soft tissue was apparent in employees exposed to alkaline aerosols containing potassium hydroxide and other surface active agents used in industry.

Several methods have been suggested to minimise dental damage, emphasising oral hygiene such as daily brushing preferably with fluoride-containing toothpaste, flossing, oral irrigation and regular dental examinations. Other methods include intervention through external factors such as through water fluoridation, altering the nature and pH of the food/drink and calcium enrichment (Meurman *et al.* 1990; Grenby 1996). The replacement of normal sugar with sweeteners in several drinks has been considered commercially and promoted as 'tooth-friendly'.

The effect of the addition of peptides, mainly those derived from milk proteins such as caseinophosphopeptides (CPP), glycomacropeptide (GMP) and total whey proteins in different food and drink formulations, has been considered as cariostatic compounds and have been investigated by several researchers (Moynihan 1999; Cairns *et al.* 2002; Cheng *et al.* 2009). Their addition to products such as toothpaste, chewing gum and soft drinks to minimise the effect of caries and direct acid erosion has been proposed (Adamson & Reynolds 1995). Other milk-derived peptides such as the peptide fractions from casein hydrolysis and different types of cheese whey have also been investigated (Grenby *et al.* 2001; Andrews *et al.* 2005, 2006; Kanekanian *et al.* 2008).

In the following sections, the role of peptides such as CPP and GMP in the protection of tooth enamel from direct acid attack is discussed in more detail.

7.5.1 *Dental erosion*

Dental erosion is the progressive loss of the dental hard surface by the direct action of acids, usually from foods and beverages with pH around 4.0 and below. Dental erosion has therefore been described as mineral dissolution, solubilisation or demineralisation of the ions from the dental hard surfaces due to the direct exposure to acids present in acidic foods and drinks. This is different to the acid produced as a by-product of fermentation by the bacteria in the plaque, which is classified as dental caries (Meurman *et al.* 1987; Cheng *et al.* 2009).

In modern Western societies, diet can be regarded as the most important factor in dental erosion and is one of the major dental health issues in recent years (Lussi *et al.* 1993; Cairns *et al.* 2002; Low & Alhuthali 2008). It has been reported that the incidence of dental caries has been declining in developed countries but, on the other hand, other dental lesions such as erosion and tooth wear have seen a rising pattern. Increasing consumption of acidic beverages, such as carbonated soft drinks and fruit juices (especially among youth), is one of the main reasons for dental erosion (Wongkhanteea *et al.* 2006; Lussi & Jaeggi 2008). For example, fruits, vegetables and especially fruit juices which are regarded as healthy foods may cause significant tooth erosion due to their acidity (Zero 1996; Lussi *et al.* 2004). Citric acid in citrus fruits and vegetables, acetic acid in vinegar and vinegar-preserved pickled vegetables, tartaric acid in wine and grapes, malic acid in apples and apple drinks and lactic acid in fermented dairy products represent the main acids in foods and drinks. Organic acids, in particular citric, malic and tartaric acids, are deemed to be more erosive due to their high-chelating nature (Meurman *et al.* 1987).

7.5.2 *Assessment of dental erosion*

Several methods have been used to measure the level of solubilisation/demineralisation which is based on chemical and/or physical means. Information about how erosion occurs and the effect of various erosive solutions, such as acids in food and beverages, in dissolving the tooth enamel is useful to assess and discover the means to reduce the prevalence of erosion. Such assessment also helps to provide further information on the chemical and physical processes of erosion and methods to minimise or prevent erosion by direct or indirect interventions. There are two main techniques to study dental erosion: either indirectly by chemical analysis to measure the mineral loss from the hard enamel surface of the tooth; or directly by studying the surface pattern of the eroded enamel.

Chemical methods

Chemical analysis is one of the major techniques which have been used in dental erosion research (Grenby *et al.* 1990; Meurman *et al.* 1990). The concentration of the dissolved calcium and the phosphate ions in the erosive solution could be measured chemically as an indication of their dissolution (Barbour & Rees 2004). Several researchers have been using hydroxyapatite (HA) as a model system for the tooth enamel. Both the model system and tooth enamel are composed of calcium phosphate, but in HA it is present mainly in its amorphous phase compared to the more defined crystalline and compact form in tooth enamel. Calcium analysis is usually measured using atomic absorption spectroscopy (AA) and phosphate concentration is usually determined colourimetrically using spectrophotometry at a wavelength of 540 nm (Chen *et al.* 1956).

By measuring the change in pH of a solution, it is also possible to determine the rate of release of OH^- ions and calculate the rate of hydroxyapatite or enamel dissolution. The mechanism of pH change is related to the neutralisation of H^+ by the released OH^- from HA or enamel, therefore increasing the pH (Shellis *et al.* 1993; Barbour & Rees 2004).

Physical methods

As a result of acid erosion to the tooth surface, the smooth enamel surface turns into a rough porous surface (Cairns *et al.* 2002) and facilitates further erosion by harbouring bacteria and their adhesion. This by itself gives rise to dental caries and results in larger pores and further colonisation of cariogenic bacteria with further damage to the tooth. Although chemical analysis is sensitive and accurate, it only provides information about the net concentrations of the released ions from a demineralised specimen. To obtain further information about the physical nature and the changes occurring to the enamel, some visual techniques such as scanning electron microscopy (SEM), profilometry, microradiography, confocal laser scanning microscopy, stereomicroscopy and atomic force microscopy (AFM) have been suggested to provide images of the crystalline surfaces, the position or nucleation of the new material and the colonisation of the oral bacteria onto tooth enamel surface.

Enamel surface strength and roughness were studied by Vitkov *et al.* (2008) using SEM, profilometer and AFM. They reported that SEM gave a much better reflection to the surface structure compared to the other methods; the latter were not able to represent the actual effect of dissolution, especially when acid etching gave high steepness of the enamel crystallites and gave a tightly twisted image which could lead to the level of erosion being misrepresented and underestimated. Typical imaging techniques which have been used are scanning electron microscopy and profilometry to examine and assess surface texture and determine the microhardness of surfaces (Lussi *et al.* 1995, 2000; Grando *et al.* 1996; Amaechi *et al.* 1999; Eisenburger & Addy 2001; Lupi-Pegurier *et al.* 2003).

In SEM, it is not necessary to polish the enamel surface (which is essential in profilometry). However, drying the surface and coating with metal or carbon is essential for scanning (Grenby 1996). SEM as a qualitative technique is used to investigate enamel erosion by taking images of the enamel surface. This technique is usually used as a supporting method to observe the changes in enamel surface pattern during acid dissolution (Shellis *et al.* 1993; Barbour & Rees 2004), which can also be used to observe the adsorption of the protective compounds on the tooth enamel. SEM analyses the surface of the bulk specimen with a high-energy beam of electrons formed by a thermionic gun, accelerated through various voltages between cathode and anode. Images are magnified by two or three electron lenses to provide details of the specimen surface (Reimer 1998).

The modified version of SEM is the environmental scanning electron microscope (ESEM) which can examine the sample without metal or carbon coating, in low vacuum and in wet conditions. The resolution and magnification of images are comparable to those of SEM (Prack 1993; Barbour & Rees 2004).

7.5.3 *The protective effect of peptides*

It has been indicated that adding a thin protective layer (albeit not a lasting one), onto tooth enamel surface as a barrier against acid erosion could protect the enamel from demineralisation. On this note, it was found that peptides from different food protein sources, especially those derived from milk proteins, namely casein, its hydrolysates and whey proteins, have been associated with dental protection through their interaction with the outer layer of the tooth, i.e. the enamel, offering a protection against acid erosion and the subsequent microbial attack. The two most-referred peptides in the literature are caseinophosphopeptide (CPP), which is derived from the proteolytic hydrolysis of casein, and glycomacropeptide (GMP), also referred to as caseinoglycomacropeptide, from whey as well as the proteose peptone fractions of β-casein (FitzGerald 1998; Grenby *et al.* 2001; Andrews *et al.* 2005, 2006; Kanekanian *et al.* 2008).

There is now substantial evidence that intact milk proteins (Vacca *et al.* 1994; Guggenheim *et al.* 1999) have anticariogenic properties, perhaps by selectively adsorbing onto the tooth surface (Guggenheim *et al.* 1999; Devold *et al.* 2006). Several research groups, especially the pioneering work of Reynolds and colleagues (Reynolds 1987; Reynolds *et al.* 1995; Adamson & Reynolds 1995) have indicated that CPP, a heavily phosphorylated peptide derived from milk casein, are even more active. To reduce dental damage due to caries or erosion some studies and experiments have been performed using milk as a source of bioactive peptides and minerals (Reynolds 1987, 1998).

From this point of view, tooth enamel erosion by acids present in food and the role of some peptides such as CPP and GMP as bioactive peptides in reducing the erosion of the enamel are further investigated through chemical and physical means. More emphasis is placed on the role of GMP in the following sections, as it is an important by-product of the cheese industry.

Caseinophosphopeptide (CPP)

CPP is produced by the action of proteolytic enzymes or micro-organisms that hydrolyse casein into small parts (Korhonen & Pihlanto 2006). Different enzymes can hydrolyse casein in different manners and produce peptides with various properties. For example, glutamic-acid-specific endopeptidase (GSE) produces CPP with fewer Ser (P) clusters and shorter chains (Park *et al.* 1998) while trypsin produces CPPs with the cluster sequence of Ser(P)-Ser(P)-Ser(P)-Glu-Glu and the ability to stabilise calcium phosphate in solution

(Poggio *et al.* 2009). Most CPPs are prepared by tryptic hydrolysis from β-casein and αs1-casein providing fraction residue 1–25 and 59–79, respectively. A few other CPP fractions, f(1–21) and f(46–70), have also been isolated from αs2-casein (Cross *et al.* 2006). In addition, the phosphate contents of the CPPs differ according to the source of casein (FitzGerald 1998; Aimutis 2004; Cross *et al.* 2006). αs2-casein contains 10–13 phosphoseryl residues, αs1-casein 8–9 residues, β-casein 5 residues and κ-casein contains only 1 (FitzGerald 1998; Park *et al.* 1998). The phosphoseryl residues in αs-casein and β-casein are followed by glutamic acid residues Ser(P)-Ser(P)-Ser(P)-Glu-Glu which is thought to have a positive effect on the bioactivity of CPP (Ono *et al.* 1994; Holt *et al.* 1996; Reynolds 1998). Glutamic acid residues after phosphoseryl sequence in casein act as mineral binding sites (Park & Allen 1998) due to the negatively charged side chains. These negative charges and the amino acid sequence can introduce calcium binding onto the enamel.

Glycomacropeptide (GMP)

GMP, a C-terminal peptide of κ-casein with 64 amino acids, is produced during cheese making by the action of chymosin. Its presence in whey is accompanied by other major whey proteins such as β-lactoglobulin (β-lg) and α-la which have also been used as a valuable component for their bioactive properties and nutritional value (Abd El-Salam *et al.* 1996; Tolkach & Kulozik 2005; Thomä-Worringer *et al.* 2006). GMP is a heterogeneous peptide due to its different content of carbohydrate, phosphorus and genetic variants due to variations in κ-casein (Abd El-Salam *et al.* 1996).

It has been reported that GMP has 11 genetic variants with two major variants A and B formed mainly from bovine milk; the other variants can be found in ovine and caprine milk (Thomä-Worringer *et al.* 2006). The difference between variant A and B is due to alteration in the available amino acid at position 136 and 148. Variant A contains Thr at position 136 and Asp at 148 while variant B contains Ile and Ala at these positions, respectively (Taylor & Woonton 2009).

On the basis of glycosylation, GMP can be divided into two major fractions. One is glycosylated and phosphorylated, referred to as glyco GMP (gGMP), comprising about 50% of the total bovine GMP; the second fraction is phosphorylated and unglycosylated, referred to as aglyco GMP (aGMP; Kreuß *et al.* 2008). The carbohydrate moiety in GMP is composed of N-acetyl galactosamine (GalNAc), N-acetyl neuraminic acid (NeuAc) or sialic acid and galactose (Gal), providing five different carbohydrate chains. The biological function of GMP is thought to be related to the gGMP and the amount of sialic acid. However, it is not clear which of these two fractions of the GMP has the most impact regarding its functional properties (Thomä-Worringer *et al.* 2006).

GMP has shown to inhibit the adherence of oral bacteria to the salivary coated tooth enamel as well as binding directly to the cell wall of some bacteria. GMP could also be incorporated into salivary pellicle and reduce the adhesion of cariogenic bacteria such *Streptococcus sobrinus* and *S. mutans* (Neeser *et al.* 1988; Thomä-Worringer *et al.* 2006) resulting in reduction in caries. It has also been shown that it might have a remineralisation property and/or inhibit demineralisation of the tooth surface, offering anti-erosive protection (Cairns *et al.* 2002; Thomä-Worringer *et al.* 2006; Sandström *et al.* 2008; Setarehnejad *et al.* 2010).

7.5.4 *Assessment of dental protection*

Chemical methods

Chemical assessment of the actual protective effect of CPP and GMP against mineral dissolution of tooth enamel has been studied by many researchers using either hydroxyapatite (as a tooth model system), tooth enamel in crushed form or as small intact pieces of enamel (Grenby *et al.* 2001; Setarehnejad *et al.* 2010).

In a study by Setarehnejad *et al.* (2009) a detailed comparison was made between three peptide samples – a laboratory-prepared CPP, a commercial CPP (cCPP) and a commercial GMP (cGMP) – for their protective role by measuring the reduction in calcium and phosphate dissolution. Commercial samples are usually available in the market from different dairy companies at different purities; however, Setarehnejad *et al.* used samples from Arla Foods Ingredients Group (Denmark). The purity of the cCPP was only 20% and the purity of cGMP was 85%. All three experimental peptides were used as a solution containing 5 mg mL^{-1} (taking into account their purity) and mixed with HA to give it a protective layer. The coated HA samples were then exposed to citrate buffer (as the erosive solution) at pH values of 3.0, 3.5, 4.0 and 4.5, which resembled the pH of most foods and beverages. The amount of dissolved calcium and phosphate in the acidic solution was measured and compared with the untreated HA, which was considered as the control. Treatment with cGMP demonstrated higher protection than cCPP (Fig. 7.4). At pH 3.0, the protection level of cGMP was 38% compared with only 30% for cCPP. It was however obvious that the protection level increases with the increase in the pH, and the protection level of cGMP increased to almost 50% at pH 4.5.

The inhibitory effect of the GMP and CPP in reducing mineral dissolution from HA surfaces has been suggested to be due to the formation of a thin layer of these peptides which acts as a barrier limiting acid access to HA, hence a reduction in mineral dissolution. In addition, these peptides may bind to the active sites of HA and prevent ion detachment from the surface. Similar observations for the protective effect of CPP and GMP have also been reported by Barbour *et al.* (2008). Although the protective effect is apparent, the actual mechanism of protection is still debatable (Andrews *et al.* 2006).

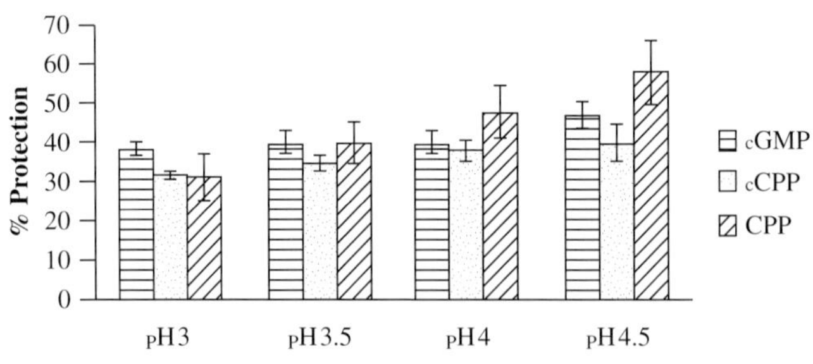

Fig. 7.4 Protective effect of the peptides against calcium and phosphate dissolution from hydroxyapatite using 0.1M citrate buffer as the erosive solution at different pH values. cGMP, commercial glycomacropeptide; cCPP, commercial caseinophosphopeptide and CPP, laboratory prepared caseinophosphopeptide. By Setarehnejad *et al.* 2009 (reproduced with permission of Springer Science + Business Media).

The presence of N-acetyl neuraminic acid (NANA), proline and acidic residues in GMP could contribute to its attachment on HA surface. The NANA can affect the net charge of GMP and the stability of the formed layer on HA (Setarehnejad *et al.* 2010). Some *in vitro* studies have indicated that proline-rich protein (PRP) binds to HA and aids remineralisation. It has also been shown that some of these PRPs are present in saliva, which could provide protective environment inside the oral cavity (Lendenmann *et al.* 2000; van Nieuw Amerongen *et al.* 2004).

On the other hand, with its multiple phosphoserine residues CPP stabilises amorphous calcium phosphate (ACP) in the solution by binding through its phosphoserine residues, forming a CPP–ACP complex. This complex can suppress mineral dissolution, increase remineralisation or possibly both. It has been suggested that the CPP–ACP complex is able to localise calcium phosphate on the tooth surface which is more resistant to acid, can buffer the calcium and phosphate ions and help to provide a saturated environment (Walker *et al.* 2006). The presence of CPP–ACP may also help the rapid return of calcium to the affected area and provide early remineralisation, replacing the lost calcium from enamel (Reynolds 1997; Oshiro *et al.* 2007), and providing better protection at an acidic pH of 4.5 where the rate of dissolution is slower compared to more acidic conditions (pH 3.0; Larsen & Nyvad 1999).

Physical methods

The chemical assessment of the protective effect of GMP, in particular the glycosylated fraction gGMP (Section 7.5.2) was investigated by Setarehnejad *et al.* (2009), and its physical examination was also investigated using SEM (Setarehnejad *et al.* unpublished data). In the latter case, SEM images were used to assess the protective effect of gGMP against acid erosion using 0.1 M citrate buffer at pH 2.5 as the erosive solution. The overall examination of the SEM images showed a distinct difference in the nature and the structure of the enamel surface of the control (untreated) and the treated enamel samples after acid exposure and after treatment with gGMP (protective agent) followed by acid exposure.

The SEM image of the untreated enamel obtained by Setarehnejad *et al.* (2009) shows a smooth surface with no visible pores (Fig. 7.5). This is in direct comparison to the

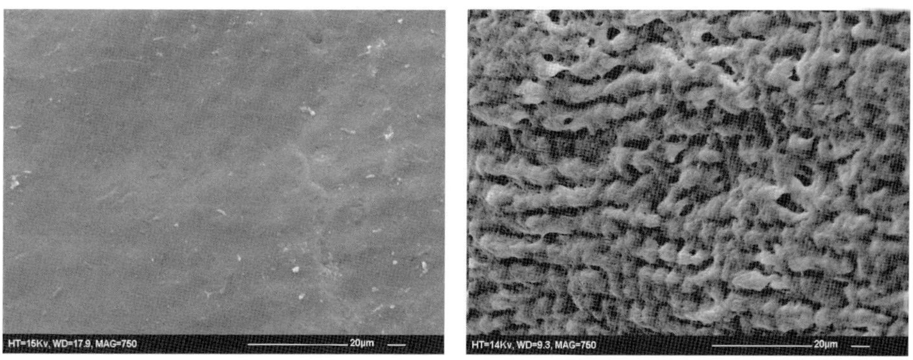

(a) (b)

Fig. 7.5 Scanning electron microscopy of (a) intact enamel and (b) the intact enamel after treatment with 0.1 M citrate buffer at pH 2.5, showing eroded surface of the enamel.

citrate-treated enamel, where the smooth surface of the untreated enamel has completely eroded to give a rough porous structure; this is an indication of substantial dissolution of the enamel, which could easily harbour micro-organisms onto the surface.

Similar observations of acid erosion have also been reported by other researchers (Grando *et al.* 1996; Hannig *et al.* 2005; Vitkov *et al.* 2008; Taubee *et al.* 2010). To eliminate such a drastic change in the outer structure of the enamel, scientists have been applying different organic and inorganic materials to prevent the erosion and strengthen the enamel.

One of these applications which were used by Setarhenejad *et al.* was the treatment of the enamel with gGMP prior to acid treatment, in order to examine the level of protection offered by an aqueous gGMP solution against acid erosion. The SEM image of the enamel treated with gGMP (Fig. 7.6) shows an additional thin deposited layer on top of the enamel (not evenly deposited but clearly visible). To study the effectiveness of this protective layer of gGMP, the sample was exposed to the same conditions used for the erosion where the sample was treated with 0.1 M citric acid. The SEM image (Fig. 7.7) clearly demonstrated that the surface remained intact and maintained its integrity by minimising

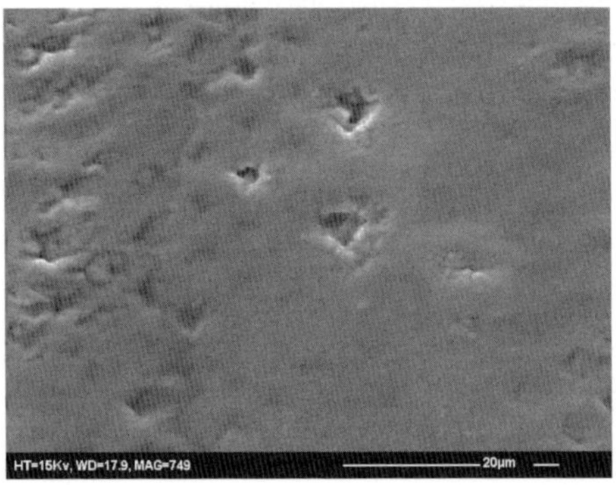

Fig. 7.6 Scanning electron microscopy of intact enamel treated with gGMP 5 mg mL^{-1}, showing a layer of the peptide attached onto the surface.

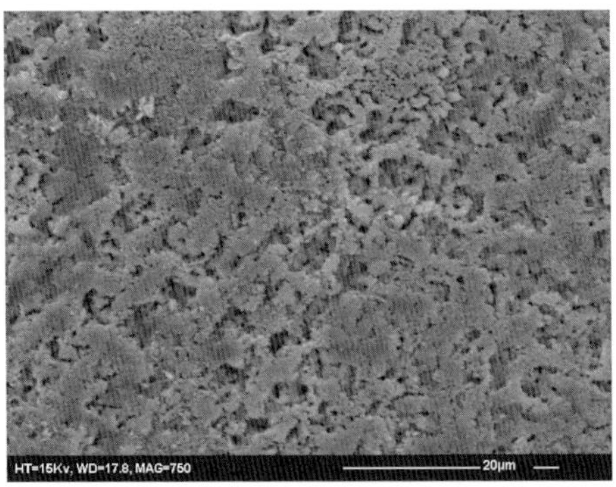

Fig. 7.7 Scanning electron microscopy of the protected enamel treated with gGMP 5 mg mL^{-1} followed by treatment with 0.1 M citrate buffer at pH 2.5, showing most of the protective layer still attached to the surface of the enamel.

the effect of acid dissolution. Such treatment proved to be functional in providing adequate protection, with the commercial application of using gGMP as protective agent within beverages.

Nakanishi *et al.* (2001) indicated that the adsorption of proteins on a solid surface varied due to their charge, size, amino acid composition and type of surface. They found that peptides rich in serine and threonine had a high affinity to be absorbed onto a solid surface such as the enamel, and indicated that acidic amino acid contributes to the attachment of the peptide. Vitorino *et al.* (2004) also indicated that proline contributed to the affinity of salivary proteins onto HA. It is therefore likely that GMP and gGMP with their specific structure and amino acid composition (Martinez *et al.* 2009) bind to HA in a similar way. The adsorbed peptides could provide a barrier against acid erosion and minimise the diffusion of calcium and phosphate ions into the exposed buffer (Hannig *et al.* 2005).

7.5.5 *Summary*

HA (a model system of tooth enamel) treated with CPP, cCPP, cGMP and gGMP showed a protective effect of 30–50% against calcium and phosphate dissolution by citrate buffer at different pH levels. The chemical analysis of the erosive citrate buffer after treatment with HA samples indicated that cGMP could have a better protective effect at lower pH compared to the other experimental peptides at the same pH levels. Expectedly, the protective effect of CPP improved with increasing pH. Peptide adsorption and mineral dissolution of specimens were examined physically by scanning electron microscopy (SEM) and chemically by analysing the calcium and phosphate content in erosive acidic citrate buffer solution. SEM images of enamel treated with the experimental peptide gGMP revealed a thinly deposited layer on the enamel surface, indicating good adhesion properties of these peptides. Such adhesion/adsorption to the surface forms a barrier against mineral dissolution from the surface of the enamel by the action of the erosive acidic condition. An enamel surface lacking such a protection layer undergoes erosion and the formation of a rough and a porous surface that could harbour spoilage bacteria, leading to dental caries. It appears that these peptides could be added to some acidic beverages to reduce their erosive effect on tooth enamel.

7.6 Conclusion

Milk ingredients exhibit numerous physiological effects in the areas of infant, clinical and sports nutrition and in functional foods. Within infant nutrition, specific proteins and components from whey fractions have been shown to protect the infant gut and may help support cognitive development in infants. In clinical nutrition, whey proteins may have a beneficial effect for immobility-caused muscle loss, Type 2 diabetes and inflammatory bowel disease. Glycomacropeptide is well suited as a protein substitute in phenylketonuria diets due to a low level of phenylalanine. In sports nutrition, the beneficial effects of whey proteins include muscle synthesis, recovery and performance. Lastly, in functional foods milk ingredients have shown improvements concerning weight management, cognitive performance and natural defence. As a functional ingredient, glycomacropeptide has also been shown to have a protective effect against tooth enamel erosion at low pH values. It was apparent that glycomacropeptide could form a protective layer over the tooth enamel and prevent enamel erosion through the direct attack of acidic foods and drinks.

Acknowledgements

The authors would like to thank Ulrich Kulozik and Markus Kreuß of the Technische Universität München (Freising, Germany) for supplying the glyco-CMP and a-glyco-CMP; Alistair Sloan of the School of Dentistry, Cardiff University for supplying tooth enamel and the use of SEM; Kelly BéruBé, Tim Jones and Wendy Rowe of Cardiff University for their assistance in using the SEM; and Arla Foods Ingredients Group P/S (Redhill, UK) for providing the cCMP cCCP for this study.

References

Abd El-Salam, M.H., El-Shibiny, S. & Buchheim, W. (1996) Characteristics and potential uses of the casein macropeptide. *International Dairy Journal*, **6**, 327–341.

Adamson, N.J. & Reynolds, E.C. (1995) Characterisation of multiple phosphorylated peptides precipitated from pancreatic digest. *Journal of Dairy Science*, **78**, 2653–2659.

Aimutis, W.R. (2004) Bioactive properties of milk proteins with particular focus on anticariogenesis. *Journal of Nutrition*, **134**, 989S–995S.

Amaechi, B.T., Higham, S.M. & Edgar, W.M. (1999) Factors influencing the development of dental erosion in vitro: enamel type, temperature and exposure time. *Journal of Oral Rehabilitation*, **26**, 624–630.

Anderson, J.W., Johnstone, B.M. & Remley, D.T. (1999) Breastfeeding and cognitive development: a meta analysis. *American Journal of Clinical Nutrition*, **70**, 525–535.

Andrews, A.T., Williams, R.J.H. & Kanekanian, A.D. (2005) beta-CN-5P and beta-CN-4P components of milk proteose-peptone: large-scale preparation and influence on the growth of cariogenic microorganisms. *Food Chemistry*, **96**(2), 234–241.

Andrews, A.T., Williams, R.J.H., Isgrove, F.H., Jenkins, K. & Kanekanian, A. (2006) β-CN-5P and β-CN-4P components of bovine milk proteose-peptones: large-scale preparation and influence on the growth of cariogenic microorganisms. *Food Chemistry*, **96**, 234–241.

Astrup, A., Chaput, J., Gilbert, J. & Lorenzen, J.K. (2010) Dairy beverages and energy balance. *Physiology & Behavior*, **100**(1), 67–75.

Azuma, N., Yamauchi, K. & Mitsuoka, T. (1984) Bifidus growth-promoting activity of a glyco-macropeptide derived from human k-casein. *Agricultural and Biological Chemistry*, **48**, 427–437.

Baer, D.J., Stote, K.S., Paul, D.R., Harris, G.K., Rumpler, W.V. & Clevidence, B.A. (2011) Whey protein but not soy protein supplementation alters body weight and composition in free-living overweight and obese adults. *Journal of Nutrition*, **141**(8), 1489–1494.

Ballevre, O., Finot, P.A. & Breuille D. (2003) Method for maintaining or improving the synthesis of mucins. United States patent application publication. US 2003/0157186 A1.

Barbour, M.E. & Rees, J.S. (2004) The laboratory assessment of enamel erosion: a review. *Journal of Dentistry*, **32**, 591–602.

Barbour, M.E., Shellis, R.P., Parker, D.M., Allen, G.C. & Addy, M. (2008) Inhibition of hydroxyapatite dissolution by whole casein: the effect of pH, protein concentration, calcium and ionic strength. *European Journal of Oral Sciences*, **116**, 473–478.

Belobrajdic, D.P., McIntosh, G.H. & Owens J.A. (2004) A high-whey-protein diet reduces body weight gain and alters insulin sensitivity relative to red meat in Wistar rats. *Journal of Nutrition*, **134**, 1454–1458.

Bettler, J. & Kullen, M.J. (2007) Infant formula enriched with alpha-lactalbumin has a prebiotic effect in healthy term infants. *Journal of Pediatric Gastroenterology and Nutrition*, **45**, E1–E90.

Beucher, S., Levenez, F., Yvon, M. & Corring, T. (1994) Effects of gastric digestive products from casein on CCK release by intestinal cells in rat. *Journal of Nutritional Biochemistry*, **5**, 578–584.

Boirie, Y., Dangin, M., Gachon, P., Vasson, M.P., Maubois, J.L. & Beufrere, B. (1997) Slow and fast dietary proteins differently modulate postprandial protein accretion. *Proceedings of National Academy of Science*, **94**, 14930–14935.

Borsheim, E., Tipton, K.D., Wolf, S.E. & Wolfe, R.R. (2002) Essential amino acids and muscle protein recovery from resistance exercise. *American Journal of Physiology: Endocrinology and Metabolism*, **283**(4), E648–E657.

Borsheim, E., Aarsland, A. & Wolfe, R.R. (2004) Effect of an amino acid, protein, and carbohydrate mixture on net muscle protein balance after resistance exercise. *International Journal of Sport Nutrition and Exercise Metabolism*, **14**(3), 255–271.

Bouhallab, S., Favrot, C. & Maubois, J.L. (1993) Growth-promoting activity of tryptic digest of caseinomacropeptide for Lactococcus lactis subsp lactis. *Lait*, **73**, 73–77.

Bouthegourd, J.C., Roseau, S.M., Makarios-Lahham, L., Leruyet, P.M., Tome, D.G. & Even, P.C. (2002) A preexercise -lactalbumin-enriched whey protein meal preserves lipid oxidation and decreases adiposity in rats. *American Journal of Physiology: Endocrinology and Metabolism*, **283**, E565–E572.

Bovee-Oudenhoven, I.M.J., Termont, D.S.M.L., Heidt, P.J. & Van der Meer, R. (1997) Increasing the intestinal resistance of rats to the invasive pathogen Salmonella enteritids: additive effects of dietary lactulose and calcium. *Gut*, **40**, 497–504.

Bovee-Oudenhoven, I.M.J., Wissink, M.L., Wouters, J.T. & Van der Meer, R. (1999) Dietary calcium phosphate stimulates intestinal Lactobacilli and decreases the severity of a Salmonella infection in rats. *Journal of Nutrition*, **129**(3), 607–612.

Bovee-Oudenhoven, I.M.J., Wissink, M.L., Van Doesburg, W., Witteman, B.J.M. & Van der Meer, R. (2003) Diarrhea caused by enterotoxigenic Escherichia coli infection of humans is inhibited by dietary calcium. *Gastroenterology*, **125**, 469–476.

Boza, J.J. (2000) Protein hydrolysates vs free amino acid-based drinks on the nutritional recovery of the starved rat. *European Journal of Nutrition*, **39**, 237–243.

Brody, E.P. (2000) Biological activities of bovine glycomacropeptide. *British Journal of Nutrition*, **84**(suppl1), S39–S46.

Brück, W.M., Graverholt, G. & Gibson, G.R. (2002) Use of batch culture and a two-stage continuous culture system to study the effect of supplemental alpha-lactalbumin and glycomacropeptide on mixed populations of human gut bacteria. *FEMS Microbiology Ecology*, **41**, 231–237.

Brück, W.M., Kelleher, S.L., Gibson, G.R., Nielsen, K.E., Chatterton, D.E. & Lönnerdal, B. (2003a) rRNA probes used to to quantify the effects of glycomacropeptides and alpha lactalbumin supplementation on the predominant groups of intestinal bacteria of infant rhesus monkeys challanged with enteropathogenic E. coli. *Journal of Pediatric Gastroenterology and Nutrition*, **37**, 273–280.

Brück, W.M., Graverholt, G. & Gibson, G.R. (2003b) A two-stage continuous culture system to study the effect of supplemental alpha-lactalbumin and glycomacropeptide on mixed cultures of human gut bacteria challanged with enteropaphogenic E. coli and S. serotype typhimurium. *Journal of Applied Microbiology*, **95**, 444–453.

Brück, W.M., Redgrave, M., Tuohy, K.M., Lönnerdal, B., Graverholt, G., Hernell, O. & Gibson, G.R. (2006) Effects of bovine alpha-lactalbumin and casein glycomacropeptide-enriched infant formulae on faecal microbiota in healthy term infants. *Journal of Pediatric Gastroenterology and Nutrition*, **43**, 673–679.

Buckley, J.D., Thomson, R.L., Coates, A.M., Howe, P.R., Denichilo, M.O. & Rowney M.K. (2010) Supplementation with a whey protein hydrolysate enhances recovery of muscle force-generating capacity following eccentric exercise. *Journal of Science and Medicine in Sport*, **13**(1), 178–181.

Burling, H. & Graverholt, G. (2008) Milk – a new source for bioactive phospholipids for use in food formulations. *Lipid Technology*, **20**, 229–231.

Burton-Freeman, B.M. (2008) Glycomacropeptide (GMP) is not critical to whey-induced satiety, but may have a unique role in energy intake regulation through cholecystokinin (CCK). *Physiology and Behavior*, **93**(1–2), 379–387.

Cairns, A.M., Watson, M., Creanor, S.L. & Foye, R.H. (2002) The pH and titratable acidity of diluting drinks and their potential effect on dental erosion. *Journal of Dentistry*, **30**, 313–317.

Calbet, J.A. & MacLean, D.A. (2002) Plasma glucagon and insulin responses depend on the rate of appearance of amino acids after ingestion of different protein solutions in humans. *Journal of Nutrition*, **8**(8), 2174–2182.

Calbet, J.A.L. & Holst, J.J. (2004) Gastric emptying, gastric secretion and enterogastrone response after administration of milk proteins or their polypeptide hydrolysates in humans. *European Journal of Nutrition*, **43**, 127–139.

Calder, P.C., Albers, R., Antoine, J.M *et al.* (2009) Inflammatory disease processes and interactions with nutrition. *British Journal of Nutrition*, **101**, S2–S34.

Carruth, B.R. & Skinne, J.D. (2001) The role of dietary calcium and other nutrients in moderating body fat in preschool children. *International Journal of Obesity*, **25**, 559–566.

Chen, P.S., Toribara, T.Y. & Warner, H. (1956) Microdetermination of phosphorus. *Analytical Chemistry*, **28**, 1756–1758.

Cheng, R., Yang, H., Shao, M., Hu, T. & Zhou, X. (2009) Dental erosion and severe tooth decay related to soft drinks: a case report and literature review. *Journal of Zhejiang University Science B*, **10**, 395–399.

Clare, D.A., Zheng, Z., Hassan, H.M., Swaisgood, H.E. & Gatignani, G.L. (2008) Antimicrobial properties of milk fat globular membrane fractions. *Journal of Food Protection*, **71**(1), 126–133.

Conn, J.W. & Newburgh, L.H. (1936) The glycemic response to isoglucogenic quantities of protein and carbohydratye. *Journal of Clinical Investigation*, **15**(6), 665–671.

Cooke, M., Cribb, P.J. & Hayes, A. (2004) The effects of short-term supplementation on muscle force recovery on eccentrically induced muscle damage in healthy individuals. In *Whey Proteins in Sports Nutrition* (ed. Cribb, P.J.) USDEC Monograph. (Presented at *The Australian Association for Exercise and Sports Science Inaugural National Conference*.)

Cribb, P. (ed.) (2004) *Whey Proteins and Immunity*. Vancouver: US Dairy Export Council.

Cribb, P.J., Williams, A.D., Hayes, A. & Carey, M.F. (2002) Effect of whey isolate and resistant training on strength, body composition and plasma glutamine. *American College of Sports Medicine*, **34**, S299.

Cribb, P.J., Williams, A.D., Carey, M.F. & Hayes, A. (2006) The effect of whey isolate and resistance training on strength, body composition, and plasma glutamine. *International Journal of Sport Nutrition and Exercise Metabolism*, **16**, 494–506.

Cross, K.J., Huq, N.L. & Reynolds, E.C. (2006) Anticariogenic peptides. In *Nutraceutical Proteins and Peptides in Health & Disease* (eds Mine, Y *et al.*), pp. 335–352. London: Taylor and Francis Group.

Daddaoua, A., Puerta, V., Zarzuelo, A., Suarez, M.D., Medina, F.S. & Martinez-Augustin, O. (2005). Bovine glycomacropeptide is anti-inflammatory in rats with hapten-induced colitis. *Journal of Nutrition*, **135**, 1164–1170.

Davies, K.M., Heaney, P.R., Recker, R.R., Lappe, J.M., Barger-Lux, M.J., Rafferty, K. & Hinders, S. (2000) Calcium intake and body weight. *Journal of Clinical Endocrinology and Metabolism*, **85**, 4635–4648.

Davis, A.M., Harris, B.J, Lien, E.L., Pramuk, K. & Trabulsi, J. (2007) alpha-Lactalbumin-rich infant formula fed to healthy term infants in a multicenter study: plasma essential amino acids and gastrointestinal tolerance. *European Journal of Clinical Nutrition*, **62**(11), 1294–3011.

Devold, T.G., Rykke, M., Isbey, D., Sorensen, E.S., Christensen, B., Langsrud, T., Svenning C., Borch-Iohnsen, B., Karlsen, J. & Vegarud, G. (2006) In vitro studies of adsorption of milk proteins onto tooth enamel. *International Dairy Journal*, **16**, 1013–1017.

Dewettinck, K., Rombaut, R., Thienpont, N., Trung Le, T., Messens, K. & Van Camp, J. (2007) Nutritional and technological aspects of milk globule membrane material. *International Dairy Journal*, **18**(5), 436–457.

Eisenburger, M. & Addy, M. (2001) Evaluation of pH and erosion time on demineralisation. *Clinical Oral Investigations*, **5**, 108–111.

Ferrando, A.A., Lane, H.W., Stuart, C.A., Davis-Street, J. & Wolfe, R.R. (1996) Prolonged bed rest decreases skeletal muscle and whole body protein synthesis. *American Journal of Physiology*, **270**(33), E627–E633.

Ferrando, A.A., Paddon-Jones, D., Hays, N.P., Kortebein, P., Ronson, O., Williams, R.H., McComb, A., Symons, T.B., Wolfe, R.R. & Evans, W. (2010) EAA supplementation to increase nitrogen intake improves muscle function during bed rest in the elderly. *Clinical Nutrition*, **29**, 18–23.

FitzGerald, R. J. (1998) Potential uses of caseinophosphopeptides. *International Dairy Journal*, **8**, 451–457.

Frid, A.H., Nilsson, M., Holst, J.J. & Bjorck, I.M.E. (2005) Effect of whey on blood glucose and insulin response to composite breakfast and lunch meals in type 2 diabetic subjects. *American Society of Clinical Nutrition*, **82**, 69–75.

Gale, C.R., Mariott, L.D., Martyn, M.N., Limond, J., Inskip, H.M., Godfey, K.M., Law, C.M., Copper, C., West, C. & Robinson, S.M. (2010) Breastfeeding, the use of docosahexaenoic acid fortified formulas in infancy and neuropsychological function in childhood. *Archives of Disease in Childhood*, **95**(3), 174–179.

Gannon, M.C., Nuttall, F.Q., Krezowski, P.A., Billington, C.J. & Parker S. (1986) The serum insulin and plasma glucose responses to milk and fruit products in Type 2 (non-insulin-dependent) diabetic patients. *Diabetologia*, **28**, 784–791.

Grando, L.J., Tames, D.R., Cardoso, A.C. & Gabilan, N.H. (1996) In vitro study of enamel erosion caused by soft drinks and lemon juice in deciduous teeth analysed by SEM. *Caries Research*, **30**(5), 373–378.

Grenby, T.H. (1996) Lessening dental erosive potential by product modification. *European Journal of Oral Sciences*, **104**, 221–228.

Grenby, T.H., Mistry, M. & Desai, T. (1990) Potential dental effects of infant's fruit drinks studied in vitro. *American Journal of Nutrition*, **64**, 273–283.

Grenby, T.H., Andrews, A.T., Mistry, M. & Williams, R.J.H. (2001) Dental caries-protective agents in milk and milk products: investigation in vitro. *Journal of Dentistry*, **29**, 83–92.

Guggenheim, B., Schmid, R., Aeschilman, J. & Neeser J. (1999) Powdered milk micellar casein prevent oral colonisation by Streptococcus sobrinus and dental caries in rat: A basis for caries-protective effect of dairy products. *Caries Research*, **33**, 446–454.

Hannig, C., Hamkens, A., Becker, K., Attin, R. & Attin, T. (2005) Erosive effects of different acids on bovine enamel: release of calcium and phosphate in vitro. *Archives of Oral Biology*, **50**, 541–552.

Hayes, A. & Cribb, P.J. (2008) Effect of whey protein isolate on strength, body composition and muscle hypertrophy during resistance training. *Current Opinion in Clinical Nutrition and Metabolic Care*, **11**, 40–44.

Hellhammer, J., Fries, E., Buss, C., Engert, V., Tuch, A., Rutenberg, D. & Hellhammer, D. (2004) Effects of Soy Lecithin phosphatic acid and phosphatidylseine complex(PAS) on the endocrine and physical response to mental stress. *Stress*, **7**, 119–126.

Hero, T., Schult, J. & Hellhammer, J. (2009) Three week intake of milk phosphatidylserine: beneficial effect on memory and psychological stress response. *Poster presentation at Nutrition Event, Lille, France, 17–18* June.

Hoffman, J.R. & Falvo, M.J. (2004) Protein: which is best? *Journal of Sports Science and Medicine*, **3**, 118–130.

Holt, C., Wahlgren, N.M. & Drakenberg, T. (1996) Ability of a casein phosphopeptide to modulate the precipitation of calcium phosphate by forming amorphous dicalcium phosphate nanoclusters. *Biochemical Journal*, **314**, 1035–1039.

Idota, T., Kawakami, H. & Nakajima, I. (1994) Growth promoting effect of N-acetylneuraminic acid containing substances on Bifidobacteria. *Bioscience, Biotechnology and Biochemistry*, **58**(9), 1720–1722.

Ingle, B.M., Hay, S.M., Bottjer, H.M. & Eastell, R. (1999) Changes in bone mass and bone turnover following ankle fracture. *Osteoporosis International*, **10**, 408–415.

Ivy, J.L., Res, P.T., Sprague, R.C. & Widzer, M.O. (2003) Effect of a carbohydrate-protein supplement on endurance performance during exercise of varying intensity. *International Journal of Sport Nutrition and Exercise Metabolism*, **13**, 382–395.

Jacobsen, R., Lorenzen, J.K., Toubro, S., Krog-Mikkelsen, I. & Astrup, A. (2005) Effect of shortterm high dietary calcium intake on 24 h energy expenditure, fat oxidation and fecal fat excretion. *International Journal of Obesity*, **29**, 292–301.

Kanekanian, A.D, Williams, R.J.H., Brownsell, VL. & Andrews, A.T. (2008) Caseinophosphopeptides and dental protection: concentration and pH studies. *Food Chemistry*, **107**, 1015–1021.

Korhonen, H. & Pihlanto, A. (2006) Bioactive peptides: production and functionality. *International Dairy Journal*, **16**, 945–960.

Kreuß, M., Krause, I. & Kulozik, U. (2008) Separation of a glycosylated and non-glycosylated fraction of caseinomacropeptide using different anion-exchange stationary phases. *Journal of Chromatography A*, **1208**, 126–132.

Kvistgaard, A.S., Pallesen, L.T., Arias, C.F., Lopez, S., Petersen, T.E., Heegaard, C.W. & Rasmussen, J.T. (2004) Inhibitory effects of human and bovine milk constituents on Rotavirus infections. *Journal of Dairy Science*, **87**, 4088–4096.

Lands, L.C., Grey, V.L. & Smountas, A.A. (1999) Effect of supplementation with a cysteine donor on muscular performance. *Journal of Applied Physiology*, **87**, 1381–1385.

Larsen, M.J. & Nyvad, B. (1999) Enamel erosion by some soft drinks and orange juices relative to their pH, buffering effect and contents of calcium phosphate. *Caries Research*, **33**, 81–87.

Layman, D.K. (2002) Role of leucine in protein metabolism during exercise and recovery. *Canadian Journal of Applied Physiology*, **27**, 646–663.

Lendenmann, U., Grogan, J. & Oppenheim, F.G. (2000). Saliva and dental pellicle - a review. *Advances in Dental Research*, **14**, 22–28.

Liljeberg, H. & Bjorck, I. (2001) Milk as a supplement to mixed meals may elevate postprandial insulinaemia. *European Journal of Clinical Nutrition*, **55**, 994–999.

Lim, K., Van Calcar, S.C., Nelson, K.L., Gleason, S.T. & Ney, D.M. (2007) Acceptable low-phenylalanine foods and beverages can be made with glycomacropeptide from cheese whey for individuals with PKU. *Molecular Genetics and Metabolism*, **92**, 176–178.

Loimarante, V., Laine, M., Soderling, E., Rokka, S., Marnila, P *et al.* (1999) Effect of bovine immune- and non-immune whey preparation on the composition and pH response of human dental plaque. *European Journal of Oral Science*, **107**, 244–250.

Lönnerdal, B. (2003) Nutritional and physiologic significance of human milk proteins. *American Journal of Clinical Nutrition*, **7**(suppl), 1537S–1543S.

Lönnerdal, B. (2010) Bioactive proteins in human milk: mechanisms of action. *Journal of Pediatrics*, **156**(2 suppl), S26–30.

Lorenzen, J.K. (2009) The effect of dietary calcium on energy balance and blood lipid profile. PhD thesis. University of Copenhagen.

Low, I.M. & Alhuthali, A. (2008) In-situ monitoring of dental erosion of tooth enamel when exposed to soft drinks. *Material Science and Engineering*, **28**, 1322–1325.

Lupi-Pegurier, L., Muller, M., Leforestier, E., Bertrand, M.F. & Bolla, M. (2003) In vitro action of Bordeaux red wine on the microhardness of human dental enamel. *Archives of Oral Biology*, **48**, 141–145.

Lussi, A. & Jaeggi, T. (2008) Erosion-diagnosis and risk factors. *Clinical Oral Investigations*, **12**(suppl1), S5–S13.

Lussi, A., Jäggi, T. & Schärer, S. (1993) The influence of different factors on in vitro enamel erosion. *Caries Research*, **27**, 387–393.

Lussi, A., Jaeggi, T. & Jaeggi-Scharer, S. (1995) Prediction of the erosive potential of some beverages. *Caries Research*, **29**, 349–354.

Lussi, A., Kohler, N., Zero, D., Schaffner, M. & Megert, B. (2000) A comparison of the erosive potential of different beverages in primary and permanent teeth using an in vitro model. *European Journal of Oral Science*, **108**, 110–114.

Lussi, A., Jaeggi, T. & Zero, T. (2004) The role of diet in the aetiology of dental erosion. *Caries Research*, **38**, 34–44.

MacDermid, P.W. & Stannard, S.R. (2006) A whey supplemented, high protein diet versus a high carbohydate diet: effects on en endurance cycling performance. *International Journal of Sport Nutrition and Exercise Metabolism*, **16**, 65–77.

Mariotti, F., Simbelie, K.L., Makarios-Lahham, L., Huneau, J.F., Laplaize, B., Tome, D. & Even, P.C. (2004) Acute ingestion of dietary proteins improves post-Eexercise liver glutathione in rats in a dose-dependent relationship with their cysteine content. *Journal of Nutrition*, **134**(1), 128–131.

Martinez, M.J., Carrera Sanchez, C., Rodríguez Patino, J.M. & Pilosof, A.M.R. (2009) Bulk and interfacial behaviour of caseinoglycomacropeptide (GMP). *Colloids and Surfaces B: Biointerfaces*, **71**, 230–237.

May, M.E. & Buse, M.G. (1989) Effects of branched-chain amino acids on protein turnover. *Diabetes/Metabolism Reviews*, **5**, 227–245.

McAuley, J.L., Linden, S.K., Png, C.W., King, R.M., Pennington, H.L., Gendler, S.J., Florin, T.H., Hill, G.R., Korolik, V. & McGuckin, M.A. (2007) MUC1 cell surface mucin is a critical element of the mucosal barrier to infection. *Journal of Clinical Investigations*, **117**(8), 2313–2324.

McDaniel, M.A., Maier, S.F. & Einstein, G.O. (2003) 'Brain-specific' nutrients: a memory cure? *Nutrition*, **19**(11–12), 957–975.

McJarrow, P., Schnell, N., Jumpsen, J. & Clandinin, T. (2009) Influence of dietary gangliosides on neonatal brain development. *Nutrition Reviews*, **67**(8), 451–463.

McKenzie, S., Phillips, S.M., Carter, S.L., Lowther, S., Gibala, M.J. & Tarnopolsky, M.A. (2000) Endurance exercise training attenuates leucine oxidation and BCOAD activation during exercise in humans. *American Journal of Physiology: Endocrinology and Metabolism*, **278**(4), E580–E587.

Meurman, J.H., Rytomaa, I., Kari, K., Laakso, T. & Murtomaa, H. (1987) Salivary pH and glucose after consuming various beverages, including sugar-containing drinks. *Caries Research*, **21**, 353–359.

Meurman, J.H., Harkonen, M., Naveri, H., Koskinen, J., Torkko, H. & Rytomaa, I. (1990) Experimental sports drinks with minimal dental erosion effect. *Scandinavian Journal of Dental Research*, **98**, 120–128.

Middleton, N., Jelen, P. & Bell G. (2004) Whole blood and mononuclear cell glutathione response to dietary whey protein supplementation in sedentary and trained male human subjects. *International Journal of Food Science and Nutrition*, **55**, 131–141.

Monteleone, P., Beinat, L., Tanzillo, C., Maj, M. & Kemali, D. (1990) Effects of phosphatidylserine on the neuroendocrine response to physical stress in humans. *Neuroendocrinology*, **52**, 243–248.

Morifuji, M., Ishizaka, M., Baba, S., Fukuda, K., Matsumoti, H., Koga, J., Kanegae, M. & Higichi, M. (2010) Comparison of different sources and degrees of hydrolysis of dietary protein: effect on plasma amino acids, dipeptides, and insulin responses in human subjects. *Journal of Agricultural and Food Chemistry*, **58**, 8788–8797.

Motouri, M., Matsuyama, H., Yamamura, J., Tanaka, M., Aoe, S., Iwanaga, T. & Kawakami, H. (2003) Milk sphingomyelin accelerates enzymatic and morphological maturation of the intestine in artificially reared rats. *Journal of Pediatric Gastroenterology and Nutrition*, **36**, 241–247.

Moynihan, P.J., Ferrier, S. & Jenkins, G.N. (1999) The cariostatic potential of cheese:cooked cheese-containing meals increase plaque calcium concentration. *British Dental Journal*, **187**, 664–667.

Nair, K.S. & Short, K.R. (2005) Hormonal and signalling role of branched-chain amino acids. *Journal of Nutrition*, **135**, 1547S–1552S.

Nakajima, K., Tamura, N., Kobayashi-Hattori, K., Yoshida, T., Hara-Kudo, Y., Ikedo, M., Sugita-Konishi, Y. & Hattori, M. (2005) Prevention of intestinal infection by glyco-macropeptide. *Bioscience, Biotechnology and Biochemistry*, **69**, 2294–3015.

Nakanishi, K., Sakiyama, T. & Imamura, K. (2001) On the adsorption of proteins on solid surfaces, a common but very complicated phenomenon. *Journal of Bioscience and Bioengineering*, **91**(3), 233–244.

Neeser, J.R., Chambaz, A., Del Vedovo, S., Prigent, M.J. & Guggenheim, B. (1988) Specific and non-specific inhibition of adhesion of oral actinomyces and streptococcii to erythrocytes and polystyrene by caseinoglycopeptide derivatives. *Infection and Immunity*, **56**, 3201–3208.

Ney, D.M., Hull, A.K., van Calcar, S.C., Liu, X. & Etzel, M.R. (2008) Dietary glycomacropeptide supports growth and reduces the concentrations of phenylalanine in plasma and brain in a murine model of Phenylketonuria. *Journal of Nutrition*, **138**, 316–322.

Nilsson, M., Stenberg, M., Frid, A.H., Holst, J.J. & Bjorck I.M.E. (2004) Glycemia and insulinemia in healthy subjects after lactose-equivalent meals of milk and other food proteins: the role of plasma amino acids and incretins. *American Journal of Clinical Nutrition*, **80**, 1246–1253.

Nilsson, M., Holst, J.J. & Bjorck, I.M.E. (2007) Metabolic effects of amino acid mixtures and whey protein in healthy subjects: studies using glucose-equivalent drinks. *American Journal of Clinical Nutrition*, **85**, 996–1004.

Novo Nordisk (2009) *A Focused Health Care*. Scientific Publications.

Nuttall, F.Q., Mooradian, A.D., Gannon, M.C., Billington, C. & Krezowski P. (1984) Effect of protein ingestion on the glucose and insulin response to a standardized oral glucose load. *Diabetes Care*, **7**(5), 465–470.

Ono, T., Ohotowa, T. & Takagi, Y. (1994) Complexes of caseinophosphopeptide and calcium phosphate prepared from casein micelles by tryptic digestion. *Bioscience, Biotechnology and Biochemistry*, **58**, 1378–1380.

Oshiro, M., Yamagouchi, K., Takamizawa, T., Inage, H., Watanabe, T., Irokawa, A., Ando, S. & Miyazaki, M. (2007) Effect of CPP-ACP paste on tooth mineralization: an FE-SEM study. *Journal of Oral Sciences*, **49**, 115–120.

Ostman, M.E., Liljeberg Elmstahl, H.G.M. & Bjorck, I.M.E. (2001) Inconsistency between glycemic and insulinemic response to regular and fermented milk products. *American Journal of Clinical Nutrition*, **74**, 96–100.

Paddon-Jones, D., Sheffield-Moore, M., Urban, R.J., Sanford, A.P., Aarsland, A., Wolfe, R.R. & Ferrando, A.A. (2004) Essential amino acid and carbohydrate supplementation ameliorates muscle protein loss in humans during 28 days bedrest. *Journal of Clinical Endocrinology and Metabolism*, **89**(9), 4351–4358.

Paddon-Jones, D., Sheffield-Moore, M., Cree, M.G., Hewlings, S.J., Aarsland, A., Wolfe, R.R. & Ferrando, A.A. (2006) Atrophy and impaired muscle protein synthesis during prolonged inactivity and stress. *Journal of Clinical Endocrinology and Metabolism*, **91**, 4836–4841.

Pal, S., Ellis, V. & Dhaliwal, S. (2010) Effects of whey protein isolate on body composition, lipids, insulin and glucose in overweight and obese individuals. *British Journal of Nutrition*, **104**, 716–723.

Park, E.J., Suh, M., Ramanujam, K., Steiner, K., Begg, D. & Clandinin, M.T. (2005) Diet-induced changes in membrane gangliosides in rat intestinal mucosa, plasma and brain. *Journal of Pediatric Gastroenterology and Nutrition*, **40**(4), 487–495.

Park, O. & Allen, J.C. (1998) Preparation of phosphopeptides derived from αs-casein and β-casein using immobilized glutamic acid-specific endopeptidase and characterization of their calcium binding. *Journal of Dairy Science*, **81**, 2858–2865.

Park, O., Swaisgood, H.E. & Allen, J.C. (1998) Calcium binding of phosphopepides derived from hydrolysis of α_s-casein or ß–casein using immobilized trypsin. *Journal of Dairy Science*, **81**, 2850–2857.

Petschow, B.W. & Talbott, R.D. (1991) Response of bifidobacterium species to growth promoters in human and cow milk. *Pediatric Research*, **29**(2), 208–213.

Pilvi, T.K., Harala, S., Korpela, R. & Mervaala, E.M. (2009) Effects of high-calcium diets with different whey proteins on weight loss and weight regain in high-fat-fed C578BL/6J mice. *British Journal of Nutrition*, **102**, 337–341.

Poggio, C., Lombardini, M., Dagna, A., Chiesa, M. & Bianchi, S. (2009) Protective effect on enamel demineralization of a CCP-ACP past: an AFM in vitro study. *Journal of Dentistry*, **37**, 949–954.

Power, O., Hallihan, A. & Jakeman, P. (2009) Human insulinotropic response to oral ingestion of native and hydrolysed whey protein. *Amino Acids*, **37**, 333–339.

Prack, E.R. (1993) An introduction to process visualization capabilities and considerations in the environmental scanning electron microscope (ESEM). *Microscopy Research and Technique*, **25**, 487–492.

Reimer, L. (1998) *Scanning Electron Microscopy Physics of Image Formation and Microanalysis*. Berlin, Heidelberg, New York: Springer-Verlag.

Requena, P., Daddaoua, A., Martinez-Plata, E., Gonzalez, M., Zarzuelo, A., Suarez, M.D., Medina, F.S. & Martinez-Augustin, O. (2008) Bovine glycomacropeptide ameliorates experimental rat ileitis by mechanisms involving downregulation of interleukin 17. *British Journal of Pharmacology*, **154**, 825–832.

Reynolds, E.C. (1987) The prevention of sub-surface demineralization of bovine enamel and change in plaque composition by casein in an intra-oral model. *Journal of Dental Research*, **66**, 1120–1127.

Reynolds, E.C. (1997) Remineralisation of enamel surface lesions by caseinophosphopeptides stabilised calcium phosphate. *Journal of Dental Research*, **76**(9), 1587–1595.

Reynolds, E.C. (1998) Anticariogenic complexes of amorphous calcium phosphate stabilized by casein phosphopeptides: a review. *Special Care in Dentistry*, **18**(1), 8–16.

Reynolds, E.C., Cain, C.J., Webber, F.L., Black, C.L., Riley, P.F., Johnson, I.H. & Perich, J.W. (1995) Anticariogenicity of calcium phosphate complexes of tryptic casein phosphopeptides in the rat. *Journal of Dental Research*, **74**, 1272–1279.

Rhoades, J.R., Gibson, G.R., Formentin, K., Beer M., Greenberg, N. & Rastall, R.A. (2005) Caseino glycomacropeptide inhibits adhesion of pathogenic Escherichia coli to human cells in culture. *Journal of Dairy Science*, **88**, 3455–3459.

Rios, D., Honório, H.M., Magalhães, A.C., Bonifácio da Silva, S.M., Delbem, A.C.B., Machado, M.A. & Buzalaf, M.A.R. (2008) Scanning electron microscopic study of the *in situ* effect of salivary stimulation on erosion and abrasion in human and bovine enamel. *Brazilian Oral Research*, **22**(2), 132–138.

Rueda, R., Maldonado, J., Narbona, E. & Gil, A. (1998) Neonatal dietary gangliosides. *Early Human Development*, **53**, S135–S147.

Russell, M.W., Childers, N.K., Michalek, S.M., Smith, D.J. & Taubman, M.A. (2004) A caries vaccine? The state of the science of immunization against dental caries. *Caries Research*, **38**, 230–235.

Sando, L., Pearson, R., Gray, C., Parker, P., Hawken, R., Thomson, P.C., Meadows, J.R., Kongsuwan, K., Smith, S. & Tellam, R.L. (2009) Bovine Muc1 is a highly polymorphic gene encoding an extensively glycosylated mucin that binds bacteria. *Journal of Dairy Science*, **92**(10), 5276–5291.

Sandström, O., Lönnerdal, B., Graverholt, G. & Hernell, O. (2008) Effect of α- lactalbumine-enriched formula containing different concentrations of glycomacropeptide on infant nutrition. *The American Journal of Clinical Nutrition*, **87**(4), 921–928.

Sanjurjo, P., Aldamiz, L., Georgi, G., Jelinek, J., Ruiz, J.I. & Boehm, G. (2003) Dietary threonine reduces plasma phenylalanine levels in patients with hyperphenylalanemia. *Journal of Pediatric Gastroenterology and Nutrition*, **36**, 23–26.

Schaafsma, G. (2000) The protein digestibility-corrected amino acid score. *Journal of Nutrition*, **130**(7), 1865S–1867S.

Schrezenmeir, J. & Jagla, A. (2000) Milk and diabetes. *Journal of the American College of Nutrition*, **19**(2), 176S–190S.

Schubert, M., Contreras, C., Franz, N. & Hellhammer, J. (2011) Milk-based phospholipids increase morning cortisol availability and improve memory in chronically stressed men. *Journal of Nutrition Research*, **31**(6), 413–420.

Sen C.K., Atalay M. & Hanninen O. (1994) Exercise induced oxidative stress: glutathione supplementation and deficiency. *Journal of Applied Physiology*, **77**(5), 2177–2187.

Setarehnejad, A., Kanekanian, A. & Tatham, A. (2009) The inhibitory effect of glycomacropeptide on dental erosion. *Dairy Science & Technology*, **89**, 233–239.

Setarehnejad, A., Kanekanian, A., Tatham, A. & Abedi, A.H. (2010) The protective effect of caseinomacropeptide against dental erosion using hydroxyapatite as a model system. *International Dairy Journal*, **20**(9), 652–656.

Shellis, R.P., Wahab, F.K. & Heywood, B.R. (1993) The hydroxyapatite ion activity product in acidic solutions equilibriated with human enamel at 37°C. *Caries Research*, **27**, 365–372.

Sprong, C., Hulstein, M.F.E. & Meer, R. (2001) Bactericidal activities of milk lipids. *Antimicrobial Agents and Chemotherapy*, **45**(4), 1298–1301.

Stehno-Bittel, L., Reddy, G.K., Gum, S. & Enwemeka, C.S. (1998) Biochemistry and biomechanics of healing tendon. I Effects of rigid plaster casts and functional casts. *Medicine and Science in Sports and Exercise*, **30**, 788–793.

Stein, T.P., Schluter, M.D., Leskiw, M.J. & Boden, G. (1999) Attenuation of the protein wasting associated with bed rest by branched-chain amino acids. *Nutrition*, **15**, 656–660.

Stevens, J.E., Pathare, N.C., Tillman, S.M., Scarborough, M.T., Gibbs, C.P., Shah, P., Jayaraman, A., Walter, G.A. & Vandenborne, K. (2006) Relative contributions of muscle activation and muscle size to plantar flexor torque during rehabilitation after immobilization. *Journal of Orthopaedic Research*, **24**, 1729–1736.

Stuart, C.A., Shangraw, R.E., Peters, E.J. & Wolfe, R.R. (1990) Effect of dietary protein on bed-rest-related changes in whole-body-protein synthesis. *American Journal of Clinical Nutrition*, **52**, 509–514.

Taubee, F., Steiniger, F., Nietzsche, S. & Norén, J.G. (2010) Scanning electron microscopic and X-ray micro analysis on tooth enamel exposed to alkaline agents. *Swedish Dental Journal*, **34**(3), 129–137.

Taylor, C.M. & Woonton, B.W. (2009) Quantity and carbohydrate content of glycomacropeptide fractions isolated from raw and heat-treated milk. *International Dairy Journal*, **19**, 709–714.

Thomä-Worringer, C., Sørensen, J. & López-Fandiño, R. (2006) Health effects and technological features of caseinomacropeptide. *International Dairy Journal*, **16**, 1324–1333.

Tipton, K.D., Biolo, G. & Wolfe, R.R. (1995) Protein metabolism in human muscle during hyperaminoacidemia following resistance exersice (abstract). *FASEB Journal*, **99**, A864.

Tipton, K.D., Gurki, B.E., Matin, S. & Wolfe, R.R. (1999) Nonessential amino acids are not necessary to stimulate net muscle protein synthesis in healthy volunteers. *Journal of Nutritional Biochemistry*, **10**, 89–95.

Tolkach, A. & Kulozik, U. (2005) Fraction of whey proteins and caseinomacropeptide by means of enzymatic crosslinking and memberane separation techniques. *Journal of Food Engineering*, **67**, 13–20.

Vacca-Smith, A.M., Van Wuyckhuyse, B.C., Tabak, L.A. & Bowen, W.H. (1994) The effect of milk and casein proteins on the adherence of Streptococcus mutans to saliva-coated hydroxyapatite. *Archives of Oral Biology*, **39**(12), 1063–1069.

van Calcar, S.C., MacLeod, E.L., Gleason, S.T., Etzel M.R., Clayton, M.K., Wolff, J.A. & Ney, D.M. (2009) Improved nutritional management of phenylketonuria by using a diet containing glycomacropeptide compared to amino acids. *American Journal of Clinical Nutrition*, **89**(4), 1068–1077.

Van der Strate, B.W.A., Beljaars, L., Molema, G., Harmansen, M.C. & Meijer, D.K.F. (2001) Anti viral activities of lactoferrin. *Antiviral Research*, **52**, 225–239.

van Loon, L.J.C., Saris W.H.M., Verhagen H. & Wagenmakers A.J.M. (2000) Plasma insulin responses after ingestion of different amino acid or protein mixtures with carbohydrate. *American Journal of Clinical Nutrition*, **72**, 96–105.

van Loon, L.J.C., Kruijshoop, M., Menheere, P.P.C.A., Wagenmakers, A.J.M., Saris, W.H.M. & Keizer, H.A. (2003) Amino acid ingestion strongly enhances insulin secretion in patients with long-Tterm Type 2 diabetes. *Diabetes Care*, **26**, 625–630.

van Nieuw Amerogen, A., Bolscher, J.G.M. & Veerman, E.C.I. (2004) Salivary proteins: protective and diagnostic value in cariology? *Caries Research*, **38**, 247–253.

Veldhorst, M.A., Nieuwenhuizen, A.G., Hochstenbach-Waelen, A., Westerterp, K.R., Engelen, M.P., Brummer, R.J., Deutz, N.E. & Westerterp-Plantenga, M.S. (2009) A breakfast with alpha-lactalbumin, gelatin, or gelatin + TRP lowers energy intake at lunch compared with a breakfast with casein, soy, whey, or whey-GMP. *Clinical Nutrition*, **28**, 147–155.

Vitkov, L., Kastner, M., Kienberger, F., Hinterdorfer, P., Schilcher, K., Grunert, I., Dumfahurt, H. & Krautgartner, W. (2008) Correlation between AFM and SEM imaging of acid-etched tooth enamel. *Ultrastructural Pathology*, **32**(1), 1–4.

Vitorino, R., Lobo, M.C., Duarte, J., Ferrer-Correia, A.J., Tomer, K.B., Dublin, J.R., Domingues, P.M. & Amado, F.M.L. (2004) In vitro hydroxyapatite adsorbed salivary proteins. *Biochemical and Biophysical Research Communications*, **320**, 342–346.

Volpi, E., Kobayashi, H., Sheffield-Moore, M., Mittendorfer, B. & Wolfe, R.R. (2003) Essential amino acids are primarily responsible for the amino acid stimulation of muscle protein anabolism in healthy elderly adults. *Journal of Clinical Nutrition*, **78**(2), 250–258.

Walker, G., Cai, F., Shen, P., Reynolds, C., Ward, B., Fone, C., Koganei, M., Oda, M. & Reynolds, C. (2006) Increased remineralisation of tooth enamel by milk containing added casein phosphopeptide-amorphous calcium phosphate. *Journal of Dairy Research*, **73**, 74–78.

Wang, B. (2009) Sialic acid is an essential nutrient for brain development and cognition. *Annual Review of Nutrition*, **29**, 177–222.

Wang, B. & Brand-Miller, J. (2003) The role and potential of sialic acid in human nutrition. *European Journal of Clinical Nutrition*, **57**(11), 1351–1369.

Wang, B., Brand-Miller, J., McVeagh, P. & Petocz, P. (2001) Concentration and distribution of sialic acid in human milk and infant formulas. *American Journal of Clinical Nutrition*, **74**(4), 510–515.

Wang, B., Yu, B., Karim, M., Hu, H., Sun, Y., McGreevy, P., Petocz, P., Held, S. & Brand-Miller, J. (2007) Dietary sialic acid supplementation improves learning and memory in piglets. *American Journal of Clinical Nutrition*, **85**(2), 561–569.

Weigle, D.S., Breen, P.A., Matthys, C.C., Callahan, H.S., Meeuws, K.E., Burden, V.R. & Purnell, J.Q. (2005) A high-protein diet induces sustained reductions in appetite, ad libitum caloric intake, and body weight despite compensatory changes in diurnal plasma leptin and ghrelin concentrations. *American Journal of Clinical Nutrition*, **82**, 41–48.

Westerterp-Plantenga, M.S., Nieuwenhuizen, A., Tome, D., Soenen, S. & Westerterp, K.R. (2009) Dietary protein, weight loss and weight maintenance. *Annual Review of Nutrition*, **29**, 21–41.

Wild, S., Roglic, G., Green, A., Sicree R. & King H. (2004) Global prevalence of diabetes. Estimates for the year 2000 and projections for 2030. *Diabetes Care*, **27**, 1047–1053.

Wongkhanteea, S., Patanapiradejb, V., Maneenutb, C. & Tantbirojnb, D. (2006) Effect of acidic food and drinks on surface hardness of enamel, dentine, and tooth-coloured filling materials. *Journal of Dentistry*, **34**, 214–220.

Wu, G., Fang, Y.Z., Yang, S., Lupton, J.R. & Turner, N.D. (2004) Glutathione metabolism and its implications for health. *Journal of Nutrition*, **134**, 489–492.

Zavorsky, G.S., Kubow, S., Grey, V., Riverin, V. & Lands, L.C. (2007) An open-label dose-response study of lymphocyte gluthathione levels in healthy men and women receiving pressurized whey protein isolate supplements. *International Journal of Food Science and Nutrition*, **58**, 429–436.

Zemel, M.B. (2003) Mechanisms of dairy modulation of adiposity. *Journal of Nutrition*, **133**, 252S–256S.

Zero, D.T. (1996) Etiology of dental erosion-extrinsic factors. *European Journal of Oral Science*, **104**, 162–77.

Zimeckl, M. & Kruzel, M.L. (2007) Milk derived proteins and peptides of potential therapeutic and nutritive value. *Journal of Experimental Therapeutics and Oncology*, **6**, 89–106.

8 Milk-derived Bioactive Components from Fermentation

O.A. Alhaj[1] and A. Kanekanian[2]

[1]Department of Food Science & Nutrition, College of Food and Agricultural Sciences, King Saud University, Kingdom of Saudi Arabia
[2]Centre for Nutrition, Dietetics and Food Science, Cardiff Metropolitan University, Cardiff, Wales, United Kingdom

8.1 Introduction

Nowadays, foods are no just considered by consumers as an energy supply to satisfy hunger and provide some nutritional needs, but are also expected to offer additional benefits beyond the traditional nutritional needs such as the prevention and minimisation of nutrition-related diseases and improvement of physical and mental well-being of the consumer (Menrad 2003; Mosilhey 2003) Milk is a complete food for newborn mammals during the early stages of rapid development (Shah 2000). Milk from different mammals including bovine (Ozer & Kirmaci 2010), camel (Alhaj & Alkanhal 2010), goat (Park & Haenlein 2006) and sheep (Park *et al.* 2007) were considered to be important functional sources and could provide a number of particular health benefits due to the presence of bioactive components.

Milk and dairy products contain a complex mixture of carbohydrates, fats, proteins, vitamins and minerals; these components were found to have bioactive properties with potential to improve human and animal's health. Bioactive components could refer to the 'compounds either naturally existing in food or ones formed and/or formulated during food processing that may have physiological and biochemical functions when consumed by humans' (Park 2009).

Milk contains 3.3–3.6% protein, of which casein is the major part (comprising about 80% of the protein). Casein exists in colloidal form with an average diameter of about 120 nm containing 93% proteins and 7% inorganic salts, mainly calcium and phosphate (Léonil *et al.* 2000). Casein micelles are classified into four major caseins: α-casein, constituting 40–45% of the casein complex with several variants, αs1, αs2, αs3, αs4 and αs5; β-casein, constituting 30–35% of the casein complex and classified into variants A, B and C; γ-casein is part of the β-casein chain with varying chain length and is classified into more than three variants; and κ-casein constitutes only 11–15% of the casein complex. These percentages differ between animal species such as cow and camel milk (Alhaj & Alkanhal 2010). Whey protein is the second major protein which is the soluble part of milk proteins that represent 0.6% of the milk

Milk and Dairy Products as Functional Foods, First Edition. Edited by A. Kanekanian.

proteins, in which 0.3% is β-lactoglobulin (β-lg), and is divided into the two variants A and B, and 0.07% is α-lactalbumin (α-la). The rest represent the other serum proteins such as bovine serum albumin (BSA), immunoglobulins, lactoperoxidase, lysozyme and lactoferrin.

8.2 Bioactive components in milk fat

A high consumption of milk and milk products is known to be one of the causes of coronary heart disease due to their high fat content, cholesterol and saturated fatty acids (SFA). Recent studies reported that out of 12 major fatty acids (FA), milk contains only three SFAs that are expected to be associated with raising total cholesterol in plasma (German & Dillard 2006). Numerous studies have reported that milk fat contains a number of bioactive components with certain biological function affecting human health which include sphingolipids and their metabolites, butyric acid and essential FAs, including conjugated linoleic acid (CLA), vitamins A, D, E and K, β-carotene (pro-vitamin A) and etheric lipids which have been shown to have a role in brain development (Ebringer *et al.* 2008). These components have been shown to have beneficial activities include anticancer (Parodi 1999; Vesper *et al.* 1999; Parodi 2001), antimicrobial (Possemiers *et al.* 2005), anti-inflammatory (German & Dillard 2006), anti-atherogenic (Mozaffarian *et al.* 2006), reduction of serum low-density lipoprotein (LPL) cholesterol (Vesper *et al.* 1999) and enhancement of immune system (Sugano *et al.* 1999).

8.2.1 *Conjugated linoleic acid (CLA) as a bioactive component*

CLA is the most important bioactive component in milk fat and has received particular interest from researchers. The concentration of CLA in camel milk fat was reported to be higher (4.56 mg g^{-1} fat) than that in bovine milk fat (Cardak *et al.* 2003; Alkhudair 2009). This compound has recorded a high content of up to 30 mg g^{-1} fat in dairy products (O'Shea *et al.* 1998). A higher content of up to 53 mg g^{-1} fat has also been reported (Collomb *et al.* 2006). The variation in CLA content in milk and dairy products could be attributed to several factors such as forage type (Chilliard *et al.* 2000; Bouattour *et al.* 2008), seasonal variation (Lock & Garnsworthy 2003), lactation stage (Jahreis *et al.* 1999) and animal species (cow, goat, camel, ewe, buffalo, etc.).

Among mammals, buffalo milk fat recorded the highest concentration of CLA of 4.83 mg g^{-1} fat due to the high fat content in buffalo milk (Van Nieuwenhove *et al.* 2004). In this case, CLA content is believed to decrease according to the following order: buffalo > camel > sheep > ewe > cow > goat milk fat (Jahreis *et al.* 1999; Van Nieuwenhove *et al.* 2004; Alkhudair 2009).

CLA was found to be naturally occurring in milk through two different pathways. The primary pathway is through endogenous synthesis in the mammary gland which involves Δ^9-desaturase of vaccenic acid in the lactating cows (Griinari *et al.* 2000). The second pathway is by bioconversion of polyunsaturated fatty acids in the rumen by some lactic acid bacteria such as *Lactobacillus delbrueckii* ssp *bulgaricus* and *Lactobacillus acidophilus* (Lin *et al.* 2005). In another study, five strains of lactic acid bacteria (*Lactobacillus acidophilus* DSM 9126, *L. delbrueckii* subsp. *bulgaricus* DSM 20080, *Bifidobacterium longum* subsp.

Infantis DSM 20088, *B. angulatum* DSM 20098, *Propionibacterium freudenreichii* subsp. *freudenreichii* DSM 20271) were able to increase the concentration of CLA in camel and cow milk (Alkhudair 2009). The predominant CLA in milk and dairy products were reported to belong to the CLA isomer 9*cis*, 11*trans* octadecadienoic acid, which comprise more than 80% of the total CLA intake in the human diet (Lawson *et al.* 2001; Collomb *et al.* 2006).

CLA is reported to possess a number of health benefits in animals and human (Bhattacharya *et al.* 2006). Studies reported that butter supplemented with CLA has shown to lower serum cholesterol concentration in rodents (Lock *et al.* 2005) and reduce body fat mass, body weight, plasma total cholesterol and LDL in human subjects (Gaullier *et al.* 2005). Similar studies have been shown to reduce cancer risk in rats (Ip *et al.* 1999; Parodi 1999; Corl *et al.* 2003) as well as having an antitumour activity in human breast cancer cells (Tanmahasamut *et al.* 2004); the possible mechanism of actions were reported by Kelley *et al.* (2007). Long-term feeding (14 weeks) of CLA may enhance immune system and increase the mucosal IgA immune response in rats (Ramírez-Santana *et al.* 2009). CLA was also found to work as anti-obesity supplement to reduce body weight in mice (Kanaya & Chen 2010). In another study on mice, dietary CLA was found to reduce allergic airway inflammation (Jaudszus *et al.* 2008) and reduce inflammation associated with collagen-induced arthritis (Huebner *et al.* 2010).

In contrast, a diet rich in CLA and butter was reported not to have an anti-inflammation effect (Raff *et al.* 2008) and have no effect on blood pressure and isobaric arterial elasticity on healthy young men (Raff *et al.* 2006).

8.3 Oligosaccharides as bioactive components in milk

Lactose is known to be the main source of sugar in milk; other minor forms of carbohydrates were also found in milk including neutral and acidic oligosaccharides, glucose, fructose and glucosamine (Fox & McSweeney 1998). Oligosaccharides are a class of carbohydrates and constitute 3–10 monosaccharide units (Mussatto & Mancilha 2007). A large amount and variety of neutral and acidic oligosaccharide were isolated from milk and colostrum of human (Kunz *et al.* 2000, 2001) and domestic farm animals (Urashima *et al.* 2001; Nakamura & Urashima 2004; Fukuda *et al.* 2010). Differences were reported between the oligosaccharide structure of human milk and domestic animal milk (Urashima *et al.* 1997). Similarly, significant homology and heterogeneity changes were also found between bovine and camel milk oligosaccharide (Fukuda *et al.* 2010).

For example, Sialyl -3'-galactosyllactose and sialyllacto-N-novopentaose-a have been found in camel milk but not in human milk/colostrum (Fukuda *et al.* 2010; Alhaj *et al.* 2013). On the other hand, 3'-GL, LNnH, 3'-SL, 6'-SL and MSLNnH oligosaccharides are present in both dromedary camel milk and human milk (Alhaj *et al.* 2013). These oligosaccharides, Sialyllacto-N-novopentaose-a, LNnH and MFLNnH, containing Gal(b1–4)GlcNAc (N-acetyllactosamine), are categorised as Type II oligosaccharides (Alhaj *et al.* 2013). However, only type II oligosaccharides were found in camel's milk. Hence, the presence or absence of Type I oligosaccharide is believed to be the main significant difference between human and camel or bovine milk oligosaccharide such as those based on a mixture of monosaccharides with β 1–3 and β 1–4 glycosidic bonds and other compounds (Fukuda *et al.* 2010).

The number of oligosaccharides identified in human milk was about 50 compared to that in bovine and caprine milk of about 30 (Saito *et al.* 1984; Martînez-Férez *et al.* 2006). Recently, thirteen oligosaccharides were characterized in Bactrian camel colostrums/milk (Fukuda *et al.* 2010), and Seven oligosaccharides Dromedary camel milk (Alhaj *et al.* 2013). Recently, 10, 3, 7 oligosaccharides were characterised in Bactrian camel colostrums/milk (Fukuda *et al.* 2010), as well as dromedary camel milk (Alhaj *et al.* 2013). The concentration of oligosaccharide in animal's milk was found to decrease during lactation compared to human milk (Finke 2000; Martin *et al.* 2001; Fukuda *et al.* 2010) however, although oligosaccharides in late lactation were reported to increase in bovine milk (Martin *et al.* 2001) and decrease in human milk (Kunz *et al.* 2001).

Human milk oligosaccharides are thought to be beneficial to the gut of breast-fed infants in terms of being important growth-promoting factors for the bifidus flora (Gori Cambrodon and Gudial-Urbano 2001). The 3'-GL oligosaccharide is present in dromedary camel milk and considered as a prebiotic component, which may suggest that dromedary camel milk oligosaccharides contains prebiotic and could be used as a food additive with this function (Urashima *et al.* 2009; Alhaj *et al.* 2013). The potential health benefits of milk oligosaccharides and their mechanisms are reported by Kunz & Rudloff (2006) and Qiang *et al.* (2009). These benefits are summarised as follows: stimulation of the growth of bifidobacteria and lactobacilli in the gastrointestinal tract; inhibition of the growth of pathogenic micro-organisms in the vagina; reducuction of obesity and diabetes Type 2; reducuction of cardiovascular risk; improvement of immune system; protection against cancer and tumours in man; influence availability and absorption of minerals; and reduction of diarrhoea.

8.4 Milk proteins as a source of bioactive peptides

Bioactive peptides have been defined as 'peptides produced in vivo or in vitro by enzymatic hydrolysis of food proteins with biological functions or physiological effects' (Gill *et al.* 1996). Later, Meisel (2004) defined bioactive substances from food as being 'components (genuine or generated) of consumption ready food which may exert regulative activities in human beyond basic nutrition'. The most recent definition of bio-functional or bioactive peptides was introduced by FitzGerald & Murray (2006) as 'peptides with hormone-or drug-like activity that eventually modulate physiological function through binding interactions to specific receptors on target cells leading to induction of physiological responses'.

Various foods were found to contain bioactive peptides encrypted in their primary structure such as dairy product, wheat gluten, eggs, rice, corn, soybean proteins and fish (Kitts & Weiler 2003). Moreover, milk is the most important source of bioactivities because most of the known bioactive peptides are derived from milk protein fragments due to minor proteins and peptides secreted into milk in active form by the mammary gland (Meisel 2004). For a long time, milk proteins were considered to only provide nutrition components such as essential amino acids for young mammals (Hambræus 1992). In the last decades, several studies have shown that milk proteins are an important nutritional and functional source and could possess a range of biological functions, depending on the amino acid sequence of the bioactive peptide. Milk proteins are not only important because of their high nutritional value but because of their availability in the markets in large amounts at moderate cost (Léonil *et al.* 2000).

Bioactive peptides have received special attention from researchers. A number of natural bioactive peptides have already been found in bovine milk reported and were discussed by Xu (1998) including: 'epidermal growth factor (EGF), transforming growth factor (TGF), nerve growth factor (NGF), insulin and insulin like growth factors I and II (IGF- I and IGF- II)'. Similarly, camel milk was also reported to contain naturally occurring antibodies (Hamers-Casterman *et al.* 1993). However, bovine milk proteins are currently recognised as the main source of bioactive peptide, specially caseins since they provide a large number of bioactive peptides (Meisel 2004).

Bioactive peptides could be produced in large quantities by the proteolysis of individual caseins and whey proteins (Meisel 1998). In general, most of the milk-protein-derived bioactive peptides contain 3–20 amino acid residues per molecule (Korhonen & Pihlanto 2003). Some angiotensin converting enzyme (ACE) -inhibitory peptides were found to contain 23 amino acids (Otte *et al.* 2007b). These bioactive peptides have single or multifunctional properties and naturally exist in the milk (Meisel & Bockelmann 1999). Some of these bioactive peptides are fully active through protein structure, while others are usually inactive in their precursor state and can only be liberated *in vivo* by gastrointestinal digests or *in vitro* by digestive or microbial proteolytic enzymes (Meisel & Bockelmann 1999; Korhonen & Pihlanto 2003).

These bioactive peptides can also be generated during the manufacturing process of milk by naturally occurring enzymes (Gobbetti *et al.* 2002). Some types of ripened cheeses are good for example for 'naturally occurring functional foods' because some bioactive peptides could be naturally generated during their manufacturing process by the microbial proteolysis of the cheese culture (Meisel 2004). However, the liberated bioactive peptides may become inactive after further proteolysis by the cheese culture during long ripening periods (Meisel *et al.* 1997). For example, medium-aged Gouda has double the ACE-inhibitory activity than that found in longer-ripened Gouda (Smacchi & Gobbetti 2000). Milk could also be the best carrier of these bioactive peptides because feeding fermented milk containing these bioactive peptides demonstrated more bioactivity effect than feeding them purely water. For example, long-term feeding of milk fermented with strains of *Lactobacillus helveticus* containing Val-Pro-Pro and Ile-Pro-Pro peptides to spontaneously hypertensive rats (SHR) attenuated blood pressure more than feeding them purely in water. This is due to the presence of other factors in fermented milk that affect blood pressure such as mineral availability (Sipola *et al.* 2001).

8.4.1 *Health benefits of milk proteins and their derived bioactive peptides*

It has been reported that bioactive peptides derived from milk proteins could provide several health benefits including: ACE-inhibitory action (Yamamoto *et al.* 1994; Clare & Swaisgood 2000; Lv *et al.* 2003; Minervini *et al.* 2003; Gobbetti *et al.* 2004; Quan *et al.* 2008; Alhaj *et al.* 2012); opioids peptides (Fiat *et al.* 1993; Rokka *et al.* 1997; Clare & Swaisgood 2000; FitzGerald & Meisel 2003, p. 676); metal binding (Holt 1994); reducuction of the risk of obesity and development of diabetes Type 2 (Erdman *et al.* 2008; Warensjo *et al.* 2010) and diabetes Type 1 (Agrawal *et al.* 2003); immunostimulating peptides (Migliore-Samour *et al.* 1989; Coste *et al.* 1992); antimicrobial peptides (El-Agamy *et al.* 1992; Liepke *et al.* 2001; Benkerroum *et al.* 2004; McCann *et al.* 2006); inhibitory effect against hepatitis C virus (Almahdy *et al.* 2011); peptides inhibitory to HIV-1 Proteinase (Gobbetti *et al.* 2002); antithrombotic peptides (Jollès & Henschen 1982; Mazoyer *et al.* 1992; Fiat *et al.* 1993); bone formation (Narva *et al.* 2004); anticancer (LeBlanc *et al.* 2005); and lowering of cholesterol level in blood (Nagaoka *et al.* 2001), *in vivo* in rats (Elayan *et al.* 2008) and *in vitro* (Alhaj *et al.* 2010). The mechanisms of these health benefits were reviewed by Gobbetti *et al.* (2002) and Alhaj & Alkanhal (2010).

The most-favoured bioactive peptides applied in foodstuffs formula to provide health benefits are ACE-inhibitory peptides, caseinophosphopeptides and immunomodulating peptides (Meisel & Bockelmann 1999).

8.4.2 *Peptide synthesis*

Peptide synthesis has been applied to the preparation of bioactive peptides; the selection of the most suitable method mainly depends on the length and the quantity of the desired peptide (Korhonen & Pihlanto 2003). Methods used for peptides synthesis have been reviewed by Gill *et al.* (1996). Three approaches are currently available for peptide synthesis. (1) Total enzymatic synthesis, which is relatively inexpensive and is currently limited to short peptides. (2) In contrast, recombinant DNA technology is the most suitable method for large peptides and proteins and could also be used for products consisting of up to several hundred amino acids (Wetzel & Goeddel 1987; Harford 1988). The application of this method requires an expensive research and development phase but, once the system is established, large quantities of the product could be obtained from very inexpensive starting materials via fermentation (Gill *et al.* 1996). (3) Chemical synthesis is the most widely used method on a laboratory scale (Korhonen & Pihlanto 2003). Two types of this method are currently available: liquid phase synthesis, which is used for the large-scale synthesis of relatively short peptides; and solid-phase synthesis, an expensive method because its instruments and reagents are restricted to this method (Gill *et al.* 1996). However, the latter method is still the most powerful method for synthesis of peptides composed of about 10 to >100 residues on a small scale and used for the rapid production of peptide libraries for screening purposes (Valembois *et al.* 1992).

8.4.3 *Milk protein hydrolysate*

The determination of casein and whey proteins fractions and their degradation products was a major task in the field of dairy science for several years and can provide valuable information (Molina *et al.* 2000; Veloso *et al.* 2002; Tanabe *et al.* 2003). It is known that protein hydrolysate can be made by breaking down the peptide bonds in proteins by individual or combined proteolytic enzymes to form a mixture of different length of peptide chains and some amino acids. However, some regions in the primary structure of casein are partially protected from proteolytic breakdown, therefore considered as strategic zones (Fiat & Jollès 1989). These regions exhibit different biological effects because they contain overlapping peptide sequences (Gobbetti *et al.* 2002). Peptides from casein hydrolysates were found to be more effective as an ACE-inhibitor than whey-protein-derived hydrolysate peptides

(Pihlanto-Leppälä *et al.* 1998; Otte *et al.* 2007a). However, the inhibition of ACE by milk protein peptides is the most-studied mechanism among scientists.

Currently, protein hydrolysates can be classified into three major groups according to their degree of hydrolysis, determining their applications: hydrolysates with a low degree of hydrolysis are used to improve the functional properties of the product; hydrolysates with a variable degree of hydrolysis are mostly used to enhance taste; and hydrolysates with a high degree of hydrolysis are mostly used as nutritional supplements and nutraceutical compounds in functional food production (Pedroche *et al.* 2004).

Biologically active peptides were reported by FitzGerald & Meisel (2003, p.675) and Korhonen & Pihlanto (2003) to be released from milk proteins *in vivo* and *in vitro* as: (1) *in vivo* digestion by gastrointestinal proteinase/peptidases; (2) *in vitro* enzymatic hydrolysis with digestive enzymes such as proteinase and exopeptidases; and (3) proteinase and peptidase enzymes produced by bacteria during the processing of dairy products including fermentation.

Bioactive peptide formation in the gastrointestinal tract

The formation of bioactive peptides from proteins and oligopeptides in dairy products can be produced by digestion (Hernández-Ledesma *et al.* 2004). In the gastrointestinal tract, long oligopeptides are further degraded by digestive enzymes, leading to the release of possible bioactive peptides (Korhonen & Pihlanto 2003). On the other hand, the biological activity of partially hydrolysed bioactive peptides could be decreased or eliminated if further digestion takes place by intestinal proteinases or brush border peptidases (Meisel & Bockelmann 1999). Various bioactive peptides were detected *in vivo* in the human and animal intestine after ingestion of milk, yogurt or bovine casein. In humans, caseinophosphopeptides (CPPs) were found in the stomach and duodenum of adults after ingestion of milk or yogurt (Chabance *et al.* 1998). Meisel *et al.* (2001) reported the presence of CPPs in the distal small intestine (ileum) of adults after milk and CPP meals ingestion. Similarly, β-Casomorphins were found in the intestinal contents of humans after bovine casein ingestion (Svedberg *et al.* 1985). In animals, the opioid peptide β-Casomorphins were found in the intestinal contents of minipigs (Meisel 1986) and of pre-ruminant calves after milk feed (Scanff *et al.* 1992).

Bioactive peptide absorption in the gastrointestinal tract

The health benefits of bioactive peptides that are either already present in food or produced *in vivo* by the action of intestinal proteinases could be exhibited locally at the luminal side of the gastrointestinal tract (Meisel & Bockelmann 1999), or they reach their organ target to offer the health benefit. In the latter case, bioactive peptides should survive/remain intact during their passage through the gastrointestinal proteinases/peptidases, luminal intestinal peptidases and serum peptidases (FitzGerald & Meisel 2003, p. 687).

The route for providing health effect is explained in Figure 8.1). When bioactive peptides pass through the gastrointestinal tract, they should be absorbed through the gastrointestinal epithelial membrane into the bloodstream to end up in the target organ to exhibit their biological effect (Korhonen & Pihlanto 2003). However, it seems that the number and size of

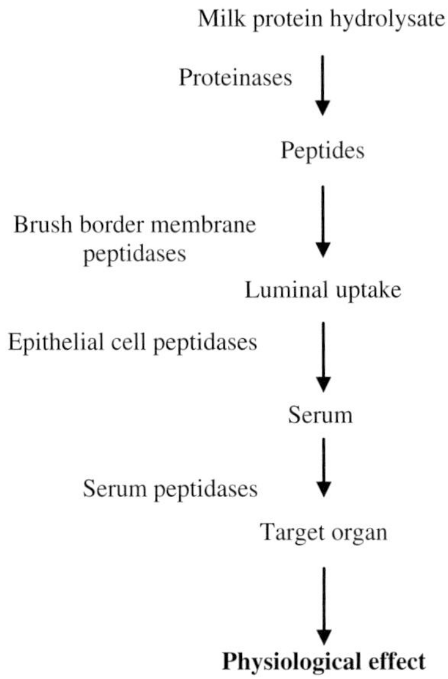

Fig. 8.1 Route of health effect from intact protein to target organ (adapted from FitzGerald & Meisel (2003). Reproduced with permission of Springer Science + Business Media.

the bioactive peptides derived from casein after milk and yogurt digestion decrease in adult humans between the stomach and the end of duodenum (Chabance *et al.* 1998).

It has been reported that peptides with a short chain such as di- and tripeptides were found to be easily absorbed in the intestine and reach the target organ without further hydrolysis by digestive enzymes (Meisel & Bockelmann 1999; Shah 2000), especially peptides containing proline (Kim *et al.* 1972). For example, small ACE-inhibitory peptides (Val-Pro-Pro and Ile-Pro-Pro) can pass through the intestine to reach peripheral target sites without being decomposed by digestive enzymes (Yamamoto 1997). The absorption of these two peptides through the gastrointestinal tract was detected *in vitro* (Masuda *et al.* 1996) and *in vivo* (Satake *et al.* 2002). ACE-inhibitory peptides were found to remain stable in commercial products (Hernández-Ledesma *et al.* 2004) and in yak milk casein hydrolysate (Mao *et al.* 2007). These short ACE-inhibitory peptides were shown to remain intact, and no further hydrolysis by digestive enzymes in SHR intestine will take place; they are absorbed directly and exhibit ACE-inhibitory activity in the abdominal aorta (Masuda *et al.* 1996). However, peptides with longer-sequence (12) amino acids such as CGP and other antithrombotic peptides were found to be absorbed through the adult gastrointestinal tract after yogurt and milk intake (Chabance *et al.* 1998).

Several studies have shown that amino acids produced by protein hydrolysates are absorbed faster and more completely than the intact proteins (Lopes *et al.* 2005). The absorption of small peptides have been reported by Gill *et al.* (1996), Tholstrup (2006) and reviewed by Fricker & Drewe (1996). However, after protein digestion, peptides could be absorbed through the gastrointestinal membranes by different mechanisms such as passive diffusion through the enterocytes and paracellular pathway or through cytotic mechanisms via dipeptide carrier or secretion back into the intestinal lumen by *p*-glycoprotein (Fricker & Drewe

1996). Not all bioactive peptides are absorbed or pass through to the bloodstream in human adults due to the prevention of digestion processes and intestinal barriers (Chabance *et al.* 1998). Some peptides could express their biological activity in the gastrointestinal tract without being absorbed, such as caseinophosphopeptides (CPPs) (Narva *et al.* 2003). Other researchers reported the potential effect of a small quantity of casein hydrolysate peptides on lipid metabolism when entering the circulation system through the lumen (Asato *et al.* 1994).

Production of bioactive peptides in milk by enzymatic hydrolysis

Basically, bioactive peptides can be produced not only by starter cultures but also by enzymatic hydrolysis during the manufacture of cheese, yogurt and other dairy products. Enzymatic hydrolysis is used as a successful method to produce active hydrolysate. This method is considered to be the most common way to produce bioactive peptides and has been shown to improve the nutritional and functional properties of milk proteins (Léonil *et al.* 2000; Gauthier & Pouliot 2003; Korhonen & Pihlanto 2006). Controlled protein hydrolysis can also be used to produce bioactive peptides (Mao *et al.* 2007). The physiochemical properties of protein hydrolysate have been reviewed by Gauthier & Pouliot (2003).

However, several studies have reported on the use of numerous proteolytic enzymes *in vitro* to release the bioactive peptides from their parent milk protein by hydrolysis with pancreatic proteinases, especially trypsin. Nagaoka *et al.* (2001) was the first to identify novel hypocholesterolemic peptides Ile-Ile-Ala-Glu-Lys from β-lg by tryptic hydrolysis. The antioxidant activity of camel α-la by enzymatic hydrolysis was recently studied by Salami *et al.* (2009) and shown to be greater than that of bovine α-la. Some peptide length and sequences such as Ile-Val-Gly-Arg-Pro-Arg-His-Gln-Gly (9 amino acids) were found not to possess ACE-inhibitory activity when administered intravenously to spontaneously hypertensive rats (SHR). However, when subjected to trypsin hydrolysis and converted into shorter peptides e.g. Ile-Val-Gly-Arg-Pro-Arg, it showed ACE-inhibitory activity after oral administration to SHR (Li *et al.* 2004), indicating that trypsin hydrolysis was essential to release the ACE-inhibitory peptide from protein chain. Moreover, the level of bioactivity was found to essentially depend on the degree of hydrolysis of protein (Alhaj *et al.* 2010).

Trypsin is synthesised in the pancreas as trypsinogen; when this inactive proenzyme enters the small intestine it is converted into its active form as trypsin by enterokinase to hydrolyse proteins and polypeptides into smaller peptides and amino acids (Lehninger 1982, p. 688). Trypsin has a narrow specificity and cleaves proteins at bonds, preferentially the carboxylic side of the basic amino acid residues of lysine and arginine (Antal *et al.* 2001; Pedroche *et al.* 2004). The optimum pH of trypsin is 7–8 but it functions best at pH 7.7.

Other endoproteinases enzymes were reported to be used individually or in combinations to generate bioactive peptides, including: pepsin, chymotrypsin, papain, thermolysin, actinase, pancreatin, elastase, alcalase and carboxypeptidase, and proline specific endopeptidase are also used (Pihlanto-Leppälä *et al.* 1994; Abubakar *et al.* 1998; McDonagh & FitzGerald 1998; Gobbetti *et al.* 2002; Gauthier & Pouliot 2003; Mizuno *et al.* 2005; McCann *et al.* 2006).

In general, digestive enzymes are preferred for producing bioactive peptides for traditional batch hydrolysis on a large scale than proteinases produced by micro-organisms through fermentation because the process is rapid and cheap and a very small quantity of enzyme is required to hydrolyse large amount of proteins in a short time. Some of the disadvantages of the digestive enzymes are being difficult to control, the narrow specificity and the need to add more enzyme due to their inactivation during the process. It has been stated by Amiot *et al.* (2004) that during hydrolysis, enzymes could be inhibited by peptides (Alder-Nissen 1986) and by peptide–peptide interactions (Pouliot *et al.* 2000) and this could also affect the overall hydrolysis level.

In contrast, proteinases produced by cultures can be costly, time consuming to produce and show low activity because the culture is nutritionally fastidious and the need to release their nutrients from the media to keep their growth. This type of hydrolysis is mainly controllable and produces favourable peptides, and their hydrolysis is gradual and non-specific because the culture produces more than one enzyme.

Formation of bioactive peptides in milk by microbial proteolysis

Biologically active peptides can be isolated in large numbers from bacterial, fungal, plant and animal origins (Gill *et al.* 1996). A number of starters, non-starter bacteria and yeasts were used to produce bioactive peptides from milk proteins. Researchers have reported the ability of some lactic acid bacteria, but not all tested strains, to produce bioactive peptides from milk proteins. The most-used lactic acid bacteria are the probiotic ones which produce bioactive peptides from different dairy products including *Lactobacillus helveticus* + *Saccharomyces cerevisiae*, *Lactobacillus acidophilus*, *Lactobacillus casei* TMC0409 + *Streptococcus thermophilus* TMC 1543, *Lactobacillus delbrueckii* subsp. *bulgaricus* SS1 + *Lactococcus lactis* subsp *cremoris* FT4, *Lactobacillus* strain GG, *Bifidobacterium longum* BB536, *B. breve* and *Bifidobacterium animalis* subsp. *lactis* (Bb-12).

Bioactive peptides can be produced using single or combined cultures such as the formation of ACE-inhibitory peptides after milk fermentation using a single strain such as *Lactobacillus helveticus* (Yamamoto *et al.* 1994) or two dairy strains such as *Lactobacillus delbrueckii* subsp *bulgaricus* and *Lactococcus lactis* subsp *cremoris* (Gobbetti *et al.* 2000). Yogurt and cheese cultures were also shown to produce different bioactive peptides in milk such as ACE-inhibitory peptides during fermentation and ripening (Gomez-Ruiz *et al.* 2002; Fuglsang *et al.* 2003; Donkor *et al.* 2007; Ong *et al.* 2007; Alhaj *et al.* 2012).

Combining cultures with different digestive enzymes has also been used to produce bioactive peptides. For example, UHT milk fermented with *Lactobacillus* GG leads to the production of several oligopeptides, resulting in the formation of bioactive peptides only after further digestion with pepsin or trypsin (Rokka *et al.* 1997).

Fermentation or co-fermentation is the most effective way to produce bioactive peptides in large quantities by microbial proteolysis, such as in milk-based fermented dairy product. For example, the production of the product Yakult® undertakes fermentation process for 7 days to cultivate the number of *Lactobacillus casei* Shirota bacteria and increase the concentration of bioactive peptides in the product (Yakult, personal communication, 2007). The long storage of 3–4 weeks with the microbial culture might result in the disappearance of some peptides of interest, and new peptides may appear due to the microbial proteolysis (Gobbetti *et al.* 2000).

However, several peptides with bioactivity derived from milk proteins released by microbial proteolysis have been reported by researchers, include immunostimulatory, ACE-inhibitory and antioxidative peptides (FitzGerald & Meisel 2003, p. 675; Korhonen

& Pihlanto 2006). In another study, the identification of two ACE-inhibitory peptides Val-Pro-Pro and Ile-Pro-Pro from milk fermented with strains of *Lactobacillus helveticus* and *Saccharomyces cerevisiae* have been reported by Nakamura *et al.* (1995). Also, the identification of ACE-inhibitory dipeptide (Tyr-Pro) from yogurt-like product fermented by *Lactobacillus helveticus* CPN 4 strain were reported by Yamamoto *et al.* (1999).

Alhaj *et al.* (2012) identified nine ACE-inhibitory peptides from fermented camel milk using two strains of *Lactobacillus* spp. *Lactobacillus helveticus* solely released three peptides of the sequences 1 (LSLSQF, SLSQF or SQF and KVLPVPQ), three peptides of the sequences 2 (TDLEN, DLEN or LEN and LHLPLPL) and one peptide of the sequence 3 (KVLPVVPQQMVPYPQ). *Lactobacillus acidophilus*, on the other hand, solely released two peptides FQEPFPDPVR and VLPFQEPVPDPVRG. All identified peptides correspond to β-casein of *Camelus dromedarius* milk. The level of bioactivity was found to essentially depend on the degree of hydrolysis of protein (Alhaj *et al.* 2010). Lactic acid bacteria including lactococci and lactobacilli (26 strains) were shown to have a proportional relationship between ACE-inhibitory activity and peptides release (Fuglsang *et al.* 2003).

Proteolytic activity of lactic acid bacteria

The purpose of adding any lactic acid bacteria (LAB) to milk is mainly based on producing lactic acid as a major end-product and, to a lesser degree, to degrade caseins and whey proteins into smaller peptides and amino acids. The presence of nitrogen in several food ecosystems is therefore considered as a limiting growth factor for lactic acid bacteria (Laan *et al.* 1989). Some lactic acid bacteria such as *Lactococcus lactis* could use oligopeptides as the main source of nitrogen during their growth in milk (Juillard *et al.* 1995b). On the other hand, *Lactobacilli* use some amino acids as an alternative source of energy to enhance their growth, such as arginine which is already produced from the primary breakdown of milk casein during cheese ripening (Kask *et al.* 1999; Laht 2003). In milk, the concentrations of essential amino acids are very low (Mills & Thomas 1981). Caseins are therefore the main source of amino acids for lactic acid bacteria in milk (Juillard *et al.* 1995a) and cheese (El-Soda *et al.* 1992). In the case of yogurt, when *Lactobacillus delbrueckii* subsp. *bulgaricus* grows this strain produces peptides and amino acids from milk proteins to encourage the growth of *Streptococcus salivarius* subsp. *thermophilus* (Tamime & Robinson 1985, p. 277). The formic acid produced by *S. salivarius* subsp. *thermophilus* and formed by milk heat treatment could accelerate the growth of *L. delbrueckii* subsp. *bulgaricus* (Tamime & Robinson 1985, p. 279; Adams & Moss 2000, p. 326), so both yogurt cultures utilise each other's metabolites for more efficient growth and acid production (Adams & Moss 2000, p. 326).

Although peptides are more favourable for the growth of lactic acid bacteria than mixtures of pure amino acids, the available amounts of free amino acids and small peptides in milk are only enough for the growth of 5–25% of the full grown culture (Juillard *et al.* 1995a). LAB have the ability to use milk casein as a source of amino acids and nitrogen because of their complex system (Shihata & Shah 2000). For example, *Lactococcus lactis* could cleave casein into amino acids and make it available for their growth through a three-step process (Wohlrab & Bockelmann 1992; Mierau *et al.* 1997). First, casein is degraded into peptides by extracellular proteinase (PrtP). In the second step, several of these released peptides are incorporated into the cell via an oligopeptide transport system (Opp) and finally, the released amino acids from these peptides could be utilised for other processes such as protein synthesis and the generation of metabolic energy.

As mentioned earlier, LAB has a complex proteolytic system to release the required nutrients by the synthesis of various enzymes such as a cell-wall-bound proteinase and also produce a number of intracellular peptidases including one or two endopeptidases. Some of these enzymes include aminopeptidases, iminopeptidase, dipeptidase, tripeptidase, prolidase, carboxypeptidase and prolinase (Bockelmann *et al.* 1995; Law & Haandrikman 1997; Christensen *et al.* 1999; Shihata & Shah 2000). At least 16 peptidases responsible for the conversion of peptides to free amino acids have been identified and biochemically characterised from LAB (Christensen *et al.* 1999).

The proteinase and peptidase enzymes of lactic acid bacteria are either bound to the cell wall or are intracellular and can only contribute after cell lysis (Law & Haandrikman 1997). Endopeptidases gained much attention because of their ability to hydrolyse peptide bonds within the oligopeptide (Janer *et al.* 2005). Lactic acid bacteria require significant amounts of essential amino acids and peptides for their growth and therefore need to hydrolyse casein and whey proteins into smaller parts to enhance maximum cell growth in milk.

Yogurt culture (*Lactobacillus delbrueckii* subsp *bulgaricus* and *S. thermophilus*) was found to have higher proteolytic activity than *Lactobacillus acidophilus* and *Bifidobacterium* spp. (Shihata & Shah 2000) and found to produce more β-D-galactosidase than probiotic bacteria (Shah & Jelen 1990; Shah 1994). This could explain 'why yogurt bacteria grow faster than probiotic bacteria and why yogurt bacteria are used as the main starter bacteria and probiotic bacteria as an adjunct starter' (Shah 2001). It has been shown that proteolysis enhances the growth of probiotic bacteria in fermented products. For example, it was reported that the presence of traditional yogurt starter cultures shortens the fermentation time, enhances the growth of probiotics, increases the population number, increases their proteolytic activities and reduces the oligosaccharide content; this is due to the supply of exogenous small peptides and essential amino acids, which are usually absent or present in very low concentration and do not support their optimal growth in milk (Law *et al.* 1976; Dave & Shah 1997a; Gomes *et al.* 1998; Shihata & Shah 2000; Wang *et al.* 2006) or could be due to the equipollence relationship between each other.

Nevertheless, lactic acid bacteria are capable of degrading different proteins and peptides to supply the required nutrients, while end-products depend on the proteolysis system of the bacterial species or enzyme capacity and specificity. However, any changes in lactic acid bacteria metabolism or growth conditions result in different-end products. Some lactic acid bacteria proteinases such as *Lactobacillus* sp. and *Lactococcus* sp. could cleave more than 40% of peptide bonds of β-casein, leading to the release of more than 100 different oligopeptides (Juillard *et al.* 1995a; Mierau *et al.* 1997). Similar hydrolysis was obtained with α-casein (Kunji *et al.* 1996). These oligopeptides will be further degraded into smaller peptides by LAB peptidase and become the main source of bioactive peptides in fermented milk. When these bioactive peptides or peptides related to the bioactive peptides sequence are liberated, the biochemical activity of microbial communities might be influenced (Gobbetti *et al.* 2002). These peptides should be carefully conditioned because they have the ability to inhibit the microbial proteolysis during ripening (Gobbetti *et al.* 1995). For example, peptides from fresh short-ripened and long-ripened cheeses have been shown to inhibit the endopeptidases of *Lactobacillus*, *Streptococcus* and *Lactococcus* strains, but have less effect on aminopeptidase activity (Gobbetti *et al.* 1995; Smacchi & Gobbetti 1998).

In general, *Lactobacilli* have complex nutritional requirements for organic substrates, amino acids and peptides, and subsequently possess stronger proteolytic activity than *Lactococcus* (Thomas & Mills 1981; Kandler & Weiss 1986) and *Bifidobacterium* spp. (Shihata & Shah 2000) and yet the released peptides are further hydrolysed by some peptidases (Meyer & Jordi 1987). On the other hand, *lactobacilli* have very limited capacity to synthesise amino acids from inorganic nitrogen sources, depending on amino acids present in the growth environment (Axelsson 1998; Christensen *et al.* 1999).

It has been reported that *Lactobacillus helveticus* ATCC 15009 requires all amino acids with the exception of four (alanine, cysteine, glycine and serine; Morishita *et al.* 1981). *Lb. helveticus* is therefore considered a good example of high proteolytic activity. This strain has a very efficient proteolytic system including general aminopeptidase, X-prolyl-dipeptidyl aminopeptidase, endopeptidase-proteinase and probably dipeptidase, tripeptidase, proline iminopeptidase and carboxypeptidase (Sasaki *et al.* 1995; Christensen *et al.* 1999).

Proteolytic activity of Bifidobacterium spp.

Bifidobacterium strains have low proteolytic activity when compared to other lactic acid bacteria (Klaver *et al.* 1993; Shihata & Shah 2000). *Bifidobacterium animalis* subsp. *Lactis*, as for many other *Bifidobacterium* species, are unable to synthesise some amino acids. In addition, milk does not contain sufficient amounts of free amino acids and small peptides for their growth (Abu-Taraboush 1996). They therefore need to hydrolyse milk proteins and utilise their products such as amino acids and peptides for their optimum growth. As a result, the level of the amino acids and peptides increase. So far, there has been very little information available in the literature regarding the proteolytic enzyme systems of *Bifidobacterium* spp. (Janer *et al.* 2005). *Bifidobacterium* spp., as with other lactic acid bacteria, do have some general aminopeptidase activity including a number of dipeptidase and possible iminopeptidase and tripeptidase (El-Soda *et al.* 1992; Shihata & Shah 2000). Cheng & Nagasawa (1983, 1984) isolated aminopeptidase and iminopeptidase activity from a cell-free extract of *B. breve*. The exopeptidase system of several *Bifidobacterium* strains was demonstrated by Minagawa *et al.* (1985) and El-Soda *et al.* (1992). Three kinds of aminopeptidase and carboxypeptidase were measured in five strains of *Bifidobateria* (Minagawa *et al.* 1985). The presence of aminopeptidase, dipeptidase, tripeptidase and carboxypeptidase activities from *B. longum*, *B. infantis* and *B. adolescentis* were demonstrated (El-Soda *et al.* 1992).

Recently, the proteolytic system of *Bifidobacterium animalis* subsp. *lactis* was studied by Janer *et al.* (2005), who demonstrated that this strain is able to produce intracellular endopeptidase (PepO), general aminopeptidase, X-prolyl-dipeptidyl aminopeptidase and proline iminopeptidase. Their study also showed that *Bifidobacterium animalis* subsp. *lactis* PepO enzyme mostly cleave peptides which bond on the N-terminal side of phenylalanine residues and were shown to have a post-proline secondary cleavage site. The number and specifications of the released bioactive peptides therefore depends on the selection of micro-organism species and growth conditions and the micro-organism's proteolytic activity, bioactive peptide composition and dairy product type.

8.5 Production and purification of bioactive peptides

Researchers have developed several techniques for the production and purification of specific biologically active peptides from milk protein hydrolysates (FitzGerald & Meisel 2003, p. 686). The lack of suitable technologies has limited the industrial-scale production of bioactive peptides from food proteins (Korhonen & Pihlanto 2006). At the same time, the method applied for the industrial purification or enrichment should be carefully chosen as it can play an important role in maintaining their biological activities (Chatterton *et al.* 2006). For example, heat treatment applied during processing can change the biological activity of the proteins by altering their structure and contributing to the lowering of their digestive properties (Chatterton *et al.* 2006). Nowadays, the technologies that are available for large-scale production has led to the introduction of some peptides with specific bioactivities to the market be used commercially. Some of these products have already been launched and are marketed internationally such as Calpis (reduction of blood pressure); BioPURE-GMP (prevention of dental caries, protection against viruses and bacteria); Capolac (helps mineral absorption); PeptoPro (improves athletic performance of muscle recovery); Vivinal Alpha (acids relaxation and sleep); and Recaldent (anticariogenic) (Korhonen 2009).

Membrane separation is the most common method for the isolation and purification of bioactive peptides from protein hydrolysate using microbial enzymes (Jelen & Lutz 1998); it is the best available technology used for peptide enrichment with a specific molecular weight range (Korhonen & Pihlanto 2007, p. 17). This technique has frequently been used for the enrichment of the ACE-inhibitory peptides from food proteins hydrolysate (Meisel *et al.* 2006, p. 285).

There are two available methods to perform the enzymatic hydrolysis however: either conventional batch hydrolysis or continuous hydrolysis using ultrafiltration membrane (Korhonen & Pihlanto 2003). Researchers have indicated several limitations for the traditional batch processing method compared to the continuous process, which are mainly related to their high use of enzyme and labour costs (Rios *et al.* 2004), their inefficiency for the frequency of start-up and shutdown procedures and need to recover and prepare the enzyme after each batch (Prazeres & Cabral 1994).

Enzymatic membrane reactors (EMR) have been used widely to produce hydrolysates with improved nutritional and/or functional properties through the total conversion of various food proteins (Prazeres & Cabral 1994; Martin-Orue *et al.* 1999). EMRs were developed to integrate several operations into a single operation; these operations include the enzymatic hydrolysis, product separation and/or concentration and enzyme recovery (Prazeres & Cabral 1994).

Ultrafiltration membrane reactors have been investigated for the continuous production of bioactive peptides, replacing the conventional batch method (FitzGerald & Meisel 2003, p. 687). This method has been shown to vastly improve productivity, efficiency and yield of the product (Deeslie & Cheryan 1981; Cheryan & Deeslie 1983) and shown to be easily scaled up (Korhonen & Pihlanto 2007, p. 17). This system was applied successfully for milk proteins for the continuous production of antithrombotic peptides from hydrolysed caseinomacropeptides (Bouhallab *et al.* 1992) and the continuous production of permeate enriched in CCPs or casomorphin precursors (Righetti *et al.* 1997). Bouhallab *et al.* (1992) reported the requirements for the continuous production of specific peptides sequences using membrane reactors as follows: 'i) highly specific enzymes; ii) better control of the enzymatic reaction and iii) high selective membranes to separate the desired peptides from the reaction mixture'.

Bordenave *et al.* (2000) suggested the technique of combining ultrafiltration (UF) techniques and enzymatic hydrolysis to produce a hydrolysate with a specified molecular size distribution. Lieske & Konrad (1996) reported the improvement of foaming and emulsifying properties of a papain hydrolysate when a selective UF step was added with a molecular weight cut-off (MWCO) of 1 kDa to remove peptide fragments from whey protein concentrates. Other methods of ultrafiltration were reported for separating out small peptides from high molecular mass residues and remaining enzymes using stepwise ultrafiltration with low MWCO membrane (Korhonen & Pihlanto 2007, p. 17). A two-step ultrafiltration process has also been used successfully to produce mixtures of polypeptides with molecular mass above 5 kDa and fractions rich in small components containing small peptides and amino acids having molecular mass below 2 kDa (Turgeon & Gauthier 1990). The first UF (MWCO of 30 kDa) is to remove the whole unhydrolysed proteins and any remaining enzymes, while the second UF (cut-off 1 kDa) is to remove small peptides and free amino acids (Lieske & Konrad 1996). Similarly, a two-step ultrafiltration membrane (30 kDa and 1 kDa) was used successfully for the enrichment of opioid peptides α-lactorphin and β-lactorphin (Pihlanto-Leppälä *et al.* 1996) and ACE-inhibitory peptides (Pihlanto-Leppälä *et al.* 2000) from whey proteins.

Microfiltration (MF) is a membrane technique also used in the dairy industry. This technique is successfully used for the removal of pathogenic or spoilage bacteria from liquid milk (Kelly & Tuohy 1997). The Tetra Pak Company (Pully, Switzerland) has developed a new microfiltration machine for commercial use called Bactocatch process; this is the most common commercially used machine and is able to remove bacteria from skim milk (Elwell & Barbano 2006). Two limitations should be considered when using this technique (Rysstad & Kolstad 2006). First, this machine does not microfilter whole milk because milk fat globules have a similar particle size distribution to that of the cells and spores; milk fat must be therefore removed first in order to microfilter the skim milk. Second, the particle size distribution of cells and spores could overlap with casein micelles; the pore size should therefore be compromised to minimise the changes in milk composition. Despite these limitations, this method could be used to replace the severe high-temperature pasteurisation and retain the sensitive physiological compounds which are usually damaged during heat treatment. This method could also be used for milk proteins separation (Jelen & Lutz 1998). and to extend the shelf life of the processed milk (Rysstad & Kolstad 2006). This method could also be ideal for camel milk, as its proteins are sensitive to severe high-temperature heat treatment (Alhaj *et al.* 2011).

Ion exchange chromatography has been used for the isolation and enrichment of peptide fractions from their protein hydrolysate. Ellegård *et al.* (1999) have developed new processing methods to isolate high-purity CPPs from tryptic hydrolysate of caseinate followed by acid precipitation, diafiltration and then using anion exchange chromatography. However, this method of separation is quite costly and uneconomical for large-scale operations (Korhonen & Pihlanto 2006). On the other hand, ion exchange chromatography was used to isolate and purify caseinomacropeptides (CMP) from whey proteins (Thomä-Worringer *et al.* 2006). The isolation was performed by passing whey containing the CMP at low pH through a cationic exchange column. The major whey proteins are usually adsorbed onto the cation exchanger because of their positively charged molecules, while CMP pass through because of their highly negative charge (Doultani *et al.* 2003).

The isolation could also be carried out by passing the acidic whey containing CMP through an anionic exchanger; however, the CMP this time will be adsorbed onto the anion exchanger instead of whey proteins due to the charges on the protein (Tek *et al.* 2005). Using a similar technique, an anion exchange membrane system has been used for the *in situ* isolation and enrichment of cationic antibacterial peptides from lactoferrin (Recio & Visser 1999). This method, described by Korhonen & Pihlanto (2007, p. 18), is based on concentrating the protein of interest within the chromatographic membrane and hydrolysing *in situ* with the desired enzyme. The resulting active peptides are retained on the column, whereas the fraction containing inactive peptides is washed out. Finally, the fraction containing the active peptides is obtained from the chromatographic membrane in a single elution step. The advantages of this process are that the isolation of the precursor protéin is unnecessary and the enzyme used in the process can be recovered, reducing the cost of the final product.

8.6 Probiotic as bioactive component in milk

Consumer acceptance of probiotic fermented milk products has grown in Europe, North American and the Middle Eastern countries (IDF 1995). In the USA, the food industry has become the largest manufacturing industry (Klaenhammer 2000). There are now a variety of dairy products containing many different blends of lactic acid bacteria and probiotic cultures including yogurts, fermented milks, sour cream, cottage cheese, ice cream and frozen dessert. One of the earliest recognised functional foods was fermented milk (Metchnikoff 1908). In use for over 5000 years, their production enhances shelf life and improves the quality of milk as well as providing health benefits.

Dairy fermentations rely on a range of micro-organisms including lactic acid bacteria, propionic acid bacteria, yeasts and moulds to produce a very wide range of products from liquid fermented products to hard cheeses matured for several years before consumption. The addition of certain genera of probiotics such as *Bifidobacterium* and *Lactobacillus* was traditionally included in probiotic products to protect the human body against harmful effects (Vaughan & Mollet 1999) and improve the quality of the product itself by extending the shelf life, as was the case with producing different fermented products using mixed cultures of *bifidobacteria* and normal yogurt culture.

The consumption of probiotic products, especially drinks, is on the increase in many developed countries. The first fermented health drink containing probiotic bacteria was introduced by Yakult and has been available to the consumer since 1935 in Japan (Karimi & Pena 2003). Japan is now the largest market for functional foods and considered as the world leader in probiotic and prebiotic products, with more than 53 different types of milk products containing probiotic organisms on the market (Shah 2001). Probiotic products are restricted to fermented milk products in Europe, especially yogurt (Hilliam 2000). Australia, North America and many European countries are the other large markets for functional foods containing probiotics, prebiotics and synbiotics (mixtures of probiotic and prebiotic) that aim to improve the immune system and intestinal function. Interest in probiotic products has been growing in some parts of the Middle East recently, mainly for their associated health claims. It is apparent that the growth in probiotics and functional foods are growing worldwide and faster than other health-promoting products; they have also been the focus of intensive research activity in recent years.

8.6.1 *What are probiotics?*

The most widely used method to increase the number of beneficial and health-promoting bacteria in the intestinal tract is by the direct consumption of food containing live bacteria. Such bacteria are called probiotics (Fuller 1989; Salminen *et al.* 1998b). The word 'probiotic' (meaning 'pro-life') is derived from the Greek language (Fuller 1989) and has been used in several different ways over the years. Their use goes back to the Roman period where Pliny the Elder in 76 AD advocated the use of fermented milks for treating gastrointestinal disturbances (Stanton 2005).

Probiotics have different definitions depending on the interpretation of the researcher in understanding their mechanisms of action or their effect on health and well-being of the host (Salminen *et al.* 1999). The term probiotics was first introduced in 1965 to describe the growth-promoting factors produced by the fermentative micro-organisms (Lilly & Stillwell 1965). However, early attempts to classify probiotics were not generally accepted (Sperti 1971). In 1974 Parker proposed a definition for the probiotics as 'organisms and substances that influenced the intestinal microflora and had beneficial effects on animals' (Parker 1974). The term 'substances' is imprecise and could include antibiotics or the promotion of microbial growth (Gibson & Roberfroid 1995). In 1989 Fuller modified the definition of probiotics to 'a live microbial feed supplement which beneficially affects the host animal by improving its intestinal microbial balance'.

According to this definition, the term 'probiotics' was restricted to feed supplements beneficial to the intestinal tract of animals, and could not be used for living micro-organisms administered in food or for location other than gastrointestinal tract (Šušković *et al.* 2001). In 1992 Havenaar and Huis in't Veld proposed the widening of Fuller's definition to 'a probiotics is a mono- or mixed culture of live micro-organisms which applied to animal or man, affect beneficially the host by improving the properties of the indigenous microflora' (Havenaar & Huis in't Veld 1992). The formal and most-used definition of probiotics was adopted by FAO/WHO in 2001 as: 'live microorganisms which when administered in adequate amounts confer a health benefit on the host'.

The most recent approach for describing probiotics was introduced by Schrezenmeir and de Vrese (2001) as 'a preparation of or a product containing viable, defined microorganisms in sufficient numbers, which alter the microflora (by implantation or colonisation) in a compartment of the host and by that exert beneficial health effects in this host'. Probiotics means any live micro-organisms, bacteria or yeast that are capable of providing any health benefits to the host and can be administrated as a component of functional food or as food supplement e.g. capsules or tablets.

The second generation of probiotics are genetically modified micro-organisms providing the host with some necessary components. For example, research has been undertaken to increase the immune response induced by genetically modified lactobacilli to produce antigens for Helicobacter or rotavirus (Mercenier *et al.* 2004). *Lactococcus lactis* strains were also genetically engineered to produce immunomodulatory cytokines (e.g. interleukins) at mucosal surfaces (Steidler & Neirynck 2003). These are not commercial products, probably due to consumers' resistance to genetically modified micro-organisms.

8.6.2 *Probiotics concept*

The concept of probiotics was first proposed by the Russian Nobel Prize winning scientist Elie Metchnikoff (Metchnikoff 1908) who suggested that longevity of Bulgarian peasants was due to the consumption of large quantities of fermented milk products. Metchnikoff has introduced his intoxication theory and stated that intestinal putrefaction and fermentation are the main cause of aging (toxicants). He believed that the consumption of a product containing bacillus can decrease the toxic effect of colonic microflora by influencing the microflora of the colon. The organism he identified was called Bulgarian bacillus and later *Bacillus bulgaricus*, and was probably what became known as *Lactobacillus bulgaricus* and is now called *Lactobacillus delbrueckii* subsp. *Bulgaricus*. This strain, together with *Streptococcus salivarius* subsp. *thermophilus*, is now a widely used culture to produce yogurt. In 1935, Rettger and colleagues became interested in the mechanisms of probiotics effects and began to investigate the use of intestine-derived species for medicinal use (Rettger *et al.* 1935). Because the products containing probiotics had not been subjected to large-scale trials of efficacy, its use was however not considered by the pharmaceutical industry (Tannock 2003).

Medication and treatment with antibiotics is effective and has been successfully used to treat many diseases for decades. In some cases, the emergence of antibiotic-resistant strains of some bacteria, such as the methicillin-resistant *Staphylococcus aureus* (MRSA), has been a challenge to the pharmaceutical industry. However, antibiotics are no longer developed by the pharmaceutical industry at a sufficient rate to be able to compete with the emergence of new stains of resistance bacteria (Bengmark 1998). Antibiotics are generally known to delay the recolonisation of the normal colonic flora (Rolfe 2000); probiotic products certainly cannot replace antibiotic treatments, but they can restore some of the lost microflora in the body and provide health benefits. These factors may support the concept of having fermented milk containing beneficial micro-organisms as a preventive approach rather than a curative method.

8.6.3 *Characteristics of probiotics*

Hundreds of species of probiotic strains have been identified as commensal organisms in the human intestinal tract, but there is no one set list of criteria for classifying a viable bacterial strain as a probiotic (Kailasapathy & Chin 2000). A joint report by both the Food and Agriculture Organization and World Health Organization Working Group (FAO/WHO 2002) pointed out the main criteria for probiotic strains *in vitro* (Table 8.1). Within probiotic bacteria, large variations can be found in probiotic species or even in its strains (Chung *et al.* 1999). The characteristics of probiotic bacteria should therefore match the human gut environment according to the claimed health benefit. Roy (2005) suggested that 'the performance of strains should be evaluated individually prior to commercial use in new dairy fermented products'. Nevertheless, *Lactobacillus acidophilus*, *Bifidobacterium* spp. and *L. casei* can be classified as probiotics because they possess beneficial health effects in the host by modulating the intestinal microflora (Schrezenmeir & De Vrese 2001).

Table 8.1 The main criteria for probiotic strains in vitro. FAO/WHO (2002). Reproduced with permission of Food and Agriculture Organization of the United Nations.

Resistance to gastric acidity
Bile acid resistance
Adherence to mucus and/or human epithelial cells and cell lines
Antimicrobial activity against potentially pathogenic bacteria
Ability to reduce pathogens adhesion to surfaces
Bile salt hydrolase activity
Resistance to spermicides (applicable to probiotics for vaginal use)

Yogurt is another type of fermented milk usually produced by adding the traditional lactic acid culture (*Lactobacillus delbrueckii* ssp. *bulgaricus* and *Streptococcus salivarius* subsp. *thermophilus*) to milk to convert the lactose in milk into lactic acid. The fermentation time of yogurt is short (approximately 4 hours) depending on the incubation temperature and the viability of the culture, compared to that for probiotic bacteria which is up to 24 hours (Dave & Shah 1997a). The survival of yogurt culture through the human gastrointestinal tract is still controversial because it is known that the two main traditional yogurt starter cultures (*L. delbrueckii* subsp *bulgaricus* and *S. salivarius* subsp. *thermophilus*) do not survive in the acidic conditions in the stomach or during their journey through the intestinal tract (Gilliland *et al.* 1984; Salminen *et al.* 1998a). Consequently, no health benefits can be gained by adding these cultures to the product. Studies in the last two decades have reported the inability of these yogurt cultures to survive for a long time, not adhering or surviving through the intestinal tract because they are not bile-resistant bacteria (Gilliland *et al.* 1984; Salminen *et al.* 1998a). These results partly agree with Garvie *et al.* (1984) who reported that when yogurt was fed continuously to rats, the *Lactobacillus delbrueckii* subsp. *bulgaricus* and *Streptococcus salivarius* subsp. *thermophilus* counts decreased which indicates that yogurt culture can only survive but not multiply in the gut of the rats.

Researchers have recently reported the viability of yogurt cultures during their passage through the gastrointestinal tract and their presence in human feces (Campo *et al.* 2005; Mater *et al.* 2005; Elli *et al* 2006) and their ability to enhance lactose intolerance (FAO/WHO 2002). Yogurt culture was considered as probiotic because they 'clearly fulfil the current concept of probiotic at least for its beneficial effect on lactose digestion in vivo' (Guarner *et al.* 2005) and was shown to have measurable health benefits (Adolfsson *et al.* 2004). Interpretation of these results must take into account the different experimental protocols such as sample size and duration of study, associated with several methodological problems. It can be concluded that the probiotic nature of yogurt culture cannot be ignored for their health benefits.

However, a list of 56 species of *Lactobacillus* and 30 species of *Bifidobacteria* has been reported by Gomes & Malcata (1999). The main probiotic organisms known worldwide include: *Lactobacillus lactis*, *L. delbrueckii* subsp *bulgaricus*, *L. brevis*, *L. johnsonii*, *L. reuteri*, *L. Cellobiosus*, *L. curvatus*, *L. fermentum*, *L. plantarum*, *Lactococcus lactis* ssp. *cremoris*, *Lactococcus lactis* subsp. *lactis*, *Bifidobacterium bifidum*, *B. animalis* subsp. *lactis* (Bb-12), *B. adolescentis*, *B. infantis*, *B. longum*, *B. thermophilum*, *Streptococcus salivarius* subsp. *thermophilus*, *S. ceremoris* and *S. diacetylactis*. Other microbial species besides lactic acid bacteria were used as probiotics such as *Pediococcus acidilactici*, *Enterococcus faecium*, *E. faecalis Bacillus subtilis* and *Propionibacterium* ssp. Some yeast has also been used as probiotic, for example *Saccharomyces boulardii* (Tamime *et al.* 2005, p. 39).

Several strains of probiotics have already been used commercially for the production of fermented milk products including *Lactobacillus acidophilus*, *L. casei* Shirota, *L. casei* Imunitass, *B. bifidum*, *B. infantis*, *B. longum*, *B. breve* and *B. animalis* subsp. *Lactis* (Reuter *et al.* 2002; Tamime 2002). A list of 19 fermented milk products containing *Bifidobacteria* in different markets worldwide was reported by Tamime *et al.* (1995), and has by now tremendously increased in number. Among *Bifidobacterium* species, *Bifidobacterium animalis* subsp. *lactis* is the most widely used species in probiotic applications (Masco *et al.* 2005). Most *Bifidobacterium* strains used in probiotic fermented milk are expected to belong to *B. animalis* but are declared on the label as *B. longum* by manufacturer (Klein *et al.* 1998; D'Aimmo *et al.* 2007). However, commercially available milk-based probiotic health drinks could contain a single strain such as Yakult® health drink (Japan) containing *L. casei* Shirota, or contain two strains such as Muller® health drink (Germany) containing *Bifidobacterium* sp. and *Lactobacillus acidophilus*. Other products could contain more than two strains such as food supplement VSL-3® (Italy) containing 8 lactic acid bacteria species.

Before delivering a product containing probiotic bacteria to the consumer, there are a number of common criteria used for isolating and defining probiotic and specific strains which include the following:

- non-toxic, non-allergic, non-pathogenic and preferably unaffected by antibiotics;
- beneficial to the host (human or animal), in other words able to provide one or more clinically documented health benefits;
- an intake of 300–400 g per week contains $10^6 - 10^8$ CFU g^{-1} of the probiotic;
- able to survive in or colonise the gut;
- viable and stable during processing, storage and even beyond the consume-by date;
- enhance the shelf life and storage stability of fermented products;
- demonstrate stability and functionality after freeze- and spray-drying methods;
- able to adhere and compete with the normal microflora;
- amenable to cultivation on an industrial scale;
- do not produce harmful and putrefactive substances but are able to produce antimicrobial metabolites (including bacteriocins, carbon dioxide, hydrogen peroxide and organic acids such as lactic and acetic acids);
- remain viable when exposed to acid, bile and oxygen;
- resist degradation caused by intestinal enzymes such as lysozymes;
- maintain a good flavour and aroma profile; and
- are genetically stable.

These probiotic criteria are adopted from Fuller (1989); Macfarlane & Macfarlane (1995); Dave & Shah (1997a); Salminen *et al.* (1998a); Gismondo *et al.* (1999); Kailasapathy & Chin (2000); Shah (2001); Šušković *et al.* (2001) and Hernández *et al.* (2005).

Bifidobaterium spp

Bifidobacterium spp. are anaerobic, non-spore forming, Gram-positive, non-motile and catalase-negative with a range of cell morphologies including short curved rods, club-shaped rods and Y–shaped branches (Gomes & Malcata 1999). The optimum growth temperature is in the range 37–41°C. *Bifidobacterium* spp. are known to grow at temperatures of 20–45°C, although growth outside this range might also occur in some strains (Marks 2003, p. 38) depending on the habitat and the origin/source of the bacteria (Biavati & Mattarelli 2001). For example, *Bifidobacterium animalis* subsp. *lactis* from animal origin has been shown to have high thermal resistance, which makes it a good choice for yogurt production because a higher incubation temperature during milk fermentation can shorten yogurt production time (D'Aimmo *et al.* 2007); it is found to have slower growth rate at 30°C (Østlie *et al.* 2005). However, it was also shown that these bacteria would survive during incubation and storage with normal yogurt culture (Lamoureux *et al.* 2002; Matsumoto *et al.* 2004). *Bifidobacterium* spp. grows best at pH 6–7, but the vast majority are unable to sustain growth at low pH valuesof 4.5–5.0 or at higher pH values of 8.0–8.5 (Gomes & Malcata 1999). The survival of *Bifidobacteria* at certain pH values varies, for example, *B. breve*, *B. longum*, *B. infantis* and *B. pseudolongum* have all been shown to be able to grow at pH 4.0 in the presence of 0.3% of bile salts or in the presence of 300 μg mL^{-1} of lysozyme, which is usually acts on Gram-positive bacteria by destroys the integrity of bacterial cell walls (Šušković *et al.* 1997). They are also resistant to at least 10 μg mL^{-1} of H_2O_2 (Gagnon *et al.* 2004) and to a limited number of antibiotics (Temmerman *et al.* 2002).

However, the beneficial effect of probiotics is usually linked to their survival and the adverse effect they impose on Gram-negative and pathogenic micro-organisms by inhibiting their growth, such as the case with *Escherichia coli* (Yusof *et al.* 2000; Ibrahim *et al.* 2003; Gagnon *et al.* 2004) and *Salmonella typhimurium* S-285 (Yusof *et al.* 2000).

In order to maintain good confidence in the use of probiotic strain, it is important to demonstrate the metabolism and characteristics of any particular strain during their production, shelf life and their ability to survive through the gastrointestinal tract. One of the Bifidobacterium strains from animal origin such as *Bifidobacterium animalis* subsp. *lactis* (Bb-12) produced and marketed by Chr. Hansen Ltd has been used in many fermented milk products. This strain already has eight clinically proven health benefits (Playne 2002) and has been shown to have some useful characteristics suitable for commercial product application such as resistance to freeze-drying (Modesto *et al.* 2004). It is stronger and more temperature tolerant compared with other *Bifidobacteria* strains from human origin such as *B. bifidum*, *B. breve*, *B. infantis* and *B. longum* (Lamoureux *et al.* 2002). This strain is bile-tolerant (Charteris *et al.* 1998) and resistant at pH 3 for 180 min, but declined slowly at pH 2 and no survival was found after 60 min at pH 1 (Pochart *et al.* 1992). This strain was found to have remarkably higher acid resistance than *B. infantis* and *B. longum* at pH 2 for 90 min in a dynamic model system of *in vitro* human upper GI tract (Mainville *et al.* 2005). This is in agreement with Vernazza *et al.* (2006), who reported that *B. lactis* (Bb-12) was more acidic resistant at pH 2 than *B. adolescentis* and *B. infantis* and two strains of *B. longum*. It has been reported (Alander *et al.* 2001; Gopal *et al.* 2003; Mättö *et al.* 2006) that Bb-12 has the ability to survive during passage through the gastrointestinal tract in human (79% of study subjects).

Experiments involving human subjects reported the survival of 23.5% of the administrated dose of *Bifidobacterium animalis* subsp. *lactis* in the samples taken from ileum (Pochart *et al.* 1992). In a similar study, this strain was detected in the faeces of all subjects after the second week of ingestion, whereas the total number was increased during the consumption of these bacteria and remained viable in the intestinal tract for 4–5 weeks (Collado *et al.* 2006). The acid tolerance of *Bifidobacterium animalis* is expected to belong to the action of H^+-ATPase activity. Their resistance to antibiotics were reported by Zhou *et al.* (2005), Antunes *et al.* (2007), D'Aimmo *et al.* (2007) and Saarela *et al.* (2007), which makes it a successful strain for being unaffected by antibiotic therapy (Rolfe 2000). Other studies showed the high sensitivity of *Bifidobacteria* to oxygen, although their oxygen tolerance is species dependent (De Vries & Stouthamer 1969; Shimamura *et al.* 1992; Talwalkar *et al.* 2001). *Bifidobacterium animalis* subsp. *lactis* has been shown to be oxygen tolerant, explaining its tendency to survive in the presence of oxygen (Meile *et al.* 1997; Bolduc *et al.* 2006). The growth and metabolite production of *B. animalis* has been studied by Østlie *et al.* (2003, 2005) who showed that *B. animalis* has a heterofermentative nature and is able to produce acetaldehyde, ethanol, acetoin, acetic, lactic, pyruvate, orotic and succinic acids in different concentrations depending on temperature and fermentation time, but has low CO_2 production ability. The above characteristics make *Bifidobacterium animalis* subsp. *lactis* a better choice for commercial probiotic formulations (Meile *et al.* 1997; Gueimonde *et al.* 2004; Masco *et al.* 2004; Bolduc *et al.* 2006).

Lactobacillus spp.

Lactobacillus spp are anaerobic, Gram-positive, non-sporulating, non-motile and rods or coccobacilli (Gomes & Malcata 1999). Several research works have been published on the sensitivity and the level of survival of some of the known probiotics such as *Lactobacillus casei* and *L. acidophilus*. Most probiotic species seem to resist low pH and other external factors, making them ideal for inclusion in fermented dairy products. They remain viable with varying degrees of survival in various food products and in human gut. Mishra & Prasad (2005) reported that *Lactobacillus casei* showed tolerance at pH 3.0, resistance to pH 2.0 for up to 90 minutes and 12 hours in the presence of 2% of bile salts. Coeuret *et al.* (2004) have also shown that *L. paracasei/casei* is resistant to 100 μg mL^{-1} lysozyme after 90 minutes, resistant to 12–14 out of 23 antibiotics and has an inhibitory effect on the growth of *S. aureus*, *Listeria monocytogenes*, *Salmonella spp.* and *E coli*. *L. casei* Shirota strain has also been shown to have the ability to survive and resist the tough conditions through the passage of the gastrointestinal tract after the consumption of fermented milk containing 10^{10} of the live strain (Yuki *et al.* 1999). This strain is also shown to have the ability to modulate the composition and metabolic activity of the intestinal flora (Spanhaak *et al.* 1998). It has also been shown that *L. acidophilus* has a high tolerance to phenol, up to 0.5%, a good survival rate at pH 4 (Šuŝković *et al.* 1997) and resistance towards certain antibiotics (Temmerman *et al.* 2002; Danielsen & Wind 2003). *L. acidophilus* was also reported to grow at a temperature of 45°C, but its optimum growth temperature is 35–40°C (Gomes & Malcata 1999). Currently, ingredient suppliers offer ambient stable probiotic cultures packaged in gel capsules that promote survival and resistance to pH, oxygen, moisture and temperature (Cruce & Goulet 2001).

Some *Lactobacillus* species such as *L. acidophilus*, *L. johnsonii* and *L. helveticus* were found to produce predominantly lactic acid, while other species such as *Lactobacillus fermentum*, *L. plantarum*, *L. casei* and *L. rhamnosus* were found to produce lactic acid, acetic acid, ethanol and formic acid (Lee *et al.* 1999). Lactic acid was found to be the main product of *Lactobacillus casei* Shirota (Sgouras *et al.* 2004). Varieties in metabolism production depend upon the homo or heterofermentative nature of the strain.

Coeuret *et al.* (2004) reported that *Lactobacillus paracasei* and *L. casei* inhibit the growth of *Staphylococcus aureus*, *Listeria monocytogenes*, *Salmonella* spp. and *Escherichia coli* at 15 and 37°C. *Lactobacillus acidophilus* M92 have been shown to inhibit the growth of *Staphylococcus aureus* 3048, *S. aureus* K-144, *Escherichia coli* 3014, *Salmonella mumum*, *Bacillus cereus*, *B. subtilis* ATCC 6633, *Enterococcus faecium* and *Candida tropicalis* (Šuŝković *et al.* 1997; Rolfe 2000). Furthermore, *Lactobacillus casei* Shirota were found to inhibit *Helicobacter pylori* and *E. coli in vivo* and *in vitro* (Sgouras *et al.* 2004).

8.6.4 *Health benefits associated with fermented milk product consumption*

The health benefits of consuming fermented milk containing lactic acid bacteria have been acknowledged for hundreds of years. Since that time a growing number of studies have confirmed that yogurt culture and probiotic bacteria really do offer several important health benefits, especially for specific groups of patients when regularly consumed in a large numbers (Macfarlane & Cummings 2002; Ouwehand *et al.* 2002). The final products should also have acceptable sensory attributes such as taste, aroma, colour and texture during their shelf life (Holzapfel *et al.* 1998).

The health benefits of consuming viable probiotic and yogurt cultures have been reviewed by Gill & Guarner (2004) and Adolfsson *et al.* (2004), respectively, while the consumption of fermented milk products containing specifically bifidobacteria and the potential health benefits were reviewed by Leahy *et al.* (2005). The health benefits of the regular intake of fermented probiotic drinks and yogurt are claimed to be the following.

1. Improves the intestinal microbial balance (Bergogne-Berezin 2000).
2. Produces lactase that prevents symptoms of lactose intolerance (Pettoello *et al.* 1989; Marteau *et al.* 2001) and reduces some forms of food allergies (Pelto *et al.* 1996).
3. Strengthens the immune system (Fiat *et al.* 1993; Marteau *et al.* 1997).
4. Reduces the risk of colon cancer (Rafter 2002) and protects against breast cancer in humans (Van't Veer *et al.* 1991) and their anti-tumour effects in rats (Abd El-Gawad *et al.* 2004).
5. Lowers the blood cholesterol levels (Noh *et al.* 1997; Lin & Chen 2000; Liong & Shah 2005; Alhaj *et al.* 2010).
6. Lowers blood pressure in rats (Sipola *et al.* 2002) and in human subjects (Hata *et al.* 1996; Seppo *et al.* 2003).
7. Plays a key role in prevention and treatment of diarrhoea (Saavedra *et al.* 1994; Scarpignato & Rampal 1995; Hilton *et al.* 1997).
8. Inhibits the growth of some pathogenic bacteria (Anand *et al.* 1984; Meurman *et al.* 1995; Yusof *et al.* 2000; Ibrahim *et al.* 2003; Sgouras *et al.* 2004).
9. Antioxidative activities in fermented soy milk (Wang *et al.* 2006).
10. Yogurt consumption improves lactose digestion, enhances the immune responce, prevents allergic disorder and alleviates acute diarrhoeal disorder (Adolfsson *et al.* 2004; Guarner *et al.* 2005).
11. More recently, yogurt cultures (*L. delbrueckii* ssp. *bulgaricus* Lb1466 and *S. thermophilus* St1342) were reported to have ACE-inhibitory activity *in vitro* with or without adding probiotic bacteria such as *L. acidophilus* L10, *L. casei* L26 and *B. lactis* B94 (Donkor *et al.* 2007).

Some health benefits could be obtained by both viable and non-viable cells including lactose intolerance treatment which reduces acute gastroenteritis duration, anti-*Candida* activity (Ouwehand & Salminen 1998) and cholesterol (Hosono & Tono-oka 1995). No firm conclusion has been drawn on the efficacy of viable cells against non-viable cells. However, Ouwehand & Salminen (1998) indicated that the viability of probiotic micro-organisms could show a better effect than the non-viable micro-organisms. Nevertheless, non-viable probiotics still have some economic advantages in terms of longer shelf life and less requirements for refrigerated transport and storage, expanding the potential use of probiotic products in the strict handling conditions in developing countries (Ouwehand *et al.* 2000b).

8.6.5 *Factors stimulating the growth and survival of* Bifidobacterium *spp.*

All definitions since 1989 (Fuller 1989) have emphasised the importance of viability of probiotic bacteria, and considered them as an essential factor for the health benefits of probiotics. This emphasis comes from the importance of maintaining the probiotic bacteria not only during its shelf life in the product but also in the gut environment until reaching the target site and delivering the health benefit(s). In general, the growth of *Bifidobacterium* spp. in milk were reported to be poor, which might be due to the lack of growth factors such as free amino acids and small peptides in milk (Gomes *et al.* 1998; Tamime *et al.* 2005, p. 57) or lack of proteolytic activity of *bifidobacteria* strains (Klaver *et al.* 1993). However, the minimum nutritional requirement for bifidobacteria to grow is 'a semi-synthetic medium containing only lactose, three amino acids including cysteine, glycine and tryptophan, several vitamins and nucleotides and some minerals' (Gomes & Malcata 1999). The growth of *Bifidobacterium* in milk might be species- and milk-dependent, whereas the growth of *B. longum* has been reported to be higher in camel milk than in bovine milk and vice versa for *B. angulatum* 27535 (Abu-Taraboush *et al.* 1998). According to Modler (1994), the growth of *Bifidobacterium* spp. was found to be enhanced by two types of factors – bifidogenic and growth – as described in the following.

Bifidogenic factors

Bifidogenic factors were defined by Modler (1994) and later modified by Gomes & Malcata (1999) as 'compounds, usually of a carbohydrate nature that survive direct metabolism by the host and reach the large bowel or cecum where they are preferentially metabolised by bifidobacteria as source of energy'. This factor is now known as prebiotic and defined by Gibson & Roberfroid (1995) as 'a non-digestible food ingredient that beneficially affects the host by selectively stimulating the growth and/or activity of one or a limited number of bacteria in the colon'. According to this definition any dietary component that reaches the large intestine are candidate prebiotics such as non-digestible carbohydrates, some peptides and proteins as well as certain lipids (Gibson & Roberfroid 1995).

The influence of bifidobacteria viability by the addition of prebiotic has been reported in many studies (e.g. Rycroft *et al.* 2001; Bruno *et al.* 2002; Martínez-Villaluenga *et al.* 2006; Vernazza *et al.* 2006) where the viability, growth and activity of *Bifidobacterium* spp. were found to be increased in milk containing oligosaccharides (Chick *et al.* 2001).

It was also reported (Vernazza *et al.* 2006) that *Bifidobacterium* spp. include *B. lactis* Bb-12 (Chr. Hansen), *B. adolescentis*, *B. infantis* and two strains of *B. longum* were able to grow and utilise a number of prebiotic substrates in different degrees, especially galacto-oligosaccharide (GOS) and isomalto-oligosaccharide (IMO).

Growth factors

These factors are also referred to as bifidus factors (Klaver *et al.* 1993) and were defined by Modler (1994) as 'compounds which promote the growth of bifidobacteria in vitro but cannot be delivered to the large bowel or cecum to selectively promote proliferation of bifidobacteria'. These factors could include many substrates such as: milk proteins and their tryptic hydrolysates; milk protein hydrolysate from microbial proteolysis; and other substrates, as described in the following.

Milk proteins and their tryptic hydrolysates. It has been reported that major whey proteins (α-la and β-lg) and casein fractions from human and bovine milk are excellent growth promoters for *Bifidobacterium* spp. (Petschow & Talbott 1990). Similarly, the viability of *Bifidobacterium* spp. in yogurt was improved by adding casein hydrolysate and whey protein concentrate containing amino acids and peptides such as tryptone (Dave & Shah 1998). The addition of whey peptide fractions containing caseinomacropeptide (CMP) and whey protein concentrate (WPC) were found to enhance the growth of *Bifidobacterium animalis* subsp. *lactis* (Janer *et al.* 2004), which was in agreement with Ibrahim & Bezkorovainy (1994) who indicated that the presence of α-la and β-lg enhanced the growth of *B. longum*. Whey proteins act as a source of amino acids and peptides and are a rich source in sulphur-containing amino acids, which are released during heat treatment, resulting in low redox potential (Dave & Shah 1998). It has also been reported (Poch & Bezkorovainy 1991) that κ-casein digested by trypsin was able to enhance the growth of *Bifidobacteria* spp. due to the presence of disulphide/sulphydryl residues (cysteine-containing peptides). Bovine casein hydrolysate and yeast extract were considered to be the best growth promoters for all tested *Bifidobacterium* spp. and were better than human and bovine whey proteins and bovine serum albumin hydrolysate (Poch & Bezkorovainy 1988). Furthermore, *Bifidobacterium* spp. preferred the peptides produced by enzymatic hydrolysis of casein as a source of nitrogen more than free amino acids (Proulx *et al.* 1992; Juillard *et al.* 1995a). The length of these peptides plays a very important role for the growth and acid production of LAB (St-Gelais *et al.* 1993). For example, the growth of bifidobacteria was not found to be stimulated by peptides produced by tryptic hydrolysis of casein having molecular mass higher than 5000 Da (Proulx *et al.* 1994).

Milk protein hydrolysate from microbial proteolysis. The growth of bifidobacteria in milk was found to improve when co-cultured with proteolytic bacterial species to provide sufficient nitrogenous compounds compared with milk without the presence of these cultures (Klaver *et al.* 1993). For example, co-culturing bifidobacteria with *L. casei* could stimulate their growth and acid production because *Lactobacillus casei* hydrolyse milk proteins into peptides and make it available for the growth of bifidobacteria (Cheng & Nagasawa 1983, 1984). Similar results were obtained by Gomes *et al.* (1998) where the growth and acid production of *B. lactis* was found to be enhanced when cocultured with *L. acidophilus*.

Other substrates. Many other compounds have also been regarded as growth factors which enhance the growth of probiotics, especially *Bifidobacterium* spp. (bifidogenic) as reported by Modler (1994) and Gomes & Malcata (1999). Compounds such as cysteine (Desjardins *et al.* 1990; Dave & Shah 1997b; Ravula & Shah 1998), threonine (Marshall *et al.* 1982), honey (Chick *et al.* 2001; Kajiwara *et al.* 2002; Haddadin *et al.* 2007), yeast extract (Poch & Bezkorovainy 1988; Ibrahim & Bezkorovainy 1994), casitone or a mixture of amino acids (Klaver *et al.* 1993), enzyme-treated chlorella, peptone and trypticase (Modler 1994), dextrin, maltose and extracts from carrots (coenzyme A) and ascorbic acid (Collins & Hall 1984; Klaver *et al.* 1993) have all been reported as bifidogenic.

8.6.6 *The direct health benefits of probiotics: Mechanisms of action*

Several mechanisms of action have been proposed regarding the health benefits of probiotics but are still not clearly understood. These mechanisms have been linked to the direct effect on the biological function of the human body and/or by inhibiting the growth of some pathogens. These proposed mechanisms could be categorised as in the following.

Probiotic population

Information about the minimum intake/dose of probiotic bacteria required to deliver the reported health claims is still unclear. However, it is known that probiotic strain(s) should be of a high population which is typically between 10^6 and 10^9 CFU g^{-1} to exert the claimed health benefits. EU guidelines propose setting the minimum level of probiotic bacteria in such a product at 10^7 CFU mL^{-1} during the shelf life of the product (EU 2003). It is known that probiotic concentration in a product is not the only critical factor for providing health benefits; daily intake also has an important effect in maintaining such health benefits. Nowadays, consumers are aware of the probiotic properties of cultured milk and dosage specifications (Mosilhey 2003).

Microflora in the large intestine can be classified according to their biological nature, which includes proteolytic species, saccharolytic organisms and organisms that can metabolise gases (Fooks *et al.* 1999), or according to the dominance in the intestine. The majority are beneficial including the dominant bacteria such as *Bifidobacterium*, *Lactobacillus*, *Bacteroides*, *Eubacterium* and *Peptostreptococcus*. The less-dominant bacteria, but still beneficial, include *Streptococcus*, *Enterococcus*, *Enterobacteriaceae* family and some *Clostridium* species. A number of aerobic and anaerobic bacteria such as *Veillonellae*, *Clostridium*, *Staphylococcus*, *Proteus* and *Pseudomonas* bacteria were also found to produce harmful nitrogen waste products such as amines, ammonia, phenols, urea, indoles, nitrites, hydrogen sulphide and secondary bile acids (Macfarlane & Macfarlane 1995; Lennart-Cedgård 1998; Gill & Rowland 2002). A reduction in the number of beneficial bacteria might allow injury to the intestine directly or contribute to diseases in the case of absorption (Clinton *et al.* 1988; Kikugawa & Kato 1986; Macfarlane & Macfarlane 1997; Aslam *et al.* 1992; Hawrelak & Myers 2004). Interestingly, the daily ingestion of *B. lactis* Bb-12 at a concentration of approximately 1–3 × 10^{10} CFU g^{-1} was found to increase the bifidobacterial number in faeces (Schiffrin *et al.* 1995; Alander *et al.* 2001). Recently, a double-blind feeding study was undertaken on 18 healthy elderly individuals using a symbiotic preparation containing a high number of *Bifidobacterium lactis* BL-01 and *B. bifidum* BB-02 combined with an inulin-based prebiotic (Bartosch *et al.* 2005). In their study, the number of *Bifidobacterium bifidum*, total bifidobacteria and total lactobacilli increased significantly in the symbiotic group during feeding when compared to a placebo.

Nutrient availability

The availability of nutrients plays an important role in the survival and growth of micro-organisms. Nutrients are an important factor in determining gut flora composition; in healthy individuals, there is a natural balance between beneficial and harmful bacteria (Gopal *et al.* 2003). In fact, frequent drinking of probiotic products will reinforce the presence and the number of probiotic bacteria in the gut and consequently reduce the chance of pathogen survival due to the lack of nutrients that have mainly been utilised by probiotics, specifically the carbohydrates in the large intestine as reported by Freter *et al.* (1983). On the other hand, adding some prebiotic substrates such as oligosaccharides and inulin promote the growth of a limited number of bacteria (Gibson *et al.* 1995). In recent studies, the consumption of fermented milk containing GOS, *B. lactis* Bb-12 (Chr. Hansen) and syrup (Alander *et al.* 2001) or reconstituted milk powder containing GOS (Gopal *et al.* 2003) was found to significantly increase the faecal count of bifidobacteria in human subjects.

Adhesion to the intestinal mucosa

Probiotic bacteria that are capable of competing for limited nutrients and adhere to the intestinal mucosa and epithelial cell surfaces are the only bacteria that can stay in the intestine. In doing so, they compete successfully to eliminate other bacteria from the ecosystem (Macfarlane & Gibson 1994) through the production of antimicrobial substances, including short chain fatty acids and antimicrobial peptides, to create beneficial conditions in the gastrointestinal tract; these conditions also decrease the luminal pH that inhibit the growth of some bacteria. Consequently, surviving probiotic bacteria can compete with other bacterial populations to occupy the adherence sites in the intestinal mucosa.

In vitro studies reported that the adhesion of probiotic bacteria was found to depend on the concentration of the ingested probiotic bacteria (Tuomola & Salminen 1998); the normal microflora does not greatly affect probiotic bacteria adhesion (Ouwehand *et al.* 1999b). Furthermore, ingestion of probiotic bacteria is able to improve the binding of human gastrointestinal microflora and stabilise it (Fuller 1991; Ouwehand *et al.* 2000a; Gopal *et al.* 2003). *In vivo* studies reported that most probiotic bacteria can only colonise the human gut temporarily, unfortunately (Ouwehand *et al.* 1999a; Tannock 1999). The administered probiotic bacteria can be detected for days or few weeks after cessation of probiotic feeding in the faeces and will thereafter disappear gradually, indicating the transient colonisation of probiotic bacteria in the gut (Bouhnik *et al.* 1992; Ouwehand *et al.* 1999a; Brigidi *et al.* 2000). For example, *Lactobacillus casei* Shirota were found not to colonise the gut epithelium permanently and were only detected in the microflora for the time of administration only (Sgouras *et al.* 2004).

However, the probiotic strains have been shown to have a strong adhesion when compared to common dairy strains (Salminen *et al.* 1998a). *Bifidobacterium lactis* Bb-12 was found to adhere well to the immobilised human ileostomy glycoproteins, human intestinal mucosa and bovine mucosa *in vitro* (Ouwehand *et al.* 2000a; He *et al.* 2001). The binding of this strain to the immobilised human ileostomy glycoproteins were found to be doubled when incubated simultaneously with *Lactobacillus* GG and *Lactobacillus delbrueckii* subsp. *bulgaricus* due to co-aggregation between strains (Ouwehand *et al.* 2000a). In the same study, other strains of probiotic did not significantly affect the adhesion of *Bifidobacterium lactis* Bb-12 to the immobilised human ileostomy glycoproteins, including *Lactobacillus johnsonii* La1 and *Lactobacillus acidophilus* La5. Furthermore, *B. adolescentis*, *B. animalis*, *B. breve*, *B. bifidum* and *B. longum* were also found to have the ability to adhere to human intestinal mucosa and bovine mucosa *in vitro* (He *et al.* 2001).

However, probiotic micro-organisms can adhere to gut epithelium specifically and non-specifically. The procedures of these adhesions have been described by Šušković *et al.* (2001) and Salminen *et al.* (1998b). The specific adhesion is usually defined as a lock and key function which occurs when the bacterial cell protein binds to a receptor on the epithelial cell. Non-specific adhesion is more general and based on physico-chemical factors such as hydrophobic or electrostatic interaction. However, *in vivo* non-specific adhesion may have no significant role in the colonisation of the epithelia, but it might have an important role in the colonisation of luminal contents (Šušković *et al.* 2001). For example, non-specific adhesion may enhance substrate uptake, consequently enforcing growth (Jonsson & Conway 1992). Some strains of *lactobacilli* including *L. casei* (BIO®, Danone), *L. casei* Imunitass (Actimel®, Danone), *L. casei* 01 (Starter culture, Chr. Hansen Ltd), *L. casei* Shirota (Yakult®, Yakult) and *L. casei* var. *rhamnosus* (Lactophilus®, Laboratoires Lyocentre) were reported to be considered as non-specific binding strains when compared to negative control *E. coli* B44 as an indicator of non-specific binding (Tuomola & Salminen 1998).

Some studies however reported the ability of non-viable probiotic bacteria to adhere to tissue culture cells, concluding that viability is not necessary for adhesion (Hood & Zottola 1988; Coconier *et al.* 1993). In general, inactivation of probiotic bacteria by heat treatment was found to decrease the ability of many probiotic strains to adhere to intestinal mucosa, but few of them were found to increase their ability to adhere to intestinal mucosa after heat treatment. Ouwehand *et al.* (2000b) suggested that 'this may be due to the changes in the cell envelop of the inactivated strains rather than due to the fact that they are dead'. They showed that the inactivation by heat treatment at 80°C and 100°C for 10 min were shown to increase the adhesion of *Bifidobacterium lactis* Bb-12 and *Propionibacterium freudenreichii* subsp. *shermanii* JS to immobilised intestinal mucosa due to the increased interaction between them after heat treatment. On the other hand, this treatment has reduced the adhesion of *Lactobacillus rhamnosus* GG and *Lactobacillus johnsonii* La1 and significantly reduced the adhesion of *Lactocoocus lactis* subsp. *cremoris* and *Lactobacillus casei* Shirota. In another study, heat treatment at 100°C for only 30 sec reduced the adhesion of *Bifidobacterium bifidum* by 40% compared with viable bacteria (Fontaine *et al.* 1994).

8.6.7 *Indirect health benefits of probiotics: Biogenic effect*

This effect occurs as a result of metabolites produced by probiotics during the fermentation process, and can be divided into antimicrobial substances, intestinal bacterial enzyme activities, immune enhancement and bioactive peptides, as described in the following.

Antimicrobial substances

Most living micro-organisms need to produce antimicrobial compounds against each other to prevent any onset of infection that would inhibit or stop the growth of other micro-organisms. However, the antipathogenic effect of probiotic bacteria could be through decreasing the luminal pH in the human gut and by the secretion of numerous antimicrobial compounds that kill or inhibit the growth of many Gram-positive pathogens such as *Staphylococcus aureus* and *Clostridium perfringens* (Hernández *et al.* 2005) and Gram-negative pathogens such as *Salmonella typhimurium* and *Escherichia coli* (Sinha 1986; Yusof *et al.* 2000). Similarly, the shelf life of a product can be extended on the same principle by producing numerous antimicrobial compounds, and some of these components such as nisin have been used as food preservatives (Qiao *et al.* 1995). The most widely produced antimicrobial compounds by probiotic bacteria are the organic acids, especially acetic and lactic acids, which comprise 90% of the produced acids (Shah 2001). In addition to these, other acids are produced such as citric, butyric, hippuric, pyruvic, succinic, orotic and uric acids in various and small quantities by probiotic bacteria (Lankaputhra & Shah 1998a; Ouwehand *et al.* 1999a; Østlie *et al.* 2003) and yogurt culture (Fernandez-Garcia & McGregor 1994).

Other studies suggest that some *Bifidobacteria* strains could inhibit a number of pathogens through a non-pH related effect such as bacteriocins (Meghrous *et al.* 1990) and hydrogen peroxide (Gibson & Wang 1994). Carbon dioxide is also produced widely by probiotic strains; some of these components can be produced if the metabolism system of the bacteria is active during their stay or passage through the intestinal tract (Ouwehand *et al.* 1999a). These inhibitory compounds do not only reduce the viable number of bacteria, but also play an important role on affecting bacterial metabolism or toxin production (Rolfe 2000). However, effectiveness levels of these antimicrobial components on the growth of the pathogens vary, whereas combined components have more effect against pathogens than individual components. For instance the presence of hydrogen peroxide together with lactic acid has a greater inhibitory effect than hydrogen peroxide or lactic acid alone (Lankaputhra & Shah 1998b). Meanwhile, hydrogen peroxide produced by the yogurt culture (*Lactobacillus delbrueckii* subsp. *bulgaricus* and *Streptococcus salivarius* subsp. *thermophilus*) is the main cause for the reduction of probiotic bacteria such as *Lactobacillus acidophilus* (Gilliland & Speck 1977; Dave & Shah 1997a) as well as reducing the population of Gram-negative *Pseudomonas* sp. (Price & Lee 1970) and Gram-positive *Staphylococcus aureus* (Dahiya & Speck 1968).

As mentioned in Section 8.6.3, *Bifidobaterium animalis* was shown to survive and maintain its viable count during incubation with yogurt culture for 28 day of storage (Lamoureux *et al.* 2002). Regardless of the beneficial effect of *Bifidobacteria* to some lactic acid bacteria when simultaneously grown, bacteriocins produced by lactic acid bacteria were also found to possess an antimicrobial effect and provide undesirable conditions against probiotic bacteria as well as pathogens (Gomez *et al.* 1997; Miteva *et al.* 1998; Hernández *et al.* 2005). However, some bacteria have self-protective mechanisms to limit the production of bacteriocin, as in the case of *Lactococcus lactis* producing nisin (Qiao *et al.* 1995).

In general, antimicrobial peptides (not casein peptides) showed a broad spectrum of antibacterial activity in addition to showing a rapid killing effect, often within minutes *in vitro*, against various targets including Gram-negative and Gram-positive bacteria, fungi, enveloped viruses, parasites and even tumour cells (Maloy & Kari 1995; Hancock & Lehrer

1998; Hiemstra 2006). It has been reported (Kappeler 1998; Kappeler *et al.* 1999) that Peptidoglycan recognition protein (PGRP) was isolated from camel milk but does not exist in cow milk; this protein inactivates pathogens by binding to peptidogly can structures in the cell wall. It is expected that camel's PGRP and its hydrolysate may have potent antibacterial properties similar to lactoferrin and lactoferricin. Higher amounts of lysozyme, lactoferrin and immunoglobulins were found in dromedary camel milk than bovine or buffalo milk (Kappeler *et al.* 1999; El-Agamy 2000; Konuspayeva *et al.* 2007; Alhaj & Alkanhal 2010). PGRP was found in greater amounts in camel milk than other antibacterial proteins such as lactoferrin, lactoperoxidase or lysozyme (Kappeler *et al.* 1999).

Intestinal bacterial enzyme activities

The metabolic activity of the colonic microflora has received considerable attention from researchers because of their pathological role in exhibiting conditions such as pseudomembranous colitis and colonic carcinogenesis (Marteau *et al.* 1990). Their other physiological functions include fermentation, which is considered to be one of the most important functions since it enlarges the surface area for enzyme action (Mahé *et al.* 1994). Probiotic bacteria alter the microflora metabolism in a beneficial manner and decrease the mutagenicity of the intestinal cells (Lidbeck *et al.* 1992). A disorder in intestinal microflora causes dysfunction of intestinal metabolism, consequently increasing the effect of harmful enzymes that are thought to be the cause of several diseases such as rheumatoid arthritis and lower colon cancer (Goldin & Gorbach 1984; Salminen *et al.* 1998a; Gill & Rowland 2002).

Information about faecal enzyme activities showed that probiotic bacteria are also able to increase the activity of some beneficial enzymes such as β-galactosidase (lactase), the deficiency or lack of which causes lactose intolerance (Martini *et al.* 1991; Sanders 1993; Rolfe 2000; Marteau *et al.* 2001). Moreover, yogurt culture (*Lactobacillus delbrueckii* subsp. *bulgaricus* and *Streptococcus salivarius* subsp. *thermophilus*) have also demonstrated a similar improvement in lactose digestion due to an increase in β-galactosidase activity in rats and humans (Garvie *et al.* 1984; Lin *et al.* 1991; Martini *et al.* 1991; Sanders 2000). Available evidence suggests that lactase activity varies between the available commercial yogurt brands, and some of these brands showed improvement in lactose digestion and tolerance (Wytock & DiPalma 1988). However, these variations in lactase activity could be due to differences in the lactase activity of the individual culture or temperature changes during product distribution from manufacturer to retailer (Wytock & DiPalma 1988). Alternatively, the disruption of microbial cell structure eventually increases lactase release by gastric acid (Martini *et al.* 1987).

Probiotic bacteria are also able to reduce the activity of some harmful enzymes such as β-glucuronidase, β-glucosidase, azoreductase, nitroreductase, dehydroxylase and steroid-7-α- dehydroxylase (Goldin *et al.* 1980; Goldin & Gorbach 1984; Rowland 1992; Pedrosa *et al.* 1995; Bengmark 1998; Gill & Rowland 2002). Hydrogen and methane production and faecal β-galactosidase and β-glucosidase activities could be used as fermentation capacity indicators (Marteau *et al.* 1990). Clinical studies have proved the metabolic activity alterations of lactic acid bacteria, whereas the faecal activities of β-glucuronidase, azoreductase and nitroreductase were significantly declined when humans were fed *L. acidophilus* of human origin (Goldin *et al.* 1980; Goldin & Gorbach 1984).

Similar results were obtained in another study where a significant reduction in all three enzyme activities was reported in healthy elderly subjects after they were fed with *Lactobacillus gasseri* for 12 days (Pedrosa *et al.* 1995). In another human study, no change was reported in the activity of faecal β-galactosidase, β-glucuronidase and azoreductase while nitroreductase activity was significantly decreased and faecal β-glucosidase increased after 3 weeks of feeding human volunteers fermented dairy products containing *Lactobacillus acidophilus* and *Bifidobacterium bifidum* (Marteau *et al.* 1990). In contrast, drinking unfermented milk alone by healthy adults showed no significant alterations in the metabolic activity (Goldin & Gorbach 1984).

Immune enhancement

The immune system is a complex system with different functions, but the main roles of this system are to control the growth of the harmful micro-organisms and distinguish them from familiar bacteria and to expel any foreign bodies such as damaged tissue and tumour cells by producing antibodies and chemical agents (Meydani & Ha 2000; Saxelin 2004). The immune system is the first line of host defence by producing two protective responses. Firstly, innate immunity does not require stimulation and is not enhanced by repeated exposure; this type of immunity has a rapid (minutes rather then days) and non-specific response, particularly to pathogens and other foreign antigens because it is based on the pattern and receptor recognition. Secondly, adaptive humoral and cellular immune functions respond to specific antigens or pathogens and are enhanced by repeated exposure 4–7 days after the beginning of an infection (Szabo 1997; Meydani & Ha 2000). Probiotic bacteria play an important role in both acquired (specific) and innate (non-specific) immunity. For example, oral administration of probiotic bacteria can improve the immunity by increasing the natural killer lymphatic cells (NK cells), phagocytic activity, macrophages, cytokine production, enhancing the circulating IgA antibody and enhance B and T cell productions (Fuller 1989; Schiffrin *et al.* 1995; Gill 1998; Arunachalam *et al.* 2000; Borregaard *et al.* 2000; Kailasapathy & Chin 2000; Meydani & Ha 2000; Simmering & Blaut 2001).

Bioactive peptides

The proteolytic activity of probiotic bacteria on milk proteins gives rise to a range of peptides, some of which are bioactive. The range of biological functions depends on the amino acid composition and sequence of the peptide. Different probiotic bacteria have been added to milk or dairy products, individually or in combination, to provide several health benefits. As part of their metabolic products, these bacteria have also been shown to hydrolyse the major components of milk proteins and increase the number of peptides available for their growth (Alhaj *et al.* 2007). Some of these released peptides from caseins and whey proteins by bacterial proteolysis were found to provide several health benefits such as hypocholesterolemic effect (Nagaoka *et al.* 2001; Alhaj *et al.* 2010) and ACE-inhibitory peptides, the major regulator of blood pressure (Minervini *et al.* 2003; Gobbetti *et al.* 2004). Other bioactivities such as antioxidants, anticarcinogenic, anticariogenic, immunmodulating and other effects are covered in Chapters 1, 2, 4 and 7.

8.7 Conclusion

Functional foods are currently receiving interest globally and consumers now expect food to provide specific benefits above their basic nutritional value. Milk from different mammalian sources is an excellent nutritional source for human adults and infants. Research has reported that milk and dairy products provide particular health benefits to the consumer depending on the bioactive component in milk. Many bioactive components are found naturally in milk such as conjugated linoleic acid (CLA), oligosaccharide and some bioactive peptides. The production of bioactive peptides can be achieved from protein hydrolysis by digestive enzymes or through fermentation when using lactic acid bacteria, including probiotic.

Research into the identification, isolation and health benefits of bioactive components in bovine milk have been given extensive coverage during the last 20 years, while non-bovine milk including camel, buffalo, human, goat, sheep and yak milk need more investigation to confirm their proposed health benefits. Some health claims of milk bioactive components have been extensively investigated *in vitro* and *in vivo* such as ACE-inhibitory activity, cholesterol reduction and antimicrobial effect. Further research is needed for the *in vitro* and *in vivo* studies to provide firm evidence of the health benefits and claims associated with bovine and non-bovine milks.

References

Abd El-Gawad, A., El-Sayed, E.M., Hafez, S.A., El-Zeini, H.M. & Saleh, F.A. (2004) Inhibitory effect of yoghurt and soya yoghurt containing *bifidobacteria* on the proliferation of ehrlich ascites tumour cells *in vitro* and *in vivo* in a mouse tumour model. *British Journal of Nutrition*, **92**, 81–86.

Abubakar, A., Saito, T., Kitazawa, H., Kawai, Y. & Itoh, T. (1998) Structural analysis of new antihypertensive peptides derived from cheese whey protein by proteinase K digestion. *Journal of Dairy Science*, **81**, 3131–3138.

Abu-Taraboush, H.M. (1996) Comparison of associative growth and proteolytic activity of yoghurt starters in whole milk from camels and cows. *Journal of Dairy Science*, **79**, 366–371.

Abu-Taraboush, H.M., Al-Dagal, M.M. & Al-Royli, M.A. (1998) Growth, viability and proteolytic activity of *Bifidobacteria* in whole camel milk. *Journal of Dairy Science*, **81**, 354–361.

Adams, M.R. & Moss, M.O. (2000) *Food Microbiology*, 2nd edn. Cambridge: The Royal Society of Chemistry.

Adolfsson, O., Meydani, S.N. & Russell, R.M. (2004) Yogurt and gut function. *American Journal of Clinical Nutrition*, **80**, 245–256.

Agrawal, R.P., Swami, S.C., Beniwal, R., Kochar, D.K., Sahani, M.S., Tuteja, F.C. & Ghorui, S.K. (2003) Effect of camel milk on glycemic control, lipid profile and diabetes quality of life in type-1 diabetes: a randomised prospective controlled cross over study. *Indian Journal of Animal Science*, **73**, 1105–1110.

Alander, M., Mättö, J., Kneifel, W., Johansson, M., Kögler, B., Crittenden, R., Mattila-Sandholm, T. & Saarela, M. (2001) Effect of galacto-oligosaccharide supplementation on human faecal microflora and on survival and persistence of *Bifidobacterium lactis* Bb-12 in the gastrointestinal tract. *International Dairy Journal*, **11**, 817–825.

Alder-Nissen, J. (1986) *Enzymatic Hydrolysis of Food Proteins*. London: Elsevier Applied Science Publishers.

Alhaj, O.A. & Alkanhal, H.A. (2010) Compositional, technological and nutritional aspects of Dromedary camel milk: a review. *International Dairy Journal*, **20**, 811–821.

Alhaj, O.A., Kanekanian, A. & Peters, A. (2007) Invistagation on whey proteins profile of commercially aviabale milk-based probiotics health drinks using fast protein liquid chromatography (FPLC). *British Food Journal*, **109**, 469–480.

Alhaj, O.A., Kanekanian, A., Peters, A. & Tatham, A. (2010) Hypocholesterolemic effect of *Bifidobacterium animalis* subsp. *lactis* (Bb12) and trypsin casein hydrolysate. *Food Chemistry*, **123**, 430–435.

Alhaj, O.A., Metwalli, A.A.M. & Ismail, E.A. (2011) Heat stability of camel milk proteins after sterilization process. *Journal of Camel Practice & Research*, **18**, 277–282.

Alhaj, O.A., Brückner, H. & Al-Khalifa, A.S. (2012) Sequences of ACE-inhibitory precursor peptides from bacterial fermented milk of *Camelus dromedarius*. *Journal of Peptide Science*, **18**, S146–S147.

Alhaj, O.A., Taufik, E., Handa, Y., Fukuda, K., Saito, T. & Urashima, T. (2013) Chemical characterization of oligosaccharides in commercially pasteurized dromedary camel (*camelus dromedarius*) milk. *International Dairy Journal*, **28**, 70–75.

Alkhudair, A.M. (2009) A comparative study on the ability of some starter cultures in increasing the level of conjugated linoleic acid in cow and camel milks. M.Sc. thesis. King Saud University, Kingdom of Saudi Arabia.

Almahdy, O., EL-Fakharany, E.M., EL-Dabaa, E., Bun Ng, T. & Redwan, E.M. (2011) Examination of the activity of camel milk casein against Hepatitis C virus (Genotype-4a) and its apoptotic potential in hepatoma and hela cell lines. *Hepatitis Monthly*, **11**, 724–730.

Amiot, J., Germain, L., Turgeon, S., Lemay, M., Ory-Salam, C. & Auger, F.A. (2004) Peptides from milk protein hydrolysates to improve the growth of human keratinocytes in culture. *International Dairy Journal*, **14**, 619–626.

Anand, S.K., Srinivasan, R.A. & Rao, L.K. (1984) Antimicrobial activity associated with *Bifidobacterium bifidum*. *Cultured Dairy Products Journal*, **19**, 6–8.

Antal, J., Pál, G., Asbóth, B., Buzás, Z., Patthy, A. & Gráf, L. (2001) Specificity assay of serine proteinases by reverse-phase high-performance liquid chromatography analysis of competing oligopeptide substrate library. *Analytical Biochemistry*, **288**, 156–167.

Antunes, A.E.C., Grael, E.T., Moreno, I., Rodrigues, L.G., Dourado, F.M., Saccaro, D.M. & Lerayer, A.L.S. (2007) Selective enumeration and viability of *Bifidobaterium animalis* subsp. *Lactis* in a new fermented milk product. *Brazilian Journal of Microbiology*, **38**, 173–177.

Arunachalam, K., Gill, H.S. & Chandra, R.K. (2000) Enhancement of natural immune function by dietary consumption of *Bifidobacterium lactis* (HN019). *European Journal of Clinical Nutrition*, **54**, 263–267.

Asato, L., Kina, T., Sugiyama, M., Shimabukuro, T. & Yamamoto, S. (1994) Effect of dietary peptides on plasma lipids and its mechanism studied in rats and mice. *Nutrition Research*, **14**, 1661–1669.

Aslam, M., Batten, J.J., Florin, T.H.J., Sidebotham, R.L. & Baron, J.H. (1992) Hydrogen sulphide induced damage to the colonic mucosal barrier in the rat. *Gut*, **33**, S69.

Axelsson, L. (1998) Lactic acid bacteria: classification and physiology. In: *Lactic acid bacteria* (eds Salminen, S. & von Wright, A.), pp. 1–72. New York: Marcel Dekker.

Bartosch, S., Woodmansey, E.J., Paterson, J.C.M., McMurdo, M.E. & Macfarlane, G.T. (2005) Microbiological effects of consuming a synbiotic containing *Bifidobacterium bifidum*, *Bifidobacterium lactis*, and oligofructose in elderly persons, determined by real-time polymerase chain reaction and counting of viable bacteria. *Clinical Infectious Diseases*, **40**, 28–37.

Bengmark, S. (1998) Probiotics and prebiotics in prevention and treatment of gastrointestinal diseases. *Gastroenterology International*, **11**, 4–7.

Benkerroum, N., Mekkaoui, M., Bennani, N. & Kamal, H. (2004) Antimicrobial activity of camel's milk against pathogenic strains of *Escherichia coli* and *Listeria monocytogene*. *International Journal of Dairy Technology*, **57**, 39–43.

Bergogne-Berezin, E. (2000) Treatment and prevention of antibiotic associated diarrhoea. *International Journal of Antimicrobial Agents*, **16**, 521–526.

Bhattacharya, A., Banu, J., Rahman, M., Causey, J. & Fernandes, G. (2006) Biological effects of conjugated linoleic acids in health and disease. *The Journal of Nutritional Biochemistry*, **17**, 789–810.

Biavati, B. & Mattarelli, P. (2001) The family Bifidobacteriaceae, In: *The Prokaryotes. An evolving electronic resource for microbiological community*, 3rd edn. (eds Dworkin, M. *et al.*). New York: Springer-Verlag.

Bockelmann, W., Beuck, H.-P., Lick, S. & Heller, K. (1995) Purification and characterization of a new tripeptidase from *Lactobacillus delbrueckii* ssp. *bulgaricus B14. International Dairy Journal*, **5**, 493–502.

Bolduc, M.-P., Raymond, Y., Fustier, P., Champagne, C.P. & Vuillemard, J.-C. (2006) Sensitivity of bifidobacteria to oxygen and redox potential in non-fermented pasteurized milk. *International Dairy Journal*, **16**, 1038–1048.

Bordenave, S., Sannier, F., Ricart, G. & Piot, J.-M. (2000) Characterization of a goat whey peptic hydrolysate produced by an ultrafiltration membrane enzymic reactor. *Journal of Dairy Research*, **67**, 551–559.

Borregaard, N., Elsbach, P., Ganz, T., Garred, P. & Svejgaard, A. (2000) Innate immunity: from plants to humans. *Immunology Today*, **21**, 68–70.

Bouattour, M.A., Casals, R., Albanell, E., Such, X. & Caja, G. (2008) Feeding soybean oil to dairy goats increases conjugated linoleic acid in milk. *Journal of Dairy Science*, **91**, 2399–2407.

Bouhallab, S., Mollé, D. & Léonil, J. (1992) Tryptic hydrolysis of caseinomacropeptide in a membrane reactor, preparation of bioactive peptides. *Biotechnology Letters*, **14**, 805–810.

Bouhnik, Y., Pochart, P., Marteau, P., Arlet, G., Goderel, I. & Rambaud, J.C. (1992) Fecal recovery in humans of viable *Bifidobacterium* sp. ingested in fermented milk. *Gastroenterology*, **102**, 875–878.

Brigidi, P., Vitali, B., Swennen, E., Altomare, L., Rossi, M. & Matteuzzi, D. (2000) Specific detection of *Bifidobacterium* strains in a pharmaceutical probiotic product and in human feces by polymerase chain reaction. *Systematic and Applied Microbiology*, **23**, 391–399.

Bruno, F.A., Lankaputhra, W.E.V. & Shah, N.P. (2002) Growth, viability and activity of *Bifidobacterium* spp. in skim milk containing prebiotics. *Journal of Food Science*, **67**, 2740–2744.

Campo, R.D., Bravo, D., Cantón, R., Ruiz-Garbajosa, P., Garcia-Albiach, R., Montesi-Libois, A., Yuste, F.-J., Abraira, V. & Baquero, F. (2005) Scarce evidence of yogurt lactic acid bacteria in human feces after daily yogurt consumption by healthy volunteers. *Applied and Environmental Microbiology* ,**71**, 547–549.

Cardak, A.D., Yetismeyen, A. & Bruckner, H. (2003) Quantitative comparison of free fatty acids in camel, goat and cow milk. *Milchwissenschaft*, **58**, 127–130.

Chabance, B., Marteau, P., Rambaud, J.C., Migliore-Samour, D., Boynard, M., Perrotin, P., Guillet, R., Jollès, P. & Fiat, A.M. (1998) Casein peptide release and passage to the blood in humans during digestion of milk or yogurt. *Biochimie*, **80**, 155–165.

Charteris, W.P., Kelly, P.M., Morell, L. & Collins, J.K. (1998) Development and application of an *in vitro* methodology to determine the transit tolerance of potentially probiotic *Lactobacillus* and *Bifidobacterium* species in the upper human gastrointestinal tract. *Journal of Applied Microbiology*, **84**, 759–768.

Chatterton, D.E.W., Smithers, G., Roupas, P. & Brodkorb, A. (2006) Bioactivity of β-lactoglobulin and α-lactalbumin-technological implications for processing. *International Dairy Journal*, **16**, 1229–1240.

Cheng, C.C. & Nagasawa, T. (1983) Associative relationships between *bifidobacteria* and *lactobacilli* in milk. *Japanese Journal of Zootechnical Science*, **54**, 740–747.

Cheng, C.C. & Nagasawa, T. (1984) Effect of peptides and amino acids produced by *Loctobacillus casei* in milk on the acid production of *bifidobacteria*. *Japanese Journal of Zootechnical Science*, **55**, 339–349.

Cheryan, M. & Deeslie, W.D. (1983) Soy protein hydrolysis in membrane reactors. *Journal of American Oil Chemists Society (JAOCS)*, **60**, 1112–1115.

Chick, H., Shin, H.S. & Ustunol, Z. (2001) Growth and acid production by lactic acid bacteria and Bifidobacteria grown in skim milk containing honey. *Journal of Food Science*, **66**, 478–481.

Chilliard, Y., Ferlay, A. & Doreau, M. (2000) Effect of different types of forage, animal fat or marine oils in cow's diet on milk fat secretion and composition, especially conjugated linoleic acid (CLA) and polyunsaturated fatty acids. *Livestock Production Science*, **70**, 31–48.

Christensen, J.E., Dudley, E.G., Pederson, J.A. & Steele, J.L. (1999) Peptidases and amino acid catabolism in lactic acid bacteria. *Antonie van Leeuwenhoek*, **76**, 217–246.

Chung, H.S., Kim, Y.B., Chun, S.L. & Ji, G.E. (1999) Screening and selection of acid and bile resistant bifidobacteria. *International Journal of Food Microbiology*, **47**, 25–32.

Clare, D.A. & Swaisgood, H.E. (2000) Bioactive milk peptides: a prospectus. *Journal of Dairy Science*, **83**, 1187–1195.

Clinton, S.K., Bostwick, D.G., Olson, L.M., Mangian, H.J. & Visek, W.J. (1988) Effects of ammonium acetate and sodium cholate on N-methyl-N-nitro-N-nitrosoguanidine induced colon carcinogenesis of rats. *Cancer Research*, **48**, 3035–3039.

Coconier, M.-H., Bernet, M.-F., Chauviére, G. & Servin, A.L. (1993) Adhering heat-killed human *lactobacillus acidophilus*, strain LB, inhibits the process of pathogenicity of diarrhoeagenic bacteria in cultured human intestinal cells. *Journal of Diarrhoeal Disease Research*, **11**, 235–242.

Coeuret, V., Gueguen, M. & Vernoux, J.P. (2004) In vitro screening of potential probiotic activities of selected lactobacilli isolated from unpasteurized milk products for incorporation into soft cheese. *Journal of Dairy Research*, **71**, 451–460.

Collado, M.C., Moreno, Y., Cobo, J.M., Mateos, J.A. & Hernández, M. (2006) Molecular detection of *Bifidobacterium animalis* DN-173010 in human feces during fermented milk administration. *Food Research International*, **39**, 530–535.

Collins, E.B. & Hall, B.J. (1984) Growth of Bifidobacteria in milk and preparation of *Bifidobacterium infantis* for a dietary adjunct. *Journal of Dairy Science*, **67**, 1376–1380.

Collomb, M., Schmid, A., Sieber, R., Wechsler, D. & Ryhänen, E.-L. (2006) Conjugated linoleic acids in milk fat: Variation and physiological effects. *International Dairy Journal*, **16**, 1347–1361.

Corl, B.A., Barbano, D.M., Bauman, D.E. & Ip, C. (2003) *Cis* - 9, *trans* - 11 CLA derived endogenously from *trans* -11 18:1 reduces cancer risk in rats. *Journal Nutrition*, **133**, 2893–2900.

Coste, M., Rochet, V., Léonil, J., Mollé, D., Bouhallab, B. & Tomé, D. (1992) Identification of the C-terminal peptides of bovine β-casein that enhance proliferation of rat lymphocytes. *Immunology Letters*, **33**, 41–46.

Cruce, P.S. & Goulet, J. (2001) Improving probiotics survival rates. *Food Technology*, **55**, 36–42.

Dahiya, R.S. & Speck, M.L. (1968) Hydrogen peroxide formation by lactobacilli and its effect on Staphylococcus aureus, *Journal of Dairy Science*, **51**, 1068–1072.

D'Aimmo, M.R., Modesto, M., Biavati, B. (2007) Antibiotic resistance of lactic acid bacteria and Bifidobacterium spp. isolated from dairy and pharmaceutical products. *International Journal of Food Microbiology*, **115**, 35–42.

Danielsen, M. & Wind, A. (2003) Susceptibility of *Lactobacillus* spp. to antimicrobial agents. *International Journal of Food Microbiology*, **82**, 1–11.

Dave, R.I. & Shah, N.P. (1997a) Viability of yogurt and probiotic bacteria in yogurt made form commercial starter culture. *International Dairy Journal*, **7**, 31–41.

Dave, R.I. & Shah, N.P. (1997b) Effect of cysteine on the viability of yoghurt and probiotic bacteria in yoghurts made with commercial starter cultures. *International Dairy Journal*, **7**, 537–545.

Dave, R.I. & Shah, N.P. (1998) Ingredient supplementation effects on viability of probiotic bacteria in yogurt. *Journal of Dairy Science*, **81**, 2804–2816.

Deeslie, W.D. & Cheryan, M. (1981) Continuous enzymatic modification of proteins in an ultrafiltration reactor. *Journal of Food Science*, **46**, 1035–1042.

De Vries, W. & Stouthamer, A.H. (1969) Factors determining the degree of anaerobiosis of *Bifidobacterium* strains.*Archiv fur Mikrobiologie*, **65**, 275–287.

Desjardins, M.-L., Roy, D. & Goulet, J. (1990) Growth of Bifidobacteria and their enzyme profiles. *Journal of Dairy Science*, **73**, 299–307.

Donkor, O.N., Henriksson, A., Singh, T.K., Vasiljevic, T. & Shah, N.P. (2007) ACE-inhibitory activity of probiotic yoghurt. *International Dairy Journal*, **17**, 1321–1331.
Doultani, S., Turhan, K.N. & Etzel, M.R. (2003) Whey protein isolate and glycomacropeptide recovery from whey using ion exchange chromatography. *Journal of Food Science*, **68**, 1389–1395.
Ebringer, L., Ferencrk, M. & Krajĉoviĉ, J. (2008) Beneficial health effects of milk and fermented dairy products: review. *Folia Microbiology*, **53**, 378–394.
El-Agamy, E.I. (2000) Effect of heat treatment on camel milk proteins with respect to antimicrobial factors: a comparison with cows' and buffalo milk proteins. *Food Chemistry*, **68**, 227e232.
El-Agamy, E., Ruppanner, R., Ismail, A., Champagne, C.P. & Assaf, R. (1992) Antibacterial and antiviral activity of camel milk protective proteins. *Journal of Dairy Research*, **59**, 169–175.
Elayan, A.A., Sulieman, A.E. & Saleh, F.A. (2008) The hypocholesterolemic effect of *Gariss* and *Gariss* containing Bifidobacteria in rats fed on a cholesterol-enriched diet. *Asian Journal of Biochemistry*, **3**, 43–47.
Ellegård, K.H., Gammelgård-Larsen, C., Sørensen, E.S. & Fedosov, S. (1999) Process scale chromatographic isolation, characterization and identification of tryptic bioactive casein phosphopeptides. *International Dairy Journal*, **9**, 639–652.
Elli, M., Callegari, M.L., Ferrari, S., Bessi, E., Cattivelli, D., Soldi, S., Morelli, L., Feuillerat, N.G. & Antoine, J.-M. (2006) Survival of yogurt bacteria in the human gut. *Applied and Environmental Microbiology*, **72**, 5113–5117.
El-Soda, M., Macedo, A. & Olson, N.F. (1992) The peptide hydrolase system of *Bifidobacterium* species. *Milchwissenschaft*, **47**, 87–90.
Elwell, M.W. & Barbano, D.M. (2006) Use of microfiltration to improve fluid milk quality. *Journal of Dairy Science*, **89**, (E. Suppl.) E20–E30.
Erdmann, K., Belinda, W.Y., Schröder, C.H. (2008) The possible roles of food-derived bioactive peptides in reducing the risk of cardiovascular disease. *The Journal of Nutritional Biochemistry*, **19**, 643–654.
EU (2003) proposal for a council regulation laying down quality characteristics and requirements for the composition, labelling and control of yogurt and yogurt-like products, AGRI/38743/2003, Rev. 3 of 19 August 2003.
FAO/WHO (Food and Agriculture Organization of the United Nations and World Health Organization) (2001) Health and nutritional properties of probiotics in food including powder milk with live lactic acid bacteria. Available at http://www.who.int/foodsafety/publications/fs_management/en/probiotics.pdf (accessed October 2013).
FAO/WHO (Food and Agriculture Organization of the United Nations and World Health Organization) (2002) Guidelines for the evaluation of probiotics in food. Available at http://www.who.int/foodsafety/fs_management/en/probiotic_guidelines.pdf (accessed October 2013).
Fernandez-Garcia, E. & McGregor, J.U. (1994) Determination of organic acids during the fermentation and cold storage of yogurt. *Journal of Dairy Science*, **77**, 2934–2939.
Fiat, A.-M. & Jollès, P. (1989) Caseins of various origins and biologically active casein peptides and oligosaccharides: Structural and physiological aspects. *Molecular and Cellular Biochemistry*, **87**, 5–30.
Fiat, A.M., Migliore-Samour, D., Jolles, P., Drouet, L., Sollier, C.B. & Caen, J. (1993) Biologically active peptides from milk with emphasis on two examples concerning antithrombotic and immunomodulating activities. *Journal of Dairy Science*,**76**, 301–310.
Finke, B. (2000) Isolierung und charakterisierung von oligosacchariden aus humanen und tierischen milchen. Ph.D. thesis, University of Giessen, Shaker Verlag, Aachen, Germany.
FitzGerald, R.J. & Meisel, H. (2003) Milk protein hydrolysates and bioactive peptides. In *Advanced Dairy Chemistry* Vol. **1**: Proteins, 3rd edn. (eds Fox, P.F. & McSweeney, P.L.H.), pp. 675–678. Kluwer Academic/Plenum Publisher.
FitzGerald, R.J. & Murray, B.A. (2006) Bioactive peptides and lactic fermentations. *International Journal of Dairy Technology*, **59**, 118–125.

Fontaine, I.F., Aissi, E.A. & Bouquelet, S.J.-L. (1994) *In vitro* binding of *Bifidobacterium bifidum* DSM 20082 to mucosal glycoproteins and hemagglutinating activity. *Current Microbiology*, **28**, 325–330.
Fooks, L.J., Fuller, R. & Gibson, G.R. (1999) Prebiotics, probiotics and human gut microbiology. *International Dairy Journal*, **9**, 53–61.
Fox, P.F. & McSweeney, P.L.H. (1998) Lactose chapter. In: *Dairy Chemistry and Biochemistry*, 1st edition. Thomson Science.
Freter, R., Stauffer, E., Cleven, D., Holdeman, L.V. & Moore, W.E. (1983) Continuous-flow cultire as in vitro models of the ecology of large intestinal flora. *Infection and Immunology*, **39**, 666–675.
Fricker, G. & Drewe, J. (1996) Current concepts in intestinal peptide absorption. *Journal of Peptide Science*, **2**, 195–211.
Fuglsang, A., Rattray, F.P., Nilsson, D. & Nyborg, N.C.B. (2003) Lactic acid bacteria: Inhibition of angiotensin converting enzyme *in vitro* and *in vivo*. *Antonie van Leeuwenhoek*, **83**, 27–34.
Fukuda, K., Yamamoto, A., Ganzorig, K., Khuukhenbaatar, J., Senda, A., Saito, T. & Urashima, T. (2010) Chemical characterization of the oligosaccharides in Bactrian camel (*Camelus bactrianus*) milk and colostrums. *Journal of Dairy Science*, **93**, 5572–5587.
Fuller, R. (1989) Probiotics in man and animals. *Journal of Applied Bacteriology*, **66**, 365–378.
Fuller, R. (1991) Probiotics in human medicine. *Gut*, **32**, 439–442.
Gagnon, M., Kheadr, E., La Blay, G. & Fliss, I. (2004) *In vitro* inhibition of *Escherichia coli* 0157:H7 by bifidobacterial strains of human origin. *International Journal of Food Microbiology*, **92**, 69–78.
Garvie, E.I., Cole, C.B., Fuller, R. & Hewitt, D. (1984) The effect of yoghurt on some components of the gut microflora and the metabolism of lactose in the rat. *Journal of Applied Bacteriology*, **56**, 237–245.
Gaullier, J.-M., Halse, J., Høye, K., Kristiansen, K., Fagertun, H., Vik, H. & Gudmundsen, O. (2005) Supplementation with conjugated linoleic acid for 24 months is well tolerated by and reduces body fat mass in healthy, overweight humans. *Journal of Nutrition*, **135**, 778–784.
Gauthier, S.F. & Pouliot, Y. (2003) Functional and biological properties of peptides obtained by enzymatic hydrolysis of whey proteins. *Journal of Dairy Science*, **86**, E78–E87.
German, J.B. & Dillard, C.J. (2006) Composition, structure and absorption of milk lipids: a source of energy, fat-soluble nutrients and bioactive molecules. *Critical Reviews in Food Science and Nutrition*, **46**, 57–92.
Gibson, G.R. & Wang, X. (1994) Regulatory effects of bifidobacteria on the growth of other colonic bacteria. *Journal of Applied Bacteriology*, **77**, 412–420.
Gibson, G.R. & Roberfroid, M.B. (1995) Dietary modulation of the human colonic microbiota: introducing the concept of prebiotics. *Journal of Nutrition*, **125**, 1401–1412.
Gibson, G.R., Beatty, E.B., Wang, X. & Cummings, J.H. (1995) Selective stimulation of Bifidobateria in the human colon by oligofructose and inulin. *Gastroenterology*, **108**, 975–982.
Gill, C.I.R. & Rowland, I.R. (2002) Diet and cancer: assessing the risk. *British Journal of Nutrition*, **88**(Suppl1), S73–S87.
Gill, H.S. (1998) Stimulation of the immune system by lactic cultures. *International Dairy Journal*, **8**, 535–544.
Gill, H.S. & Guarner, F. (2004) Probiotics and human health: a clinical perspective. *Postgraduate Medical Journal*, **80**, 516–526.
Gill, I., López-Fandiño, R., Jorba, X. & Vulfson, E.N. (1996) Biologically active peptides and enzymatic approaches to their production. *Enzyme Microbial Technology*, **18**, 162–183.
Gilliland, S.E. & Speck, M.L. (1977) Instability of *Lactobacillus acidophilus* in yogurt. *Journal of Dairy Science*, **60**, 1394–1398.
Gilliland, S.E., Staley, T.E. & Bush, L.J. (1984) Importance of bile tolerance of *lactobacillus acidophilus* used as a dietary adjunct. *Journal of Dairy Science*, **67**, 3045–3051.
Gismondo, M.R., Drago, L. & Lombardi, A. (1999) Review of probiotics available to modify gastrointestinal flora. *International Journal of Antimicrobial Agents*, **12**, 287–292.

Gobbett, M., Stepaniak, L., Fox, P.F., Sørhaug, T. & Tobiassen, R. (1995) Inhibition of endoandamino-peptidase activities in cytoplasmic fractions of Lactococcus, Lactobacillus and Propionibacterium by peptides from different cheeses. *Milchwissenschaft*, **50**, 565–570.

Gobbetti, M., Ferranti, P., Smacchi, E., Goffredi, F. & Addeo, F. (2000) Production of Angiotensin I-converting-enzyme inhibitory peptides in fermented milks started by *Lactobacillus delbrueckii* sub sp. *bulgaricus* SS1 and *Lactococcus lactis* subsp. *ceremoris* FT4. *Applied and Environmental Microbiology*, **66**, 3898–3904.

Gobbetti, M., Stepaniak, L., De Angelis, M., Corsetti, A. & Di, R. (2002) Latent bioactive peptides in milk proteins: proteolytic activation and significance in dairy processing. *Critical Reviews in Food Science and Nutrition*, **42**, 223–239.

Gobbetti, M., Minervini, F. & Rizzello, C.G. (2004) Angiotensin I-converting-enzyme inhibitory and antimicrobial bioactive peptides. *International Journal of Dairy Technology*, **57**, 173–188.

Goldin, B.R. & Gorbach, S.L. (1984) The effect of milk and lactobacillus feeding on human intestinal bacterial enzyme activity. *American Journal of Clinical Nutrition*, **39**, 756–761.

Goldin, B.R., Swenson, L., Dwyer, J., Sexton, M. & Gorbach, S.L. (1980) Effect of diet and *Lactobacillus acidophilus* supplements on human faecal bacterial enzymes. *Journal of National Cancer Institute*, **64**, 255–260.

Gomes, A.M.P. & Malcata, X.F. (1999) *Bifidobacterium* spp. and *Lactobacillus acidophilus*: biological, biochemical, technological and therapeutical properties relevant for use as probiotics. *Trends in Food Science and Technology*, **10**, 139–157.

Gomes, A.M.P., Malcata, X.F. & Klaver, F.A.M. (1998) Growth enhancement of *Bifidobacterium lactis* Bo and *Lactobacillus acidophilus* Ki by milk hydrolysates. *Journal of Dairy Science*, **81**, 2817–2825.

Gomez, S., Cosson, C. & Deschamps, A.M. (1997) Evidence for a bacteriocin-like substance produced by a new strain of *Streptococcus* sp., inhibitory to Gram-positive food-borne pathogens. *Research Microbiology*, **148**, 757–766.

Gomez-Ruiz, J.A., Ramos, M. & Recio, I. (2002) Angiotensin converting enzyme- inhibitory peptides in Manchego cheeses manufactured with different starter cultures. *International Dairy Journal*, **12**, 697–706.

Gopal, P.K., Prasad, J. & Gill, H.S. (2003) Effect of the consumption of *Bifidobacterium lactis* HN019 (DR10^{TM}) and galaco- oligosaccarides on the microflora of the gastrointestinal tract in human subjects. *Nutrition Research*, **23**, 1313–1328.

Gori Cambrodon, I. & Gudial-Urbano, M. (2001) Oligosaccharides supplementation in infant formula: natural prebiotics. *Alimenteria*, **38**, 77–81.

Griinari, J.M., Corl, B.A., Lacy, S.H., Choinard, P.Y., Nurmela, K.V.V. & Bauman, D.E. (2000) Conjugated linoleic acid is synthesized endogenously in lactating dairy cows by Δ^9- desaturase . *Journal of Nutrition*, **130**, 2285–2291.

Guarner, F., Perdigon, G., Corthier, G., Salminen, S., Koletzko, B. & Morelli, L. (2005) Should yoghurt cultures be considered probiotic? *British Journal of Nutrition*, **93**, 783–786.

Gueimonde, M., Delgado, S., Mayo, B., Ruas-Madiedo, P., Margolles, A. & De los Reyes-Gavilán, C.G. (2004) Viability and diversity of probiotic *Lactobacillus* and *Bifidobacterium* populations included in commercial fermented milks. *Food Research International*, **37**, 839–850.

Haddadin, M.S.Y., Nazer, I., Abu Raddad, S. & Robinson, R.K. (2007) Effect of honey on the growth and metabolism of two bacterial species of intestinal origin. *Pakistan Journal of Nutrition*, **6**, 693–697.

Hambræus, L. (1992) Nutritional aspects of milk proteins. In *Advanced Dairy Chemistry-1. Proteins*, 2nd edn (ed. Fox, P.F.), pp. 57–490. London: Elsevier Applied Science.

Hamers-Casterman, C., Atarouch, T., Muyldermans, S., Bendolman, N. & Hamers, R. (1993) Naturally occurring antibodies devoid of light chains. *Nature*, **363**, 446–448.

Hancock, R.E.W. & Lehrer, R. (1998) Cationic peptides: a new source of antibiotics. *Trends in Biotechnology*, **16**, 82–88.

Harford, S. (1988) Genetic engineering and the pharmaceutical industry. In: *Molecular Biology and Biotechnology* (eds Walker, J.M. & Gingold, E.B.), pp. 149–158. London: Royal Society of Chemistry.

Hata, Y., Yamamoto, M., Ohni, M., Nakajima, K., Nakamura, Y. & Takano, T. (1996) A placebo-controlledstudy of the effect of sour milk on blood pressure in hypertensive subjects. *American Journal of Clinical Nutrition*, **64**, 767–771.

Havenaar, R. & Huis in't Veld, J.H.J. (1992) *A General View in the Lactic Acid Bacteria in Health end Disease* (ed. Wood, B.J.B.), pp. 151–170. London: Elsevier Applied Science.

Hawrelak, J.A. & Myers, S.P. (2004) The causes of intestinal dysbiosis: a review. *Alternative Medicine Review*, **9**, 180–197.

He, F., Ouwehand, A.C., Hashimoto, H., Isolauri, E., Benno, Y. & Salminen, S. (2001) Adhesion of *Bifidobacterium spp.* to human intestinal mucus. *Microbiology and Immunology*, **45**, 259–262.

Hernández, D., Cardell, E. & Zárate, V. (2005) Antimicrobial activity of lactic acid bacteria isolated from Tenerife cheese: initial characterization of plantaricin TF711, a bacteriocin-like substance produced by Lactobacillus plantarumTF711. *Journal of Applied Microbiology*, **99**, 77–84.

Hernández-Ledesma, B., Amigo, L., Ramos, M. & Recio, I. (2004) Angiotensin converting enzyme inhibitory activity in commercial fermented products. Formation of peptides under simulated gastrointestinal digestion. *Journal of Argicultual and Food Chemistry*, **52**, 1504–1510.

Hiemstra, P.S. (2006) Defensins and cathelicidins in inflammatory lung disease: beyond antimicrobial activity. *Biochemical Society Transactions*, **34**, 276–278.

Hilliam, M. (2000). Functional food: How big is the market? *World Food Ingredients*, **12**, 50–53.

Hilton, E., Kolakowski, P., Singer, C. & Smith, M. (1997) Efficacy of Lactobacillus GG as a diarrhoeal prevention in travellers. *Journal of Travel Medicine*, **4**, 41–43.

Holt, C. (1994) The biological function of casein. *In* Yearbook 1994, The Hannah Research Institute, Ayr, Scotland, pp. 60–68.

Holzapfel, W.H., Haberer, P., Snel, J., Schillinger, U. & Huis in't Veld, J.H.J. (1998) Overview of gut flora and probiotics. *International Journal of Food Microbiology*, **41**, 85 101

Hood, S.K. & Zottola, E.A. (1988) Effect of low pH on the ability of *Lactobacillus acidophilus* to survive and adhere to human intestinal cells. *Journal of Food Science*, **53**, 1514–1516.

Hosono, A. & Tono-oka, T. (1995) Binding of cholesterol with lactic acid bacterial cells. *Milchwissenschaft*, **50**, 556–560.

Huebner, S.M., Campbell, J.P., Butz, D.E., Fulmer, T.G., Gendron-Fitzpatrick, A. & Cook, M.E. (2010) Individual isomers of conjugated linoleic acid Reduce inflammation associated with established collagen-induced arthritis in DBA/1 Mice. *Journal of Nutrition*, **140**, 1454–1461.

Ibrahim, S.A. & Bezkorovainy, A. (1994) Growth promoting factors for *Bifidobacterium longum*. *Journal of Food Science*, **59**, 189–191.

Ibrahim, S.A., Dharmavaram, S.R., Seo, C.W. & Shahdazi, G. (2003) Antimicrobial activity of bifidobacterium longum (NCFB) as influnced by spices. *Internet Journal of Food Safety*, **2**, 6–8.

IDF (International Dairy Federation) (1995) Consumption statistics for milk and milk Products 1993. Document no 301, pp. 4–6. International Dairy Federation, Brussels.

Ip, C., Banni, S., Angioni, E., Carta, G., McGinley, J., Thompson, H.J., Barbano, D. & Bauman, D. (1999) Conjugated linoleic acid-enriched butter fat alters mammary gland morphogenesis and reduces cancer risk in rats. *Journal of Nutrition*, **129**, 2135–2142.

Jahreis, G., Fritsche, J. & Kraft, J. (1999) Species-dependent, seasonal, and dietary variation of conjugated linoleic acid in milk. In: *Advances in Conjugated Linoleic Acid Research*, Vol. **1** (eds Yurawecz, M.P., Mossoba, M.M., Kramer, J.K.G., Pariza, M.W. & Nelson, G.J.), pp. 215–225. Champaign, IL: AOCS Press.

Janer, C., Peláez, C. & Requena, T. (2004) Caseinomacropeptide and whey protein concentrate enhance *Bifidobacterium lactis* growth in milk. *Food Chemistry*, **86**, 263–267.

Janer, C., Arigoni, F., Lee, B.H., Peláez, C. & Requena, T. (2005) Enzymatic ability of *Bifidobacterium animalis* subsp. *lactis* to hydrolyze milk proteins: identification and characterization of Endopeptidase O. *Applied and Environmental Microbiology*, **71**, 8460–8465.

Jaudszus, A., Krokowski, M., Möckel, P., Darcan, Y., Avagyan, A., Matricardi, P., Jahreis, G. & Hamelmann, E. (2008) *Cis*-9,*trans*-11-conjugated linoleic acid inhibits allergic sensitization and airway inflammation via a PPARγ-related mechanism in mice. *Journal of Nutrition*, **138**, 1336–1342.

Jelen, P. & Lutz, S. (1998) Functional milk and dairy products. In Functional Foods, Biochemical and Processing Aspects (ed. Mazza, G.), pp. 357–380. Lancaster: Technomic Publishing Co. Inc.

Jollès, P. & Henschen, A. (1982) Comparison between the clotting of the blood and milk. *Trends in Biochemical Science*, **7**, 325–330.

Jonsson, E. & Conway, P. (1992) Probiotics for pigs. In *Probiotics. The Scientific Basis*, (ed. Fuller, R.), pp. 260–316. London: Chapman & Hall.

Juillard, V., Laan, H., Kunji, E.R.S., Jeronimus-Stratingh, C.M., Bruins, A.P. & Konings, W.N. (1995a) The extracellular pI-type proteinase of *Lactococcus lactis* hydrolyzes β-casein into more than one hundred different oligopeptides. *Journal of Bacteriology*, **177**, 3472–3478.

Juillard, V., Le Bars, D., Kunji, E.R.S., Konings, W.N., Gripon, J.C. & Richard, J. (1995b) Oligopeptides are the main source of nitrogen for *Lactococcus lactis* during growth in milk. *Applied and Environmental Microbiology*, **61**, 3024–3030.

Kailasapathy, K. & Chin, J. (2000) Survival and therapeutic potential of probiotic organisms with reference to *Lactobacillus acidophilus* and *Bifidobacterium* spp. *Immunology and Cell Biology*, **78**, 80–88.

Kajiwara, S., Gandhi, H. & Ustunol, Z. (2002) Effect of honey on the growth of and acid production by human intestinal *Bifidobacterium* spp.: An *In Vitro* comparison with commercial oligosaccharides and inulin. *Journal of Food Protection*, **65**, 214–218.

Kanaya, N. & Chen, S. (2010) Conjugated linoleic acid reduces body weight gain in ovariectomized female C57BL/6J mice. *Nutrition Research*, **30**, 714–721.

Kandler, O. & Weiss, N. (1986) Regular non-sporing Gram-positive rods. In *Bergey's Manual of Systematic Bacteriology*, vol. 2 (eds Sneath, P.H.A., Mair, N.S. & Sharpe, M.E.), pp. 1209–1234. Baltimore: Williams and Wilkins.

Kappeler, S. (1998) Compositional and structural analysis of camel milk proteins with emphasis on protective proteins. Ph.D. thesis, Swiss Federal Institute of Technology, Zürich, Switzerland.

Kappeler, S., Farah, Z. & Puhan, Z. (1999) Alternative splitting of lactophorin mRNA from lactating mammary gland of the camel(*Camelus dromedarius*). *Journal of Dairy Science*, **82**, 2084–2093.

Karimi, O. & Pena, A.S. (2003) Probiotics: isolated bacteria strain or mixtures of different strains. *Drugs of Today*, **39**, 565–597.

Kask, S., Laht, T.-M., Pall, T. & Paalme, T. (1999) A study on growth characteristics and nutrient consumption of *Lactobacillus plantarum* in A-stat culture. *Antonie van Leeuwenhoek*, **75**, 309–320.

Kelley, N.S., Hubbard, N.E. & Erickson, K.L. (2007) Conjugated linoleic acid isomers and cancer. *Journal of Nutrition*, **137**, 2599–2607.

Kelly, P.M. & Tuohy, J.J. (1997) The effectiveness of microfiltration for the removal of microorganisms. *International Dairy Federation Bulletin*, **320**, 26–31.

Kikugawa, K. & Kato, T. (1986) Formation of a mutagenic diazoquinone by interaction of phenol with nitrite. *Food Chemistry and Toxicology*, **26**, 209–214.

Kim, Y.S., Birtwhistle, W. & Kim, Y.W. (1972) Peptide hydrolases in the brush border and soluble fractions of small intestinal mucosa of rat and man. *The Journal of Clinical Investigation*, **51**, 1419–1430.

Kitts, D. & Weiler, K. (2003) Bioactive proteins and peptides from food sources. Applications of bioprocesses used in isolation and recovery. *Current Pharmaceutical Design*, **9**, 1309–1323.

Klaenhammer, T.R. (2000). Probiotic bacteria: today and tomorrow. *Journal of Nutrition*, **130**, 415S–416S.

Klaver, F.A.M. & Van der Meer, R. (1993) The assumed assimilation of cholesterol by Lactobacilli and *Bifidobacterium bifidum* is due to their bile salt-deconjugation activity. *Applied and Environmental Microbiology*, **59**, 1120–1124.

Klaver, F.A.M., Kingma, F. & Weerkamp, A.H. (1993) Growth and survival of bifidobacteria in milk. *Netherlands Milk Dairy Journal*, **47**, 151–164.

Klein, G., Pack, A., Bonaparte, C. & Reuter, G. (1998) Taxonomy and physiology of probiotic lactic acid bacteria. *International Journal of Food Microbiology*, **41**, 103–125.

Konuspayeva, G., Faye, B., Loiseau, G. & Levieux, D. (2007) Lactoferrin and immunoglobulin contents in camel's milk (Camelus bactrianus, Camelus dromedarius, and hybrids) from Kazakhstan. *Journal of Dairy Science*, **90**, 38e46.

Korhonen, H. (2009) Milk-derived bioactive peptides: from science to applications. *Journal of Functional Foods*, **1**, 177–187.

Korhonen, H. & Pihlanto, A. (2003) Food-derived bioactive peptides – opportunities for designing future foods. *Current Pharmaceutical Design*, **9**, 1297–1308.

Korhonen, H. & Pihlanto, A. (2006) Bioactive peptides: Production and functionality. *International Dairy Journal*, **16**, 945–960.

Korhonen, H. & Pihlanto, A. (2007) Bioactive peptides from food proteins. In *Handbook of Food Products Manufacturing*, Volume **2** (ed. Hui, Y.H.), pp. 5–37. New Jersey: John Wiley& Sons, Inc.

Kunji, E.R.S., Mierau, I., Hagting, A., Poolman, B. & Konings, W.N. (1996) The proteolytic systems of lactic acid bacteria. *Antonie van Leeuwenhoek*, **70**, 187–221.

Kunz, C. & Rudloff, S. (2006) Health promoting aspects of milk oligosaccharides. *International Dairy Journal*, **16**, 1341–1346.

Kunz, C., Rudloff, S., Baier, W., Klien, N. & Strobel, S. (2000) Oligosaccharides in human milk: Structural, functional, and metabolic aspects. *Annual Review of Nutrition*, **20**, 699–722.

Kunz, C., Rudloff, S. & Michaelsen, K.F. (2001) Concentration of human milk oligosaccharides during late lactation. *The Journal of the Federation of American Societies for Experimental Biology (FASEB)*, **15**, A986.

Laan, H., Smid, E.J., Tan, P.S.T. & Konings, W.N. (1989) Enzymes involved in the degradation and utilization of casein in *Lactococcus lactis*. *Netherlands Milk and Dairy Journal*, **43**, 327–345.

Laht, T.-M. (2003) Influence of microstructure of the curd on enzymatic and microbiological processes in Swiss-type cheese. Dissertation, Tallinn University.

Lamoureux, L., Roy, D. & Gauthier, F. (2002) Production of oligosaccharides in yoghurt containing Bifidobacteria and yoghurt cultures. *Journal of Dairy Science*, **85**, 1058–1069.

Lankaputhra, W.E.V. & Shah, N.P. (1998a) Adherence of probiotic bacteria to human colonic cells. *Bioscience Microflora*, **17**, 105–113.

Lankaputhra, W.E.V. & Shah, N.P. (1998b) Antimutagenic properties of probiotic bacteria and organic acids. *Mutation Reseach*, **397**, 169–182.

Law, B.A., Seggin, E. & Sharpe, M.E. (1976) Amino acid nutrition of some commercial cheese starters in relation to their growth in peptone-supplemented whey media. *Journal of Dairy Research*, **43**, 291–300.

Law, J. & Haandrikman, A. (1997) Proteolytic enzymes of lactic acid bacteria. *International Dairy Journal*, **7**, 1–11.

Lawson, R.E., Moss, A.R. & Givens, D.I. (2001) The role of dairy products in supplying conjugated linoleic acid to man's diet: A review. *Nutrition Research Reviews*, **14**, 153–172.

Leahy, S.C., Higgins, D.G., Fitzgerald, G.F. & Van Sinderen, D. (2005) Getting better with Bifidobacteria: A review. *Journal of Applied Microbiology*, **98**, 1303–1315.

LeBlanc, A., Matar, C., LeBlanc, N. & Perdigón, G. (2005) Effects of milk fermented by *Lactobacillus helveticus* R389 on a murine breast cancer model. *Breast Cancer Research*, **7**, R477–R486.

Lee, Y.-K., Nomoto, K., Salminen, S. & Gorbach, S.L. (1999) *Handbook of Probiotics*. New York, NY: John Wiley & Sons, Inc.

Lehninger, A.L. (1982) *Principles of Biochemistry*. New York: Worth Publishers, INC.

Lennart-Cedgård, M.D. (1998) Probiotics: The link between health and disease. *Positive Health Magazine*, **33**.
Léonil, J., Gagnaire, V., Mollé, D., Pezennec, S. & Bouhallab, S. (2000) Application of chromatography and mass spectrometry to the characterization of food proteins and derived peptides. *Journal of Chromatography A*, **881**, 1–21.
Li, G.H., Le, G.W., Shi, Y.H. & Shrestha, S. (2004) Angiotensin I-converting enzyme inhibitory peptides derived from food proteins and their physiological and pharmacological effects. *Nutrition Research*, **24**, 469–486.
Lidbeck, A., Oevervik, E., Rafter, J., Nord, C.E. & Gustafsson, J.A. (1992) Effect of *Lactobacillus acidophilus* supplements on mutagen excretion in faeces and urine in humans. *Microbial Ecology in Health and Disease*, **5**, 59–67.
Liepke, C., Zucht, H.-D., Forssmann, W.-G. & Ständker, L. (2001) Purification of novel peptide antibiotics from human milk. *Journal of Chromatography B*, **752**, 369–377.
Lieske, B. & Konrad, G. (1996) Physico-chemical and functional properties of whey protein as affected by limited papain proteolysis and selective ultrafiltration. *International Dairy Journal*, **6**, 13–31.
Lilly, D.M. & Stillwell, R.H. (1965) Probiotics, growth promoting factors produced by microorganisms. *Science*, **147**, 747–748.
Lin, M.Y. & Chen, T.-W. (2000) Reduction of cholesterol by *Lactobacillus acidophilus* in culture broth. *Journal of Food and Drug Analysis*, **8**, 97–102.
Lin, M.Y., Savaiano, D. & Harlander, S. (1991) Influence of non fermented dairy products containing bacterial starter cultures on lactose Maldigestion in humans. *Journal of Dairy Science*, **74**, 87–95.
Lin, T.Y., Hung, T.H. & Cheng, T.S.J. (2005) Conjugated linoleic acid production by immobilized cells of *Lactobacillus delbrueckii* ssp *bulgaricus* and *Lactobacillus acidophilus*. *Food Chemistry*, **92**, 23–28.
Liong, M.T. & Shah, N.P. (2005) Acid and bile tolerance and cholesterol removal ability of lactobacilli strains. *Journal of Dairy Science*, **88**, 55–66.
Lock, A.L. & Garnsworthy, P.C. (2003) Seasonal variation in milk conjugated linoleic acid and Δ^9-desaturase activity in dairy cows. *Livestock Production Science*, **79**, 47–59.
Lock, A.L., Home, C.A.M., Bauman, D.E. & Salter, A.M. (2005) Butter naturally enriched in conjugated linoleic acid and vaccenic acid alters tissue fatty acids and improves the plasma lipoprotein profile in cholesterol - fed hamsters. *Journal of Nutrition*, **135**, 1934–1939.
Lopes, D.C.F., Delvivo, F.M. & Silvestre, M.P.C. (2005) Hydrolysates of skim milk peptide profiles using two proteolytic enzymes. *British Food Journal*, **107**, 42–53.
Lv, G.S., Huo, G.C. & Fu, X.Y. (2003) Expression of milk-derived antihypertensive peptide in *Escherichia coli*. *Journal of Dairy Science*, **86**, 1927–1931.
Macfarlane, G.T. & Gibson, G.R. (1994) Metabolic activities of the normal colonic flora. In *Human Health: The Contribution of Microorganisms* (ed. Gibson, S.A.W.), pp. 17–52. London: Springer-Verlag.
Macfarlane, G.T. & Macfarlane, S. (1997) Human colonic microbiota: ecology, physiology and metabolic potential of intestinal bacteria. *Scandinavian Journal of Gastroenterology*, **222** (Suppl), 3–9.
Macfarlane, G.T. & Cummings, J.H. (2002) Probiotics, infection and immunity. *Current Opinion in Infectious Diseases*, **15**, 506–510.
Macfarlane, S. & Macfarlane, G.T. (1995) Proteolysis and amino acid fermentation. In Human Colonic Bacteria: Role in Nutrition, Physiology and Pathology (eds Gibson, G.R. & Macfarlane, G.T.), pp. 75–100. Boca Raton FL: CRC Press.
Mahé, S., Marteau, P., Huneau, J., Thuillier, F. & Tomé, D. (1994) Intestinal nitrogen and electrolyte movements following fermented milk ingestion in man. *British Journal of Nutrition*, **71**, 169–180.
Mainville, I., Arcand, Y. & Farnworth, E.R. (2005) A dynamic model that simulates the human upper gastrointestinal tract for the study of probiotics. *International Journal Food Microbiology*, **99**, 287–296.

Maloy, W.L. & Kari, U.P. (1995) Structure-acitvity studies on magainins and other host defense peptides. *Biopolymers*, **37**, 105–122.

Mao, X.-Y., Ni, J.-R., Sun, W.-L., Hao, P.-P. & Fan, L. (2007) Value-added utilization of yak milk casein for the production of angiotensin-I-converting enzyme inhibitory peptides. *Food Chemistry*, **103**, 1282–1287.

Marks, J. (2003) Successful probiotic bifidobacteria. In *Handbook of Functional Dairy Products* (eds Shortt, C. & O'Brien, J.), pp. 33–69. Florida: CRC Press.

Marshall, V.W., Cole, W.E. & Mabbit, L.A. (1982) Fermentation of specially formulated milk with single strains of bifidobacteria. *International Journal of Dairy Technology*, **35**, 143–144.

Marteau, P., Pochart, P., Flourié, B., Pellier, P., Santos, L., Desjeux, J.F. & Rambaud, J.C. (1990) Effect of chronic ingestion of a fermented dairy product containing *Lactobacillus acidophilus* and *Bifidobacterium bifidum* on metabolic activities of the colonic flora in humans. *American Journal of Clinical Nutrition*, **52**, 685–688.

Marteau, P., Vaerman, J.P., Dehnnin, J.P., Bord, S., Brassart, D., Pochart, P., Desjeux, J.F. & Rambaud, J.C. (1997) Effects of intrajejunal perfusion and chronic ingestion of *Lactobacillus johnsonii* strain LA1 on serum concentrations and jejunal secretions of immunoglobulins and serum proteins in healthy humans. *Gastroenterology and Clinical Biology*, **21**, 293–298.

Marteau, P.R., Verse, M.D., Cellier, C.J. & Schrezenmeir, J. (2001) Protection from gastrointestinal diseases with the use of Probiotics. *American Journal of Clinical Nutrition*, **73**(Suppl2), 430S–436S.

Martin, M.J., Martin-Sosa, S., Garcia-Pardo, L.A. & Hueso, P. (2001) Distribution of bovine sialoglycoconjugates during lactation. *Journal of Dairy Science*, **84**, 995–1000.

Martin-Orue, C., Henry, G. & Bouhallab, S. (1999) Tryptic hydrolysis of κ-caseinomacropeptide: control of the enzymatic reaction in a continuous membrane reactor. *Enzyme and Microbial Technology*, **24**, 173–180.

Martînez-Férez, A., Rudloff, S., Guadix, A., Henkel, C.A., Pohlentz, G., Boza, J.J., Gaudix, E.M. & Kunz, C. (2006) Goats' milk as a natural source of lactose-derived oligosaccharides: isolation by membrane technology. *International Dairy Journal*, **16**, 173–181.

Martínez-Villaluenga, C., Frías, J., Gómez, R. & Vidal-Valverde, C. (2006) Influence of addition of raffinose family oligosaccharides on probiotic survival in fermented milk during refrigerated storage. *International Dairy Journal*, **16**, 768–774.

Martini, M.C., Bollweg, G.L., Levitt, M.D. & Savaiano, D.A. (1987) Lactose digestion by yogurt β-galactosidase: influence of pH and microbial cell integrity. *American Journal of Clinical Nutrition*, **45**, 432–436.

Martini, M.C., Lerebours, E.C., Lin, W.J., Harlander, S.K., Berrada, N.M., Antoine, J.M. & Savaiano, D.A. (1991) Strains and species of lactic acid bacteria in fermented milks (yogurts): effect on in vivo lactose digestion. *American Journal of Clinical Nutrition*, **54**, 1041–1046.

Masco, L., Ventura, M., Zink, R., Huys, G. & Swings, J. (2004) Polyphasic taxonomic analysis of *Bifidobacterium animalis* and *Bifidobacterium lactis* reveals relatedness at the subspecies level: reclassification of *Bifidobacterium animalis* as *Bifidobacterium animalis subsp. animalis* subsp. nov and *Bifidobacterium lactis* as *Bifidobacterium animalis* subsp. *lactis* subsp. nov. *International Journal of Systematic and Evolutionary Microbiology*, **54**, 1137–1143.

Masco, L., Huys, G., De Brandt, E., Temmerman, R. & Swings, J. (2005) Culture-dependent and culture-independent qualitative analysis of probiotic products claimed to contain bifidobacteria. *International Journal of Food Microbiology*, **105**, 221–230.

Masuda, O., Nakamura, Y. & Takano, T. (1996) Antihypertensive peptides are present in aorta after oral administration of sour milk containing these peptides to spontaneously hypertensive rats. *Journal of Nutrition*, **126**, 3063–3068.

Mater, D.D.G., Bretign,Y.L., Firmesse, O., Flores, M.-J., Mogenet, A., Bresson, J.-L. & Corthier, G. (2005) *Streptococcus thermophilus* and *Lactobacillus delbrueckii subsp. bulgaricus* survive gastrointestinal transit of healthy volunteers consuming yogurt. *FEMS Microbiology Letters*, **250**, 185–187.

Matsumoto, M., Ohishi, H. & Benno, Y. (2004) H+-ATPase activity in Bifidobacterium with special reference to acid tolerance. *International Journal of Food Microbiology*, **93**, 109–113.
Mättö, J., Fondén, R., Tolvanen, T., Wright, A., Vilpponen-Salmela, T., Satokari, R. & Saarela, M. (2006) Intestinal survival and persistence of probiotic *Lactobacillus* and *Bifidobacterium* strains administered in triple-strain yogurt. *International Dairy Journal*, **16**, 1174–1180.
Mazoyer, E., Bal Dit Sollier, C., Drouet, L., Fiat, A.M., Jollès, P. & Caen, J. (1992) Active peptides from human and cow's milk proteins effect on platelets function and vessel wall. In *Food Nutrition and Immunity*, Vol. 1 (eds Paubert-Braquet, M., Dupont, C. & Paoletti, R.) Basel: Dynamic Nutrition Research, Karger.
McCann, K.B., Shiell, B.J., Michalski, W.P., Lee, A., Wan, J., Roginski, H. & Coventry, M.J. (2006) Isolation and characterisation of a novel antibacterial peptide from bovine αs1-casein. *International Dairy Journal*, **16**, 316–323.
McDonagh, D. & FitzGerald, R.J. (1998) Production of caseinophosphopeptides (CPPs) from sodium caseinate using a range of commercial protease preparations. *International Dairy Journal*, **8**, 39–45.
Meghrous, J., Euloge, P., Junelles, A.M., Ballongue, J. & Petitdemange, H. (1990) Screening of *Bifidobacterium* strains for bacteriocin production. *Biotechnology Letters*, **12**, 575–580.
Meile, L., Ludwig, W., Rueger, U., Gut, C., Kaufmann, P. & Dasen, G. (1997) Bifidobacterium lactis sp. nov., a moderately oxygen tolerant species isolated from fermented milk. *Systematic and Applied Microbiology*, **20**, 57–64.
Meisel, H. (1986) Chemical characterisation and opioid activity of an exorphin isolated from *in vivo* digests of casein. *FEBS Letters*, **196**, 223–227.
Meisel, H. (1998) Overview on milk protein-derived peptides. *International Dairy Journal*, **8**, 363–373.
Meisel, H. (2004) Multifunctional peptides encrypted in milk proteins. *BioFactors*, **21**, 55–61.
Meisel, H. & Bockelmann, W. (1999) Bioactive peptides encrypted in milk proteins: proteolytic activation and thropho-functional properties. *Antonie van Leeuwenhoek*, **76**, 207–215.
Meisel, H., Goepfert, A. & Günther, S. (1997) ACE-inhibitory activities in milk products. *Milchwissenschaft*, **52**, 307–311.
Meisel, H., Bernard, H., Fairweather-Tait, S., FitzGerald, R.J., Hartmann, R., Lane, C.N., McDonagh, D., Teucher, B. & Wal, J.M. (2001) Nutraceutical and functional food ingredients for food and pharmaceutical applications. *British Journal of Nutrition*, **85**, 635 (1 page).
Meisel, H., Walsh, D.J., Murray, B. & FitzGerald, R.J. (2006) ACE inhibitory peptides. Chapter 13. In *Nutraceutical Proteins and Peptides in Health and Disease* Vol. **4** (eds Mine, Y. & Shahidi, F.), pp. 269–315. Boca Raton, Florida: Taylor & Francis Group Publisher.
Menrad, K. (2003) Market and marketing of functional food in Europe. *Journal of Food Engeneering*, **56**, 181–188.
Mercenier, A., Hols, P., Roussel, Y. *et al.* (2004) Screening and construction of probiotic strains with enhanced protective properties against intestinal disorders. *Microbial Ecology in Health and Disease*, **16**, 86–95.
Metchnicoff, E. (1908) *The Prolongation of Life*. New York, NY: Putmans Sons.
Meurman, J.H., Antila, H., Korhonen, A. and Salminen, S. (1995) Effect of *lactobacillus rhamnosus strain* GG (ATCC 53103) on the growth of *streptococcus sobrinus in vitro*. *The European Journal of Oral Sciences*, **103**, 253–258.
Meydani, S.N. & Ha, W.K. (2000) Immunologic effects of yogurt. *American Journal of Clinical Nutrition*, **71**, 861–872.
Meyer, J. & Jordi, R. (1987) Purification and characterization of X-prolyl-dipeptidyl-aminopeptidase from *Lactobacillus lactis* and *Streptococcus thermophilus*. *Journal of Dairy Science*, **70**, 738–745.
Mierau, I., Kunji, E.R.S., Venema, G. & Kok, J. (1997) Casein and peptide degradation in lactic acid bacteria. *Biotechnology Genetic Engineering Review*, **14**, 279–301.
Migliore-Samour, D., Floch, F. & Jolles, P. (1989) Biologically active casein peptides implicated in immunomodulation. *Journal of Dairy Research*, **56**, 357–362.

Mills, O.E. & Thomas, T.D. (1981) Nitrogen sources for growth of lactic streptococci in milk. *New Zealand Journal of Dairy Science and Technology*, **16**, 43–55.

Minagawa, E., Kaminogawa, S., Tsukasaki, F., Motoshima, H. & Yamauchi, K. (1985) Exopeptidase profiles of bifidobacteria. *Journal of Nutritional Science and Vitaminology*, **31**, 599–606.

Minervini, F., Algaron, F., Rizzello, C.G., Fox, P.F., Monnet, V. & Gobbetti, M. (2003) Angiotensin I-Converting-Enzyme Inhibitory and Antibacterial Peptides from *Lactobacillus helveticus* PR4 Proteinase Hydrolyzed Caseins of Milk from Six Species. *Applied and Environmental Microbiology*, **69**, 5297–5305.

Mishra, V. & Prasad, D.N. (2005) Application of in vitro methods for selection of *Lactobacillus casei* strains as potential probiotics. *International Journal of Food Microbiology*, **103**, 109–115.

Miteva, V., Lvanova, I., Budakov, I., Pantev, A., Stefanova, T., Danova, S., Moncheva, P., Mitev, V., Dousset, X. & Boyaval, P. (1998) Detection and characterization of a novel antibacterial substance produced by a Lactobacillus delbrueckii strain 1043. *Journal of Applied Microbiology*, **85**, 603–614.

Mizuno, S., Matsuura, K., Gotou, T., Nishimura, S., Kajimoto, O., Yabune, M., Kajimoto, Y. & Yamamoto, N. (2005) Antihypertensive effect of casein hydrolysate in a placebo-controlled study in subjects with high-normal blood pressure and mild hypertension. *British Journal of Nutrition*, **94**, 84–91.

Modesto, M., Mattarelli, P. & Biavati, B. (2004) Resistance to freezing and freeze-drying storage processes of potential probiotic bifidobacteria. *Annals of Microbiology*, **54**, 43–48.

Modler, H.W. (1994) Bifidogenic factors-sources, metabolism and applications-review paper. *International Dairy Journal*, **4**, 383–407.

Molina, E., Frutos, M. & Romos, M. (2000) Capillary electrophoresis characterization of the casein fraction of cheeses made from cows', ewes' and goats' milks. *Journal of Dairy Research*, **67**, 209–216.

Morishita, T., Deguchi, Y., Yajima, M., Sakurai, T. & Yura, T. (1981) Multiple nutritional requirements of Lactobacilli: genetic lesions affecting amino acid biosynthetic pathways. *Journal of Bacteriology*, **148**, 64–71.

Mosilhey, S.H. (2003) Influence of different capsule materials on the physiological properties of microencapsulated *Lactobacillus acidophilus*. Dissertation, Bonn University

Mozaffarian, D., Katan, M.B., Ascherio, A., Stampfer, M.J. & Willett, W.C. (2006) Trans fatty acids and cardiovascular disease. *New England Journal of Medicine*, **354**, 1601–1613.

Mussatto, S.I. & Mancilha, I.M. (2007) Non-digestible oligosaccharide: A review. *Carbohydrate Polymers*, **68**, 587–597.

Nagaoka, S., Futamura, Y., Miwa, K., Awano, T., Yamauchi, K., Kanamaru, Y., Tadashi, K. & Kuwata, T. (2001) Identification of novel hypocholesterolemic peptides derived from bovine milk β-lactoglobulin. *Biochemical and Biophysical Research Communications*, **281**, 11–17.

Nakamura, T. & Urashima, T. (2004) The milk oligosaccharides of domestic farm animals. *Trends in Glycoscience and Glycotechnology*, **16**, 135–142.

Nakamura, Y., Yamamoto, N., Sakai, K., Okubo, A., Yamazaki, S. & Takano, T. (1995) Purification and characterization of angiotensin I converting enzyme inhibitors from sour milk. *Journal of Dairy Science*, **78**, 777–783.

Narva, M., Kärkkäinen, M., Poussa, T., Lamberg-Allardt, C. & Korpela, R. (2003) Caseinphosphopeptides in milk and fermented milk do not affect calcium metabolism acutely in postmenopausal women. *Journal of American College of Nutrition*, **22**, 88–93.

Narva, M., Halleen, J., Väänänen, K. & Korpela, R. (2004) Effects of *Lactobacillus helveticus* fermented milk on bone cells *in vitro*. *Life Sciences*, **75**, 1727–1734.

Noh, D.O., Kim, S.H. & Gilliland, S.E. (1997) Incorporation of cholesterol into the cellular membrane of *Lactobacillus acidophilus* ATCC 43121. *Journal of Dairy Science*, **80**, 3107–3113.

Ong, L., Henriksson, A. & Shah, N.P. (2007) Angiotensin converting enzyme inhibitory activity in Cheddar cheeses made with the addition of probiotic *Lactobacillus casei* sp. *Lait*, **87**, 149–165.

O'Shea, M., Lawless, F., Stanton, C. & Devery, R. (1998) Conjugated linoleic acid in bovine milk fat: A food-based approach to cancer chemoprevention. *Trend Food Science and Technology*, **9**, 192–196.

Østlie, H.M., Helland, M.H. & Narvhus, J.A. (2003) Growth and metabolism of selected strains of probiotic bacteria milk. *International Journal of Food Microbiology*, **87**, 17–27.

Østlie, H.M., Treimo, J. & Narvhus, J.A. (2005) Effect of temperature on growth and metabolism of probiotic bacteria milk. *International Dairy Journal*, **15**, 989–997.

Otte, J., Shalaby, S.M., Zakora, M., Pripp, A.H. & El-Shabrawy, S.A. (2007a) Angiotensin-converting enzyme inhibitory activity of milk protein hydrolysates: Effect of substrate, enzyme and time of hydrolysis. *International Dairy Journal*, **17**, 488–503.

Otte, J., Shalaby, S.M.A., Zakora, M. & Neilsen, M.S. (2007b) Fractionation and identification of ACE-inhibitory peptides from α-lactalbumin and β-casein produced by thermolysin-catalysed hydrolysis. *International Dairy Journal*, **17**, 1460–1472.

Ouwehand, A.C. & Salminen, S.J. (1998) The health effects of cultured milk products with viable and non-viable bacteria. *International Dairy Journal*, **8**, 749–758.

Ouwehand, A.C., Kirjavainen, P.V., Shrott, C. & Salminen, S. (1999a) Probiotics: mechanisms and established effects. *International Dairy Journal*, **9**, 43–52.

Ouwehand, A.C., Niemi, P. & Salminen, S.J. (1999b) The normal faecal microflora does not affect the adhesion of probiotic bacteria in vitro. *FEMS Microbiology Letters*, **177**, 35–38.

Ouwehand, A.C., Isolauri, E., Kirjavainen, P.V., Tölkkö, S. & Salminen, S.J. (2000a) The mucus binding of *Bifidobacterium lactis* Bb12 is enhanced in the presence of *Lactobacillus* GG and *Lactobacillus delbrueckii* subsp. *bulgaricus*. *Letters in Applied Microbiology*, **30**, 10–13.

Ouwehand, A.C., Tölkkö, S., Kulmala, J., Salminen, S. & Salminen, E. (2000b) Adhesion of inactivated probiotic strains to intestinal mucus. *Letters in Applied Microbiology*, **31**, 82–86.

Ouwehand, A.C., Salminen, S. & Isolauri, E. (2002) Probiotics; an overview of beneficial effects. *Antonie Van Leeuwenhoek*, **82**, 279–289.

Ozer, B.H. & Kirmaci, H.A. (2010) Functional milks and dairy beverages. *International Journal of Dairy Technology*, **63**, 1–15.

Park, Y.W. (2009) Overview of bioactive components in milk and dairy products. In *Bioactive Components in Milk and Dairy Products*, Chapter 1 (ed. Park, Y.W.) Iowa: Blackwell Publishing.

Park, Y.W. & Haenlein, G.F.W. (2006) Therapeutic and hypoallergenic values of goat milk and implication of food allergy. In Handbook of Milk on Non-bovine Mammals, *Chapter 2* (eds Park, Y.W. & Haenlein, G.F.W.), pp. 121–135. Iowa: Blackwell Publishing.

Park, Y.W., Juárez, M., Ramos, M. & Haenlein, G.F.W. (2007) Physico-chemical characteristics of goat and sheep milk. *Small Ruminant Research*, **68**, 88–113.

Parker, R.B. (1974) Probiotics, the other half of the antibiotics story. *Animal Nutrition and Health*, **29**, 4–8.

Parodi, P.W. (1999) Conjugated linoleic acid and other anticarcinogenic agents of bovine milk fat. *Journal of Dairy Science*, **82**, 1339–1349.

Parodi, P.W. (2001) Cow's milk components with anti-cancer potential. *Australian Journal of Dairy Technology*, **56**, 65–73.

Pedroche, J., Yust, M.M., Lqari, H., Girón-Calle, J., Vioque, J., Alaiz, M. & Millán, F. (2004) Production and characterization of casein hydrolysates with a high amino acid Fischer's ratio using immobilized proteases. *International Dairy Journal*, **14**, 527–533.

Pedrosa, M.C., Golner, B.B., Goldin, B.R., Barakat, S., Dalla, G.E. & Ruseel, R.M. (1995) Survival of yogurt-containing organisms and *Lactobacillus gasseri* (ADH) and their effect on bacterial enzyme activity in the gastrointestinal tract of healthy and hypochlorhydric elderly subjects. *American Journal of Clinical Nutrition*, **61**, 353–358.

Pelto, L, Salminen, S.J. & Isolauri, E. (1996) Lactobacillus GG modulates milk induced immune inflammatory response in milk-hypersensitive adult. *Nutrition Today*, **31**(Suppl), 45S–46S.

Petschow, B.W. & Talbott, R.D. (1990) Growth promotion of *Bifidobacterium* species by whey and casein fractions from human and bovine milk. *Journal of Clinical Microbiology*, **28**, 287–292.

Pettoello, M.M., Guandalini, S., Ecuba, P., Corvino, C. & di Martino, L. (1989) Lactose malabsorption in children with symptomatic Giardia lamblia infection: feasibility of yoghurt supplementation. *Journal of Pediatric Gastroenterology and Nutrition*, **9**, 230–295.

Pihlanto-Leppälä, A., Antila, P., Mäntsälä, P. & Hellman, J. (1994) Opioid peptides produced by in-vitro proteolysis of bovine caseins. *International Dairy Journal*, **4**, 291–301.

Pihlanto-Leppälä, A., Koskinen, P., Paakkari, I., Tupasela, T. & Korhonen, H. (1996) Opioid whey protein peptides obtained by membrane filtration. *International Dairy Federation Bulletin*, **311**, 36–38.

Pihlanto-Leppälä, A., Rokka, T. & Korhonen, H. (1998) Angiotensin I converting enzyme inhibitory peptides derived from bovine milk proteins. *International Dairy Journal*, **8**, 325–331.

Pihlanto-Leppälä, A., Koskinen, P., Piilola, K., Tupasela, T. & Korhonen, H. (2000) Angiotensin I-converting enzyme inhibitory properties of whey protein digests: concentration and characterization of active peptides. *Journal of Dairy Research*, **67**, 53–64.

Playne, M.J. (2002) The health benefits of probiotics. *Food Australia*, **54**, 71–74.

Poch, M. & Bezkorovainy, A. (1988) Growth-enhancing supplements for various species of the genus *Bifidobacterium*. *Journal of Dairy Science*, **71**, 3214–3221.

Poch, M. & Bezkorovainy, A. (1991) Bovine milk κ-casein trypsin digest is a growth enhancer for the genus *Bifidobacterium*. *Journal of Agriculture and Food Chemistry*, **39**, 73–77.

Pochart, P., Marteau, P., Bouhnik, Y., Goderel, I., Bourlioux, P. & Rambaud, J.C. (1992) Survival of bifidobacteria ingested via fermented milk during their passage through the human small intestine: an in vivo study using intestinal perfusion. *American Journal of Clinical Nutrition*, **55**, 78–80.

Possemiers, S., Van Camp, J., Bolca, S. & Verstraete, W. (2005) Characterization of the bactericidal effect of dietary sphingosine and its activity under intestinal conditions. *International Journal of Food Microbiology*, **105**, 59–70.

Pouliot, Y., Gauthier, S.F. & L'Heureux, J. (2000) Effect of peptide distribution on the fractionation of whey hydrolysates by nanofiltration membranes. *Lait*, **80**, 113–122.

Prazeres, D.M.F. & Cabral, J.M.S. (1994) Enzymatic membrane bioreactors and their applications. *Enzyme and Microbial Technology*, **16**, 738–750.

Price, R.J. & Lee, J.S. (1970) Inhibition of *Pseudomonas* species by hydrogen peroxide producing lactobacilli. *Journal of Milk and Food Technology*, **33**, 13–18.

Proulx, M., Gauthier, S.F. & Roy, D. (1992) Utilization of casein enzymatic hydrolysates for the growth of bifidobacteria. *Lait*, **72**, 393–404.

Proulx, M., Ward, P., Gauthier, S.F. & Roy, D. (1994) Comparison of bifidobacterial growth-promoting activity of ultrafiltered casein hydrolyzate fractions. *Lait*, **74**, 139–152.

Qiang, X., YongLie, C. & QianBing, W. (2009) Health benefit application of functional oligosaccharides. *Carbohydrate Polymers*, **77**, 435–441.

Qiao, M., Immonen, T., Koponen, O. & Saris, P.E.J. (1995) The cellular location and effect on nisin immunity of the NisI protein from *Lactococcus lactis* N8 expressed in *Escherichia coli* and *L. lactis*. *FEMS Microbiology Letters*, **131**, 75–80.

Quan, S., Tsuda, H. & Miyamoto, T. (2008) Angiotensin I-converting enzyme inhibitory peptides in skim milk fermented with *Lactobacillus helveticus* 130B4 from camel milk in Inner Mongolia, China. *Journal of the Science of Food and Agriculture*, **88**, 2688–2692.

Raff, M., Tholstrup, T., Sejrsen, K., Straarup, E.M. & Wiinberg, N. (2006) Diets rich in conjugated linoleic acid and vaccenic acid have no effect on blood pressure and isobaric arterial elasticity in healthy young men. *Journal of Nutrition*, **136**, 992–997.

Raff, M., Tholstrup, T., Basu, S., Nonboe, P., SØrensen, M.T. & Straarup, E.M. (2008) A diet rich in conjugated linoleic acid and butter increases lipid peroxidation but does not affect atherosclerotic, inflammatory, or diabetic risk markers in healthy young men. *Journal of Nutrition*, **138**, 509–514.

Rafter, J. (2002) Lactic acid bacteria and cancer: mechanistic perspective. *British Journal of Nutrition*, **88**(Suppl1), S89–S94.

Ramírez-Santana, C., Castellote, C., Castell, M., Rivero, M., Rodríguez-Palmero, M., Franch, A. & Pérez-Cano, F.J. (2009) Long-term feeding of the *cis*-9,*trans*-11 isomer of conjugated linoleic acid reinforces the specific immune response in rats. *Journal of Nutrition*, **139**, 76–81.

Ravula, R.R. & Shah, N.P. (1998) Effect of acid casein hydrolysate and cysteine on the viability of yogurt and probiotic bacteria in fermented frozen dairy desserts. *The Australian Journal of Dairy Technology*, **53**, 175–179.

Recio, I. & Visser, S. (1999) Two ion-exchange chromatographic methods for the isolation of antibacterial peptides from lactoferrin *in situ* enzymatic hydrolysis on an ion-exchange membrane. *Journal of Chromatography A*, **831**, 191–201.

Rettger, F., Levy, M.N., Weinstein, K. & Weiss, J.E. (1935) Lactobacillus acidophilus *and its Therapeutic Application*. Yale University Press New Haven, CT.

Reuter, G., Klein, G. & Goldberg, M. (2002) Identification of probiotic cultures in food samples. *Food Research International*, **35**, 117–124.

Righetti, P.G., Nembri, F., Bossi, A. & Mortarino, M. (1997) Continuous enzymatic hydrolysis of β-casein and isoelectric collection of some of the biologically active peptides in an electric field. *Biotechnology Progress*, **13**, 258–264.

Rios, G.M., Belleville, M.P., Paolucci, D. & Sanchez, J. (2004) Progress in enzymatic membrane reactors-a review. *Journal of Membrane Science*, **242**, 189–196.

Rokka, T., Syväoja, E.-L., Tuominen, J. & Korhonen, H. (1997) Release of bioactive peptides by enzymatic proteolysis of *Lactobacillus* GG fermented UHT milk. *Milchwissenschaft*, **52**, 675–678.

Rolfe, R.D. (2000) The role of probiotic cultures in the control of gastrointestinal health. *Journal of Nutrition*, **130**, 396S–402S.

Rowland, I.R. (1992) Metabolic interactions in the gut. In *Probiotics* (ed. Fuller, R.), pp. 29–53. London, UK: The Scientific Basis, Chapman & Hall.

Roy, D. (2005) Technological aspects related to the use of bifidobacteria in dairy products, review. *Lait*, **85**, 39–56.

Rycroft, C.E., Jones, M.R., Gibson, G.R. & Rastall, R.A. (2001) A comparative *in vitro* evaluation of the fermentation properties of prebiotic oligosaccharides. *Journal of Applied Microbiology*, **91**, 878–887.

Rysstad, G. & Kolstad, J. (2006) Extended shelf life milk-advances in technology. *International Journal of Dairy Technology*, **59**, 85–96.

Saarela, M., Maukonen, J., Wright, A., Vilpponen-Salmela, T., Patterson, A.J., Scott, K.P., Hämynen, H. & Mättö, J. (2007) Tetracycline susceptibility of the ingested *Lactobacillus acidophilus* LaCH-5 and *Bifidobacterium animalis* subsp. *lactis* Bb-12 strains during antibiotic/probiotic intervention *International Journal of Antimicrobial Agents*, **29**, 271–280.

Saavedra, J.M., Bauman, N.A., Oung, I., Perman, J.A. & Yolken, R.H. (1994) Feeding of bifidobacterium bifidum and streptococcus thermophilus to infants in hospital for prevention of diarrhoea and shedding of rotavirus. *Lancet*, **344**, 1046–1049.

Saito, T., Itoh, T. & Adachi, S. (1984) Presence of two neutral disaccharides containing N – acetylhexosamine in bovine colostrum as free forms. *Biochimica et Biophysica Acta*, **801**, 147–150.

Salami, M., Yousefi, R., Ehsani, M.R. *et al.* (2009) Enzymatic digestion and antioxidant activity of the native and molten globule states of camel a-lactalbumin: Possible significance for use in infant formula. *International Dairy Journal*, **19**, 518–523.

Salminen, S., Ouwehand, A.C. & Isolauri, E. (1998a) Clinical application of probiotic bacteria. *International Dairy Journal*, **8**, 563–572.

Salminen, S., Wright, A.V., Morelli, L., Marteau, P., Brassart, D., de Vos, W.M., Fonden, R., Saxelin, M., Collins, K., Mogensen, G., Birkeland, S.E. & Mattila-Sandholm, T. (1998b) Demonstration of safety of probiotics-a review. *International Journal of Food Microbiology*, **44**, 93–106.

Salminen, S., Ouwehand, A.C., Benno, Y. & Lee, Y.K. (1999) Probiotics: how should they be defined? *Trends in Food Science & Technology*, **10**, 107–110.

Sanders, M.E. (1993) Effects of consumption of lactic cultures on human health. *Advances in Food Nutrition Research*, **37**, 67–130.

Sanders, M.E. (2000) Considerations for use of probiotic bacteria to modulate human health. *Journal of Nutrition*, **130**, S384–S390.

Sasaki, M., Bosman, B.W. & Tan, P.S.T. (1995) Comparison of proteolytic activities in various lactobacilli. *Journal of Dairy Research*, **62**, 601–610.

Satake, M., Enjoh, M., Nakamura, Y., Takano, T., Kawamura, Y., Arai, S. & Shimizu, M. (2002) Transepithelial transport of the bioactive tripeptide, Val-Pro-Pro, in human intestinal Caco-2 cell monolayers. *Bioscience, Biotechnology and Biochemistry*, **66**, 378–384.

Saxelin, M. (2004) *Health Benefits of Lactobacillus GG*. Finland: Valio Ltd.

Scanff, P., Yvon, M., Thirouin, S. & Pelissier, J.P. (1992) Characterization and kinetics of gastric emptying of peptides derived from milk proteins in the pre-ruminant calf. *Journal of Dairy Research*, **59**, 437–447.

Scarpignato, C. & Rampal, P. (1995) Prevention and treatment of traveler's diarrhea: a clinical pharmacological approach. *Chemotherapy*, **41**(Suppl1), S48–S81.

Schiffrin, E.J., Rochat, F., Link-Amster, H., Aeschlimann, J. & Donnet-Hughes, A. (1995) Immunomodulation of Human Blood Cells Following the Ingestion of Lactic acid bacteria. *Journal of Dairy Science*, **78**, 491–497.

Schrezenmeir, J. & de Vrese, M. (2001) Probiotics, prebiotics and symbiotics approaching a definition. *American Journal of Clinical Nutrition*, **73**, 361S–364S.

Seppo, L., Jauhiainen, T., Poussa, T. & Korpela, R. (2003) A fermented milk high in bioactive peptides hasa blood pressure-lowering effect in hypertensive subjects. *American Journal of Clinical Nutrition*, **77**, 326–330.

Sgouras, D., Maragkoudakis, P., Petraki, K., Martinez-Gonzalez, B., Eriotou, E., Michopoulos, S., Kalantzopoulos, G., Tsakalidou, E. & Mentis, A. (2004) In Vitro and In Vivo Inhibition of *Helicobacter pylori* by *Lactobacillus casei* Strain Shirota. *Applied and Environmental Microbiology*, **70**, 518–526.

Shah, N.P. (1994) Lactobacillus acidophilus and lactose intolerance: A review. *ASEAN Food Journal*, **9**, 47–54.

Shah, N.P. (2000) Effects of milk-derived bioactives: an overview. *British Journal of Nutrition*, **84**(Suppl1), S3–S10.

Shah, N.P. (2001) Functional foods from probiotics and prebiotics. *Food Technology*, **55**, 46–53.

Shah, N.P. & Jelen, P. (1990) Survival of lactic acid bacteria and their lactases under acidic conditions. *Journal of Food Science*, **55**, 506–509.

Shihata, A. & Shah, N.P. (2000) Proteolytic profiles of yogurt and probiotic bacteria. *International Dairy Journal*, **10**, 401–408.

Shimamura, S., Abe, F., Ishibashi, N., Miyakawa, H., Yaeshima, T.A. & Tomita, M. (1992) Relationship between oxygen sensitivity and oxygen metabolism of *Bifidobacterium* species. *Journal of Dairy Science*, **75**, 3296–3306.

Simmering, R. & Blaut, M. (2001) Pro- and prebiotics-the tasty guardian angels. *Applied Mircobiology and Biotechnology*, **55**, 19–28.

Sinha, R.P. (1986) Toxicity of organic acids for repair-deficient strains of *Escherichia coli*. *Applied and Environmental Microbiology*, **51**, 1364–1366.

Sipola, M., Finckenberg, P., Santisteban, J., Korpela, R., Vapaatalo, H. & Nurminen, M.-L. (2001) Long term intake of milk peptides attenuates development of hypertension in spontaneously hypertensive rats. *Journal of Physiology and Pharmacology*, **52**, 745–754.

Sipola, M., Finckenberg, P., Korpela, R., Vapaatalo, H. & Nurminen, M.L. (2002) Effect of long-term intake of milk products on blood pressure in hypertensive rats. *Journal of Dairy Research*, **69**, 103–111.

Smacchi, E. & Gobbetti, M. (1998) Peptides from several Italian cheeses inhibitory to proteolytic enzymes of lactic acid bacteria, *Pseudomonas fluorescens* ATCC 948 and to the angiotensin-I converting enzyme. *Enzyme Microbial Technology*, **22**, 687–694.

Smacchi, E. & Gobbetti, M. (2000) Bioactive peptides in dairy products: synthesis and interaction with proteolytic enzymes. *Food Microbiology*, **17**, 129–141.

Spanhaak, S., Havenaar, R. & Schaafsma, G. (1998) The effect of consumption of milk fermented by *Lactobacillus casei* strain Shirota on the intestinal microflora and immune parameters in humans. *European Journal of Clinical Nutrition*, **52**, 899–907.

Sperti, G.S. (1971) *Probiotics*. West Points, CT: Avi Publishing Co.

Stanton, C., Ross, R.P., FitzGerald, G.F. & Sinderen, D.V. (2005) Fermented functional foods based on probiotics and their biogenic metabolites. *Current Opinion in Biotechnology*, **16**, 198–203.

Steidler, L. & Neirynck, S. (2003) *In situ* delivery of therapeutic proteins by recombinant *Lactococcus lactis*. *The Journal of Microbiology*, **41**, 63–72.

St-Gelais, D., Roy, D., Haché, S. & Desjardins, M.L. (1993) Growth of non-proteolytic *Lactococcus lactis* in culture medium supplemented with different casein hydrolysates. *Journal of Dairy Science*, **76**, 3327–3337.

Sugano, M., Yamasaki, M., Yamada, K. & Huang, Y.S. (1999) Effect of conjugated linoleic acid on polyunsaturated fatty acid metabolism and immune functions. In *Advances in Conjugated Linoleic Acid Research* (eds Yurawecz, M.P., Mossaba, M.M., Kramer, J.K.G., Pariza, M.W. & Nelson, G.J.), pp. 327–339. Champaign, IL: AOCS Press.

Šuŝković, J., Brkic, B., Matoŝić, S. & Maric, V. (1997) *Lactobacillus acidophilus* M92 as potential probiotic strain. *Milchwissenschaft*, **52**, 430–435.

Šušković, J., Kos, B., Goreta, J. & Matoŝić, S. (2001) Lactic acid Bacteria and Bifidobacteria in symbiotic effect. *Food Technology and Biotechnology*, **39**, 227–235.

Svedberg, J., de Haas, J., Leimenstoll, G., Paul, F. & Teschemacher, H. (1985) Demonstration of β-casomorphin immunoreactive materials in *in vitro* digests of bovine milk and in small intestine contents after bovine milk ingestion in adult humans. *Peptides*, **6**, 825–830.

Szabo, G. (1997) Alcohol's contribution to compromised immunity. *Alcohol Health and Research World*, **21**, 30–41.

Talwalkar, A., Kailasapathy, K., Peiris, P. & Arumugaswamy, R. (2001) Application of RBGRa simple way for screening of oxygen tolerance in probiotic bacteria. *International Journal of Food Microbiology*, **71**, 245–248.

Tamime, A.Y. (2002) Fermented milks: a historical food with modern applications-a review. *Eurepean Journal of Clinical Nutrition*, **56**, S2–S15.

Tamime, A.Y. & Robinson, R.K. (1985) *Yoghurt Science and Technology*. Oxford: Pergamon Press Ltd.

Tamime, A.Y., Marshall, V.E. & Robinson, R.K. (1995) Microbiological and technological aspects of milks fermented by bifidobacteria. *Journal of Dairy Research*, **62**, 151–187.

Tamime, A.Y., Saarela, M., Korslund Søndergaard, A., Mistry, V.V. & Shah, N.P. (2005) Production and maintenance of viability of probiotic micro-organiss in dairy products. In *Probiotic Dairy Products* (ed. Tamime, A.Y.), pp. 39–72. Oxford: Blackwell Publishing Ltd.

Tanabe, S., Tesaki, S., Watanabe, J., Fukushi, E., Sonoyama, K. & Kawabata, J. (2003) Isolation and structural elucidation of a peptide derived from edam cheese that inhibit β-lactoglobulin transport. *Journal of Dairy Science*, **86**, 464–468.

Tanmahasamut, P., Liu, J., Hendry, L.B. & Sidell, N. (2004) Conjugated linoleic acid blocks estrogen signaling in human breast cancer cells. *Journal of Nutrition*, **134**, 674–680.

Tannock, G.W. (ed.) (1999) Introduction. In *Probiotics: A critical review*. Norfolk, UK: Horizon Scientific Press, pp. 1–4.

Tannock, G.W. (2003) Probiotics: time for a dose of realism. *Current Issues in Intestinal Microbiology*, **4**, 33–42.

Tek, H.N., Turhan, K.N. & Etzel, M.R. (2005) Effect of conductivity, pH, and elution buffer salinity on glycomacropeptide recovery from whey using anion exchange chromatography. *Journal of Food Science*, **70**, E295–E300.

Temmerman, R., Pot, B., Huys, G. & Swings, J. (2002) Identification and antibiotic susceptibility of bacterial isolates from probiotic products. *International Journal of Food Microbiology*, **81**, 1–10.

Tholstrup, T. (2006) Dairy products and cardiovascular disease. *Current Opinion in Lipidology*, **17**, 1–10.

Thomä-Worringer, C., Sørensen, J. & López-Fandiño, R. (2006) Health effects and technological features of caseinomacropeptide. *International Dairy Journal*, **16**, 1324–1333.

Thomas, T.D. & Mills, O.E. (1981) Proteolytic enzymes of starter bacteria. *Netherland Milk Dairy Journal*, **35**, 255–273.

Tuomola, E.M. & Salminen, S.J. (1998) Adhesion of some probiotic and dairy *Lactobacillus* strains to Caco-2 cell cultures. *International Journal of Food Microbiology*, **41**, 45–51.

Turgeon, S.L. & Gauthier, S.F. (1990) Whey peptide fractions obtained with a two-step ultrafiltration process: production and characterization. *Journal of Food Science*, **55**, 106–110.

Urashima, T., Murata, S. & Nakamura, T. (1997) Structural determination of monosialyl trisaccharides obtained from caprine colostrum. *Comparative Biochemistry Physiology*, **116B**, 431–435.

Urashima, T., Saito, T., Nakamura, T. & Messer, M. (2001) Oligosaccharides of milk and colostrums in non-human mammals. *Glycoconjugate Journal*, **18**, 357–371.

Urashima, T., Kitaoka, M., Asakuma, S. & Messer, M. (2009) Milk oligosaccharides. In Advanced Dairy Chemistry. Lactose, Water, Salts and Minor Constituents, Vol. **3** (eds McSweeney, P.L.H. & Fox, P.F.), pp. 295e349. New York, NY: Springer Science Business Media.

Valembois, C., Mendre, C., Cavadore, J.C. & Calas, B. (1992) Continuous flow synthesis of a biologically active small protein and a large peptide. *Tetrahedron Letters*, **33**, 4005–4008.

Van Nieuwenhove, C., González, S., Pérez-Chaia, A. & De Ruiz Holgado, A.P. (2004) Conjugated linoleic acid in buffalo (Bubalus bubalis) milk from Argentina. *Milchwissenschaft*, **59**, 506–508.

Van't Veer, P., Van Leer, E.M., Rietdijk, A., Kok, F.J., Schouten, E.G., Hermus, R.J.J. & Sturmans, F. (1991) Combination of dietary factors in relation to breast-cancer occurrence. *International Journal of Cancer*, **47**, 649–653.

Vaughan, E.E. & Mollet, B. (1999) Probiotics in the new millennium. *Die Nahrung*, **43**, S148–S153.

Veloso, A.C.A., Teixeiria, N. & Ferreira, I.M.P.L.V.O. (2002). Separation and quantification of the major casein fractions by reverse-phase high performance liquid chromatography and urea-polyacrylamide gel electrophoresis detection of milk adulterations. *Journal of Chromatography A*, **967**, 209–218.

Vernazza, C.L., Gibson, G.R. & Rastall, R.A. (2006) Carbohydrate preference, acid tolerance and bile tolerance in five strains of *Bifidobacterium*. *Journal of Applied Microbiology*, **100**, 846–853.

Vesper, H., Schelma, E., Nikolova-karakashion, M.N., Dillehay, D.L., Lynch, D.V., Mercill, A.H. (1999) Sphingolipids in food and the emerging importance of sphingolipids to nutrition. *Journal of Nutrition*, **129**, 1239–1249.

Wang, Y.-C., Yu, R.-C. & Chou, C.-C. (2006) Antioxidative activities of soymilk fermented with lactic acid bacteria and bifidobacteria. *Food Microbiology*, **23**, 128–135.

Warensjo, E., Nolan, D. & Tapsell, L. (2010) Chapter 1-Dairy food consumption and obesity-related chronic disease. *Advances in Food and Nutrition Research*, **59**, 1–41.

Wetzel, R. & Goeddel, D.V. (1987) Synthesis of polypeptides by recombinant DNA methods. In The Peptides: Analysis, Synthesis and Biology Vol. **5**. (eds Udenfriend, S. & Meienhofer, J.), pp. 2–65. New York: Academic Press.

Wohlrab, Y. & Bockelmann, W. (1992) Purification and characterization of a dipeptidase from *Lactobacillus delbrueckii* subsp. *bulgaricus*. *International Dairy Journal*, **2**, 345–361.

Wytock, D.H. & DiPalma, J.A. (1988) All yogurts are not created equal. *American Journal of Clinical Nutrition*, **47**, 454–457.

Xu, R.J. (1998) Bioactive peptides in milk and their biological and health implications. *Food Review International*, **14**, 1–16.

Yamamoto, N. (1997) Antihypertensive peptides derived from food proteins. *Biopolymers*, **43**, 129–134.

Yamamoto, N., Akino, A. & Takano, T. (1994) Antihypertensive Effect of the Peptides Derived from Casein by an Extracellular Proteinase from *Lactobacillus helveticus* CP790. *Journal of Dairy Science*, **77**, 917–922.

Yamamoto, N., Maeno, M. & Takano, T. (1999) Purification and characterization of an antihypertensive peptide from a yogurt-like product fermented by *Lactobacillus helveticus* CPN4. *Journal of Dairy Science*, **82**, 1388–1393.

Yuki, N., Watanabe, K., Mike, A., Tagami, Y., Tanaka, R., Ohwaki, M. & Morotomi, M. (1999) Survival of a probiotic, *Lactobacillus casei* strain Shirota, in the gastrointestinal tract: Selective isolation from faeces and identification using monoclonal antibodies. *International Journal of Food Microbiology*, **48**, 51–57.

Yusof, R.M., Haque, F., Ismail, M. & Hassan, Z. (2000) Isolation of *Bifidobacterium infantis* and its antagonistic activity against ETEC 0157 and *Salmonella typhimurium* S-285 in weaning foods. *Asia Pacific Journal of Clinical Nutrition*, **9**, 130–135.

Zhou, J.S., Pillidge, C.J., Gopal, P.K. & Gill, H.S. (2005) Antibiotic susceptibility profiles of new probiotic *Lactobacillus* and *Bifidobacterium* strains. *International Journal of Food Microbiology*, **98**, 211–217.

9 Minerals and Vitamins in Milk and Dairy Products

I. Gordon

1290 Versoix, Switzerland

9.1 Introduction

9.1.1 *Mineral and vitamin needs*

It is self-evident that all foods are nutritious and thus functional. In this chapter we consider the delivery of minerals and vitamins via a range of dairy products and indicate the proportion of the recommended daily consumption. There exist a number of ways of expressing desirable intake. In this chapter the terminology taken from the US National Academy of Sciences (NAS) Institute of Medicine will be used as standard, including:

- dietary reference intake (DRI) components;
- estimated average requirement (EAR), which reflects the estimated median requirement and is particularly appropriate for applications related to planning and assessing intakes for groups of persons;
- recommended dietary allowance (RDA) or intake (RDI), derived from EAR and meets or exceeds the requirement for 97.5% of the population;
- tolerable upper intake level (UL), the highest average daily intake that is likely to pose no risk of adverse effects to almost all individuals in the general population; and
- adequate intake (AI), used when an EAR/RDA cannot be developed; average intake level based on observed or experimental intakes.

RDA is used in this chapter as the reference, the source being the Food and Nutrition Board (2004). The recommended level of consumption varies with age and sex. Factors include age, both with respect to growth in children and simply as a share of body weight; sex, with a special subset of females who are pregnant or lactating; other sources of nutrition, notably where a balanced diet may not be available. For reasons of simplification, the RDA data presented in this chapter (with the exception of that of calcium) are limited to that of male adults.

The functional benefits offered by milk and milk products are especially well documented in the case of calcium, although vitamins are also referenced (especially vitamin D) and their benefits promoted to a lesser extent. Throughout the chapter the term 'mineral' is taken to

Milk and Dairy Products as Functional Foods, First Edition. Edited by A. Kanekanian.

refer to the metal ion, whether or not it be complexed. Anion interest largely concerns phosphates. As well as benefits, the negative influence of sodium chloride cannot be ignored especially in hard cheeses. In order to compare like with like across dairy products, liquid milks are assumed to be 200 mL servings, hard cheeses 30 g, soft cheeses 40 g and yogurts 150 mL.

9.1.2 *Bioavailability*

The issue of bioavailability is fraught. The concept is simple: anything ingested has to pass through a strong acid environment prior to either being absorbed into the body (being bioavailable), consumed by the intestinal flora (which may or may not be positive for the host organism) or rejected in the faeces. The two main claims regarding bioavailability relate to protection of some nutrient during passage through the gastrum or an enhanced rate of passage through the intestinal wall.

Particular claims are made for calcium bioavailability in dairy products due to the significant share of total calcium which is chelated and/or enmeshed in the casein micelles or suspended colloidally as calcium phosphate, rather than being present simply as hydrated ions. Those calcium ions complexed with casein peptides are claimed to maintain the calcium in a soluble form and protect it against precipitation in the intestine, thus facilitating absorption (Ebringer *et al.* 2008).

Smeets (1955) proposed that calcium in milk is found at about 65% protein bound and 35% diffusible at 'normal' milk pH. This has been confirmed in many later publications. A major examination was published in 2006 by Lin *et al.*

9.1.3 *Differences between minerals and vitamins*

The chapter first looks at the minerals available in milks of various species before examining in more depth mineral availability in cow milk products. It continues with a similar approach to examine vitamin contents in milk and cow milk derivatives.

There are some significant differences between the mineral and vitamin parts of the chapter: the mineral contents are found only in the aqueous phase of the milk and are unchanging in total mass with time unless precipitated; the vitamins, on the other hand, vary, some being fat- and others water-soluble. Further, many vitamins degrade with heat and/or light and so concentrations will depend on process and storage conditions.

9.1.4 *Caveat*

A major caveat concerns the variability of the raw milk feedstock from any one specie. As would be expected, milk composition is influenced by many exogenous and endogenous factors. Particularly important are: species; breed; point of collection with regard to the whole period of lactation and number of lactations of the animal; feed composition; and drinking water composition (notably for trace metal ions).

Experiments in modifying milk composition by varying feed composition have been carried out, but such manipulation is rarely viable in practice for large commercial enterprises. Consequently, significant differences appear between sources in the literature concerning specific levels of minerals and vitamins; difference of ±10% can be found. When milk from less-well-studied species is examined, the published data show an alarming range of mineral and vitamin content, partly due to the more diverse breeds from which the milk may be collected and to the greater variation in analysis over the post-partum period.

Some discrepancies will be observed even in cow's milk and when the data are drawn from a common source. As an example, full-cream cow's milk has a calcium content quoted as being 120 mg per 100 g, and yet full-cream yogurt appears at 200 mg per 100 g. Even if the yogurt is considered to have added skim milk powder, this is surprising. Differing sources also relate the supply of nutrients to differing RDA bases and legislations. Again, the original source comparisons have been used in the discussion in this chapter despite the apparent inconsistency that this may imply. This approach has been deemed preferable to the alternative of forcing a specific national view of RDA.

9.2 Minerals in dairy products

9.2.1 *Introduction*

Milk contains about 0.9% minerals. The main elements are sodium, potassium, calcium, magnesium, chloride, phosphate and sulphate. In also contains a relatively large number of essential trace elements such as iron, iodine, copper, manganese, zinc, cobalt, selenium and chromium.

The processing of milk influences the content of minerals in the various dairy ingredient fractions. Simple ions such as sodium, potassium and chloride and elements bound in soluble complexes such as iron and cobalt will concentrate in whey. Other minerals such as calcium, magnesium, manganese and zinc are associated with caseins, depending on the pH of the processing system. In cheese-making, for example, higher acidity levels will push more minerals into the whey. Heat treatment causing denaturation of ion-specific binding proteins may also influence the separation. Typically, iron and copper, bound to rather heat-stable proteins, will pass to the whey while manganese and zinc will accompany calcium and magnesium.

The major focus in the promotion of milks and dairy products when functional foods are being considered is the delivery of calcium, sometimes linked to magnesium and phosphorus in order to reflect possibly improved bioavailability. This in turn is largely related to a growing concern about the impact of osteoporosis in an increasingly aged population. The arguments are set out in the following sections.

Calcium

Calcium is one of the most important elements in the human body. It plays an important part in many body processes and regulates many of the chemical reactions that occur in the body such as contraction and relaxation of muscles, transmission of nerve impulses and clotting of the blood. About 99% of the calcium found in the body is stored in the bony skeleton and teeth. If the diet does not supply enough calcium, the body activates hormones which cause the skeleton to break down and release calcium back into the blood so normal levels of calcium in the blood are maintained. Since calcium is lost from the body continuously through the skin and normal daily excretions, everyone needs calcium every day in order to maintain a positive balance i.e. ingesting at least as much as is being lost.

Dairy foods and dairy ingredients such as whey and whey derivatives are excellent sources of calcium. The bioavailability of calcium from foods varies greatly; for example, little calcium is absorbed from spinach and the availability from certain beans is only half as good as that from milk. Numerous other dietary and non-dietary factors such as vitamin D, lactose, fructo-oligosaccharides, phosphorus, protein, sodium and other elements influence calcium metabolism. Many of these are also constituents of milk and whey products. Research also suggests that osteoporosis, hypertension, colon cancer and cardiovascular diseases may prevent intake of calcium. Further, calcium, potassium and magnesium (also present in milk and whey minerals) help to combat high blood pressure.

Osteoporosis

The impact of osteoporosis on the increasingly long-lived population of advanced economies and its potential mitigation by consumption of dairy products has been a focus of the marketing of milk for some time.

The basic tenet is that osteoporosis onset is inversely related to the maximum bone density reached in early adulthood. This is because bone mass/density is developed during the growth phase and will be related to the adequacy of the constituents in the diet and to the need for strength, itself related to muscle loads exerted on the skeleton through exercise (especially during childhood and adolescence).

Throughout life there is a constant turnover of calcium via the osteoclast/osteoblast interaction and bone density falls slowly but inexorably through later life. This is amplified in women by a sudden drop which occurs around the menopause. It is for this reason that even women who had adequate calcium input in their youth may still develop a 'widow's hump' at advanced age as the vertebrae slowly collapse.

Recommendations of appropriate daily intake of calcium throughout life vary marginally between jurisdictions. Table 9.1 provides a good overview, as proposed by the FAO in 2002.

There has been much enthusiasm for calcium supplementation in old people, not only via consumption of milk products but also via over-the-counter (OTC) calcium carbonate tablets. However, very recently several papers have discussed calcium supplementation with or without co-administration of vitamin D (which is said to enhance the bioavailability of the calcium ions). There is concern that the outcome measure usually used as a surrogate for reduced mortality (bone density) does not match outcomes as measured by fracture and actual mortality. Meanwhile, there have been suggestions that calcium supplementation may increase the risk of cardiac infarction (Bolland *et al.* 2010).

Table 9.1 Recommended calcium intake for humans at different age groups based on European and North American data. FAO/WHO 2001. Reproduced with permission of Food and Agriculture Organization of the United Nations.

Group	Recommended intake (mg day^{-1})
Infants and children	
0–6 months	
Human milk	300
Cow milk	400
7–12 months	400
1–3 years	500
4–6 years	600
7–9 years	700
Adolescents (10–18 years, in particular during growth spurt)	1300
Adults	
Females	
Pregnancy	1200
Lactation	1000
Males and females	
19–65 years	1000
>65 years	1300

9.2.2 *Milks*

The milks consumed around the world are largely of cow, buffalo, sheep, goat and, to a lesser extent camel, mare or llama in origin. Table 9.2 lists the typical contents of minerals in the milk of these species. Comparing each to cow's milk, the following observations may be drawn (bearing in mind the caveat described in Section 9.1.4 that differing sources give a wide spread of data for these milks):

- buffalo milk is somewhat higher in calcium and copper, while being lower in potassium and zinc;
- sheep milk demonstrates enhanced calcium, phosphorus and zinc;
- goat milk is very similar to cows milk;
- camel milk is lower in iron but higher in copper and zinc;
- mare milk is lower in phosphorus; and
- llama milk is higher in calcium.

Table 9.2 Levels of minerals in milk from different animals (expressed in mg or μg per 200 mL milk).

Mineral	Cow	Buffalo	Sheep	Goat	Camel	Mare	Llama	RDA adult
Source	Park & Haenlein (2000)	Ranjan *et al.* (2005)	Coni *et al.* (1999)	Park & Haenlein (2000)	Bornaz *et al.* (2009)		Morin *et al.* (1995)	
Sodium (mg)	116	88	100	82	120		60	1500
Potassium (mg)	304	214	272	362	300			4700
Calcium	250	366	390	260	224	270	400	1200
Magnesium (mg)	23	36	36	32	24	20		420
Phosphorus (mg)	238	164	300	220	180	174	240	700
Iron (ug)	160		140	120	60			8000
Copper (ug)	120	300	80	100	500			900
Zinc (ug)	700	200	1060	800	1000			8000
Chloride (mg)	200	116		300			200	2300
Selenium (ug)	3			3				55
Iodine (ug)	42	30		44				150

None of these variations make any significant difference to the functional food quality of the milk (Table 9.3). The main species only provide significant contributions to RDA in the case of calcium, phosphorus and iodine, although copper features in four and zinc in three types of milk.

The remainder of this section considers cow milks and their products, with the exception of some comparisons with sheep and goat milk cheeses.

Fresh cow milk

No changes are to be expected between raw milk intake and the supply of ions present in pasteurised milk for sale. It is self evident that a semi-skimmed or skimmed milk will contain a slightly higher percentage of all minerals as the lipids present in whole milk (though containing 20–60% aqueous phase) are progressively removed in processing.

Looking in more detail at the mineral contents of typical retail milks, these may be compared to the RDI. For simplicity, Table 9.4 shows these for the three most commonly consumed milks in the Western world in terms of adult RDI (the impact of age and sex on RDI was discussed in Section 9.1.1). Although the definitions of semi-skimmed or half-fat milks vary by jurisdiction, they are all very similar. The data are based on consumption of 200 mL milk. The National Academy of Science provide further age- and sex-dependent RDIs, as well as those for lactating females.

The very significant contribution of calcium, phosphorus and iodine to the RDI/RDA in only 200 mL of cow's milk is striking.

Table 9.3 Percentage share of adult mineral RDA found in milks (200 mL portion).

Mineral	Cow	Buffalo	Sheep	Goat	Camel	Mare	Llama
Source	Park & Haenlein (2000)	Ranjan *et al.* (2005)	Coni *et al.* (1999)	Park & Haenlein (2000)	Bornaz *et al.* (2009)		Morin *et al.* (1995)
Sodium	8	6	7	5	8		4
Potassium	6	5	6	8	6		
Calcium (mg)	21	31	33	22	19	23	33
Magnesium	5	9	9	8	6	5	
Phosphorus	34	23	43	31	26	25	34
Iron	2	0	2	2	1		
Copper	13	33	9	11	56		
Zinc	9	3	13	10	13		
Chloride	9	5		13			9
Selenium	5			5			
Iodine	28	20		29			

Table 9.4 Mineral content in mg or μg per 200 mL in milk and the % RDA provided by semi-skimmed milk.

	Whole milk[1]	Semi-skimmed[1]	Skimmed[1]	RDA adult[2]	Semi-skimmed as %RDA[2]
Sodium (mg)	89	89	91	1300	7
Potassium (mg)	319	321	334	4700	7
Calcium (mg)	243	247	258	800	31
Magnesium (mg)	23	23	23	420	5
Phosphorus (mg)	192	194	198	700	28
Iron (mg)	0.06	0.04	0.06	8	1
Copper (mg)[3]	0.1	0.1	0.1	0.9	11
Zinc (mg)	0.8	0.8	1	8	10
Chloride (mg)	183	179	179	2300	8
Selenium (ug)	2	2	2	55	4
Iodine (ug)	64	62	62	150	41

Source: [1]Dairy Council (www.milk.co.uk);
[2]Food and Nutrition Board (2004);
[3]Park & Haenlein (2000)

UHT and sterilised cow milks

The data presented above reflect the position in heat-treated milks as well, since the minerals are not degraded by these processes.

Concentrated cow milks

Mineral concentrations simply follow the degree of concentration. The situation is somewhat more complex in the case of vitamins, where the various concentrated milks are defined and considered in detail.

9.2.3 *Fermented milks*

Fermentation without change of concentration does not change the mineral picture. The most important sector within fermented milks is yogurts by far, and cow milk yogurts play the larger part.

Cow milk yogurts

Table 9.5 considers the situation for plain cow milk yogurts, both full cream and low fat. As noted for milks, calcium, phosphorus and iodine levels are such that one yogurt (typically 150 g) provides a significant share of RDA.

When fruit yogurts are considered the above quantities are diluted; rarely is any new mineral introduced. A number of yogurts are produced with the addition of skim milk powder,

Table 9.5 Mineral content in mg or μg per 150 g in full-fat and low-fat yogurt and their respective contribution to RDA.

	RDA[1]	Whole-milk yogurt[2]	RDA (%)	Low-fat yogurt[2]	RDA (%)
Sodium (mg)	1500	120.0	8	95	6
Potassium (mg)	4700	420.0	9	342	7
Calcium (mg)	1200	300.0	25	243	20
Magnesium (mg)	420	29.0	7	24	6
Phosphorus (mg)	700	255	36	215	31
Iron (μg)	8000	150	2	150	2
Copper (mg)	900	Trace	–	0.05	–
Zinc (μg)	8000	1100	14	0.9	–
Chloride (mg)	2300	255	11	353	15
Selenium (ug)	55	3	5	3.0	5
Iodine (ug)	150	95	63	51	34

Source: [1]Food and Nutrition Board (2004);
[2]Dairy Council (www.milk.co.uk)

which naturally increases mineral contents in proportion to the quantity added. In some cases milk protein concentrates may be used. In this case fewer mineral ions will be associated with this addition. Finally, 'Greek' yogurts contain added cream and the milk base is usually ultrafiltered to increase solids levels. In this case the lipid element will slightly dilute the mineral contents of the whole while the ultrafiltration (or use of milk protein concentrate) may rebalance them. The situation with drinking yogurts is similar to set and stirred yogurts.

Goat milk yogurts

Goat milk yogurts offer a similar picture with little contribution to RDA in a single portion (Table 9.6).

Mineral fortification of cow milks

In the developed world, a number of dairies promote milks with added calcium. Fortification is typically carried out to raise the level by up to 50%. A typical example is the Spanish Dairy Puleva which offers a milk branded as Calcio with such a level of enrichment. The product also is enriched with vitamin D.

9.2.4 *Cheeses*

The production of cheese implies the addition of small quantities of cultures which may dilute the source milk marginally. However, the key element in changing the proportion of any mineral present is the partition of it between the cheese and the whey. This will depend on the quantity of whey drained from the cheese and the relative occlusion of particular metal ions in the casein matrix of the cheese itself.

Table 9.6 Mineral content in mg or μg per 150 g in full-fat goat milk yogurt and its contribution to RDA. Adapted from Park & Haenlein 2006. Reproduced with permission of John Wiley & Sons.

Mineral	RDA	Plain yogurt	RDA (%)
Sodium (mg)	1500	12.7	1
Potassium (mg)	4700	24.5	1
Calcium (mg)	1200	24.2	2
Magnesium (mg)	420	2.7	1
Phosphorus (mg)	700	21.6	3
Iron (mg)	8	0.1	–
Copper (mg)	900	0.0	–
Zinc (mg)	8	0.1	–

Again, it is conceivable to fortify the cheese milk in order to enhance the mineral content of the cheese, although this is generally not seen in practice. However, cold storage of milk can deplete the ionic share of calcium to the detriment of curd formation (Ash & Wilbey 2010). This is often overcome by addition of calcium chloride (typically <0.02%).

The mineral contents of different cheeses depend upon the manufacturing processes and the specific variety of cheese. Cheeses made with a coagulating enzyme e.g. cheddar, tend to retain most of the calcium within the curd that remains after extraction of the whey. The other mineral levels of hard cheese reflect relatively low levels of retained whey, and so are somewhat higher than those of soft cheeses.

Cow milk cheeses

Hard cheese mineral contents (Table 9.7) vary naturally with the moisture content. In this section the contribution of 30 g (a typical matchbox-sized portion) is used as a standard for hard cheese and 40 g for soft cheese.

The functional contributions of note are: very high shares of calcium and phosphorus and zinc; a useful contribution of iodine; and difficulties with the salt content.

Cheeses coagulated by lactic acid alone, that is, without a coagulating enzyme (e.g. cottage or cream cheeses) tend to be lower in calcium. For example, a portion (40 g) of regular cottage cheese will provide an average adult (19–50 years old) with only approximately 7% of the recommended daily requirement for calcium, i.e. a quarter of that delivered by cheddar cheese. Most other minerals will be partitioned predominately into the whey phase. In the case of a typical soft cheese, the sodium issue disappears since heavy brining is not part

Table 9.7 Mineral content in mg per 30 g of cheddar cheese and its contribution to RDA.

Mineral	RDA[1]	In 30 g cheese[2]	RDA (%)
Sodium (mg)	1500	217	14
Potassium (mg)	4700	23	0
Calcium (mg)	1200	222	19
Magnesium (mg)	420	9	2
Phosphorus (mg)	700	152	22
Iron (mg)	8000	0.09	0
Copper (mg)	900	0.01	0
Zinc (mg)	8000	1.2	0
Chloride (mg)	2300	312	14
Selenium (ug)	55	1.8	3
Iodine (ug)	150	9	6

Source: [1]Food and Nutrition Board (2004);
[2]Adapted from Handbook of Milk of Non-Bovine Dairy Animals. Reproduced with permission of John Wiley & Sons.

Table 9.8 The mineral content in mg per 40 g of reduced-fat cottage cheese relative to RDA.

Mineral	RDA[1]	In 40 g cottage cheese[2]	RDA (%)
Sodium (mg)	1500	120	8
Potassium (mg)	4700	64	1
Calcium (mg)	1200	51	4
Magnesium (mg)	420	5	1
Phosphorus (mg)	700	68	10
Iron (mg)	8000	Trace	–
Copper (mg)	900	Trace	–
Zinc (mg)	8000	0.2	–
Chloride (mg)	2300	196	9
Selenium (ug)	55	2	4
Iodine (ug)	150	10	7

Source: [1]Food and Nutrition Board (2004);
[2]Adapted from Handbook of Milk of Non-Bovine Dairy Animals
Reproduced with permission of John Wiley & Sons.

of the process. As an example, the case of reduced-fat cottage cheese is shown in Table 9.8, where the portion is typically 40 g rather than 30 g for the hard equivalent. Cottage cheese therefore delivers only minor quantities of the mineral RDAs.

Goat milk cheeses

Again we may distinguish between hard and soft cheeses, although goat cheeses are often artisanal and exhibit a very wide range of analytical parameters (Table 9.9). Note the significant contributions per portion for only phosphorus and calcium (Table 9.10).

9.2.5 *Whey products*

A significant proportion of the minerals are drained from a cheese in the whey. As noted above, low-pH cheeses yield different mineral profiles in both cheese and whey that of rennet/chymosin-coagulated cheeses. Whey is also generated in the manufacture of casein. As would be expected, the mineral contents of wheys from different species vary greatly (Table 9.11).

Concentrations are usually enhanced in whey protein concentrates, although the partition of minerals between protein concentrate and permeate varies with process conditions. Whey powder, whey protein concentrate (WPC), whey protein isolate (WPI), whey permeate and reduced lactose whey contain calcium in concentrations of 0.5–2 g per 100 g powder. Whey itself is largely used in animal feed, although a small number of whey drinks exist (e.g. Rivella in Switzerland). Whey may also be dried, either as sweet or acid whey powders. The concentration of mineral ions is naturally related to the evaporation step.

Table 9.9 Mineral content in mg per 30 g in cheddar-type goat cheese and its respective contribution to RDA.

Mineral	RDA[1]	In 30 g cheddar cheese[2]	RDA (%)
Sodium (mg)	1500	84	6
Potassium (mg)	4700	3	–
Calcium (mg)	1200	140	12
Magnesium (mg)	420	10	2
Phosphorus (mg)	700	123	18
Iron (mg)	8	0.4	4
Copper (mg)	900	0.1	–
Zinc (mg)	8	0.2	2

Source: [1]Food and Nutrition Board (2004);
[2]Adapted from Handbook of Milk of Non-Bovine Dairy Animals. Reproduced with permission of John Wiley & Sons.

Table 9.10 A comparison of the mineral content in mg per 40 g goat's soft cheese with respect to RDA.

Mineral	RDA[1]	In 40 g soft cheese[2]	RDA (%)
Sodium (mg)	1500	66.9	4
Potassium (mg)	4700	4.2	–
Calcium (mg)	1200	27.6	2
Magnesium (mg)	420	2.3	1
Phosphorus (mg)	700	44.2	6
Iron (mg)	8	0.3	4
Copper (mg)	900	0.1	–
Zinc (mg)	8	0.1	2

Source: [1]Food and Nutrition Board (2004);
[2]Adapted from Handbook of Milk of Non-Bovine Dairy Animals. Reproduced with permission of John Wiley & Sons.

Table 9.11 Calcium and phosphorus content in mg per 100g of whey derived from different milk sources. Bornaz, Sahli, Attalah & Attia (2009). Reproduced with permission of Elsevier.

Mineral (mg per 100 g)	Cow	Sheep	Goat	Camel
Calcium	26	33	22	43
Phosphorous	22	25	20	33

Cow milk whey products

Sweet whey powder. The differences between skim milk powder and sweet whey powder (Table 9.12) emphasise the significance of the partition of ionic species between cheese and whey in the cheese-making process. Naturally the concentration of interesting ionic species in cheeses noted above implies a lower functional role for whey and whey products in mineral supply.

Whey protein concentrates. The protein content in whey protein concentrates (WPCs) falls within the range 35–80% (Table 9.13). Isolates are also available containing around 90% protein and lower mineral content. The quantity of various minerals essentially follows the degree of concentration, although process conditions (in particular the membrane specification of the reverse osmosis step or choice of electrodialysis) have a major impact.

The significant variation in ionic concentrations across the ever more concentrated whey protein entities, where two different methods of concentration are applied, illustrates the major impact of process in the production of WPCs. Again, however, the functional supply of minerals from these whey concentrates is not very significant.

Table 9.12 Comparison in mineral content in g per 100 g of skimmed milk powder (SMP) and sweet whey powder (SWP). Delaney (1976). Reproduced with permission of Elsevier.

Mineral (g per 100 g)	SMP	SWP	Difference	Difference as a percentage
Sodium	0.54	0.67	−0.13	−24
Potassium	1.6	1.06	0.54	34
Calcium	1.3	0.67	0.63	48
Magnesium	0.13	0.11	0.02	15
Phosphorus	1.02	0.63	0.39	38
Chloride	1.23	1.61	−0.38	−31

Table 9.13 Comparison in mineral content in g per 100 g of sweet whey powder (SWP) and whey protein concentrate (WPC) using both electrodialysis and ultrafiltration. Delaney (1976). Reproduced with permission of Elsevier.

Mineral (g per 100 g)	SWP	WPC 35% (electrodialysis)	% Change (SWP−WPC35) SWP	WPC 50% (ultrafiltration)	% Change (SWP−WPC50) SWP
Sodium	0.67	0.024	96	0.28	58
Potassium	1.06	0.03	97	1.01	5
Calcium	0.67	0.5	25	0.26	61
Magnesium	0.11	0.05	55	0.09	18
Phosphorus	0.63	0.05	92	0.33	48

9.2.6 *Butter fat*

None of the ionic species is partitioned into butter oil. In cream and butter, the share is related to the aqueous portion, which falls within the range 60–85% for creams and usually 12% for butter.

9.2.7 *Commercial mineral concentrates*

Several of the dairy ingredient manufacturers make whey mineral products. Brand names such as Lactoval® (DMV), XtraCal (Glanbia), Capolac® (Arla Foods) and Vitalmor Ca (Armor Proteines) are well known in the industry.

Such mineral products represent calcium and phosphorus in a bioavailable form; in addition, they may contain magnesium and potassium together with iron, zinc and copper. Milk mineral products usually contain calcium in concentrations of 23–28 g per 100 g powder. All these milk mineral products give a base for mineral fortification in a large number of application areas. The composition of milk/whey mineral products depends on the type of whey used as raw material and the process technology applied for manufacturing. The composition of typical such products is listed in Table 9.14.

Milk mineral products have a neutral taste and a bland flavour, providing excellent sensorial properties in formulations. The guaranteed milk origin, the natural food grade and a range of valuable minerals and trace elements provide label-friendliness.

Table 9.14 Chemical composition of commercial milk calcium (%). Giract, personal communication. Reproduced with permission of Giract.

Total mineral content (%)	80
Ash (%)	70
Calcium (%)	25
Phosphorus (%)	14
Ca/P	1.8
Magnesium (%)	1.5
Sodium (%)	0.7
Potassium (%)	0.8
Zinc (mg per 100 g)	27
Copper (mg per 100 g)	0.4
Iron (mg per 100 g)	1.9
Organic mineral (citrate, %)	9
Protein	5
Fat	1

Table 9.15 Global production (2010) of milk calcium in volume and value for use in food fortification. Giract, personal communication. Reproduced with permission of Giract.

Milk calcium production 2008	Volume (tons)	Value (US$ million)
Europe	600	3–6
USA	700	3.5–7
Japan	900	4.5–9
Total	2200	11–22

9.2.8 *Applications*

The claimed high bioavailability and correct calcium/phosphorus ratio make the products particularly interesting. They are primarily used in dietary supplements as well as in the fortification of cereals, bakery, dairy and soft drinks. In a commercial sense, milk minerals can essentially be narrowed down to milk calcium (as a phosphate salt, part chelated). Although today the world market is small at about 2000 tons (Table 9.15), milk calcium has experienced significant growth in recent years based on its perception as one of the best sources for product fortification. Asia is particularly interested in this respect as milk consumption per capita is low compared to Europe and USA; the Japanese market therefore outweighs both the European and US markets.

9.2.9 *Summary*

Milks and dairy products deliver significant proportions of the RDA of a number of minerals in a relatively concentrated and bioavailable form. They are therefore rightly considered as functional foods, in particular for calcium, phosphorus and iodine.

9.3 Vitamins in dairy products

9.3.1 *Introduction*

There are two significant differences between the mineral and vitamin contents of milks and dairy products: the sensitivity of some vitamins when exposed to heat and light; and the differential solubility of differing vitamins in the aqueous and lipid phases, and thus changes in relative concentrations as products have more or less of one of these phases

Data have been combined for two separate vitamins in order to simplify Tables 9.16 and 9.17. 'Vitamin A' is reported in retinol equivalents, being the sum of retinol and one-sixth of the value of carotene, while 'niacin' is reported as niacin equivalents, being the sum of niacin as such and niacin from tryptophan in all data sourced from the Dairy Council (http://www.milk.co.uk) in Tables 9.16 and 9.17.

Table 9.16 Levels of vitamins in milk from different animals, expressed as mg or μg per 200 mL milk. Adapted from Park & Haenlein (2006). Reproduced with permission of John Wiley & Sons.

Vitamin	Cow	Buffalo	Sheep	Goat	Camel	Mare	RDA adult
Thiamin (μg)	80	100	160	80	80		1400
Riboflavin (μg)	324	200	712	260	120		1600
Niacin (μg)	168	520	832	620	800		18000
Vitamin B6 (μg)	120	760	160	120	100		2000
Vitamin B12 (μg)	0.7	0.6	1.4	0.2	4.0		2.5
Folate (μg)	10	20	12	2.0	0.2		400
Pantothenate (μg)	628	300	816	820	180		5000
Biotin (μg)	4.0	5.2	10.0	6.0			30
Vitamin C (mg)	1.8	5.0	1.9	2.6		4.8	60
Vitamin A (μg)	104	140	168	88	30	68	1000
Vitamin D (μg)	0.36	0.00	0.06	0.22		0.6	5
Vitamin E (mg)	0.18	0.20	0.22	0.06	0.01		15

Table 9.17 Proportion of adult vitamin RDA found in 200 mL milk (%). Adapted from Park & Haenlein (2006). Reproduced with permission of John Wiley & Sons.

Vitamin	Cow	Buffalo	Sheep	Goat	Camel	Mare
Thiamin (μg)	6	7	11	6	6	
Riboflavin (μg)	20	13	45	16	8	
Niacin (μg)	1	3	5	3	4	
Vitamin B6 (μg)	6	38	8	6	5	
Vitamin B12 (μg)	28	24	57	8	160	
Folate (μg)	3	5	3	1		
Pantothenate (μg)	13	6	16	16	4	
Biotin (μg)	13	17	33	20		
Vitamin C (mg)	3	8	3	4		8
Vitamin A (μg)	10	14	17	9	3	7
Vitamin D (μg)	7	0	1	4		12
Vitamin E (mg)	1	1	1			

Table 9.16 shows that raw milks have varied vitamin concentrations depending on species and feed regime. Again, the share in this glass-sized portion may be compared to adult RDA. The data for buffalo milk for vitamin B6 is surprising and may not represent the norm. Several of the non-bovine species have enhanced levels, notably of niacin and vitamin A, compared to cow milk. All species offer significant shares of RDA of vitamin B12, biotin and riboflavin.

9.3.2 *Cow milks*

As was the case with minerals, the vitamin contents of typical retail milks may be compared to the RDI at various ages. For simplicity, Table 9.18 lists these for the three most commonly consumed milks in the Western world. Although the definition of semi-skimmed or half-fat milks varies by jurisdiction, they are all very similar. This data reflect pasteurised products. The key vitamins in a typical semi-skimmed cow's milk are vitamin B12, riboflavin and, to a lesser extent, pantothenate.

UHT and sterilised cow milks

As would be expected, vitamins which are heat sensitive are reduced in concentration with increasing heat treatment, and are virtually eliminated from sterilised milks as shown in Table 9.19. Vitamin C is very labile and folate and B6 are affected by UHT processing; only sterilisation conditions affect thiamin, B12 and pantothenate.

Concentrated cow milks

Vitamin concentrations follow the degree of concentration to the extent that they are not degraded by the process or by changes in butterfat ratio. The various concentrations of vitamins listed in Table 9.20 reflect the degree of concentration, with the caveat of any heating process reducing the concentrations of the more labile vitamins.

Table 9.18 Levels of different vitamins in different types of pasteurised milk with their RDAs, expressed as mg or µg per 200 mL milk. Scott & Bishop (1986). Adapted from Park & Haenlein (2006). Reproduced with permission of John Wiley & Sons.

Vitamin	Whole milk	Semi-skimmed	Skimmed milk	RDA adult	Semi-skimmed as RDA (%)
Thiamin µg	80	80	60	1400	6
Riboflavin µg	480	500	460	1600	31
Niacin µg	3200	1400	1600	18000	8
Vitamin B6 µg	120	120	120	2000	6
Vitamin B12 µg	0.8	0.8	0.9	2.5	32
Folate µg	6	6	6	400	2
Pantothenate µg	700	700	800	5000	14
Biotin µg	5	6	7	30	20
Vitamin C mg	4	4	3	60	7
Vitamin A µg	62	40	2	1000	4
Vitamin D µg	Trace	Trace	Trace	5.0	<1

Table 9.19 The effect of heat treatment on vitamin levels of different types of milk (μg g^{-1}).

Heat-treated cow's milk (μg g^{-1})	Raw	Pasteurised whole	Pasteurised semi-skimmed	UHT	Sterilised	Degradation with heat
Thiamin	0.4	0.40	0.42	0.40	0.30	only high
Riboflavin	1.5	1.7	2.0	1.8	1.5	no
Niacin	0.8	0.83	1.07	0.91	0.95	no
Vitamin B6	0.7	0.6	0.61	0.44	0.39	yes
Vitamin B12	0.005	0.0036	0.004	0.002	0.002	only high
Folate	0.05	0.057	0.056	0.008	0.002	yes
Pantothenate	2.5	3.4	3.8	3.3	2.9	only high
Biotin	0.02	0.02	0.024	0.019	0.019	no
Vitamin C	20	8.4	11	1.9	0.8	yes
Vitamin A	520					
Vitamin D	1.8					

Source: Smith (1950), Porter (1966) and Scott & Bishop (1986).

Table 9.20 The effect on vitamin concentration of different types of concentrated and evaporated milk products.

Concentrated cow's milk (μg g^{-1})	Raw	Evaporated	Condensed	Condensed skimmed	Dried skimmed
Thiamin	0.4	0.65	0.85	1.1	3.8
Riboflavin	1.5	4.2	4.6	5.2	16.3
Niacin	0.8	2.4	2.9	3	10.2
Vitamin B6	0.7	0.72	3	10.2	6
Vitamin B12	0.005	0.0014	0.0073	0.0091	0.026
Folate	0.05	0.11	0.15	0.16	0.51
Pantothenate	2.5	7.5	8.5	10.3	33.3
Biotin	0.02	0.04	0.056	0.052	0.2
Vitamin C	20	15	41	51	132
Vitamin A	9.4				
Vitamin D	1.8				

Source: Smith (1950), Porter (1966) and Scott & Bishop (1986).

9.3.3 *Fermented milks*

Within fermented milks, by far the most important sector is yogurts.

Cow milk yogurts

Table 9.21 lists the concentration of vitamins in plain yogurts, both full-cream and low-fat. Full-fat yogurts provide a significant share of the RDA of riboflavin and, to a lesser extent, that of biotin, pantothenate and B12. In low-fat yogurts the share of biotin halves, while pantothenate appears to be marginally enhanced and vitamin A is diminished. Surprisingly, thiamine content appears greater (this may be an artefact of differing samples). When fruit yogurts are considered, the above quantities are diluted and rarely is any new vitamin introduced other than vitamin C. However, some sources suggest enhancement of other vitamin groups, including thiamin and vitamin A. The situation with drinking yogurts is closer to the standards for milks. Greek yogurts have enhanced cream content, increasing the quantities of fat-soluble vitamins.

Goat milk yogurts

Goat milk yogurts are much less common that cow milk yogurts. Only a significant contribution to the RDA for riboflavin can be obtained from a 150 g pot/portion, as shown in Table 9.22.

Table 9.21 Level of different vitamins in two different types of yogurt, expressed as mg or μg per 150 g yogurt, with their RDA values.

Vitamin	RDA[1]	Whole-milk yogurt[2]	RDA (%)	Low-fat yogurt[2]	RDA (%)
Thiamin (mg)	1.4	0.09	6	0.18	13
Riboflavin (mg)	1.6	0.4	25	0.33	21
Niacin (mg)	18	2.3	12.8	1.7	9.4
Vitamin B6 (mg)	2	0.15	8	0.02	1
Vitamin B12 (mg)	0.0024	0.0003	13	0.0005	0
Folate (μg)	400	27	7	27	7
Pantothenate (mg)	6	0.75	13	0.84	14
Biotin (mg)	0.03	0.0039	13	0.0023	8
Vitamin C (mg)	60	1.2	2	2	3
Vitamin A (μg)	800	47	6	12	2
Vitamin D (μg)	5	0	0	0.2	4

Source: [1]Food and Nutrition Board (2004);
[2]Dairy Council (www.milk.co.uk)

Table 9.22 Levels of different vitamins in goat milk yogurt, expressed as mg or μg per 150 g yogurt, and RDA values.

Vitamin	RDA[1]	Plain yogurt[2]	RDA (%)
Thiamin (mg)	1.4	0.0	3
Riboflavin (mg)	1.6	0.2	11
Niacin (mg)	18	0.3	2
Vitamin B6 (mg)	2	0.1	3
Vitamin B12 (mg)	0.0024	Trace	Trace
Folate (μg)	400	7.0	Trace
Pantothenate (mg)	6	0.2	Trace
Biotin (mg)	0.03	0.001	2
Vitamin C (mg)	60	1	2
Vitamin A (μg)	800	32	4
Vitamin D (μg)	5	Trace	Trace

Source: [1]Food and Nutrition Board (2004);
[2]Scott & Bishop (1986)

9.3.4 *Vitamin fortification of cow milks*

As noted in the section on mineral fortification, a number of suppliers have chosen to fortify milk, the most frequent relevant fortification is the addition of vitamin D (usually together with calcium) with a health claim related to control of osteoporosis. The health claim permitted is inevitably dependent on local legislation.

9.3.5 *Cheeses*

The supply of vitamins in cheese naturally relates to its water (residual whey) content. Here, fat-soluble vitamins will be more favoured the harder the cheese. For hard cheeses a portion size of 30 g is again used, while for soft cheeses 40 g is the norm.

Cow milk cheeses

There are many hundreds, if not thousands, of cow milk cheeses around the world; hard and semi-hard cheeses are more prevalent. Table 9.23 lists the typical values for a cow's milk hard cheese, in this case cheddar cheese. In terms of vitamin content, a portion (30 g) of cheddar cheese only provides an important share of the RDA of vitamin B12. Soft cheeses provide similarly low levels of vitamin RDAs. Reduced-fat cottage cheese provides only minor quantities of the mineral RDAs (Table 9.24), again with vitamin B12 being the most important.

Table 9.23 Levels of different vitamins in cheddar cheese, expressed as mg or μg per 30 g portion, and RDA values.

Vitamin	RDA[1]	In 30 g cheddar[2]	RDA (%)
Thiamin (mg)	1.4	0.01	1
Riboflavin (mg)	1.6	0.12	8
Niacin (mg)	18	2.03	11
Vitamin B6 (mg)	2	0.05	2
Vitamin B12 (μg)	2.5	0.72	29
Folate (μg)	400	9.30	2
Pantothenate (mg)	5	0.15	3
Biotin (mg)	0.03	Trace	Trace
Vitamin C (mg)	60	Trace	Trace
Vitamin A (μg)	1000	116	12
Vitamin D (μg)	5	0.09	2
Vitamin E (mg)	15	0.16	1

Source: [1]Food and Nutrition Board (2004);
[2]Dairy Council (www.milk.co.uk)

Table 9.24 The vitamin content in mg per 40 g of reduced fat cottage cheese with respect to the RDA.

Vitamin	RDA[1]	In 40 g cottage cheese[2]	RDA (%)
Thiamin (mg)	1.4	0.02	–
Riboflavin (mg)	1.6	0.10	6
Niacin (mg)	18	1.5	8
Vitamin B6 (mg)	2	0.02	1
Vitamin B12 (μg)	2.5	0.24	10
Folate (μg)	400	9	2
Pantothenate (mg)	5	0.10	2
Biotin (mg)	0.03	0.002	7
Vitamin C (mg)	60	Trace	–
Vitamin A (μg)	1000	6.3	1
Vitamin D (μg)	5	Trace	–
Vitamin E (mg)	15	0.01	–

Source: [1]Food and Nutrition Board (2004);
[2]Dairy Council (www.milk.co.uk)

Goat milk cheeses

Table 9.25 lists the typical vitamin contents of a hard cheddar goat's milk cheese. Only vitamin B12 and vitamin A are present in any significant quantities in this sample. The vitamin contents of soft goat's milk cheeses are similar (Table 9.26). Only vitamin A is present in any significant quantity in terms of RDA.

9.3.6 *Whey products*

As shown in Table 9.27, there are major differences in the availability of many vitamins between sweet whey powder and skim milk powder. Equally, there are many differences between ultrafiltered WPC and sweet whey powder. Not immediately obvious are the reasons for large changes in the ratios for folate and B12.

9.3.7 *Butter fat*

The vitamins found in butterfat are A, D and E. Vitamin A levels were extensively studied by Semb *et al.* in Wisconsin as early as 1934. A wide range of results was observed, both between breeds and stage of lactation. In colostrum, levels were 5–15 times higher than those from

Table 9.25 Levels of different vitamins in hard goat (cheddar) cheese, expressed as mg or μg per 30 g, and RDA values.

Vitamin	RDA[1]	In 30 g cheese[2]	RDA (%)
Thiamin (mg)	1.4	0.009	1
Riboflavin (mg)	1.6	0.12	8
Niacin (mg)	18	0.021	–
Vitamin B6 (mg)	2	0.03	2
Vitamin B12 (μg)	2.5	0.33	13
Folate (μg)	400	9.9	2
Pantothenate (mg)	5	0.108	2
Biotin (mg)	0.03	0.0009	3
Vitamin C (mg)	60	Trace	Trace
Vitamin A (μg)	1000	97.5	10
Vitamin D (μg)	5	0.078	2
Vitamin E (mg)	15	0.159	1

Source: [1]Food and Nutrition Board (2004);
[2]Adapted from Handbook of Milk of Non-Bovine Dairy Animals. Reproduced with permission of John Wiley & Sons.

Table 9.26 Levels of different vitamins in soft goat cheese, expressed as mg or μg per 40 g and RDA values.

Vitamin	RDA[1]	In 40 g soft goat cheese[2]	% RDA
Thiamin (mg)	1.4	0.016	–
Riboflavin (mg)	1.6	0.252	2
Niacin (mg)	18	0.26	0
Vitamin B6 (mg)	2	0.048	0
Vitamin B12 (μg)	2.5	0.8	3
Folate (μg)	400	7.6	Trace
Vitamin C (mg)	60	trace	Trace
Vitamin A (μg)	1000	124	12
Vitamin D (μg)	5	0.2	4
Vitamin E (mg)	15	0.316	2

Source: [1]Food and Nutrition Board (2004);
[2]Adapted from Handbook of Milk of Non-Bovine Dairy Animals. Reproduced with permission of John Wiley & Sons.

Table 9.27 A comparison in vitamins level between skim milk powder (SMP), sweet whey powder (SWP) and whey protein concentrate (WPC). Delaney (1976). Reproduced with permission of Elsevier.

Vitamins ($\mu g\ g^{-1}$)	SMP	SWP	% Change (SMP−SWP)/ SMP	WPC ultra-filtered	% Change (SWP−WPCuf)/ SWP
Thiamin	2.9	2.6	10	3.8	−46
Riboflavin	18.9	21.4	−13	31	−45
Niacin	11.4	9.3	18	11.7	−26
Vitamin B6	2.8	4.1	−46	3.3	20
Vitamin B12	0.038	0.023	39	0.21	−813
Folate	0.044	0.72	−1536	5.9	−719
Pantothenate	1281	1356	6	1462	−8
Biotin	0.22	0.31	−41	0.4	−29

5–100-days post-parturition which ranged from 5.7–10.3 ug g^{-1}. From 100 days onwards, levels were about 25% lower than these. Interbreed differences were at their most extreme between Jersey (5.7 ug g^{-1}) and Holstein (10.3 ug g^{-1}). A significant variation in vitamin A content is observed throughout the year (Searles & Armstong 1969). The spread between high and low values are reported to be as much as +50%, between 6.8 and 9.2 μg g^{-1}.

Similar observations were made for vitamin E where the annual variation lay between 24 and 36 μg g^{-1}, the higher value again being in the summer. Vitamin D levels are mainly influenced by feed and have ten-fold ranges in a single breed. Milk is not a useful source of this vitamin.

The wide range and rather unpredictable variations in all these vitamins effectively rules out claims of functional food status for butterfat, even before the issue of saturated fat is addressed.

9.3.8 *Summary*

Vitamin claims are limited in terms of functional food positioning. Both pasteurised and UHT milks may claim useful contributions of vitamin B12, riboflavin and, to a lesser extent, pantothenate. Cheddar cheese has major quantities of vitamins B12, A, niacin and riboflavin and some folate. The fat-soluble vitamins show too great a variability, both interbreed and seasonal, to be able to make meaningful claims.

References

Ash, A. & Wilbey, A. (2010) The nutritional significance of cheese in the UK diet. *International Journal of Dairy Technology*, **63**(3), 305–319.

Bolland, M., Avenell, A., Baron, J., Grey, A., MacLennan, G. & Reid, I. (2010) Effect of calcium supplements on risk of myocardial infarction and cardiovascular events, meta-analysis. *British Medical Journal*, **341**, 289.

Bornaz, S., Sahli, A., Attalah, A. & Attia, H. (2009) Physicochemical characteristics and renneting properties of camels' milk: a comparison with goats', ewes', and cows' milks. *International Journal of Dairy Technology*, **62**(4), 505–513.

Coni, E., Bocca, B. & Caroli, A. (1999) Minor and trace element content of two typical Italian sheep dairy products. *Journal of Dairy Research*, **66**, 589–598.

Delaney, R.A.M. (1976) The composition, properties and uses of whey protein concentrates. *International Journal of Dairy Technology*, **29**(2), 91–101.

Ebringer, J., Ferencik, M. & Krajovic, J. (2008) Beneficial health effects of milk and fermented dairy products. *Folia Microbiologica*, **53**, 378–394.

Food and Nutrition Board (2004) Institute of Medicine, National Academy of Sciences. Available at http://www.fao.org/docrep/004/Y2809E/y2809e0h.htm#bm17.5 (accessed October 2013).

Lin, M.-J., Lewis, M.J. & Grandison, A.S. (2006) Measurement of ionic calcium in milk. *International Journal of Dairy Technology*, **59**(3), 192–199.

Morin, D.E., Rowan, L.L., Hurley, W.L. & Braselton, W.E. (1995) The composition of milk from llamas in the United States. *Journal of Dairy Science*, **78**, 1713–1720.

Park, Y.W. & Haenlein, G.F.W. (eds) (2000) *Handbook of Milk of Non-Bovine Dairy Animals*. Iowa: Wiley-Blackwell.

Porter, J. (1966) The nutritional aspects of milk and milk products. *International Journal of Dairy Technology*, **19**(1), 3–7.

Ranjan, P., Arora, S., Shama, G., Sindhu, J., Kansal, V. & Sangwan, R. (2005) Bioavailability of calcium and physicochemical properties of calcium-fortified buffalo milk. *International Journal of Dairy Technology*, **58**(3), 185–189.

Scott, K.J. & Bishop, D.R. (1986) Nutrient content of milk and milk products: vitamins of the B complex and vitamin C in retail market milk and milk products. *International Journal of Dairy Technology*, **39**(1), 32–35.
Searles, S.K. & Armstrong, J.G. (1969) Vitamin E, vitamin A, and carotene content of Alberta butter. *Journal of Dairy Science*, **52**(2), 150–156.
Semb, J., Bauman, C.A. & Steenbock, H. (1934) Fat-soluble vitamins: XLI. The carotene and vitamin A content of colostrum. *Journal of Biological Chemistry*, **107**, 697–703.
Smeets, W.T.G.M. (1955) The determination of the concentration of calcium ions in milk ultrafiltrate. *Netherlands Milk and Dairy Journal*, **9**, 249–360.
Smith, J. (1950) The food value of milk and milk products. *Journal of the Society of Dairy Technology*, **3**(4), 254–263.

10 Legislations and Relevant Regulations

M. Hickey

Charleville, Co., Cork, Ireland

10.1 Background and introduction

The belief that certain foods provide health benefits above and beyond their nutritional value has been recognised for centuries by various cultures in many countries; indeed, the ancient Greek physician Hippocrates, the father of Western medicine, is attributed with the quote 'Let food be thy medicine and medicine be thy food'. The first scientific claims for food providing health benefits were made by the Russian-born Nobel Laureate Élie (Ilya) Metchnikoff working at the Pasteur Institute in Paris. He referred to the observations of a Bulgarian student in Geneva called Stamen Grigoroff (1878–1945), regarding the number of centenarians to be found in Bulgaria where yogurt was a stable food (Metchnikoff 1908). In addition, Metchnikoff believed that there was potential to replace harmful bacteria in our bodies with beneficial bacteria.

At about the same time French physician Henry Tissier, who also worked at the Pasteur Institute, observed that the stools of young children with diarrhoea were characterised by low numbers of particular y- or bifid-shaped bacteria, while those of healthy children had high numbers of the same type of organisms (Tissier 1906; Anukam & Reid 2010). Tissier suggested that there was a possibility of administering such bacteria to the ill children. Somewhat later, Japanese doctor Minoru Shirota working in a microbiology laboratory at Kyoto Imperial University's School of Medicine, was the first to follow up on these ideas. During the 1930s, Shirota succeeded in isolating one such bacterium, *Lactobacillus casei* strain Shirota, which he believed could help to maintain the balance of beneficial bacteria in the gut. Over time, Shirota cultivated the beneficial micro-organism to survive digestion, and went on to incorporate it into a fermented milk beverage known as Yakult. Yakult was introduced to the Japanese market in 1935; today it is sold in over 25 countries worldwide.

Since that time, many researchers have isolated other micro-organisms that have health benefits associated with humans, and these have been commercialised. A recent scientific paper listed 17 strains of lactobacilli and five strains of bifidobacteria that are used by internationally known food manufacturers such as Nestlé, Danone, Arla, Valio, Yakult and Fonterra and major starter culture suppliers such as Chr. Hansen and Danisco (Sanders 2003). The fortification of foods with vitamins and minerals is another aspect of seeking the enhancement of a food ingredient for a functional reason, and an example is the addition of iodine to salt

Milk and Dairy Products as Functional Foods, First Edition. Edited by A. Kanekanian.

by manufacturers in the US at around 1917 to prevent the nutritional deficiency disease goitre (Platzman 1999).

From a legal perspective, while the concept of using beneficial bacteria originated early in the last century, the term 'probiotic' only appeared in the 1960s. A number of scientific definitions have appeared in the literature since then but, as yet, there is no internationally recognised legal definition. This has led to a situation where there is no clear and legally consistent way in which the term is, or can be, used in product labelling.

In addition, as an increasing number of scientific reports appear in the literature detailing the specific health benefits of probiotic micro-organisms, there has been an increasing clamour to allow the use of specific claims in the labelling, marketing and advertising of products containing probiotic microflora. These considerations arise not only with probiotic products, but also encompass the overall area of functional foods which fall into the grey area between foods and medicines. As a reaction to this, some countries developed their own legislative systems to ensure food safety and consumer protection and to address the issues raised by this relatively new category of products.

Inevitably, differences arise between the approaches used in different jurisdictions and this chapter covers the approaches in a number of regions of the world starting with Japan then going on to discuss the European Union, The United States, Canada, Australia/New Zealand and China. It shall then go on to describe the progress that has taken place at an international level and, in particular, the role and status of Codex Alimentarius is examined. Finally, some conclusions and possible future developments are also considered.

10.2 The Japanese approach

Whenever legislation on the subject of functional foods is discussed, reference is made to the Japanese system of regulation for foods with health claims and foods for special dietary uses. Many will be familiar with the category of Foods for Specified Health Use (FOSHU), though as we shall see this is but a specific category in Japan of such products. The Japanese concept of functional foods and the consequential legislation on foods with health claims, including FOSHU, have their origins in two main factors:

1. the severe malnutrition in certain sections of the Japanese population in the aftermath of World War II, leading to the schools lunch programme and permitting the addition of certain nutrients to certain staple foods such as bread and rice; and
2. concerns regarding the ageing of the Japanese population and the burden that this was likely to place on the state's finances in the future.

In 1952 the Nutrition Improvement Law (Law No. 248 enacted by the Japanese Parliament, The Diet, on 31 July 1952) created a food category of foods for special dietary uses. Article 1 of the law stated 'This law aims to contribute to promote national welfare by means of striving for the elevation of ideas of nutritional improvement of the people, the inquiry into the nutritional conditions of the people and the maintenance and the development of the health and the physical strength of the people devising necessary steps about the betterment of the nutritional conditions of the people' (Mitsuda 1958). This law, which paved the way for later legislation in this area, mainly addressed the implementation of the outcomes of the National Nutrition Surveys which had been undertaken annually since 1946: the development of a national nutrition policy and the provision of nutrient-enrichment foods for specific groups such as infants, pregnant women, medical patients and for the elderly with certain eating difficulties. Foods for special health uses (FOSHU) came somewhat later (Tomita 2007; Yamada *et al.* 2008).

In 1984, the Ministry of Education, Science and Culture carried out a statistical analysis of food nutrition (Ichikawa, 1994) and, in 1986, a similar report of this study identified three main functions of food:

1. its nutritive value;
2. its organoleptic appeal; and
3. its physiological factors, which include the regulation of bodily functions, disease prevention, promotion of recovery and good health.

The concept of functional foods grew out of the third function above, and a Functional Foods Forum was established comprising experts from food and nutrition departments of Japanese universities. In addition, the then Ministry of Health and Welfare was promoting functional foods and the industry was keen to market such products using health claims. In 1989, the Nutrition Law was enacted allowing the manufacturers of functional foods to make health claims. A Functional Foods Liaison Board was established to work with the industry, and various Functional Food Subcommittees were set up. FOSHU became the official term for such foods from 1991, replacing the original term 'functional foods', defined as 'processed foods containing ingredients that aid specific bodily functions in addition to being nutritious'. The law also aimed to prevent misleading and ill-defined health claims and listed 5 categories of Food for Special Dietary Uses:

1. milk powder for pregnant and lactating women;
2. infant formula;
3. food for elderly people who have difficulty chewing or swallowing;
4. foods for medical patients; and
5. FOSHU.

In 2003 a new category of Foods with Nutrient Function Claims (FNFC) was added to the existing category of FOSHU. In 2005, following expert consultations, the relevant Ministry (then named the Ministry for Health Labour and Welfare, MHLW) changed the existing FOSHU system to include the new subsystems – standardised FOSHU, qualified FOSHU and disease risk reduction claims for FOSHU – based on the level of scientific evidence supporting the relationship between the food or ingredient in the food and the health condition concerned (Yamada *et al.* 2008).

The Consumer Affairs Agency (CAA) was established as a new government agency on 1 September 2009 and one of the responsibilities it was given was the administration of the Japanese food labelling system, including those provisions covering FOSHU. The CAA began discussions on a review of the health foods systems in November 2009. The report on these discussions was published in late August of 2010 and included improving the FOSHU approval process with regard to speed and transparency and also the presentation of the necessary labelling information. The report also included a commitment to regulate health foods, such as dietary supplements and other health-related products, outside the FOSHU system by developing guidelines to cover false and misleading labelling; such foods are not allowed to make health or nutrient function claims. In addition to the CAA, the Consumer Commission was also established on 1 September 2009. It operates as an independent third party agency to monitor general consumer affairs administration of the government ministries and agencies, including the CAA. It is composed of a panel of independent experts and its meetings are held in public and open to consumers to attend. Certain expert consultative groups and committees of the Consumer Commission have roles in the review of FOSHU applications, as outlined in Section 10.2.5.

10.2.1 *Subsystems of FOSHU*

Qualified FOSHU refers to the approval of a food with a health function which is not substantiated with the high level of scientific evidence to meet that required of regular FOSHU, or to a food with certain effectiveness, but without the mechanism of the effective element for the function being clearly established.

Standardised FOSHU are approved where a product meets the set standards and specifications.

In the case of a reduction of disease risk FOSHU, a claim is permitted when reduction of disease risk is clinically and nutritionally established for an ingredient (Japanese Ministry of Health Labour and Welfare 2011a).

10.2.2 *Broad headings for approved FOSHU*

Approved FOSHU product claims may be grouped under seven broad headings that, together with some examples of ingredients with FOSHU approval, correspond to each of these broad categories (see Table 10.1).

10.2.3 *Disease reduction risk FOSHU*

There are two approved disease reduction claims for foods (Japanese Ministry of Health Labour and Welfare 2011a) as follows.

- Calcium and osteoporosis: the claim statement is that a proper amount of calcium contained in healthy meals, with appropriate exercise, may support healthy bones of young women and reduce the risk of osteoporosis when aged.
- Folic acid and neural tube defects: it may be claimed that the intake of the proper amount of folic acid contained in healthy meals may support women to bear a healthy baby by reducing the risk of neural tube defects, such as spondyloschisis, during foetal development. Spondyloschisis is a congenital fissure of one or more of the vertebral arches and is also known as rachischisis, or a cleft spine.

Table 10.1 Seven broad functional category headings under which FOSHU ingredients may be grouped in Japan, with examples of some of the ingredients approved under each category. Data from Consumer Affairs Agency, Japan (2011), Japanese Ministry of Health Labour and Welfare (2011a) and Japanese Ministry of Health Labour and Welfare (2011b).

Broad FOSHU category	Ingredients exhibiting health functions related to these*	Total approved to April 2011
Gastrointestinal health	Lactic acid bacteria, bifidobacteria, oligosaccharides, lactose, dietary fibre, indigestible dextrin, polydextrol, guar gum, psyllium seed coat	350
Hypertension control	Lactotripeptide, casein dodecaneptide, tochu leaf glycoside (geniposidic acid), sardine peptide (Valyl tyrosine)	120
Blood sugar control	Wheat albumin, indigestible dextrin, guava tea polyphenols, L-arabiose, catechin (the main polyphenol from green teas)	141
Triglycerides and body fat	Partially degraded sodium alginate, dietary fibre from psyllium seed husk, middle-chain-length fatty acids, isolated soya protein	70
Cholesterol and blood triglyceride control	Chitosan (produced commercially by deacetylation of chitin), isolated soya protein, partially degraded sodium alginate	142
Dental health	Erythrytol, maltitiose, paratinose, casein phosphopeptide	79
Bone formation	Soybean isoflavone, milk basic protein (MBP, the basic fraction of whey protein), casein phosphopeptide, calcium citrated malate	53

*Some ingredients exhibit health functions in more than one of these categories

10.2.4 *Foods with nutrient function claims (FNCF)*

There are also legal provisions for foods with nutrient function claims (FNFC) which can be made for foods that meet specified requirements for any of 12 named vitamins and the minerals calcium, copper, iron, magnesium and zinc (Yamada *et al.* 2008; Japanese Ministry of Health Labour and Welfare 2011b). Unlike FOSHU products, FNFC products do not require permission from or notification to the Japanese authorities, provided that it meets the established standards and specifications which specify the nutrient content required for each claim as set down, and the labelling must include any warnings as required (see Table 10.2).

10.2.5 *Essential elements for obtaining FOSHU approval*

The Japanese approval procedure differs from those of other countries such as the US, where only generic claims can be made on certain foods, and the EU where (at least in the case of general function claims) all foods that meet the specified requirements may carry the claim. Prior to September 2009 it was the norm that applications were submitted to the local or regional health authority and passed on to the Office of Health Policy on Newly Developed Foods of the Ministry of Health, Labour and Welfare, but since the formation of

Table 10.2 Specified nutrients and their requirements as regards their advisable daily intakes for which Foods with Nutrient Function Claims (FNFCs) are permitted in Japan. Data from Yamada *et al.* (2008) and Japanese Ministry of Health Labour and Welfare (2011b).

Nutrient	Nutrient function claim*	Advisable nutrient intake range for making the claim
Minerals		
Calcium	Calcium is a nutrient necessary in the development of bone and teeth.	210–600 mg
Copper	Copper is a nutrient that helps the formation of red blood cells. Copper is a nutrient that helps the proper function of many body enzymes. Copper is a nutrient that helps in bone formation.	0.18–6 mg
Iron	Iron is a nutrient necessary in red blood cell formation.	2.25–10 mg
Magnesium	Magnesium is a nutrient necessary in the development of bone and teeth. Magnesium is a nutrient that helps maintain proper blood circulation. Magnesium is a nutrient that helps proper function of many body enzymes. Magnesium is a nutrient that helps in energy generation.	75–300 mg
Zinc	Zinc is a necessary nutrient to maintain normal taste. Zinc is a nutrient that helps maintain healthy skin and mucous membranes. Zinc is a nutrient that is involved in the metabolism of protein and nucleic acids. Zinc is a nutrient that is helpful in maintaining health.	2.1–25 μg
Vitamins		
Vitamin A	Vitamin A helps to maintain vision in the dark. Vitamin A helps to keep skin and mucosa healthy.	135–600 μg
Vitamin B1 (Thiamin)	Vitamin B1/Thiamin is a nutrient that helps to produce the energy from carbohydrate. Vitamin B1/Thiamin is a nutrient that helps to maintain healthy skin and mucosa.	0.3–25 μg
Vitamin B2 (Riboflavin)	Vitamin B2/Riboflavin is a nutrient that helps to maintain healthy skin and mucosa	0.33–12 mg
Vitamin B6	Vitamin B6 is a nutrient that helps to produce the energy from protein. Vitamin B6 is a nutrient that helps to keep healthy skin and mucosa.	0.3–25 mg
Vitamin B12	Vitamin B12 is a nutrient that aids in the formation of red blood cells.	0.6–60 μg
Vitamin C	Vitamin C is a nutrient that helps to maintain skin and mucosa healthy. Vitamin C is a nutrient that has an anti-oxidising effect.	24–1000 mg
Vitamin D	Vitamin D is a nutrient that promotes the absorption of calcium in gut intestine and aids in the growth of bone.	1.5–5 μg
Vitamin E	Vitamin E is a nutrient that helps to protect fat in the body from oxidisation and to maintain the health of cells.	2.4–150 mg
Biotin	Biotin is a nutrient that helps to maintain healthy skin and mucosa.	14–500 μg
Folic acid	Folic acid is a nutrient that aids in the formation of red blood cells. Folic acid is a nutrient that contributes to the normal growth of the foetus.	60–200 μg
Niacin	Niacin is a nutrient that helps to maintain healthy skin and mucosa.	3.3–60 mg
Pantothenic acid	Pantothenic acid is a nutrient that helps to maintain healthy skin and mucosa.	1.65–30 mg

*When a FNFC claim wording is used, specified warnings are also required on the label against the consumption of more than the recommended daily intake of the nutrient and that consuming a large quantity of the product does not promote health or cure disease. The wording of nutrient function claims as given in this table are translations from the original Japanese and should be regarded as indicative only. The specified text in Japanese should be used. Based on Yamada *et al.* (2008) and Japanese Ministry of Health Labour and Welfare (2011b).

the CAA applications are now submitted to the Labelling Division of the Consumer Affairs Agency. The application should provide:

- name of the applicant company and the name of a representative;
- name and address of the head office and factory;
- name of the product;
- shelf life of the product;
- content amount;
- reason for seeking permission or approval;
- explanation of how the product contributes to the improvement of diet and the maintenance/enhancement of health of the entire population;
- health claim and nutrition information on the label of the product;
- list of ingredients and composition as percentages;
- details of manufacturing process;
- certificate of nutrient and energy analysis;
- daily intake amount recommended;
- considerations and precautions regarding intake;
- instructions for preparation, storage or intake of the product;
- a sample of the product packaging with labels and claims;
- documentation that demonstrates clinical and nutritional evidence relating to the health claim of the product or its functional components;
- documentation regarding safety of the product or its functional components, including additional human studies regarding the eating experience;
- documentation in relation to the physicochemical properties of the product and the analytical methods for the product's functional components;
- results of the quantitative and qualitative analyses of its functional components as nutrient constituents of the food, and the analytical methods;
- a copy of the applicant's organisational statutes or certificate of contribution;
- description of the production method and factory equipment and details of the quality control system;
- a copy of the consignment contract if the applicant is not the manufacturer;
- reasons for not attaching any of the above requirements if not provided; and
- other materials and documentation as appropriate to support the application.

If the application is deemed adequate and in order, the CAA then forwards the application to the Assessment and Evaluation Group for Novel Food of the Consumer Commission to review its efficacy. It is then forwarded to the Expert Assessment Group of the Food Safety Commission which reviews the food safety aspects. It is then returned to the Assessment Committee for Novel Foods of the Consumer Commission for a comprehensive review of both efficacy and food safety. Requests for additional documentation and information may be made throughout the process.

Following successful completion of the various reviews, the application is passed to the Ministry of Health, Labour and Welfare to check whether the labelling violates the

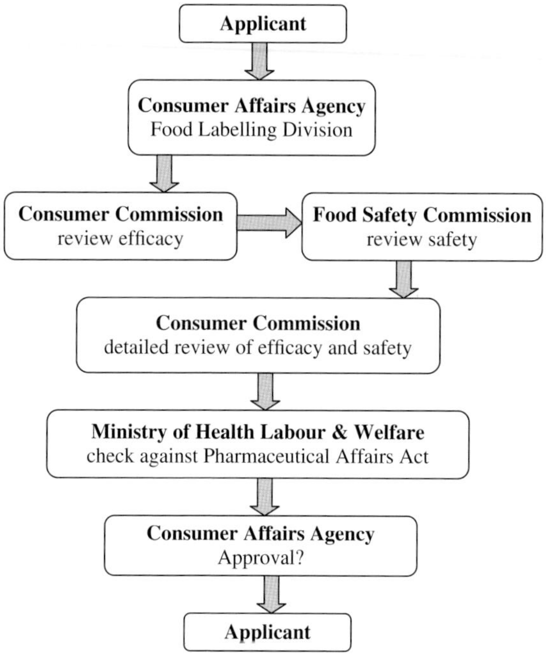

Fig. 10.1 The FOSHU approval system.

Fig. 10.2 The FOSHU logo.

Pharmaceutical Affairs Act. On completion of the entire process, the product receives FOSHU approval and receives permission to use the approved claim (see Fig. 10.1). The labelling of approved products can also carry the FOSHU logo (see Fig. 10.2) or the Qualified FOSHU logo (see Fig. 10.3) as appropriate. The examples shown are those that indicate, in Japanese characters, above the logo 消費者庁許可 showing that the products are permitted

Fig. 10.3 The qualified FOSHU logo.

or approved by the Consumer Affairs Agency (CAA). Products approved by the Ministry of Health, Labour and Welfare prior to the establishment of the CAA in September 2009 may indicate this by the use of the Japanese characters 厚生労働省許可 above the logo.

The approval process generally takes about one year from the time of submission to obtain approval. The rate of applications was initially slow; this was attributed to the length of time taken to obtain approval (in the earlier years it was reputed to take some years) and also to the costs involved. From 1991 to 1998, only 126 products received FOSHU approval; however, amendments to the FOSHU Law in 1998 reduced the requirements for the amount of scientific documentation that manufacturers must submit. The requirement for a certificate confirming that all submitted scientific documentation had been reviewed by outside scientific experts was removed and replaced by a requirement that studies had been published in a scientific journal; an industry-sponsored journal is deemed acceptable in this respect. Finally, the requirement that products should be tested by the National Health and Nutrition Laboratory has now been removed, as the manufacturer's own analytical tests are now acceptable.

The responsibility for reviewing applications was also transferred from the Ministry of Health and Welfare to the local prefectures at that time, but this is now the responsibility of the CAA. As a result of these changes, the rate of approvals increased significantly. By the end of 2003, a total of 396 products had been granted FOSHU status and by April 2011 the total had progressed to 955 products (see Fig. 10.4) (Consumer Affairs Agency (Japan) 2011).

A list of all FOSHU products is available in Excel file format but in Japanese only (Japanese Ministry of Health Labour and Welfare 2010). In terms of product approvals, the largest category is that of gastrointestinal health which consists mainly of probiotics, prebiotics and dietary fibre. By the end of 2001, such products accounted for about 60% of the 289 approved products (Arai *et al.* 2002), and probiotics accounted for about one-third of the gastrointestinal health category. In 2008 it was indicated that approvals related to gastrointestinal health still accounted for about half of all FOSHU approvals and by April 2011 the figure was 36% (Table 10.1; Fig. 10.5) (Yamada *et al.* 2008; Consumer Affairs

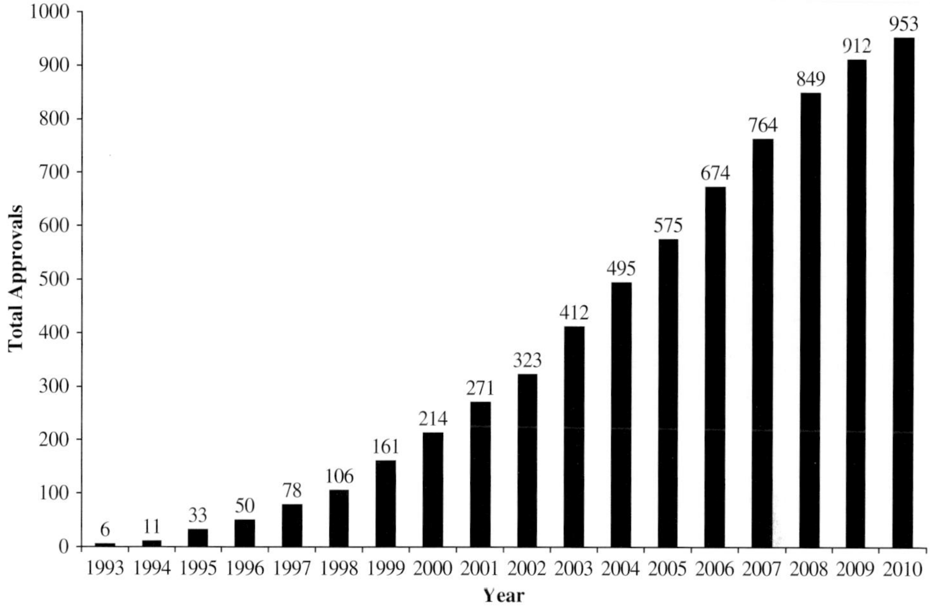

Fig. 10.4 Cumulative FOSHU approvals from 1993 to 2010. Adapted from Consumer Affairs Agency , Japan (2011).

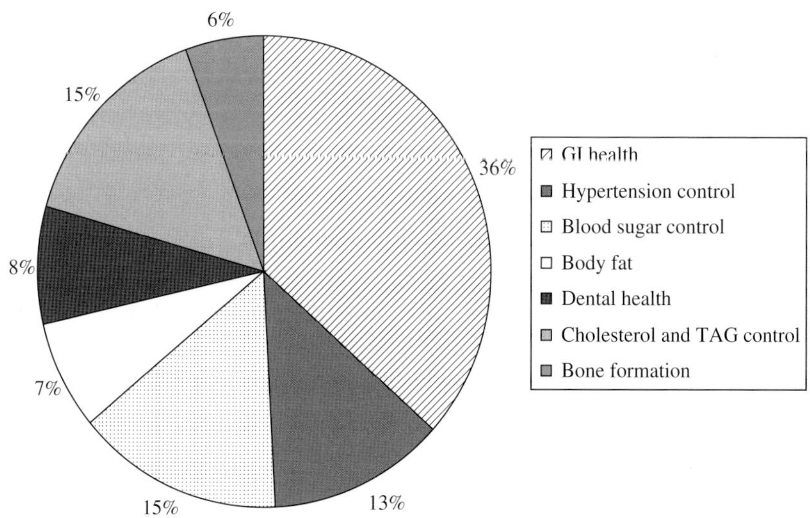

Fig. 10.5 FOSHU by health effect as a percentage of total approvals. Data from Consumer Affairs Agency, Japan (2011). (See plate section for the colour version of this figure.)

Table 10.3 Examples of health claims approved on FOSHU products containing probiotic micro-organisms.

Commercial product	Health claims
Meiji milk products (Bulgarian yogurt)	Due to the effects of *Lactobacillus* LB 81, this yogurt regulates the balance of intestinal bacteria that lead to and maintain a good intestinal condition.
Yakult	Due to the effects of the Yakult strain (*Lactobacillus casei* strain Shirota), which can reach the intestine alive, Yakult maintains the intestine in good health by increasing beneficial bacteria, decreasing harmful bacteria and by improving the intestinal environment.
Morinaga milk industry (Bifidus yogurt)	This yogurt contains living bifidobacteria (*Bifidobacterium longum* BB536). It helps to increase intestinal bifidobacteria, improve the intestinal environment and regulate the intestinal condition.
Takanashi milk products (Onaka-He-GG)	Due to the effects of *Lactobacillus rhamnosus* GG, which can reach the intestine alive, this yogurt increases beneficial bacteria and decreases harmful bacteria. It improves the intestinal environment and regulates the intestinal condition.

Agency (Japan) 2011). Some examples of health claims for milk-based probiotic products that have been approved for FOSHU products on the Japanese market are listed in Table 10.3.

The Japanese FOSHU system has a number of unique features: it is a voluntary procedure, but it has government approval; it is granted to individual products; specific health claims are approved prior to use; approval is based on documented scientific safety and effectiveness considerations; and approved products may use the FOSHU logo on their labels.

Being the first regulatory system to address functional foods, it has proved a model and reference point for the sector. It is expensive to obtain FOSHU approval, however; the cost has been estimated as being between \$850,000 and \$1.5m (Yamaguchi 2004; Patel *et al.* 2008).

10.3 The legislative situation in the European Union (EU)

At present in the EU, there is no legal definition of the term functional food nor is there specific legislation governing them. Nonetheless, a number of horizontal legislation considerations need to be taken into account, not just the regulation on nutrition and health claims (which is probably that which has most immediate impact) but also legislation addressing aspects of food safety, novel foods and food ingredients and food labelling. The latter aspects shall be addressed before going on to deal with nutrition and health claim in some detail.

10.3.1 *Relevant EU food safety legislation*

Regulation 178/2002 lays down some general principles and requirements on food safety for food in the EU (EU 2002). Article 14.1 of this regulation requires that food put on the market should be safe. The regulation goes on to specify that food safety in this context encompasses short-term, intermediate and long-term effects on consumers, and also any possible effects on subsequent generations (through teratogenic effects). Prior to its adoption, food manufacturers were required to ensure their products met the specific requirements of EU legislation but, unlike the laws of an increasing number of member states, the law did not include the broad requirement that food put on the market should be safe. While it could be held that the Product Liability Directive 85/374 (EU 1985, 1999) made it a requirement to have safe products by imposing strict liability on manufacturers whose products caused harm to individuals; it did not allow action by the authorities before the products actually caused a problem. Furthermore, the scope of the Liability Directive excluded primary agricultural products. By way of contrast, the General Food Law Regulation 178/2002 allows pre-emptive action and also removes the exemption for primary agricultural produce.

Novel Foods Regulation in the EU

The Novel Foods Regulation 258/97 (EU 1997), which came into force on 15 May 1997, requires novel foods and novel food ingredients that have not been previously used to a significant degree within the EU to undergo a safety assessment before being placed on the market. An interesting discussion could be had on the precise legal meaning of the words 'significant degree' in the definition, but doubtless clarity on this shall emerge over time. The categories of foods listed in the scope of the regulation include novel food ingredients and novel food processes, as well as novel foods themselves. It is interesting to note that between May 1997 and September 2010 a total of 118 applications were made and a further 158 notifications were received under Article 5 (substantial equivalence) of this regulation (EU 2011a).

Very few of these applications and notifications relate to milk products; the exceptions are certain fermented milks, fat spreads and other milk-based products containing added phytosterols and phytostanols. Two applications were made regarding bovine lactoferrin by Bipolar S.A., Les Isnes, Namur, Belgium in August 2007 and TNO, Zeist, The Netherlands on behalf of DMV International in March 2009. A Commission decision has not been issued on either application but the UK Advisory Committee for Novel Foods and Processes (ACNFP) considered the initial Belgian opinion on the Bipolar S.A. application by written correspondence in August 2008. Its members highlighted a number of concerns about the requirements for allergy labelling and also about the nature and extent of toxicological studies that had been carried out by the applicant company. These were reflected in the UK response that was sent to the European Commission. When discussing the TNO application on behalf of DMV International in July 2010, the same UK committee confirmed that its concerns over the previous Bipolar S.A. application for lactoferrin still applied.

Some milk-based products which could be considered as functional foods, such as milk products containing probiotics, were on the market prior to the regulation coming into force; they were therefore unaffected by this regulation because commercial foods in use in at least one member state before the regulation EU came into force were allowed on the EU market under the 'principle of mutual recognition'. One example of such prior approval of a probiotic is *Lactobacillus rhamnosus* GG, sometimes referred to simply as LGG, which was approved by the UK Advisory Committee for Novel Foods and Processes (ACNFP) in 1992 (Anon. 1992). It should be noted that its approval was as a novel starter culture and not as a novel probiotic. This was the first micro-organism to undergo a formal novel food evaluation by the UK ACNFP, and there was discussion in the committee as to whether the proposed use was actually novel. The conclusion was that, while the organism was already present in the UK diet at low levels, the proposed use could significantly increase the level of consumption and that justified its assessment as a novel food. Furthermore, the conclusion of the assessment was that there were no food safety concerns regarding its consumption in foods. In 2006 a scientific paper outlined some of the selection criteria and clinical evaluation applicable to *Lactobacillus* GG (Salminen 1996), and the specific techniques used in the selection process for the organism are described fully in US Patent 4839281 (Gorbach & Goldin 1989).

Genetic modification

Genetically modified organisms (GMOs) are organisms such as plants, animals and micro-organisms which have had their genetic characteristics modified artificially in order to give them new or improved properties. To ensure the safety of GMOs, the EU has established a detailed legal framework covering the area.

Given the controversy in recent years regarding GMOs, it is likely to be some time before functional foods using genetic modification will appear on the EU market. It has to be remembered that, as part of the selection and evaluation of probiotics, improvements of the performance of some of the strains are made. If any of these procedures involve genetic modification as defined by the regulation, then the EU legislation on GMOs must be respected. This establishes the conditions under which GMOs, or foods containing or derived from GMOs, may be developed, used or marketed.

The definition of a GMO is given in Article 2 of Directive 2001/18 (EU 2001c) as an organism in which the genetic material has been altered in a way that does not occur naturally by mating and/or by natural recombination. The particular GMO legislation that applies in any situation depends on the stage of the process involved, but the following is the list of the regulations that apply.

- Directive 90/219 is on the contained use of genetically modified micro-organisms in the case of, for example, the confined environment of laboratory research (EU 1990, 1998, 2001b).
- Directive 2001/18 on the deliberate release into the environment of genetically modified organisms, for the placing on the market of GMOs, for cultivation, import or processing into industrial products (EU 2001b).
- Regulation 1829/2003 is on genetically modified food and feed, for the placing on the market of GMOs intended for food and of food products containing, consisting of, or produced from GMOs (EU 2003a).

Where a food product contains or consists of GMOs, the applicant has a choice: either the application as a whole is subject solely to Regulation (EC) 1829/2003 or the application, or part of it, is subject to Directive 2001/18/EC and to Regulation (EC) 1829/2003 (EU 2001c, 2003a). In addition, GMOs and food products derived from GMOs placed on the market must also comply with labelling and traceability requirements. These requirements are found in Regulation 1829/2003 and in Regulation 1830/2003 on the traceability and labelling of GMOs and of food products produced from GMOs (EU 2003a, 2003b).

As regards approval of GMOs, foods containing GMOs and foods containing ingredients derived from GMOs are regarded as a particular category of novel foods and are treated as outlined in the previous section for novel foods as a whole.

10.3.2 *Claims and food labelling provisions*

Food labelling in the EU is governed by Directive 2000/13 (EU 2000, 2001a). The general principle underpinning food labelling is that purchasers must not be misled. Consequently, the information on food labels must be clear and unambiguous, and must not mislead the consumer to a material degree. The text on food labels regarding the characteristics of the food and, in particular, its nature, identity, properties, composition, quantity, durability, origin, method of manufacture or production should avoid claims such as: attributing to a particular food effects or properties which it does not possess; suggesting that a particular food possesses special characteristics when, in fact, all similar foodstuffs possess such characteristics; or attributing to a foodstuff the property of preventing, treating or curing a human disease or refer to such properties.

These principles and requirements must be respected when companies wish to make claims for their food products. A claim on a food label may be regarded as a declaration regarding the properties of that food above and beyond those that are required by legislation.

Following much discussion and debate, Regulation 1160/2001 on the provision of food information to consumers was adopted and published in November 2011 (EU 2011b). Among its purposes this repeals Directive 200/13 among other directives and regulations and amends a number of other regulations. This regulation came into force on 13 December 2011 and its provisions apply from 13 December 2014. The provisions on mandatory nutritional declaration apply from 13 December 2016 and specific requirements, such as regards minced meat, apply from 1 January 2014. This new regulation retains the provisions outlined above as regards the general principle underpinning food labelling that purchasers must not be misled and claims that should be avoided. Many more specific provisions on the subject of claims were addressed in the regulation on nutrition and health claims adopted in late 2006.

10.3.3 *Nutrition and health claims*

The regulation of health claims in the EU has proved very difficult. Efforts were made in the late 1980s and early 1990s at a time when nine of the then twelve member states had either national legislation or guidelines on claims, giving rise to a situation that had the potential to create barriers to the completion of the open market. The EU Commission produced draft proposals for a regulation, which was worked on for some years, but this effort foundered on the contentious issue of health claims. Some countries wanted a total prohibition on health claims, while others were prepared to accept them provided they were subject to strict criteria. Barriers to and difficulties in intra-community trade continued to arise as a consequence and 'health claims' continued to appear on foods, viewed by some as a problem as they were unregulated.

Eventually in July 2003 the commission published a new proposal for regulation on nutrition and health claims. The resulting regulation 1924/2006 was finally adopted and published on 30 December 2006. The original published version was corrected in early 2007, and two subsequent amendments were adopted in 2008 and another in 2010 (EU 2006, 2008b, 2008c, 2010). Full implementation of all the provisions in this regulation is not scheduled until 2021; an implementing regulation (Reg 353/2008) was adopted in April 2008 (EU 2008a).

The scope of the main regulation encompasses nutrition and health claims made on all foods marketed within the EU and includes claims made in labelling, presentation and advertising of foods (including brand names). The scope also includes food supplements, foods for particular nutritional uses (ParNuts), natural mineral waters and water intended for human consumption.

A claim is defined as referring to any message or representation which is not mandatory under EU or national legislation, including pictorial, graphical or symbolic representation in any form which states, suggests or implies that a food has particular characteristics. A nutrition claim is any claim which states, suggests or implies that a food has particular beneficial nutritional properties due to the energy, nutrients or other substances provided, not provided or provided in reduced/increased amounts. A health claim is any claim that states, suggests or implies that a relationship exists between a food category, a food or one of its constituents and health.

While nutrient content claims had been addressed to some extent already, the regulation of the field of health claims at community level was new. Health claims are only permitted if the following information is included in the labelling (or, if there is no such labelling, in the presentation and advertising): a statement indicating the importance of a varied and balanced diet and a healthy lifestyle; the quantity of the food and pattern of consumption required to obtain the claimed beneficial effect; where appropriate, a statement addressed to persons who should avoid using the food; and an appropriate warning for products that are likely to present a health risk if consumed to excess.

10.3.4 *Types of health claims*

Health claims which describe or refer to the role of a nutrient/substance in body functions are called Article 13 claims. These are further subdivided into Article 13.1 health claims and Article 13.5 health claims. Health claims that refer to a reduction of disease risk or claims referring to children's development and health are called Article 14 claims. It should be mentioned that medicinal claims, which claim that food can treat, prevent or cure any disease or medical condition, are still prohibited on foods; such medicinal claims can only be made for licensed medicines.

Article 13.1 claims

Article 13.1 claims refer to the role of a nutrient/substance in growth, development or body functions, psychological and behavioural functions and slimming, weight-control, weigh-reduction, satiety or reduction of available energy. These are based on generally accepted scientific evidence and must be well understood by the average consumer. They are sometimes referred to as 'general function' claims.

In mid-2006 the commission asked member states to submit a list of health claims (under Article 13) made by the food industry within their jurisdiction; to this end the competent authorities in each country contacted their food companies. The deadline for the submission of these national lists was the end January 2008. It is believed the commission expected to receive about 2000 such claims from member states – they actually received about 40,000.

The UK Food Standards Agency (FSA) submitted 2556 such claims (available on their website contained in five tables grouped by substance); the Food Safety Authority of Ireland (FSAI) submitted 314 submissions (including some duplicates), some of which were identified by them as borderline Article 14 claims.

This original number of Article 13.1 claims submitted to the European Commission was reduced to, or more correctly consolidated into, 4185 such claims taking into account 9720 similar claims and these were forwarded to the European Food Safety Authority (EFSA). A total of 1719 of these passed pre-screening by EFSA and 2466 were sent back for further information prior to their being considered and assessed by the EFSA panel on Dietetic Products, Nutrition and Allergies (the NDA Panel). Of the 2466 claims sent back to member states for clarification, 331 were subsequently withdrawn, clarification was received on 1515 and no clarification was received on the remaining 620. The total number of Article 13.1 claims was increased by a further 452 in 2010, mainly from a new member state, raising the total to be assessed to 4306. Of that number, 1548 claims on botanicals have been put on hold by the commission as they consider how these should be handled. This leaves a total of 2758 Article 13.1 claims to be assessed.

Based on the advice of EFSA, the commission had committed to adopt a list of such permitted claims by 31 January 2010. This deadline for completion proved to be unrealistic, as the task was not actually completed for a further 18 months. The EFSA opinions were published in 6 batches, starting in October 2009 and ending in July 2011, by which time the EFSA NDA Panel had completed its assessment of the remaining 2758 of these claims and issued 341 opinions on them. The assessments and opinions are available on the EFSA website (EFSA 2011a).

A personal review and analysis of 2719 of these EFSA opinions indicated that 478 (17.6%) received favourable opinions, that is, the claims submitted were borne out by the evidence supplied; a total of 2212 (81.3%) received unfavourable outcomes and 30 (1.1%) received mixed outcomes. The latter cases are due to the grouping of a number of claims in a single opinion; some of the health claims therein have received favourable outcomes while others received unfavourable opinions (e.g. the opinion on Article 13.1 Health Claim I.D. No 121 on health claims involving 'biotin' and 'bone, teeth, hair, skin and nail health' received a favourable opinion as regards the health claim on hair and skin, but an unfavourable opinion as regards the health claim on bone, teeth and nails).

The remaining 38 health claims relate to the roles of various vitamins, minerals and omega-3 fatty acids on various body functions, which had been evaluated in previously published EFSA opinions. Their outcomes cannot be readily analysed without identifying each of the corresponding earlier opinions and their outcomes.

Some types or categories of claims appear to have received a higher percentage of favourable outcomes (see Table 10.4), while other types and categories were notably unsuccessful. By way of example, in an analysis of approximately 338 vitamin and mineral claims on which opinions were published between October 2009 and November 2010, 208 (61.54%) received favourable opinions, 121 (35.8%) were unsuccessful and 9 (2.66%) received mixed outcomes. On the other hand, with the exception of two successful claims regarding live yogurt cultures and lactose digestion, none of the opinions on the other 285 probiotic claims were favourable. It has been indicated that EFSA is liaising with the European Commission and member states regarding the re-submission of a limited number

Table 10.4 Examples of the more successful broad categories of products where the outcomes of European Food Safety Authority (EFSA) evaluations gave favourable opinions. Data from EFSA (2011a), (2011b) and (2011c).

- Vitamins and minerals: many and various claims (e.g. calcium and/or vitamin D and their role in bone health/strength or teeth structure)
- Specific dietary fibres (e.g. oat bran) related to their role in the control of blood sugar (glucose), blood cholesterol control or weight management (it should be noted that where non-specific 'dietary fibre' was the subject of the claim, these received unfavourable opinions)
- Antioxidant effects of polyphenols in olive oil
- Specific fatty acids (such as omega-3, omega-6, DHA* etc.) and their role in heart function
- Plant sterols/plant stanols (including their esters) and their role in heart health or cholesterol levels
- Various sugar replacers (such as polyols) and their role in maintaining tooth mineralisation or lowering the increase of blood glucose levels after meals
- Certain carbohydrate-electrolyte drinks or creatine and their role in sports performance
- Live yogurt cultures and lactose digestion

Source: EFSA (2011a, b, c).
*DHA: docosahexaenoic acid

of general health claims, including a number of probiotic claims where the strains concerned were considered by EFSA to have been insufficiently characterised and also regarding some claims for which the evidence provided during the initial submission was not deemed sufficient to establish a cause and effect relationship (EFSA 2011a).

As regards milk-based foods for which Article 13.1 health claims were submitted, it should be pointed out that quite a number of milk products should meet the requirements that would allow calcium and/or vitamin D claims referring to teeth or bone growth and health (subject to any difficulties that may arise regarding nutrient profiles, which shall be discussed in Section 10.3.5). As regards Article 13.1 claims for other milk-based functional foods, Table 10.5 lists a number of examples of such claims that were unsuccessful.

The main reasons for submissions receiving unfavourable opinions included:

- insufficient information to identify or characterise the substance or ingredient on which the claim is based (examples include some claims on 'probiotics' or on 'dietary fibre' where the specific fibre was not specified);
- insufficient scientific evidence to demonstrate that the claimed effect was beneficial to the maintenance or improvement of the functions of the body (examples include foods with 'antioxidant properties' and claims on 'water elimination' by the kidneys/renal system);
- lack of precision regarding the wording of the health claim being made (examples include claims using broad terms such as 'energy' and 'vitality' and claims on 'gut health', 'women's health' or 'mental energy');
- lack of sufficient human studies containing the necessary scientific data to demonstrate the claimed health benefit;
- claims referring to food categories which were considered to be too broad to allow them to be linked to specific effects, such as 'fruits and vegetables' and 'dairy products', rather than specifying specific foods; and
- claims which did not meet the definition of a health claim or that were considered to be Article 14 health claims as contained in the regulation (e.g. a claim not meeting the definition of a health claim is that the intake of cod liver oil improves the digestion of omega-3 fatty acids).

Table 10.5 Examples of some milk-based Article 13.1 claims (other than probiotic-type claims) where the outcomes of European Food Safety Authority (EFSA) evaluations gave unfavourable opinions. Adapted from EFSA (2011a).

Ingredient and/or food product	Claimed effect
Calcium in dairy products	Weight management
Dairy calcium	Bone health, bone structure, mineralisation and density
Selenium in milk and dairy products	Antioxidant to prevent oxidative stress, proper thyroid function, maintenance of cellular redox status
Whey protein (6 out of a total of 12 claims, all of which received unfavourable opinions)	Promotes protein synthesis when taken after exercise; supports the increase in lean body mass when combined with exercise and a hypocaloric diet; supports the decrease in body fat when combined with exercise and a hypocaloric diet; supports a healthy immune system; increases satiety; muscle mass in the elderly.
Whey protein hydrolysate	Muscle recovery
Whey protein isolate (Lacprodan DI-9212)	Increased muscle synthesis
Milk fat globule membrane/milk phospholipids	Immune/immunity system
Whey protein concentrate rich in α-albumin	Cognition/memory
Special hydrolysed milk proteins	Angiotensin-converting enzyme (ACE) inhibitor
Lactotripeptides	Blood pressure
Peptides (from milk products fermented with *L. helveticus*)	Cardiovascular system
Bovine colostrum	Immune system
Lacprodan PL-20; milk protein concentrate with a high content of phospholipids (effective component phosphatidyl serine)	Stress reduction; enhanced memory function
Fermented whey	Gut health
Lactose	Calcium absorption
Carbohydrates in dairy products	Foods with a low glycemic index (GI) and a low or slow blood glucose response
Lactose naturally present in dairy products	Dental health

Based on a review of EFSA NDA Panel Opinions on Article 13.1 Health Claims. Adapted from EFSA (2011a).

Botanicals refer to plants, algae, fungi or lichens and to preparations derived from them. They are often referred to as herbal substances or herbal remedies. Over the years they have become widely available on the EU market, usually in the form of food supplements, and are particularly popular for those who favour or have a preference for alternative medicine. Such products are typically labelled as natural foods and a variety of claims are made regarding possible health benefits. They can be bought over the counter in pharmacies, supermarkets, specialist shops and even on the internet. While most of these products have a long history of use in Europe, some concerns exist with regard to their safety and quality, including issues regarding the need to ensure that concentrations of their bioactive ingredients are within safe limits. Some examples of the botanicals concerned include St John's wort, bilberry extracts, garlic, ginseng, ginko tree extracts, sage oil and guarana. Details of the Article 13.1 botanical claims may be found on the EFSA website; we shall not dwell further on these however, as botanical products are outside the scope of this chapter.

The EFSA assessment of Article 13.1 health claims, as given in its opinions as published, may be viewed as scientific 'risk assessment'. The commission, in consultation with the member states, must now undertake the 'risk management' process which in this case is to adopt the list of permitted Article 13 claims as soon as possible to complete the overall procedures. They have already started on this and it is their stated aim to present the list for adoption before the end of 2011. Once this list is adopted and fully operational, all the acceptable general function health claims on the EU market will have been substantiated by science and assessed as not misleading; it will also facilitate the national competent authorities in the enforcement of compliance with the regulation and will guarantee fair competition among operators.

Article 13.5 claims

Article 13.5 claims are of a similar nature to the Article 13.1 health claims, but are based on newly developed scientific evidence and/or applications which contain requests for the protection of proprietary data. By November 2011 the EFSA panel had received 60 applications, 14 of which were subsequently withdrawn. So far 36 scientific opinions have been adopted. Of these 36 opinions that have been published, only 5 (13.9%) were favourable to the claims proposed (EFSA 2011b). None of these favourable claims concerned milk-based products, but of those which received unfavourable opinions 7 were milk-based or enriched-milk products, including 4 probiotic-containing products.

Article 14 health claims

Article 14 health claims address the reduction of disease risk and also include health claims that refer to children's development and health. The benchmark for approval is that the highest standard of scientific validity is required. It has been indicated that large clinical studies based on human trials are necessary. By the end of October 2011 EFSA had received 275 such applications, of which 118 were subsequently withdrawn including a number of probiotic-type claims. The list of applications includes food supplements and foods for particular nutritional uses (ParNuts), in addition to foods of a general nature.

A total of 85 opinions, covering 90 applications, had been adopted by the EFSA NDA panel by mid-October 2011; it should be pointed out that Article 14 claims are assessed individually and are not grouped or consolidated. Of these 85 published opinions, 24 (28.24%) were favourable and 61 (71.76%) were unfavourable (EFSA 2011c). Table 10.6 lists examples of Article 14 claims on some 15 milk-based products and related products of relevance to the dairy sector, 7 of which received favourable and 8 of which received unfavourable opinions from the EFSA NDA panel evaluations. Some Article 14 claims would appear to be similar to some Article 13.1 claims, but because they make reference to effects on children they were dealt with under Article 14. The main reasons for submissions receiving unfavourable opinions were similar to those outlined in the section on Article 13.1 claims.

Table 10.6 Article 14 claims on some milk-based products and related products of relevance to the dairy sector that received favourable and unfavourable opinions from European Food Safety Authority (EFSA) evaluations. Adapted from EFSA (2011c).

Favourable opinions Product	Wording of the claim that reflects the scientific evidence (EFSA Opinion)*	Unfavourable opinions Product(s)	Claimed effect by applicant
Low-fat/low-trans-spreadable-fat rich in unsaturated and omega-3 fatty acids as a replacement for a fat rich in saturated/trans fatty acids (hard fat, butter, stick margarine)	Consumption of saturated fat increases blood cholesterol concentrations; consumption of mono- and/or polyunsaturated fat in replacement of saturated fat has been shown to lower/reduce blood cholesterol; blood cholesterol lowering may reduce the risk of (coronary) heart disease	A combination of a low molecular weight fraction from bovine blood and bovine lactoferrin	Increasing suppressed concentrations of secretory IgA and reducing the risk of common cold with sore throat
Fat-based foods such as fat spreads and low-fat foods such as milk and yogurt	Plant sterols have been shown to lower/reduce blood cholesterol. Blood cholesterol lowering may reduce the risk of coronary heart disease	A combination of a low molecular weight fraction from bovine blood and bovine lactoferrin	Increasing suppressed concentrations of secretory IgA and reducing the risk of influenza and common cold
Low-fat fermented milk product enriched with plant sterols/stanols (phytosterols)	Phytosterols have been shown to lower/reduce blood cholesterol. High blood cholesterol is a risk factor in the development of coronary heart disease.	Fermented milk drink containing *Lactobacillus casei* (DN-114001) and yogurt cultures	Reduction of the presence of *Clostridium difficile* toxins in the gut which reduces the incidence of acute diarrhoea
Infant and follow-on formula supplemented with DHA (at about 0.3% of total fatty acids)	DHA contributes to the visual development of infants	Infant and follow-on formulae supplemented with DHA (at about 0.3% of total fatty acids) and a ratio ARA:DHA ratio between 1.4:1 and 2:1	DHA and ARA contribute to the optimal brain development of infants and young children
Follow-on formula in compliance with the criteria of composition laid down in Directive 2006/141/EC	Alpha-linolenic acid, an essential fatty acid, contributes to brain and nerve tissue development	Follow-on formulae with a fixed combination of short-chain galacto-oligosaccharides, acidified milk, nucleotides and beta-palmitate	Aids minor intestinal ailments (such as colic, constipation, digestive symptoms)
Calcium and vitamin D in yogurt, drinking yogurt and fromage frais matrix (the claim relates to calcium and vitamin D directly and not to the food product)	Calcium and vitamin D are needed for the normal growth and development of bone in children	Dairy foods including milk, cheese and yogurt	Three portions of dairy food every day, as part of a balanced diet, may help promote a healthy body weight during childhood and adolescence
Fresh cheese (e.g. cream cheese, rocitta, quark, mozzarella) high in protein and containing significant amounts of calcium and/or vitamin D	Calcium and/or vitamin D is needed for the normal growth and development of bone in children	Dairy foods (milk and cheese)	Dairy foods (milk and cheese) promote dental health in children
		Granulated lycopene-whey complex (lycopene is the active ingredient)	Reduction of the risk of atherosclerotic plaques

*This wording is that which, in the opinion of the EFSA NDA panel, reflects (or is justified by) the scientific evidence submitted with the application; it is not that claimed by the applicant in their submission.

Based on a review of EFSA NDA Panel Decisions on Article 14 Health Claims. Adapted from EFSA (2011c).

10.3.5 *Nutrient profiles*

Article 4 of the EU Regulation 1924/2006 on the use of nutrition and health claims indicates the intention to link the use of both these categories of claim to foods or categories of foods meeting certain requirements and conditions related to their nutrient profiles. It specifies that nutrient profiles should be based on scientific knowledge on diet and nutrition, and their relation to health. Article 4 goes on to state that such nutrient profiles will take account of the following in particular: fat, saturated fat, trans fat, sugar and salt/sodium content of foods bearing claims; the role and importance of the food and its contribution to the diet of the general population or certain risk groups; and the overall composition of the food and the presence of nutrients that have been scientifically recognised as having an effect on health.

In February 2001 the commission formally requested EFSA to provide relevant scientific advice for the setting of nutrient profiles. The EFSA opinion was published in February 2008 (EFSA 2008). In that opinion the nutrient profile of the overall diet is identified as an important factor for good health and the overall nutrient profile of a balanced diet has already been defined by science taking into account recommendations for intakes of energy and specific nutrients. Because diets are composed of many component foods, an overall good dietary balance can be obtained from foods with different nutrient profiles; it is not really necessary for each individual food to meet the overall nutrient profile required for a 'balanced' diet.

On the other hand, individual foods may influence the nutrient profile of the overall diet depending in particular on the amount of its intake. This factor has to be taken into account in considering food products that should be eligible to carry health claims. The setting of nutrient profiles should therefore take into account the dietary role and importance of food groups and the contribution of nutrients to the overall diet of the population. Given the particular concerns regarding the western diet with growing levels of obesity and diet-related diseases (e.g. cardiovascular conditions, diabetes and certain cancers), particular consideration had to be given to those nutrients that, when consumed to excess, may contribute to these diseases. At the same time, other nutrients such as the minerals calcium, iron, magnesium and zinc and certain vitamins where dietary intake may be inadequate (especially if the consumption of certain food categories which are good sources of certain important nutrients are impacted adversely by their nutrient profiles) cannot be disregarded. It is therefore necessary that these seemingly conflicting factors are taken into account.

Work has started on the development of nutrient profiles but progress is slow. In the EFSA opinion it is stated that the term 'nutrient profile' refers to the nutrient composition of a food or diet. 'Nutrient profiling' is the classification of foods for specific purposes based on their nutrient composition. It is recognised that, in this particular case, it is intended solely for the regulation of nutrition and health claims made on foods (EFSA 2008).

From the limited information in the public domain there appears to be general acceptance that solid foods other than those in a specified food group category are likely to have the following threshold levels: sodium: 300 mg 100 g^{-1} or 100 mL^{-1}; saturates: 2 g 100 g^{-1} or 100 mL^{-1}; and sugars: 10 g 100 g^{-1} or 100 mL^{-1}.

The food groups that are regarded as having important roles in the diet include: cereals and cereal products (within which bread and breakfast cereals may have different threshold levels to other foods within the group); dairy products (within which cheese may require different threshold levels to other dairy foods); fish and fish products; fruits, vegetables and their products; meat and meat products; non-alcoholic beverages; ready meals, soups, sandwiches; seeds and nuts (excluding their oils); and vegetable oils and spreadable fats. These groupings are not finalised and may be further expanded or subdivided.

Information on the progress of nutrient profiling is quite limited, but some working documents are available. A working document became available in October 2008, followed by another updated document in February 2009 (Health and Consumers Directorate of the European Commission 2008b, 2009). The commission also held wide-ranging consultations with stakeholders on the subject including public health groups, consumer groups and different sectors of the food industry at two meetings of the Working Group on the Implementation of Health and Nutrition Claims of the Advisory Group on the Food Chain and Animal and Plant Health during 2008 (Health and Consumers Directorate of the European Commission 2008a). Since these were taken from working documents and state that they cannot be regarded as an official position of the European Commission, they have no legal status and are likely to be further amended before being agreed upon. It is therefore not appropriate at this time to include any details of the nutrient profile tables other than the general indications given in this section.

One issue of concern that has been raised by the dairy industry is that the sodium and saturated fat threshold levels would adversely affect the making of health or nutrition claims in cheese in particular, notwithstanding the role of many cheeses as good sources of calcium and protein in the European diet. However, a report quoting a UK Food Standards Agency letter related to the March 2009 meeting of the Working Group on the Implementation of Health and Nutrition Claims of the Advisory Group on the Food Chain and Animal and Plant Health includes suggestions that revisions include raising the threshold level for saturated fat under the general dairy category to 2.6 mg $100\,g^{-1}$ and raising those specifically for cheese for sodium to 900 mg and of saturated fat to 20 g $100\,g^{-1}$ (Jukes 2009).

While these threshold levels would allow some cheeses to carry health and nutrition claims based on their composition, especially reduced-fat cheeses, many common and popular varieties with relatively high fat levels and/or salt levels would still have problems. The work on developing nutrient profiles continues; the final outcome and timescale to complete the task is still unclear. It would appear that since mid-2009 the relevant working group has focused its attention on the completion of the assessments of the health claims themselves, together with some additional discussions on nutritional claims such as those relating to trans fats.

10.4 The US legislative situation regarding health claims and functional foods

10.4.1 *Background to US federal legislation*

Only US federal legislation is addressed in this chapter. Until 1900 there was little federal legislation addressing food and food standards; the individual states controlled domestically produced and distributed foods, but this control was markedly inconsistent from state to state. From the early 1880s the US Department of Agriculture's (USDA) Division of Chemistry (renamed the Bureau of Chemistry in 1901), under Harvey Wiley who had been appointed its chief chemist in 1883, began researching the adulteration and misbranding of food (and drugs) on the market. Although without any regulatory powers, the division published their findings in a ten-part series entitled Foods and Food Adulterants. Based on these results, Wiley started to lobby for a federal law to set standards for food and drugs in interstate trade. In this he was assisted and supported by state regulators, consumer bodies, medical doctors, pharmacists and certain journalists. Their efforts coincided with a general trend for increased federal regulations in all matters pertinent to safeguarding public health. State laws provided varying degrees of protection against practices such as misrepresenting the ingredients of food products or medicines (Swann 2008). It should also be added that in the early 1900s the food industry strongly supported national food legislation in order to obtain national uniformity in regulatory requirements and to build credibility for the food supply (Porter & Earl 1992).

Despite considerable debate on the issue of constitutionality surrounding States' rights, Congress enacted the Food and Drugs Act 1906 (Pub. L. No. 59–384 34 STAT. 768 1906), sometimes called the 'Wiley Act' in honour of its chief advocate. This act was aimed at 'preventing the manufacture, sale, or transportation of adulterated or misbranded or poisonous or deleterious foods, drugs, medicines, and liquors, and for regulating traffic therein, and for other purposes'. Congressional Acts identify and grant broad authority to federal agencies to interpret their provisions into the US Code, with the relevant enforcement agencies identified.

Under the US Food and Drugs Act 1906, responsibility for administration and examination for 'adulteration' or 'misbranding' was granted to the Wiley's USDA Bureau of Chemistry (USDABOC). Over the years the name of this body has changed to the more familiar Food and Drug Administration (FDA). The evolution and development of this organisation is discussed in Section 10.4.2.

Despite the vigour with which the USDABOC pursued its new powers and the intentions of the original Act, it did not really succeed in establishing a regulation at the federal level. Firstly, it regulated the adulteration and mislabelling of foods in interstate trade in their original packages based in their labelling. However, when bulk packages were opened and repacked, responsibility for control reverted to the individual states. Secondly, it did not have the clear mandate to develop standards for foods. Thirdly, in challenges the courts upheld State regulations that differed from, or were additional to, those imposed at the federal level (Porter & Earl 1992). Amendments were made to the original Act over the 25 years that followed 1906 but, due to ongoing problems, the federal regulators, consumer groups and the media pressed for a new act with more powers and scope in the early 1930s.

It took 5 years to be passed, but the Federal Food, Drug and Cosmetic (FDC) Act of 1938 was finally adopted (US Congress 1938). As well as extending the scope of the earlier act to cover cosmetics and therapeutic devices, this new Act repealed the Food and Drugs Act 1906 and contained the following new provisions of relevance to food:

- allowed that safe tolerances be set for unavoidable poisonous substances;
- permitted standards of identity, quality and fill-of-container to be set;
- authorised the inspection of manufacturing premises; and
- added the use of court injunctions to the previous penalties of seizures and prosecutions.

Although this law has been subject to frequent amendments in the intervening years, it remains the basis for federal regulation by the FDA to the present day. One amending act of particular relevance is the Nutrition Labeling and Education Act of 1990 (NLEA) (Pub. L 101–535) (US Congress 1990).

10.4.2 *Evolution and development of the FDA*

When the Food and Drugs Act 1906 was passed into law, responsibility for its administration and for the examination of food for 'adulteration' or 'misbranding' was granted to the USDA Bureau of Chemistry. Over time this body has changed its name to the Food and Drugs Administration and has transferred the government department under which it operates on a number of occasions, according to the following timeline:

- 1927: the non-regulatory duties of the USDA Bureau of Chemistry were transferred with the USDA and its name was changed to the Food, Drug and Insecticide Administration.
- 1930: the name was shortened to the more familiar Food and Drug Administration (FDA);
- 1940: it was transferred to the Federal Security Agency;
- 1953: the agency was transferred to the Department of Health, Education and Welfare (HEW);
- 1968: became part of the Public Health Service within HEW;
- 1980: the education function was removed from HEW to create the Department of Health and Human Services (DHSS), the current home of the FDA.

The FDA is now responsible for about 80% of the US food supply. The exceptions are the safety, wholesomeness, labelling and packaging of meat, poultry and certain egg products, which are the responsibility of the US Department of Agriculture (Swann 2008).

10.4.3 *Functional food and claims on food in the US*

As is the case with the EU, the US does not have any specific legislation on functional foods. However, the US Institute of Medicine of the Academy of Sciences has defined functional foods as 'any modified food or food ingredient that may provide a health benefit beyond the traditional nutrients it contains'; this is not a legal definition, however. The approach to regulation is again focused on the area of health claims.

In the USA the Food and Drugs Administration has defined four categories of foods: conventional foods; dietary supplements (i.e. intended to supplement the diet and marketed like conventional foods, although they also have to be labelled as dietary supplements); foods for special dietary use (i.e. intended to supply particular dietary needs for physiological conditions, overweight, food allergies, infant formula); and medicinal foods (i.e. intended for dietary management of a specific disease, under the supervision of a doctor or other health professional).

In theory, functional foods could fit into any of the above-mentioned categories but, to date, none would seem to be used in medical foods and there are very few applications of their use in foods for special dietary purposes. By way of example, a number of conventional foods contain probiotics (mainly dairy products such as yogurts, cultured milks, milk and cottage cheese). The US standard of identity for yogurt requires the use of the conventional starter culture organisms (*S. thermophilus* and *Lb. delbrueckii* subsp. *bulgaricus*) to be present, but no levels are proscribed. Certain species of lactobacilli and bifidobacteria may also be added; indeed it is estimated that over 75% of yogurts sold in the US market contain *Lactobacillus acidophilus* (Sanders 2003). However, The National Yogurt Association has introduced a voluntary Live Active Culture seal for products that contain live starter cultures, and this requires refrigerated yogurt to contain 10^8 cfu g^{-1} and frozen yogurt 10^7 cfu g^{-1} at the time of manufacture. These requirements do not apply to the levels of other cultures named on the label (Sanders 2003).

The main market for probiotics in the US is in dietary supplements where they are sold in the form of pills, capsules, powders and drinks. The reason that so many probiotics are sold as dietary supplements would appear to be that prior approval of structure/function claims is not required for this category. This aspect is further discussed in the following section.

Food safety in the US

The US food safety requirements depend on the category of foods to which a product belongs, but for foods and food supplements the onus for food safety is primarily on the manufacturers. Dietary supplements that were on the market prior to 1994 are assumed to be safe and new products may be marketed 75 days after the manufacturer provides notice and substantiation to the FDA that the product can reasonably be expected to be safe. This is much shorter and easier than the prior approval needed for food additives, where generally recognised as safe (GRAS) status cannot be demonstrated by manufacturers; however, such prior approval is expensive and lengthy. Of course, the FDA can act at any time if it has concerns about the safety of any product on the market and it can also challenge a manufacturer's notice and substantiation.

Claims and labelling

The Nutrition Labeling and Education Act (NLEA) of 1990 allowed health and disease prevention claims on food labels such as 'any substance that expressly or by implication characterises the relationship of any substance to a disease or health-related condition' (US Congress 1990). Such claims on foods are subject to prior authorisation by the FDA.

The Dietary Supplement Health Education Act (DSHEA), which was enacted in 1994, created another category of statements generally referred to as 'structure/function' claims that may be made for dietary supplements (US Congress 1994). These statements may: claim a benefit related to a nutrient deficiency disease; describe the role of a nutrient or dietary ingredient intended to affect a structure or function in humans that characterise the means by which a nutrient or dietary ingredient acts to maintain such structure or function; or describe general well-being from consumption of a nutrient or dietary ingredient. The manufacturer is responsible for ensuring the accuracy and truthfulness of these claims; the FDA does not approve them. For this reason, the law stipulates that if a dietary supplement label includes such a claim, it must state in a disclaimer that the FDA has not evaluated the claim. The disclaimer must also state that the dietary supplement product is not intended to 'diagnose, treat, cure or prevent any disease' because only a drug can legally make such a claim. Some report that this requirement is not always followed in practice (Berner & O'Donnell 1998).

In 1997, the Food and Drug Administration Modernisation Act (FDAMA) provided a second way for a health claim to be used on foods (US Congress 1997). This allows certain health claims to be made as a result of a successful submission of a notification based on an 'authoritative statement' from a scientific body of the US Government or the National Academy of Sciences. The government did not include dietary supplements in the provisions for health claims based on authoritative statements. Consequently, this method of managing health claims cannot be used for dietary supplements at this time. Qualified health claims may be made for dietary supplements only, and came about as a result of the US Court of Appeals in the case of Pearson *versus* Shalala. This court decision required the FDA to allow appropriately qualified health claims that would be misleading without such qualification. These qualified claims are based on the weight of the scientific evidence, i.e. there is more evidence for than against the relationship, but fall short of the validity standard required for foods under NLEA.

Nutrition and health claims in the US fall into three categories: health claims; nutrient content claims and structure/function claims. Health claims describe a relationship between a food, food component or dietary supplement ingredient and reducing the risk of a disease or health-related condition. Nutrient content claims describe the level of a nutrient or dietary substance in the product using terms such as 'free', 'high' or 'low' or compare the level in a food to that of another food using terms such as 'more', 'reduced' or 'lite'. Structure/function claims describe the role of a nutrient or dietary ingredient intended to affect normal structure or function in humans, e.g. 'calcium builds strong bones'. They may also characterise the means by which a nutrient or dietary ingredient acts to maintain such structure or function, e.g. 'fibre maintains bowel regularity' or 'antioxidants maintain cell integrity'. Finally, they may describe general well-being from consumption of a nutrient or dietary ingredient.

At present, the FDA has approved 12 health claims or nutrient content claims for foods that meet the Significant Scientific Agreement (SSA) standard (see Table 10.7) and 4 FDAMA (FDA Modernisation Act) Health Claims (Health Claims Authorised Based on an Authoritative Statement by Federal Scientific Bodies) (see Table 10.8); some of these apply to dietary supplements as well as conventional foods. As well as the requirements for approved claims, the FDA has detailed the requirements for the food making the claim, the food claim requirements and model claim statements. Full details of these claims can be found in the Code of Federal Regulations and in appendix C of the Food Labelling Guide (FDA 2009).

Table 10.7 Twelve health claims approved by the US Food and Drugs Administration (FDA) that meet the Significant Scientific Agreement (SSA) Standard. FDA (2009).

Calcium and osteoporosis
Dietary lipids and cancer
Sodium and hypertension
Dietary saturated fat, cholesterol and coronary heart disease
Fibre-containing grain products, fruit and vegetables and cancer
Fruits, vegetables and grain products that contain fibre, particularly soluble fibre, and risk of coronary heart disease.
Fruits and vegetables and cancer
Folate and neural tube defects
Dietary sugar alcohols and dental caries
Soluble fibre from certain foods and the risk of coronary heart disease
Soy protein and the risk of coronary heart disease
Plant sterol/stanol esters and the risk of coronary heart disease

Table 10.8 Wording of four US FDAMA health claims based on an authoritative statement by federal scientific bodies. FDA (2009).

Diets rich in wholegrain foods and other plant foods and low in total fat, saturated fat and cholesterol may reduce the risk of heart disease and some cancers
Diets containing foods that are a good source of potassium and that are low in sodium may reduce the risk of high blood pressure and stroke
Drinking fluoridated water may reduce the risk of dental caries or tooth decay
Diets low in saturated fat and cholesterol, and as low as possible in *trans* fat, may reduce the risk of heart disease

In US legislations there is no explicit recognition of any benefits of functional foods as such. Some contend that statements such as 'a probiotic helps proper digestive function' would be a structure function statement and not necessarily a health claim, but if the claim was that it 'helped reduce the risk of cancer' it would be a health-related claim and thus subject to FDA approval (Berner & O'Donnell 1998). Some products sold as dietary supplements make structural/function claims such as 'when taken daily, helps fortify your body's natural defenses and helps keep your body at its best' (Actimel, Dannon) and 'Helps create a favourable environment for the growth of beneficial flora, which dramatically influences metabolism and physical well being' (Acidophilus, Cell Tech) (Sanders 2003).

Structure/function claims have historically appeared on the labels of conventional foods and dietary supplements. The Dietary Supplement Health and Education Act of 1994 (DSHEA) established regulatory procedures for such claims for dietary supplement labels (although they can also be applied to conventional foods).

As structure/function claims are not FDA approved, there is no definitive list of such claims. However, such claims may describe or characterise: the role of a nutrient or dietary ingredient intended to affect normal structure or function in humans (e.g. 'calcium builds strong bones'); the means by which a nutrient or dietary ingredient acts to maintain such structure or function (e.g. 'fibre maintains bowel regularity' or 'antioxidants maintain cell integrity'); general well-being from consumption of a nutrient or dietary ingredient; or a benefit related to a nutrient-deficiency disease (e.g. vitamin C and scurvy), as long as the statement also tells how widespread (or otherwise) such a disease is in the US.

Although structure/function claims do not require pre-approval by FDA, they must be truthful and not misleading (the manufacturer is responsible for ensuring the accuracy and truthfulness of these claims) and the FDA must be notified of dietary supplement claims within 30 days of their first use. If a dietary supplement label includes such a claim, it must state in a disclaimer that the FDA has not evaluated the claim. Such a disclaimer is not required on conventional foods.

Structure/function (S/F) claims and disease claims for conventional foods focus on the effects derived from nutritive value, while S/F claims for dietary supplements may focus on nutritive as well as non-nutritive effects. The FDA is likely to interpret the dividing line between S/F claims and disease claims in a similar manner for conventional foods and for dietary supplements.

If this is not complicated enough, the US also has a system of qualified health claims. The FDA began considering such claims under interim procedures in September 2003. US court decisions had clarified the need to provide for health claims based on less science evidence, as long as the claims did not mislead the consumers. As with approved health claims, qualified health claims should also be based on a relationship between a substance and a health-related condition. In common with all health claims, qualified health claims require that a petition (i.e. an application) be submitted to the FDA. An enforcement discretion letter is issued by the FDA if it does not object to the use of the claim specified in the letter, provided that the products that bear the claim are consistent with the stated criteria. The FDA is committed to having all letters of enforcement discretion posted on their website. Once the letter is posted on the website, all manufacturers are informed of how the FDA intends to exercise its enforcement discretion on the use of the specific qualified health claim.

The scientific support for qualified health claims does not have to be as strong as that for health claims with SSA. Under its interim guidance, the FDA is tentatively providing for three levels of science below the SSA standard: good to moderate level of scientific agreement; low level of scientific agreement; and very low level of scientific agreement. The criteria for the scientific review are described in the FDA interim guidance (FDA 2011). As of mid-2011, the US has 15 qualified health claims (see Table 10.9).

Table 10.9 Fifteen qualified health claims recognised by the US Food and Drug Administration where the scientific support does not have to be as strong as that for health claims that meet the Significant Scientific Agreement (SSA) standard; the wording of such claims must include the term 'qualified. FDA (2009).

Dietary supplements containing a minimum of 0.8 mg folic acid and neural tube defects
Dietary supplements containing vitamins B6 and B12 and/or folic acid and vascular disease
Dietary supplements containing selenium and cancer
Dietary supplements containing vitamin E and/or vitamin C and cancer
Dietary supplements containing soy-derived phosphatidylserine and cognitive dysfunction and dementia
Specified types of whole or chopped nuts as listed and specified in detail and heart disease
Whole or chopped walnuts and heart disease
Conventional foods and dietary supplements that contain EPA and DHA omega-3 fatty acids and coronary heart disease (CHD)
Products that are essentially pure olive oil and labelled as such and coronary heart disease (CHD)
Dietary supplements containing chromium picolinates and diabetes
Dietary supplements containing calcium and colon/rectal cancer and calcium and recurrent colon/rectal polyps
Dietary supplements containing calcium and hypertension, pregnancy-induced hypertension and pre-eclampsia
Cooked, raw, dried, or canned tomatoes and tomato sauces containing at least 8.37% salt-free tomato solids and prostate, ovarian, gastric and pancreatic cancers
Unsaturated fatty acids from canola oil and reduced risk of CHD (detailed product requirements are specified)
Corn oil and corn oil-containing products and a reduced risk of CHD (detailed product requirements are specified)

10.5 The Canadian legislative situation regarding health claims and functional foods

10.5.1 *Introduction and background to Canadian federal legislation*

As is the case in the US, agriculture and food regulation in Canada is administered at both federal and individual provincial levels (based on Constitution Act 1867 and jurisprudence/case law) as follows.

- It is based on the division of powers between the provinces and federal government, sections 91 and 92 of the Constitution Act 1867.
- Along with immigration, agriculture was the only other domain defined as explicitly concurrent (i.e. as provincial and federal competency) in 1867.
- Federal government has competency/powers over trade and commerce but the provinces have jurisdiction to matters over 'property and civil rights'.
- The Provinces have competency and jurisdiction in matters related to health, but federal government plays a direct role in the licensing of food and drugs.
- National agricultural marketing boards that practise supply management are matters of both federal and provincial jurisdiction, the two orders effectively delegating their respective powers (over inter-provincial and intra-provincial trade and commerce) to the marketing boards.

For the purposes of this chapter, only federal legislation is considered. It should also be noted that Canada has a bilingual policy and all federal legislation is published in both French and English. Federal responsibility for the development of the national food labelling requirements is shared between two federal departments: Health Canada and the Canadian Food Inspection Agency (CFIA). Health Canada is responsible for the establishment of policies and standards relating to the health, safety and nutritional quality of food sold in Canada under the Food and Drugs Act. The CFIA is responsible for the administration of food labelling policies related to misrepresentation and fraud in respect to food labelling, packaging and advertising and the general food and fish labelling provisions respecting grade, quality and composition. In addition, the CFIA has responsibility for the administration of the food-related provisions of the Consumer Packaging and Labelling Act including basic food label information, net quantity, metrication and bilingual labelling.

10.5.2 *Health claims on foods in Canada*

In common with many other countries, Canada does not have specific legislation on functional foods but relies instead on its general food legislation to regulate their promotion and advertising. The Canadian authorities recognise that health claims on foods may assist consumers to make more informed decisions about their food choices but only providing that such claims are truthful, scientifically valid and not misleading. The Food Directorate of Health

Canada is responsible for the development of policies, regulations and standards that relate to the use of health claims on foods. It is also the Food Directorate that makes the necessary assessments of these claims by reviewing mandatory and voluntary pre-market submissions. The decisions reached concerning health claims which are based on Health Canada reviews are available on its website. Food products may also be subject to safety assessment if they are considered novel foods. The CFIA enforces policies, regulations and standards set by Health Canada, governing the safety and nutritional quality of all food sold in Canada.

In 1998, in response to growing consumer and marketing interest in nutrition and health, Health Canada published a Policy Paper on Nutraceuticals/Functional Foods and Health Claims on Foods (Health Canada 1998). In 2002, an Interim Guidance Document was published that outlined standards of evidence for evaluating foods with health claims (Health Canada 2002). In 2003, the Canadian Food and Drug Regulations were amended to introduce the first series of authorised health claims in Canada. In 2009, Health Canada updated the Interim Guidance Document replacing it with the Guidance Document for Preparing a Submission for Food Health Claims (Health Canada 2009a). Also in 2009, Health Canada posted the Guidance Document 'The Use of Probiotic Microorganisms in Food' and a new guidance document 'Classification of Products at the Food-Natural Health Product Interface: Products in Food Format', which was updated in 2010 (Health Canada 2009b, 2010).

The term 'health claim' is not defined in the Canadian Food and Drug Regulations, but Health Canada recognises the definition of 'any representation that states, suggests, or implies that a relationship exists between a food or a constituent of that food and health' (L'Abbé *et al.* 2008). This is the definition as included in the Codex Alimentarius, discussed in Section 10.8. In Canada health claims may be stated explicitly with words or implicitly using symbols, graphics, logos; product names, brand names and trade marks are also included.

Health claims have been classed into three main categories: disease risk reduction and therapeutic claims; function claims; and general health claims.

Disease risk reduction claims

Disease risk reduction claims are generally statements that link a food or a constituent of a food with the reduction of the risk of developing a diet-related disease or a condition when considered as part of the total diet; examples of such diet-related diseases and conditions include elevated blood pressure, certain cancers and osteoporosis. In practice, most disease risk reduction claims are in essence drug claims. A drug claim is a claim that suggests or implies that the product has the properties of a drug or that the food or ingredient thereof has effects above and beyond that which is normally associated with a food. Because of this, the Canadian Food and Drug Regulations were amended in 2002 to exempt foods with drug-like claims from the provisions of the Act and its related drug regulations in order to permit disease risk reduction and therapeutic claims; these must be specifically permitted by the Food and Drug Regulations. The composition of a food that carries the claim must also contribute to a dietary pattern associated with the claimed benefit.

A complete list of such claims, statements and conditions for their use is given in Section B.01.603 of the (Canadian) Food and Drugs Regulations (Minister of Justice Canada 2011).

The scientific data is reviewed on a regular basis, meaning that this is an evolving list. The following are the five health claim statements or claims that are listed at this time:

- A healthy diet containing foods high in potassium and low in sodium may reduce the risk of high blood pressure, a risk factor for stroke and heart disease.
- A healthy diet with adequate calcium and vitamin D combined with regular physical activity helps to achieve strong bones and may reduce the risk of osteoporosis.
- A healthy diet low in saturated and trans fat and may help to reduce the risk of heart disease.
- A healthy diet rich in a variety of vegetables and fruit may help to reduce the risk of some types of cancers.
- Statements regarding non-fermentable carbohydrates in gums and hard candies that the product: 'won't cause cavities'; 'does not promote tooth decay'; 'does not promote dental caries'; and is 'non-cariogenic'.

All of these are disease risk reduction claims. The updated Guidance Document for Preparing a Submission for Food Health Claims is intended to ensure that health claims for foods are substantiated in a systematic, comprehensive and transparent manner. When submissions are being prepared for the use of new health claims on food products, they are required to follow the format set out in this guidance document (Health Canada 2009a).

The substantiation of such health claims in Canada is based on the following nine guiding principles: a systematic approach; transparency; human evidence; comprehensiveness; a high level of certainty (i.e. the majority of high-quality human studies support a statistically significant favourable effect); a demonstration of causality; biological relevance of the claimed effect; the feasibility of the effective dose; the health claim wording; and the substantiation of one food-health relationship in a submission.

Health Canada is currently reviewing comments on its Position Paper on Five US Health Claims Considered for Use in Canada. This paper details its plan to proceed with publication of proposed regulatory amendments to allow an additional two new health claims (Health Canada 2006) relating to: a healthy diet rich in a variety of vegetables, fruit and whole grain products and reduced risk of heart disease; and a diet rich in folate along with a daily folic acid supplement and reduced risk of having a baby with a birth defect of the brain or spinal cord. However, these claims are not yet included in the Canadian Food and Drug Regulations.

Therapeutic claims

Therapeutic claims are those concerning the treatment or alleviation of a health-related disease or condition or about restoring, correcting or changing certain body functions. As of July 2011 there are no therapeutic claims approved for food in Canada.

Function claims

Function claims are claims about the specific beneficial effects that a food or a constituent of a food has on normal bodily functions. Such claims include not just the maintenance of a physiological function, but also that of physical or mental performance. Claims of this type must be clearly distinguishable from disease risk reduction or therapeutic effects as outlined above.

Function claims are permitted under Sections B.01.311(3), D.01.006 and D.02.004 of the Food and Drug Regulations. The following examples of function claims are given in the Guide to Food Labelling and Advertising: coarse wheat bran and the promotion of laxation or regularity; green tea and its antioxidant effect on blood or blood lipids; and pysllium (or more precisely the fibre from pysllium seed) and the promotion of laxation or regularity.

The term 'laxation' is not defined as such, but the Guide to Food Labelling and Advertising indicates that it is taken as referring to 'the normal softness and bulking of the stool resulting from such factors as increased undigested residue or bacterial mass, trapping of gases or water retention'. On the other hand, the food product should not refer to the food being a laxative or mention the relief of constipation on a label or advertisement, as this would characterise the product as a drug which is not permitted.

If new function claims are to be made then the supplier should have the necessary scientific evidence to validate the claim prior to its use in food labelling or advertising. This evidence may be used by the CFIA, in collaboration with Health Canada, to evaluate whether the product complies with the requirements of the regulations. Manufacturers are encouraged to contact Health Canada for advice regarding the acceptability of function claims on food products prior to their use, but they are not legally required to do so. Where new function claims are evaluated and found to be acceptable, they will be added to the relevant table in the guide. In determining the acceptability of new function claims, two factors are taken into account: the standards of the evidence, which must be applicable to the target group; and the claim(s) should clearly state a specific and scientifically supported physiological effect associated with good health or performance.

Nutrient function claims

Nutrient function claims, which were previously known in Canada as biological role claims, are a subdivision of function claims that describe the well-established roles of energy or known nutrients that are essential for the maintenance of good health or for normal growth and development. Nutrient function claims are not made for a food or foods as such; they may only be made respecting the energy value or nutrients in a food. The term 'nutrient' is not defined in the Canadian Food and Drug Regulations, but the Guide to Food Labelling and Advertising considers a substance to be a nutrient if it is recognised as such by the US Institute of Medicine of the National Academies (see www.iom.edu; Canadian Food Inspection Agency 2011).

The following two general nutrient function claims are permissible for all the recognised nutrients: energy (or name of the nutrient) is a factor in the maintenance of good health and in normal growth and development. It is important to recognise that in Canada nutrient function claims cannot be made for a food as such; rather they may only be made respecting the energy value or nutrients in the food. To illustrate the difference, the statement 'milk is an excellent source of calcium which helps build strong bones and teeth' is an *acceptable* claim under Canadian legislation, because it relates to a nutrient. On the other hand, the statement

Table 10.10 Acceptable nutrient function claims in Canada. Canadian Food Inspection Agency (2011).

Nutrient	Acceptable nutrient function claim*
Protein	Helps build and repair body tissues; helps build antibodies
Fat	Supplies energy; aids in the absorption of fat-soluble vitamins
DHA, polyunsaturated omega-3 fatty acid docosahexaenoic acid (22:6)	Supports the normal physical development of the brain, eyes and nerves primarily in children under two years of age
ARA, polyunsaturated omega-6 fatty acid arachidonic acid (20:4)	Supports the normal physical development of the brain, eyes and nerves, primarily in children under two years of age
Carbohydrate	Supplies energy; assists in the utilisation of fats
Vitamin A	Aids normal bone and tooth development; in the development and maintenance of night vision; in the development and maintenance of night vision
Vitamin D	Factor in the formation and maintenance of bones and teeth; enhances calcium and phosphorus absorption and utilisation
Vitamin E	A dietary antioxidant that protects the fat in body tissues from oxidation
Vitamin C	A factor in the development and maintenance of bones, cartilage, teeth and gums; a dietary antioxidant; a dietary antioxidant that significantly decreases the adverse effects of free radicals on normal physiological functions; a dietary antioxidant that helps to reduce free radicals and lipid oxidation in body tissues
Thiamine (vitamin B1)	Releases energy from carbohydrate; aids normal growth
Riboflavin (vitamin B2)	Factor in energy metabolism and tissue formation
Niacin	Aids in normal growth and development; factor in energy metabolism and tissue formation
Vitamin B6	Factor in energy metabolism and tissue formation
Folate	Aids in red blood cell formation; a factor in normal early foetal development; a factor in the normal early development of the foetal brain and spinal cord
Vitamin B12	Aids in red blood cell formation
Pantothenic acid	Factor in energy metabolism and tissue formation
Calcium	Aids in the formation and maintenance of bones and teeth
Phosphorus	Factor in the formation and maintenance of bones and teeth
Magnesium	Factor in energy metabolism, tissue formation and bone development
Iron	Factor in red blood cell formation
Zinc	Factor in energy metabolism and tissue formation
Iodine	Factor in the normal function of the thyroid gland
Selenium	A dietary antioxidant involved in the formation of a protein that defends against oxidative stress

*To make a nutrient function claim for a vitamin and mineral, that nutrient must have an established Recommended Daily Intake (RDI) and the food must be a recognised source (i.e. contain a minimum of 5% of the RDI) for that nutrient as specified in the Canadian Food and Drugs Regulations.

'milk helps build strong bones and teeth' is an *unacceptable* claim under Canadian legislation because it refers to the food and not a nutrient. In total, there are 23 individual nutrients for which nutrient function claims may be made (see Table 10.10).

General health claims

General health claims are broad-type claims that relate to the promotion of good health through encouraging healthy eating or that provide dietary guidance. These claims should not refer to a specific or general health effect, disease or health condition. There are no specific regulations governing the use of general health claims on foods, other than the general requirement that they must not be false, misleading or deceptive. In addition, there are no standardised nutritional criteria laid down for foods that carry these types of claims.

Probiotic claims

Canada has specifically addressed probiotic claims in its Guide to Food Labelling and Advertising, which contains the definition developed by the 2001 FAO/WHO expert consultation on health and nutritional properties of probiotics in food (FAO/WHO 2002; Canadian Food Inspection Agency 2011). There are two types of probiotic claims which can be made on food: strain-specific claims are claims about the health benefits or effects of specific strains of probiotics (at the present time there are no strain-specific claims accepted by Health Canada); and non-strain-specific claims are statements about the nature of probiotics (a list of non-strain-specific probiotic claims, acceptable without the need to conduct a detailed review of the scientific literature for the basis of the claim, is given in the guide).

Health Canada has also prepared a guidance document, 'The Use of Probiotic Microorganisms in Food', that sets out the conditions under which health claims about probiotics would be considered acceptable (Health Canada 2009b). This indicates that the term probiotic and other similar terms or representations should be accompanied by specific and validated statements about the benefits or effects of the probiotic. Health claims about the health benefits or effects of probiotics are statements that should be validated and should be supported by strain-specific evidence. When making a probiotic claim, the manufacturer should have the necessary documentation available to support the identity, safety, viability, concentration and stability of the specific probiotic strain that is used.

Where required, the manufacturer and supplier should follow all legal requirements applicable to the sale of food including any concerning the use and labelling of ingredients used in novel technology. The food should contain the amount of the probiotic micro-organism(s) that would be required to result in the claimed health benefit for the full shelf life of the product. Documentation to support the stability and viability of the probiotic strain(s) should be maintained.

The wording of acceptable claims that may be made on 16 non-strain-specific probiotics has been developed (see Table 10.11), including phrases such as: probiotic that naturally forms part of the gut flora; provides live micro-organisms that naturally form part of the gut flora; probiotic that contributes to healthy gut flora; or provides live micro-organisms that contribute to healthy gut flora.

Table 10.11 The eligible non-strain-specific species for probiotic claims in the Canadian Guide to Food Labelling and Advertising. Adapted from Canadian Food Inspection Agency 2011.

Lactobacilli spp	Bifidobacteria spp
Lactobacillus acidophilus	*Bifidobacterium adolescentis*
Lactobacillus casei	*Bifidobacterium animalis* subsp. *animalis*
Lactobacillus fermentum	*Bifidobacterium animalis* subsp. *lactis* (the synonym *B. lactis* may be used for labelling purposes)
Lactobacillus gasseri	*Bifidobacterium bifidum*
Lactobacillus johnsonii	*Bifidobacterium breve*
Lactobacillus paracasei	*Bifidobacterium longum* subsp. *infantis* comb. nov.
Lactobacillus plantarum	*Bifidobacterium longum* subsp. *longum* subsp. nov.
Lactobacillus rhamnosus	
Lactobacillus salivarius	

Adapted from Canadian Food Inspection Agency (2011).

10.6 Health claims for foods in Australia and New Zealand

10.6.1 *The evolution of health and related claims in Australia and New Zealand*

Historically Australia and New Zealand had separate legislation; that partly remains the situation to this day. However, in 1995 Australia and New Zealand signed a Joint Food Standards Setting Treaty which committed both countries to the development and implementation of a single set of food standards. The Food Standards Treaty provides for a bi-national agency, now called Food Standards Australia New Zealand (FSANZ), to undertake the relevant food standards development for both Australia and New Zealand. FSANZ has produced a Food Standards Code, which is regularly updated, that contains the joint standards developed to date (Food Standards Australia New Zealand 2011a). The Australia and New Zealand Food Regulation Ministerial Council (ANZFRMC) sets policy guidelines for the development of food standards by FSANZ. FSANZ must have regard to these policy guidelines when it develops or reviews food standards. It is not required to consult on them, nor consider their regulatory impact, but it automatically implements them in its work.

In 1989 the Food Standards Committee of the Australian National Health and Medical Research Council prohibited health claims based on the principles of public health nutrition, such as individual foods did not prevent or cause disease, that the reduction of disease risk was associated with total diet and lifestyle and that the causes of those diseases in which diet played some part were multifactorial. As a consequence, up to 1996 health claims were explicitly prohibited in Australia by Clause 19 of Standard 1A of the old Australian Food Standards Code which was then in place; a similar situation applied in New Zealand.

In 1996 the then National Food Authority, which had replaced the Food Standards Committee of the Australian National Health and Medical Research Council, reviewed this prohibition of the use of health and related claims on food labels (Preston & Lawrence 1996). This review led to the publication of a policy on nutrition, health and related claims in late 2003 by the Ministerial Council (ANZFRMC) (Australia and New Zealand Food Regulation Ministerial Council 2003); an updated version of this policy for 2011 is also available (Australia And New Zealand Food Regulation Ministerial Council 2011). FSANZ then commenced the process of developing a new standard on health claims.

In the interim, regulation is by the transitional standard on health claims: Standard 1.1A.2. This standard incorporates Clause 19 of Standard A1 of the Australian Food Standards Code, and operates pending the adoption of the more specific Standard 1.2.7 which is under development. Under the transitional standard only one health claim about a serious disease can be made; this is a claim about the benefits of maternal consumption of folate and the claim is to the effect that folate may reduce the risk of having a baby with a neural tube defect such as spina bifida (Food Standards Australia New Zealand 2008c; Tapsell 2008).

The proposals for the new standard 2.1.7 were published as 'Proposal P293 Nutrition, Health and Related Claims: the Final Assessment Report' which sets out in detail the reasons why new regulations were needed, the options that were considered during the development of the standard and a draft standard for incorporating into the Food Code as an attachment (Food Standards Australia New Zealand 2008a; Tapsell 2008). A guide to the proposed new code is also published (Food Standards Australia New Zealand 2008b). Following submission of this report, the Ministerial Council requested a review of the draft standard which included a number of issues that FSANZ were asked to reconsider, in particular the enforceability of the proposed regulatory approach for general level health claims and the application of the nutrient profiling scoring criteria to nutrition content claims. Although the initial deadline for the review was September 2008, this deadline has been extended a number of times and the latest information is that the review report was due to be submitted to the Ministerial Council in October 2011 (Food Standards Australia New Zealand 2011d).

Although not yet finalised and some changes are still possible, the aims of this proposed new FSANZ Standard 1.2.7 are to: remove ambiguity and uncertainty in the marketplace; provide a comprehensive regulatory approach for nutrition content and health claims on food and in food advertising; protect consumers from misleading or deceptive claims; assist consumers to select foods for a balanced diet; provide new opportunities for industry; ensure due regard is given to costs to consumers and industry; and develop support in the community for the new standard.

Under the new FSANZ standard on health claims, it is intended that three types of claims be regulated: nutrition content claims (statements about the presence or absence of a nutrient, energy or a biologically active substance in the food); general level health claims (claims about the effect of a nutrient or substance in a food on a health function or a non-serious disease; see Table 10.12); and high-level health claims (claims about the effect of a nutrient or substance in a food that make reference to a serious disease or a biomarker of a serious disease e.g. blood cholesterol and blood pressure).

Seven high-level health claims, where diet and disease or disease biomarker relationships were identified, were selected for individual review by external Australian and New Zealand experts engaged by FSANZ, based on reference information contained in the Australian and New Zealand Dietary Guidelines. Five of these diet–disease/biomarker relationships were assessed as reaching a convincing level of evidence by the experts, while two others were

Table 10.12 Nutrient function statements that may be used as a basis for making a general health claim in Australia and New Zealand. Adapted from Australia New Zealand Food Standards Code, Standard 1.2.7.

Nutrient	Nutrient function statement[a]
Vitamin A	Vitamin A is necessary for normal vision
Thiamin (vitamin B1)	Thiamin is necessary for the normal metabolism of carbohydrates
Riboflavin (vitamin B2)	Riboflavin contributes to the normal release of energy from food
Niacin (vitamin B3)	Niacin is necessary for the normal release of energy from food
Pantothenic acid (vitamin B5)	Pantothenic acid is necessary for the normal metabolism of fat
Vitamin B6	Vitamin B6 acid is necessary for the normal metabolism of protein
Biotin (vitamin B7)	Biotin contributes to normal fat metabolism and energy production
Folate (vitamin B9)[b]	Folate is necessary for normal blood formation
Vitamin B12	Vitamin B12 contributes to normal blood formation
Vitamin C	Vitamin C is necessary for normal structure and function of connective tissue (such as that required for normal gums, skin, healing processes, bone and cartilage)
Vitamin D	Vitamin D is necessary for the normal absorption and utilisation of calcium and phosphate
Vitamin E	Vitamin E is necessary for cell protection from the damage caused by free radicals (such as oxidation of polyunsaturated fatty acids in red blood cell membranes)
Vitamin K	Vitamin K is necessary for normal coagulation (blood clotting)
Calcium[c]	Calcium is necessary for normal structure of bones and teeth
Copper	Copper is necessary for the normal function of the immune system
Iodine	Iodine is necessary for normal production of thyroid hormones; Iodine is necessary for normal brain development in the unborn child, babies and young children
Iron	Iron contributes to normal blood formation
Magnesium	Magnesium is necessary for normal energy metabolism
Manganese	Manganese contributes to normal bone function
Phosphate	Phosphorus is necessary for the normal structure of bone and teeth
Selenium	Selenium is necessary for cell protection from some types of damage caused by free radicals
Zinc	Zinc contributes to the normal structure of skin and normal wound healing
Protein	Protein helps to build and repair body tissues
Dietary fibre	Dietary fibre contributes to regular laxation

Adapted from Australia New Zealand Food Standards Code, Standard 1.2.7.
[a]The nutrient function wording is not binding; however, the claim statement used must be consistent with the scientific basis of this wording.
[b]It is envisaged that a high-level claim may be made linking folate intake by women of child-bearing age and a reduced risk of neural tube defects (see Table 10.13)
[c]It is envisaged that a high-level claim may be made linking calcium intake, alone or in conjunction with vitamin D, with reduced osteoarthritis risk (see Table 10.13)

Table 10.13 The outcomes of seven high-level health claims involving disease-risk or disease biomarker relationships reviewed by external experts appointed by FSANZ. The individual reports of these reviews may be accessed on the FSANZ website. Adapted from Australia New Zealand Food Standards Code, Standard 1.2.7.

Food or food constituent	Disease/biomarker	Relationship	Outcome of expert review of evidence
Sodium (with or without potassium)	Hypertension	Positive	Reached convincing level
Fruit and vegetables intake	Coronary heart disease	Inverse	Reached convincing level
Saturated fat and/or trans fat	Elevated serum cholesterol or heart disease	Positive	Reached convincing level
Calcium (with or without vitamin D)	Osteoporosis	Inverse	Reached convincing level
Folate	Neural tube defects	Inverse	Reached convincing level
Wholegrains	Coronary heart disease	Inverse	Did not reach a convincing level
Omega-3 fatty acids	Cardiovascular disease	Inverse	Did not reach a convincing level

Adapted from Australia New Zealand Food Standards Code, Standard 1.2.7.

considered not to have reached the required level of evidence required (see Table 10.13). Claims for disease treatment are not allowed on foods (Food Standards Australia New Zealand 2008b). All claims have to be scientifically substantiated and not misleading.

10.6.2 *The Australia and New Zealand nutrient profiling model*

It is proposed that foods carrying general and high-level health claims will need to meet certain eligibility criteria, including those imposed under the nutritional profile scoring system. This is a fairly complex system which envisages three food categories:

- Category 1: beverages (including milk and milk-based beverages);
- Category 2: foods in general other than those in categories 1 and 3 (including cheese and processed cheese with calcium content ≤320 mg $100g^{-1}$; and
- Category 3: cheese and processed cheese with calcium content >320 mg $100g^{-1}$, including edible oil, edible oil spreads and butter.

Tables are provided to calculate the baseline points based on the sum of the energy points, saturated fat points, total sugars points and sodium points based on their content levels. Fruit and vegetable points (V), protein points (P) and fibre points (F) are then calculated, again from tables based their content included in the standard. The total points for the food product are then calculated to obtain:

$$\text{final score} = \text{baseline points} - V - P - F \qquad (10.1)$$

(however, for foods whose baseline points are ≥13, 0 protein points are given). Figure 10.6 provides a flowchart of the final score calculations and the eligibility of products in the 3 categories to make health claims.

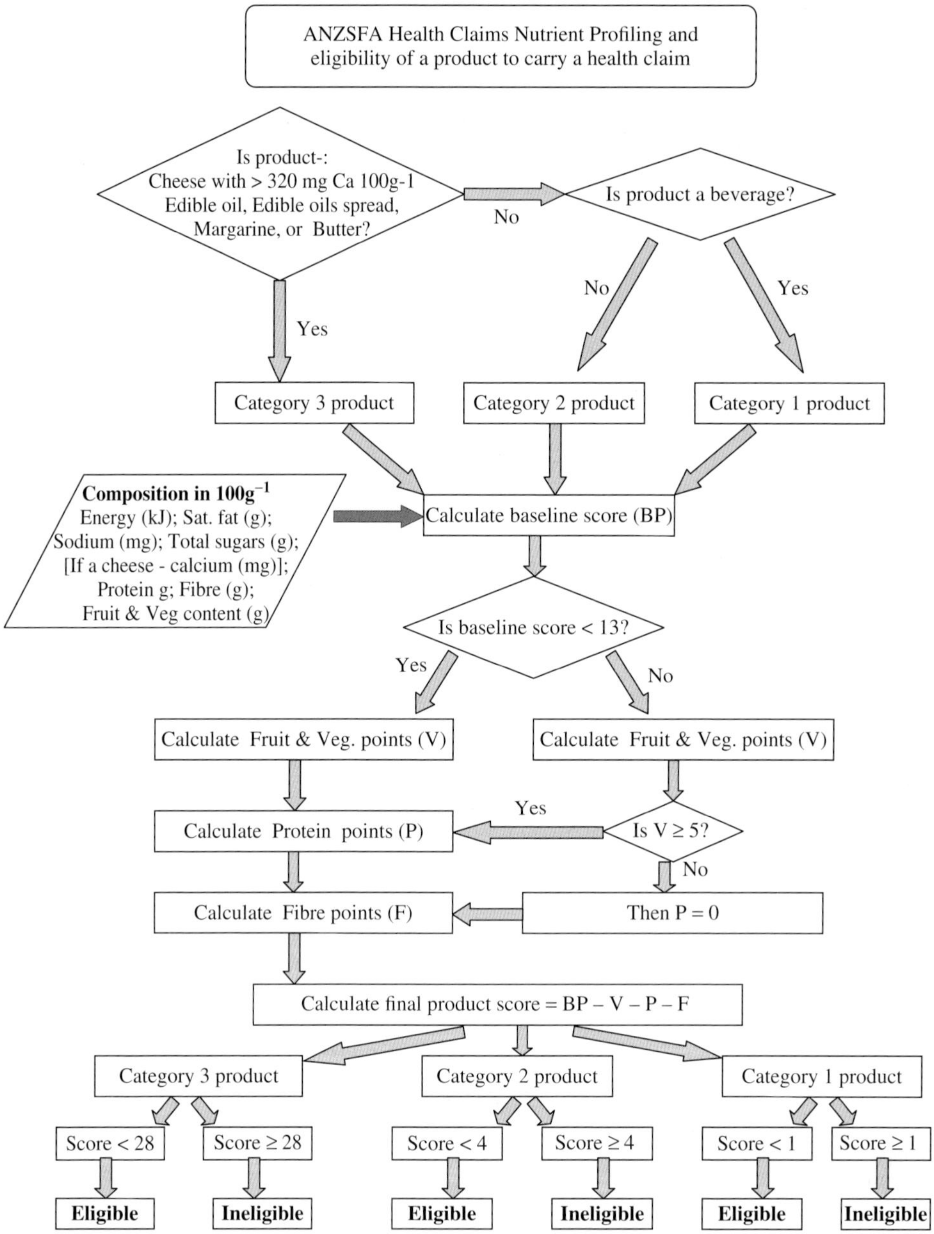

Fig. 10.6 ANZFSA health claims nutrient profiling and eligibility of a product to carry a health claim. Data from Food Standards Australia New Zealand, Food Standards Code Proposal 293.

There are levels for the final score that are based on the category into which the food fits. If it is below the level specified for its category, then that food is eligible to make a health claim (or a glycaemic index claim or a diet claim as also covered by the proposed standard). For categories 1–3, these final scores are <1, <4 and <28, respectively.

To help overcome the complexity and evaluate actual products, an online calculator and a guide to approaching the calculations (with worked examples) are available (Food Standards Australia New Zealand 2011b, 2011c). See Table 10.14 for examples of how the nutrient profile scoring system in the proposed Standard 1.2.7 might apply to some milk and milk products such as fluid whole and semi-skimmed milks for direct consumption, low-fat and full-fat yogurts both plain and with fruit and some cheeses.

As with the developments regarding nutrient profiling in the EU, cheeses are the milk products that are likely to be most affected by the system. The relevance of the calcium content of cheeses in deciding the food category is exemplified by the cottage cheese example. It probably comes as no surprise that cheddar cheese, Gouda and other full-fat varieties will not be eligible, for example, to make nutrient function claims such as that calcium is necessary for normal structure of bones and teeth. FSANZ claims that its system has been tested with over 10,000 foods using a database compiled from Australian and New Zealand foods.

Nonetheless, the nutrient profiling scoring system has its critics. Criticisms include the facts that: the scoring system is not proportionate to the health impact; there are no threshold levels set; it ignores normal consumption patterns and serving sizes; in certain

Table 10.14 Some examples of dairy product nutrient profile scores based on their composition and their consequent eligibility to make health claims. Adapted from Australia New Zealand Food Standards Code, Standard 1.2.7.

Product	Composition[a]						Total score	Eligible to make health claim[b]
	Saturated ($g\ 100\ g^{-1}$)	Protein ($g\ 100\ g^{-1}$)	Sugars ($g\ 100\ g^{-1}$)	Sodium ($g\ 100\ g^{-1}$)	Energy ($kJ\ 100\ g^{-1}$)	Fibre ($g\ 100\ g^{-1}$)		
Milk (full fat)	3.9 (2.5)	3.3	4.6	43	274	0	0	Yes
Semi-skimmed milk	1.7 (1.1)	3.5	4.7	43	195	0	−1	Yes
Low-fat yogurt (plain)	1.0 (0.65)	4.8	7.5	63	239	0	−1	Yes
Low-fat fruit yogurt[c]	1.1 (0.8)	4.2	12.7	62	331	0.2	0	Yes
Full-fat yogurt (plain)	3.0 (1.9)	5.7	7.8	80	333	0	−1	Yes
Full-fat fruit yogurt[c]	3.0 (2.0)	4.0	16.6	58	463	0.2	−3	Yes
Cheddar cheese[d]	36.6 (21.7)	25.4	0.1	723	1725	0	34[d]	No
Gouda cheese[d]	25.3 (20.3)	30.6	0.1	925	1562	0	34[d]	No
Half-fat cheese[e]	15.8 (9.9)	32.7	0.1	670	1141	0	19[e]	Yes[d]
Cottage cheese (plain)[f]	4.3 (2.3)	12.6	3.1	300	423	0	1[f]	Yes

[a]Compositional data based on the McCance and Widdowson Food Compositional Database (Food Standards Agency UK, 2010)

[b]Eligibility to make a health claim does not necessarily mean a health claim may be made; other factors should be take into account

[c]Based on a fruit content $\leq 40\ g\ 100\ g^{-1}$ or a concentrated fruit content $<25\ g\ 100\ g^{-1}$; therefore no fruit points are allowed.

[d]A product that scores ≥13 baseline points is not permitted to score points for protein unless it scores 5 or more points from fruit, vegetables, nuts and legumes. Hence the 5 points that could normally be deducted for the protein content does not apply in this case as the baseline points are 34.

[e]As for footnote (d), but the total score is <28 and the calcium content is $>320\ mg\ 100\ g^{-1}$ so health claims may be made.

[f]Cottage cheese has $127\ mg\ 100\ g^{-1}$ calcium and is therefore a Category 2 product according to Standard 1.2.7. The total score is <4 and therefore eligible to make a health claim.

Adapted from Australia New Zealand Food Standards Code, Standard 1.2.7.

cases small formulation changes can make ineligible products eligible again; and it differs from other nutrient profiling systems, such as that being developed in the EU, especially in terms of complexity. These criticisms may explain why the Ministerial Council asked the FSANZ to reconsider the application of the nutrient profiling scoring criteria to nutrition content claims.

10.6.3 *Enforcement of the health claim proposals in Australia and New Zealand*

The proposed Standard 1.2.7 envisages that the enforcement of its provisions will be at state or territorial level in Australia, which has 6 states (New South Wales, Queensland, South Australia, Tasmania, Victoria and Western Australia) and two territories (Northern Territory and Australian Capital Territory or ACT); in New Zealand enforcement will be at national level by the New Zealand Food Safety Authority (NZFSA). The latter was an independent public service body from July 2007 but on 1 July 2010 it amalgamated once again with the New Zealand Ministry of Agriculture and Forestry (MAF).

10.7 Health foods and functional foods in China

10.7.1 *Background*

In traditional Chinese medicine (TCM) there is a long-documented history going back to the West Zhou Dynasty in about 1000 BC that links the use or the avoidance of food to prevent, treat or alleviate certain diseases or their symptoms. Examples of such foods or food ingredients include both relatively common products such as honey, oysters, mussels, almonds, garlic, ginger, etc. and more exotic products such as deer horn antler, musk, snake, bear bile, ant beverage, earthworm, chrysanthemum, ginseng, etc. Foods and medicines came to be seen as coming from the same sources, and many of the herbs used in TCM were used both as medicines and as ingredients of regular foods. Theories on nutrition in TCM came to regard food as having four roles or uses (Weng & Chen 1996): food as diet; food use as a tonic; food used as a medicine; and the avoidance of food in certain cases.

The use of food in TCM is outside the scope of this chapter, except in demonstrating the historical basis of the situation that evolved over 3 millennia in China.

10.7.2 *Chinese legislative structures*

In China the highest state body and the only legislative house of government is the National People's Congress (NPC). Delegates who are elected by China's provinces, autonomous regions, municipalities and the armed forces hold office for five years, and the full congress is convened for one session each year. The 12th and current NPC, which was elected in 2013, has 2987 members. The NPC enacts the national laws and appoints the prime minister, other ministers and the president. The NPC chooses a Standing Committee from among its members, which is the permanent body of the NPC. It is the NPC Standing Committee which convenes the parliament, issues regulations, interprets laws and oversees government activities. The NPC also appoints the members of the State Council, which is the equivalent of the government in other countries. The State Council is organised as ministries based on the various sectors of the economy; these include for example the Ministry of Health and the Ministry of Agriculture (Lähteenmäki-Uutela, 2009a, 2009b; National People's Congress of the People's Republic of China 2011).

The Ministry of Health is the most relevant in regulating foodstuffs and medicines; however, the State Food and Drug Administration (SFDA), which was founded in 2003, has a major role in drafting various guidelines related to safety and efficacy of foodstuffs and at this time operates directly under the State Council. Foods such as healthy foods and novel foods that require pre-market authorisation come under the jurisdiction and competence of the SFDA (Lähteenmäki-Uutela 2009b; Information Center of SFDA 2011).

10.7.3 *The healthy (functional) foods sector in China and its regulation*

By the late 1980s healthy foods in China were regarded as fitting into one of three categories: fortified foods, special nutrition foods and foods for special health use. In 1986 the Chinese Ministry of Health enacted a Hygienic Standard and Regulation on the Use of Nutritional Fortification Substances. Later revisions resulted in 40 substances being authorised for such use including amino acids, vitamins and minerals but not any of the plants, herbs or products of animal origin used in TCM. In 1992 a National Standard for the Labelling of foods was issued which covered infant formulae, nutritionally fortified food and modified food such as high-fibre, low-sodium and low-fat foods where the labels were required to specify the target groups for whom the foods were intended. Because of the TCM link with certain foods, it was also deemed necessary to clearly differentiate between what was a food and what was a drug and to avoid confusion in their regulation. This was achieved by defining food in Article 54 of the current Food Hygiene Law as 'any finished product or raw materials to be provided for people to eat or drink, as well as any product that has traditionally served as both food and medicament, with the exception of products used solely for medical purposes' (Ministry of Health of China 1995; Kan 1996).

Drugs were defined in Article 57 of the Drug Administration Law of 1984; this was revised in 2001 and may now be found in Article 102 of the current Drug Administration Law (Ministry of Health of China 2001): 'Drugs refer to articles which are used in the prevention, treatment and diagnosis of human diseases and intended for the regulation of the physiological functions of human beings, for which indications, usage and dosage are established, including Chinese crude drugs, prepared slices of Chinese crude drugs, traditional Chinese medicine preparations, chemical drugs substances and their preparations, antibiotics, biochemical drugs, radioactive pharmaceuticals, serum, vaccines, blood products and diagnostic agents'. The main factor that determines whether a product is a food or a drug is whether it is used as a medicine or not; however, there was still a certain ambiguity as regards the use of certain traditional herbs and at that time it came down to how the product was labelled and advertised.

The development of the Chinese healthy (functional) food sector began in the early 1980s when commercially produced products started to be offered for retail sale. Some of these stated they could improve health and remedy certain diseases or conditions. By 1994 there were estimated to about 3000 factories producing such foods estimated to be worth $4 billion at that time. However, there were concerns as regards the identity, name, efficacy and possibly even the food safety aspects of these products, and it was recognised that a system of evaluation and assessment of these foods was necessary.

In March 1996 the Chinese Ministry of Health developed provisions to cover healthy foods (which began to be regarded as functional foods), which it defined as food which has special health functions. Such foods were intended for consumption by specific population groups for the regulation of certain functions of the body, but did not include therapeutic use.

The new Food Safety Law of 2009, which replaced an earlier Food Hygiene Law of 1995, contained some specific provisions relating to health foods in Article 51 requiring the state to stringently supervise foods claimed to have particular effects on human health (National People's Congress of China 1995, 2009). The relevant supervision and administration departments were obliged to perform their functions and undertake the responsibilities in accordance with the law. The new Food Safety Law has the following requirements.

- No food claimed to have particular effects on human health shall cause any acute, sub-acute or chronic harm to the human health.
- The labels and instructions of such food shall not involve the effect of prevention or treatment of any disease.
- The contents thereof shall be true and indicate applicable groups of people, inapplicable groups of people, effective ingredients or symbolic ingredients and contents thereof, etc.
- The effects and ingredients of a product shall be consistent with the indications in the labels and instructions.

There are 27 broad headings under which health claims may be made for foods or food ingredients in China. The type of scientific data required for each of these categories is specified, and indicates which should be obtained from human trials, animal trials or both (see Table 10.15). Prior to submission of the application, reports must be obtained from an authorised Chinese laboratory under the headings: toxicological safety assessment; evaluation of the functionality or efficacy of the active ingredient; analysis of the level of the active ingredient present; on the stability or shelf life of the product; and on the microbiological quality of the product.

In the case of health products manufactured in China, the required laboratory reports should be from one of a list of laboratories approved by the SFDA throughout China. For imported products, the reports should be from the Institute of Nutrition and Food Safety of the Chinese Center for Disease Control in Beijing.

Applications for product approval of Chinese-produced health foods should be made through the provincial health administrative department, which carries out the preliminary examination before passing it on to the SFDA. Applications for imported health foods are sent straight to the SFDA. The application should detail the product characteristics, the full formulation or ingredients, the analytical methods required, details of the manufacturing process, the relevant product and raw material specifications, the product label (including directions for its use), product samples and evidence from scientific literature regarding the characterisation, safety and efficacy of the food or its active ingredient.

Table 10.15 Health claims permitted in China (Yang 2008; Agriculture and Agri-Food Canada 2011; RJS Medical Technology 2011).

Health claims	Type of supporting scientific test data required
Alleviating physical fatigue	Animal
Assisting in protection against chemical injury of the liver	Animal
Assisting in protection against irradiation	Animal
Enhancement of anoxia endurance	Animal
Enhancement of the immune system	Animal
Improvement of sleep	Animal
Increase in bone density	Animal
Anti-oxidative function	Animal and human
Assisting in blood lipids reduction	Animal and human
Assisting in blood pressure reduction	Animal and human
Assisting in blood sugar reduction	Animal and human
Assisting in memory improvement	Animal and human
Assisting in weigh control (overweight or obesity control)	Animal and human
Facilitating bowel movement	Animal and human
Facilitating digestion (regularity)	Animal and human
Facilitating lead excretion	Animal and human
Improving child growth and development	Animal and human
Improving nutritional anaemia	Animal and human
Increasing milk secretion	Animal and human
Moistening and cleaning throat	Animal and human
Protection of gastric mucosa from damage	Animal and human
Regulating gastrointestinal microflora	Animal and human
Alleviating eye fatigue	Human
Eliminating acne	Human
Eliminating skin melasma (dark pigmentation patches)	Human
Improving skin ability to retain moisture	Human
Improving skin oil content	Human

While the market for such products in China is huge, the requirement for testing in government-approved agencies within China is a major challenge, especially as all the relevant information has to be provided in Chinese. This increases both the approval costs and the time taken to get to market. The testing process can take from 6 to 12 months, dependent upon whether or not human trials are required in China. Following the submission to the SFDA, and if everything thereafter progresses smoothly, the application can be approved within a further 3 months. If additional information is required this period can however be extended by 9–19 months. The application fee is relatively modest ($1200) and the main cost is in the testing. It has been estimated that the overall approval cost in China can be between $14,500 and $34,000, which is considerably cheaper than in Japan where the total cost has been estimated as being between $850,000 and $1.5 million (Yamaguchi 2004; Patel *et al.* 2008).

Approximately 8200 functional food products were approved in China by the Ministry of Health or the SFDA in the period from 1996 to mid-2007; at the end of that period it was estimated that only about 30% of the approved products were still on the market, however. The approval rate of applications by the SFDA in the period 2003–2007 was between 27% and 33%. Approximations on the top three categories under the 27 broad headings for which health claims may be made for foods or food supplements were 33% for enhancement of the immune function, 15% for alleviating fatigue and 9% for reduction in blood lipids. Supplements in the form of pills or capsules comprise >60% of the functional food products on the market, while conventional foods comprise <1% (Yang 2008). Estimates for the annual value of functional foods in China were of the order $4–6 billion in 2007 and around $14 billion in 2009; this is expected to grow to $70 billion in 2015 (Ford *et al.* 2007; Patel *et al.* 2008; Yang 2008; RedFern Associates 2010).

10.8 Codex Alimentarius

10.8.1 *Background, structure, operation and role*

At the 11th Session of the Conference of the Food and Agriculture Organisation (FAO) of the United Nations in November 1961, a resolution was passed to establish a Codex Alimentarius Commission (CAC) and a request was made for the World Health Organisation (WHO) to support the establishment of a joint FAO/WHO food standards programme. In 1962, the Joint FAO/WHO Food Standards Conference asked the CAC to implement the joint FAO/WHO foods standards programme, and to create the Codex Alimentarius. The 16th World Health Assembly approved this in May 1963 and adopted the statutes of the CAC.

The CAC is an international intergovernmental body. Its membership is open to member nations and associate members of the FAO and/or the WHO. As of July 2011, it had 184 member countries from all continents and regions. In recent years the CAC meets every year and the venue alternates between the FAO headquarters in Rome and WHO headquarters in Geneva. Nominated senior officials represent member governments at Codex meetings. National delegations may also include representatives of the industry, consumers and academia. A significant number of other international governmental organisations, for example the International Office of Epizootics (OIE) and the World Trade Organisation

(WTO), and international non-governmental organisations (NGOs) such as the International Dairy Federation (IDF), the European Dairy Association (EDA) and the Confederation of the Food and Drink Industries of the EU (CIAA), also attend in an observer capacity. In June 2011 the CIAA changed its name to FoodDrinkEurope. As of July 2009 there were 161 NGOs with Codex Alimentarius observer status. Observers are allowed to contribute to meeting at all stages except in final decisions. Final decisions are the exclusive prerogative of member governments.

The CAC has established two types of subsidiary committees: (a) Codex Committees, and (b) Co-ordinating Committees. The former type of committee is subdivided into General Subject Committees (currently 10 in number, named for the horizontal nature of their work), and Commodity Committees (currently 11 in number, 7 of which are adjourned *sine die*) which develop the standards for specific foods or classes of foods. There are six Regional Co-ordinating Committees whose role is to ensure that the CAC is responsive to regional interests and the needs of developing countries. The CAC also establishes *ad hoc* Intergovernmental Task Forces given stated tasks on specific topics. Currently, there are two such Task Forces on antimicrobial resistance and animal feeding. The Codex Alimentarius structure is depicted in Figure 10.7.

Codex Alimentarius can be translated from the Latin as 'the food code' and its three main aims are: the protection of consumer health; ensuring fair trading practices; and facilitating international trade. The Codex Alimentarius comprises: food standards; codes of hygienic or technological practice; pesticide evaluations; maximum residue limits for pesticides; guidelines for contaminants; food additive evaluations; and veterinary drug evaluations.

Food standards cover specific commodities and also general issues, which have horizontal application. For example, they encompass food labelling, food additives, food hygiene, contaminants, nutrition and foods for special dietary uses and methods of analysis and sampling.

The CAC has established a number of principles on the scientific basis for its decision-making processes (Randell & Race 1996). These principles ensure that the quality and food safety provisions are based on sound science and, that in establishing food standards, other legitimate factors relevant to consumer health may be considered and in the promotion of fair trade. Codex standards and related texts are subject to revision as and when deemed necessary by the CAC and its subsidiary bodies, in order to ensure that they are consistent with and reflect current scientific knowledge. Any member of Codex may identify and present new scientific or other information to the relevant body that may warrant a revision.

10.8.2 *Codex standards, their international relevance and their role in the WTO*

The Uruguay round of multilateral trade negotiations held under the General Agreement on Tariffs and Trade (GATT) took place between 1986 and 1994 and led to the formation of the World Trade Organisation (WTO) on the 1st of January 1995. For the first time, GATT agreements included agriculture and food in its scope; however, the Marrakesh agreement of 1994 also included the agreements on sanitary and phytosanitary (SPS) measures and on technical barriers to trade (TBT). These agreements acknowledge the need for the harmonisation of international standards to minimise the risk of sanitary, phytosanitary and other

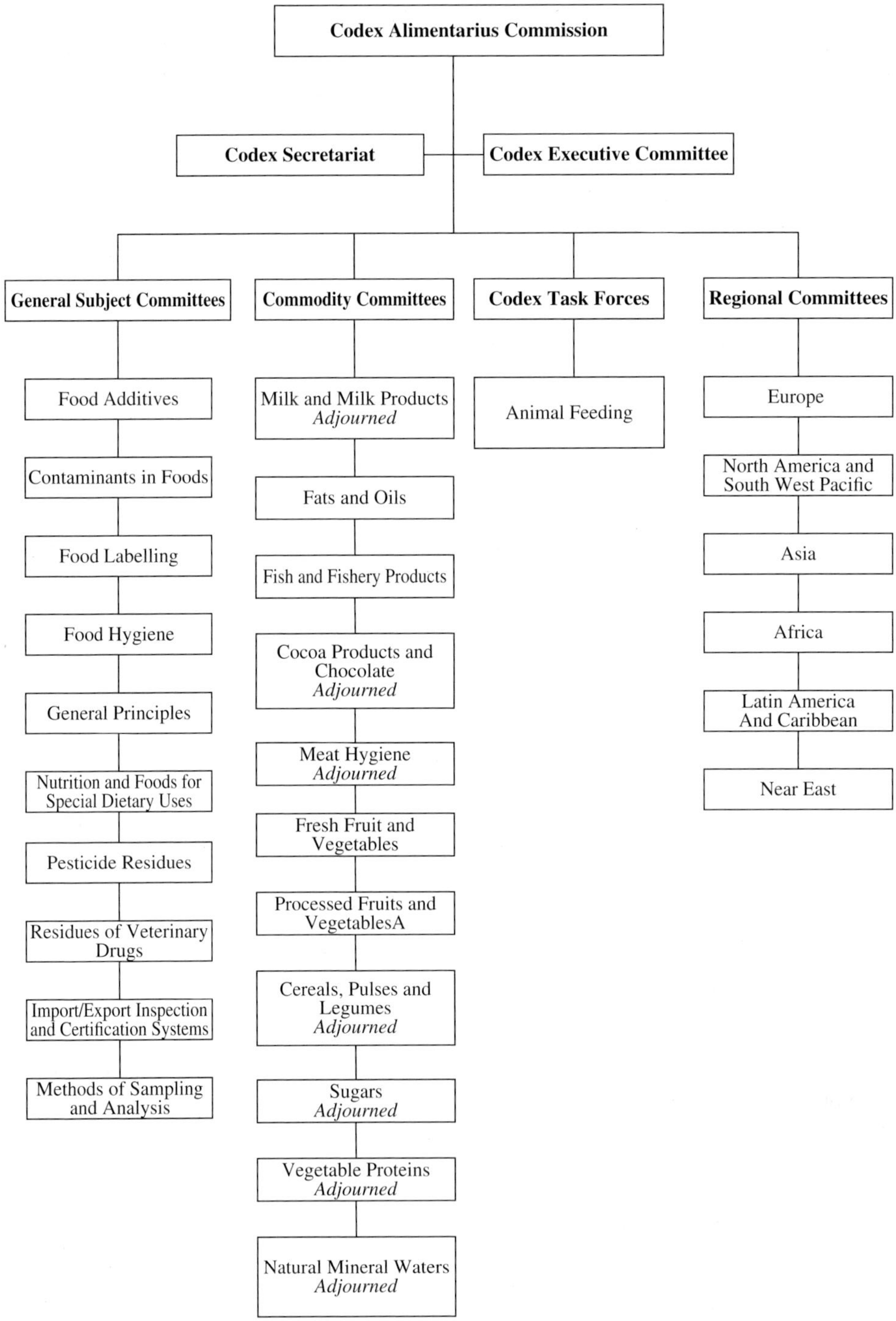

Fig. 10.7 The structure of Codex Alimentarius. Adapted from WHO/FAO (2011). Reproduced with permission of Food and Agriculture Organization of the United Nations.

technical standards becoming barriers to international trade. The SPS and TBT agreements therefore gave formal recognition to international standards, guidelines and recommendations of international organisations, including the CAC, as reference points for facilitating international trade and resolving disputes. The role of Codex Alimentarius in this regard is now well recognised.

It should be noted however that consumer groups have expressed some criticism of Codex standards on the basis of the time taken to elaborate standards and sound science; the latter basis may not necessarily take into account other considerations, such as consumer concerns (O'Rourke 1999).

10.8.3 *Codex and the issue of health and nutrition claims*

Codex has not addressed functional foods *per se*; rather, it has developed a series of standards, guidelines and recommendations that addresses the issue of labelling claims associated with such products. The Codex General Standard for the Labelling of Prepackaged Foods (GSLPF) was adopted by the CAC in 1981; it was subsequently revised in 1985 and 1991 and amended in 1999 and 2001 (FAO/WHO 2007c). This standard defines a claim as 'any representation which states, suggests or implies that a food has particular qualities relating to its origin, nutritional properties, nature, processing, composition or any other quality'.

The Codex General Standard for the Labelling of and Claims for Prepackaged Foods for Special Dietary Uses was adopted by the Codex Alimentarius Commission in 1985 (FAO/WHO 2007b). The Codex General Guidelines on Claims was originally adopted in 1979 with a revised version adopted in 1991 (FAO/WHO 2007a). This text contains the same definition of a claim as given above under the GSLPF but also prohibits claims which: (1) state that any given food will provide an adequate source of all essential nutrients (except in certain well-defined products where a Codex standard regulates such claims as admissible claims or where appropriate authorities have accepted that the product is an adequate source of all essential nutrients); (2) imply that a balanced diet or ordinary foods cannot supply adequate amounts of all nutrients; (3) cannot be substantiated; (4) state a food is suitable for use in the prevention, alleviation, treatment or cure of a disease, disorder or particular physiological condition unless it meets specific criteria further detailed in section 3.4 of these guidelines; or (5) could give rise to doubt about the safety of similar food or which could arouse or exploit fear in the consumer.

The task of developing guidelines on the use of nutrition and health claims came under the remit of the Codex Committee on Food Labelling (CCFL), but with review and endorsement of specific nutritional aspects and the development of guidelines on scientific substantiation by the Codex Committee for Foods for Special Dietary Uses (CCNFSDU). This proved a difficult task; for example, in written comments submitted to the CCFL meeting in 1994, some countries (Denmark and Finland) opposed both nutrient function and health claims while others (New Zealand and Switzerland) opposed health claims but could accept nutrient function claims. However, Australia, Sweden and the US were prepared to accept both these types of claims provided they were subject to strict criteria. Given the lack of consensus on the matter, the CCFL agreed in 1996 to delete all references to health claims in the guidelines and sent those on nutrition claims only to the CAC. These were adopted in 1997 as Guidelines for Use of Nutrition Claims CAC/GL 23–1997. Work continued on addressing the more contentious issue of health claims and, at its 32nd session in 2004, the CCFL finally agreed the provisions on health claims. These were consolidated into the existing use of nutritional claims guidelines, were renamed as Guidelines on the Use of Nutrition and Health Claims and forwarded to the CAC for adoption at its 27th session (FAO/WHO 2007d).

These guidelines are long and detailed and, at the outset, they lay down a number of principles concerning health claims, including the following.

- They should be consistent with national health policy, nutrition policy and support same as applicable.
- They should be supported by a sound and sufficient body of scientific evidence to substantiate the claim.
- They should provide truthful and non-misleading information to aid consumers in choosing healthy diets and be supported by specific consumer information.
- Their impact on consumers' eating habits and buying habits should be monitored, in general, by competent authorities.
- The prohibition of claims as outlined in Section 3.4 of the Codex General Guidelines for the Use of Claims should remain (see above).
- The Codex definition of a health claim is given as 'any representation that states, suggests, or implies that a relationship exists between a food or a constituent of that food and health' and three types of related claims are defined: (1) nutrient function claims (that describe the physiological role of the nutrient in growth, development and normal functions of the body); (2) other function claims (concerning specific beneficial effects of the consumption of foods or their constituents in the context of the total diet on normal functions or biological activities of the body, such as a positive contribution to health or to the improvement of a function or modifying or preserving health); and (3) reduction of disease risk claims (relating to the consumption of a food or food constituent, in the context of the total diet, to the reduced risk of developing a disease or health-related condition). Risk reduction means significantly altering a major risk factor(s) for a disease or health-related condition.

There is a requirement that any health claim must be accepted by, or acceptable to, the competent authorities of the country where the product is sold.

As regards substantiation of health claims, in the text as adopted in 2004 reference was made to a parallel text being developed by the CCNFSDU. At its 25th session in November 2003, the CCNFSDU considered a text with the working title of 'Recommendations on the Scientific Basis for Health Claims'. The resulting text was finally adopted by the CAC in 2009, is now entitled 'Recommendations on the Scientific Substantiation of Health Claims' and is now included as an Annex to the Guidelines on the Use of Nutrition and Health Claims. A footnote states that this document should be read in conjunction with the Working Principles for Risk Analysis for Food Safety for Application by Governments (FAO/WHO 2007e).

It should be pointed out that Codex Alimentarius does not evaluate health claims. Rather, these guidelines are intended for individual governments to facilitate their evaluation of health claims made by industry and also as a reference for industry in preparing dossiers aimed at providing substantiation of such claims. They focus on the process, the criteria that should be used for substantiation and for the systematic review of the relevant scientific evidence. The following are some of the criteria.

- Health claims should primarily be based on evidence provided by well-designed human intervention studies. It is recognised that human observational studies are not necessarily sufficient alone, but these may contribute to the totality of the evidence.
- Data from *ex vivo* or *in vitro* animal model studies are not regarded as sufficient, but again may be used to provide additional supportive information.
- The totality of the evidence, which can include appropriate unpublished data, should be identified and reviewed.
- Evidence based on human studies should demonstrate a consistent association between the food or food constituent and the claimed health effect, with little or none to the contrary.
- Substantiation can take into account specific situations or alternate processes; for example, nutrient function claims may be substantially based on generally accepted statements by recognised expert scientific bodies that have been verified and validated over time.
- Some health claims, such as those involving a relationship between a food category and a health effect, may be substantially based on observational studies. Such studies should provide a consistent body of evidence from a number of well-designed studies.
- Evidence-based dietary guidelines and authoritative statements prepared or endorsed by a competent authoritative body and meeting the same high scientific standards may also be used.

A summary of the labelling requirements under the Codex Guidelines for the use of Nutrition and Health Claims includes: a statement of the quantity of any nutrient or other constituent of the food that is the subject of the claim; the target group, if appropriate; how to use the food to obtain the claimed benefit and other lifestyle factors or other dietary sources, where appropriate; if appropriate, advice to vulnerable groups on how to use the food and advice to groups, if any, who need to avoid the food; maximum safe intake of the food or constituent where necessary; how the food or food constituent fits within the context of the total diet; and a statement on the importance of maintaining a healthy diet.

There should be periodic reviews of health claims to study new knowledge of the benefits, to look at consumption patterns of products with health claims, to ensure they are in line with the claims and to look at positive effects and possible negative effects after long-term consumption.

The above Codex guidelines and recommendations combined should provide a firmer legal basis for health claims at an international level, but leave the actual approval or acceptance of such claims to individual governments to ensure that they are in line with national dietary policies and guidelines.

10.8.4 *The Codex Standard for Fermented Milks*

At its 26th session in July 2003, the CAC adopted a new Codex Standard for Fermented Milks (CAC 2003) which replaced two former standards for (a) yogurt and sweetened yogurt, and

(b) flavoured yogurt and heat-treated products after fermentation. This standard expanded its scope to encompass a broader range of fermented milks such as kefir, acidophilus milk and koumiss. While not specifically aimed at probiotics, this new standard includes compositional requirements for the minimum level of starter culture organisms (1×10^7 cfu g^{-1}) and, where a content labelling claim is made for specific micro-organisms other than the normal starter cultures, a minimum of 1×10^6 cfu g^{-1} is required (FAO/WHO 2011). The earlier standards required only that micro-organisms should be viable and abundant, without setting specific minima.

10.9 Other international developments

It is evident from the current review that many aspects pertaining to the regulation of the category of so-called functional foods, or specific subcategories such as probiotics, still exist within a 'grey' area. As international trade in these products continues to grow, many regulators express a need to develop an appropriate regulatory framework for the control of such products to ensure the ongoing need to protect human health, prevent consumer deception and facilitate fair trade. This is reflected in some other developments at Codex level in recent years.

One such initiative was the joint FAO/WHO expert consultation on health and nutritional properties of probiotics in food held in October 2001 at Cordoba in Argentina (FAO/WHO 2001). Eleven experts from ten countries attended to evaluate the latest information (i.e. scientific evidence available on functional foods, safety aspects of probiotics and the methodologies to assess such aspects). The report of this consultation lists a number of conclusions, including that the regulatory status of probiotics is not well established on an international basis and that (at that time at least) regulatory procedures are in place or sufficiently developed in only a few countries to permit probiotic products to describe specific health benefits. In addition to the scientific recommendations, the international experts also made a number of recommendations pertaining to regulatory matters as follows.

- In order to be termed a probiotic, the micro-organism must be able to confer defined health benefits on the host in the actual product that will be made available to humans.
- Good manufacturing practices must be applied in terms of quality control, shelf-life conditions established and clear labelling to include minimum dosage and verifiable health claims.
- The regulatory status of probiotics as a component of foods should be established on an international level.
- The regulatory framework to be established should include issues related to probiotics such as efficacy, safety, labelling, fraud and claims.
- Probiotic products shown to confer defined health benefits should be permitted to describe these health benefits.
- Surveillance systems should be put in place to record and analyse any adverse events associated with probiotics in food; these systems should include trace-back and post-marketing surveillance and they should also monitor the long-term health benefits of probiotic strains.

To follow on from the consultation, a working group was established to draft guidelines for the evaluation of probiotics in food and met in London, Ontario in 2002. In the report of their discussions they concluded that, from the current state of evidence, probiotic effects are strain-specific (ftp://ftp.fao.org/es/esn/food/wgreport2.pdf). The same working group proposed a definition for probiotics as follows: 'Live micro-organisms which when administered in adequate amounts confer a health benefit on the host'. Furthermore, they also proposed that adoption of the guidelines of their report should be a prerequisite for calling a strain a probiotic.

It has to be stressed that the conclusions and recommendations of these groups are those of the participants and do not imply any opinion on the part of the organisers. In addition, the participants were all scientists and not legislators. The needs, expectations and priorities of the two groups do not necessarily concur; however, they do identify gaps and shortcomings which, if addressed, could help to clarify and remove at least some of the uncertainty from the current regulatory situation.

10.10 Summary and conclusions

As can be seen from the review of the legislative systems of the various countries and regions outlined in this chapter, two basic types of regulatory approach to functional foods are used: (1) where individual food products are approved as for example in the FOSHU approach used in Japan and also to an extent in China; and (2) through the use of health, nutrient function, nutrient content or similar type claims as is used in a wider basis worldwide such as in the EU, US, Canada Australia and New Zealand.

The absence of an internationally adopted definition of what is meant by 'functional food' in itself creates problems. While the Japanese defined the term in their legislation and working definitions exist, the latter have no particular status in international standards and in national or regional legislative systems. Furthermore, although all the various regulatory systems on functional foods and functional ingredients are based on the provision of scientific justification, the level of scientific evidence required seems to vary throughout the world. This has resulted in a situation where certain products that have achieved FOSHU status in Japan are unlikely to be allowed to make the same claim when sold in the EU, although the scientific justification used in both cases is at least similar.

It is unlikely that this situation will be resolved at an international level by the Codex Alimentarius; as it develops guidelines for claims to be used by member governments in assessing claims, their adoption is voluntary. It is possible that the WTO might become involved at some future date if the case is made that technical barriers to trade are involved in the prohibition of a certain health claim, but it is far from clear that this will happen.

Will all this result in a decline in the demand for functional foods or will the market for such products continue to grow? The answer is not clear at this time but the evidence is that existing products on the market are still increasing their market penetration and, once a brand has captured and reinforced its position in consumers' minds, it is very difficult to change those minds (Mellentin 2010). The greater challenges are for new products to achieve recognition and acceptance. As a consequence there is little doubt that the efforts for the scientific substantiation of the effectiveness of functional foods will continue and indeed probably increase, athough this will also necessitate greater and increasing expenditure (especially if larger human trials are required). Resources for such research are likely to concentrate more and more on larger multinational food companies; smaller food companies may hope to benefit however either from their innovation or from the commercialisation of their products by larger companies.

The additional challenges posed by nutrient profiling also looms large for the dairy sector, especially for the higher-fat products such as cheese and milk-fat-based products. Emerging nutritional research offers more hopeful signs that milk fat may be less of an issue than previously thought, but much work and effort remains to convince the wider medical and regulatory world of this. The message will not be a simple one, but unfortunately simple approaches are often preferred in the regulatory world as they are easier to communicate. This will probably require more education of the general public on nutritional matters. Who should do this and, even more importantly, who shall or should fund the cost? These are all matters that will continue to be discussed and debated for quite some time to come.

Acknowledgements

I gratefully acknowledge my International Dairy Federation colleague and friend Yuki Morita of Yakult Europe and her colleague Ryouichi Akahoshi of Yakult Honsha, Japan, for their generous assistance with the translation and interpretation of certain Japanese texts, including the Japanese characters used in Section 10.2.5, and in supplying the images of the FOSHU logos used in this chapter.

References

Agriculture and Agr-Food Canada (2011) Registration of Health Food in China 2010. Available at http://www.ats.agr.gc.ca/asi/5492-eng.htm (accessed November 2013).

Anon. (1992) Annual Report 1992: Advisory Committee for Novel Foods and Processes. Ministry of Agriculture Fisheries and Foods Publications, UK.

Anukam, K.C. & Reid, G. (2010) Probiotics: 100 years (1907–2007) after Elie Metchnikoff's Observation. *Communicating Current Research and Educational Topics and Trends in Applied Microbiology*, **1**,466–474.

Arai, S., Moringa, Y., Yoshikawa, T., Ichiishi, E., Kiso, Y., Yamazaki, M., Morotimo, M., Shimizu, M., Kuwata, T. Kaminogawa, S. (2002) Recent trends in functional food science and industry in Japan. *Bioscience Biotechnology & Biochemistry*, **66**, 2017–2029.

Australia and New Zealand Food Regulation Ministerial Council (2003) Policy guideline on nutrition health and related claims. Available at http://www.health.gov.au/internet/main/publishing.nsf/Content/foodsecretariat-communiques-03_12dec.htm (accessed October 2013).

Australia and New Zealand Food Regulation Ministerial Council (2011) Policy guideline on nutrition health and related claims (update 2011). Available at http://www.health.gov.au/internet/main/publishing.nsf/Content/foodsecretariat-policy-guidelines#10 (accessed November 2013).

Berner, L.A. & O'Donnell, J.A. (1998) Functional foods and health claims legislation: applications to dairy foods. *International Dairy Journal*, **8**, 355–362.

CAC (Codex Alimentarius Commission) (2003) Twenty-sixth Session, FAO Headquarters, Rome, 30 June–7 July 2003. Report, pp. 13–14 in ALINORM 03/41. Available at http://www.codexalimentarius.org/meetings-reports/en/?sortingDate=072003 (accessed November 2013).

Canadian Food Inspection Agency (2011) Health claims. In *Guide to Food Labelling and Advertising*, pp.8.1–8.46. Ottawa. Available at http://www.inspection.gc.ca/food/labelling/guide-to-food-labelling-and-advertising/eng/1300118951990/1300118996556 (accessed November 2013).

Consumer Affairs Agency (Japan) (2011) Regulatory systems of health foods in Japan. Available at http://www.caa.go.jp/en/pdf/syokuhin338.pdf (accessed October 2013).

EFSA (2008) The setting of nutrient profiles for foods bearing nutrition and health claims pursuant to Article 4 of the Regulation (EC) No 1924/2006. Scientific Opinion of the Panel on Dietetic Products, Nutrition and Allergies. *The EFSA Journal*, **6**(2), 1–44.

EFSA (2011a) 'General function' health claims under Article 13. Available at http://www.efsa.europa.eu/en/topics/topic/article13.htm (accessed November 2013).

EFSA (2011b) 'New function' health claims under Article 13.5. Available at http://www.efsa.europa.eu/en/topics/topic/article13-5.htm (accessed November 2013).

EFSA (2011c) Claims on disease risk reduction and child development or health under Article 14. Available at http://www.efsa.europa.eu/en/topics/topic/article14.htm (accessed November 2013).

EU (1985) Council Directive 85/374/EEC of 25 July 1985 on the approximation of the laws, regulations and administrative provisions of the Member States concerning liability for defective products. *Official Journal of the European Commission*, **L210**, 29–33.

EU (1990) Council Directive 90/219/EEC of 23 April 1990 on the contained use of genetically modified micro-organisms. *Official Journal of the European Commission*, **L117**, 1–14.

EU (1997) Regulation (EC) 2/97/EC of 27 January 1997 concerning novel foods and novel food ingredients. *Official Journal of the European Commission*, **L43**, 1–6.

EU (1998) Council Directive 98/81/EC of 26 October 1998 on the contained use of genetically modified micro-organisms. *Official Journal of the European Commission*, **L330**, 13–31.

EU (1999) Directive 1999/34/EC of the European Parliament and of the Council of 10 May 1999 amending Council Directive 85/374/EEC on the approximation of the laws, regulations and administrative provisions of the Member States concerning liability for defective products. *Official Journal of the European Commission*, **L141**, 20–21.

EU (2000) Directive 2000/13/EC of the European Parliament and of the Council of 20 March 2000 on the approximation of the laws of the Member States relating to the labelling, presentation and advertising of foodstuffs. *Official Journal of the European Commission*, **L109**, 29–42.

EU (2001a) Commission Directive 2001/101/EC of 26 November 2001 amending Directive 2000/13/EC of the European Parliament and of the Council on the approximation of the laws of the Member States relating to the labelling, presentation and advertising of foodstuffs. *Official Journal of the European Commission*, **L310**, 19–21.

EU (2001b) Council Decision 2001/204/EC of 8 March 2001 supplementing Directive 90/219/EEC as regards the criteria for establishing the safety, for human health and the environment, of types of genetically modified micro-organisms. *Official Journal of the European Commission*, **L73**, 32–34.

EU (2001c) Council Directive 2001/18/EC on the deliberate release into the environment of genetically modified organisms and repealing Council Directive 90/220/EC. *Official Journal of the European Commission*, **L106**, 1–39.

EU (2002) Regulation (EC) No 178/2002 of the European Parliament and of the Council of 28 January 2002 laying down the general principles and requirements of food law, establishing the European Food Safety Authority and laying down procedures in matters of food safety. *Official Journal of the European Commission*, **L31**, 1–24.

EU (2003a) Regulation (EC) No 1829/2003 of the European Parliament and of the Council of 22 September 2003 on genetically modified food and feed. *Official Journal of the European Commission*, **L268**, 1–23.

EU (2003b) Regulation (EC) No 1830/2003 of the European Parliament and of the Council of 22 September 2003 concerning the traceability and labelling of genetically modified organisms and the traceability of food and feed products produced from genetically modified organisms and amending Directive 2001/18/EC. *Official Journal of the European Commission*, **L268**, 24–28.

EU (2006) Corrigendum to Regulation (EC) No 1924/2006 of the European Parliament and of the Council of 20 December 2006 on nutrition and health claims made on foods. *Official Journal of the European Union*, **L12**, 3–18.

EU (2008a) Commission Regulation (EC) No 353/2008 of 18 April 2008 establishing implementing rules for applications for authorisation of health claims as provided for in Article 15 of Regulation (EC) No 1924/2006 of the European Parliament and of the Council. *Official Journal of the European Commission*, **L109**, 11–16.

EU (2008b) Regulation (EC) No 107/2008 of the European Parliament and of the Council of 15 January 2008 amending Regulation (EC) No 1924/2006 on nutrition and health claims made on foods as regards the implementing powers conferred on the Commission. *Official Journal of the European Commission*, **L39**, 8–10.

EU (2008c) Regulation (EC) No 109/2008 of the European Parliament and of the Council of 15 January 2008 amending Regulation (EC) No 1924/2006 on nutrition and health claims made on foods. *Official Journal of the European Commission*, **L39**, 14–15.

EU (2010) Commission Regulation (EU) No 116/2010 of 9 February 2010 amending Regulation (EC) No 1924/2006 of the European Parliament and of the Council with regard to the list of nutrition claims. *Official Journal of the European Commission*, **L37**, 16–18.

EU (2011a) Novel foods and novel food ingredients: Introduction and legislation. Available at http://ec.europa.eu/food/food/biotechnology/novelfood/index_en.htm (accessed October 2013).

EU (2011b) Regulation (EU) No 1169/2011 of the European Parliament and of the Council of 25 October 2011 on the provision of food information to consumers, amending Regulations (EC) No 1924/2006 and (EC) No 1925/2006 of the European Parliament and of the Council, and repealing Commission Directive 87/250/EEC, Council Directive 90/496/EEC, Commission Directive 1999/10/EC, Directive 2000/13/EC of the European Parliament and of the Council, Commission Directives 2002/67/EC and 2008/5/EC and Commission Regulation (EC) No 608/2004. *Official Journal of the European Commission*, **L304**, 18–63.

FAO/WHO (2001) Health and nutritional properties of probiotics in food including powder milk with live lactic acid bacteria. Available at http://www.who.int/foodsafety/publications/fs_management/en/probiotics.pdf (accessed October 2013).

FAO/WHO (2002) Guidelines for the evaluation of probiotics in food. Report of a Joint FAO/WHO Working Group on Drafting Guidelines for the Evaluation of Probiotics in Food. Available at ftp://ftp.fao.org/es/esn/food/wgreport2.pdf (accessed October 2013).

FAO/WHO (2007a) General guidelines for claims. In *Food Labelling*, pp. 21–23. Secretariat of the Codex Alimentarius Commission, Rome.

FAO/WHO (2007b) General standard for the labelling of and claims for prepackaged foods for special dietary uses (CODEX STAN 146–1985). In *Food Labelling*, pp. 17–20. Secretariat of the Codex Alimentarius Commission, Rome.

FAO/WHO (2007c) General standard for the labelling of prepackaged foods (CODEX STAN 1–1985). In *Food Labelling*, pp. 1–9. Secretariat of the Codex Alimentarius Commission, Rome.

FAO/WHO (2007d) Guidelines for the use of nutrition and health claims (CAC/GL 23–1977). In *Food Labelling*, pp. 33–41. Secretariat of the Codex Alimentarius Commission, Rome.

FAO/WHO (2007e) *Working Principles for Risk Analysis for Food Safety for Application by Governments (CAC/GL 62–2007)*. Secretariat of the Codex Alimentarius Commission, Rome.

FAO/WHO (2011) Codex standard for fermented milks. In *Milk and Milk Products*, pp. 6–16. Secretariat of the Codex Alimentarius Commission, Rome.

FDA (2009) Guidance for industry: a food labeling guide. Appendix C: Health claims. Available at http://www.fda.gov/Food/GuidanceRegulation/GuidanceDocumentsRegulatoryInformation/LabelingNutrition/ucm064919.htm (accessed November 2013).
FDA (2011) Guidance for industry: a food labeling guide. 8. Claims. Available at http://www.fda.gov/food/guidanceregulation/guidancedocumentsregulatoryinformation/labelingnutrition/ucm064908.htm (accessed October 2013).
Food Standards Agency (UK) (2010) McCance and Widdowson's The Composition of Foods integrated dataset (CoF IDS). Available at http://www.food.gov.uk/multimedia/spreadsheets/cofids.xls (accessed November 2013).
Food Standards Australia New Zealand (2008a) Final assessment report, Proposal P293: Nutrition, health and related claims. Available at http://www.fda.gov/downloads/Food/GuidanceRegulation/UCM265446.pdf (accessed November 2013).
Food Standards Australia New Zealand (2008b) Nutrition, health and related claims: a short guide to the new standard. Available at http://www.foodstandards.gov.au/code/proposals/pages/proposalp293nutritionhealthandrelatedclaims/p293finalassessmentr3889.aspx (accessed November 2013).
Food Standards Australia New Zealand (2008c) Transitional standard. Health claims: Standard 1.1A.2. Available at http://www.comlaw.gov.au/Details/F2009C00878 (accessed November 2013).
Food Standards Australia New Zealand (2011a) Australia New Zealand Food Standards Code. Available at http://www.foodstandards.gov.au/code/Pages/default.aspx (accessed November 2013).
Food Standards Australia New Zealand (2011b) Calculation method for determining foods eligible to make health claims (final). Available at http://www.foodstandards.gov.au/code/proposals/documents/P293_PFAR_Att_7%20-_Health%20Claims%20Nutrient%20Profiling%20Calculator.pdf (accessed November 2013).
Food Standards Australia New Zealand (2011c) Health claims nutrient profiling calculator. Available at http://www.foodstandards.gov.au/consumerinformation/nutritionhealthandrelatedclaims/nutrientprofilingcal3499.cfm (accessed October 2013).
Food Standards Australia New Zealand (2011d) Timeline for Proposal P293: Nutrition, health and related claims. Available at http://www.foodstandards.gov.au/consumerinformation/nutritionhealthandrelatedclaims/healthclaimsstandard5081.cfm (accessed October 2013).
Ford, K., Konishi, Y., Rajalahti, R. & Pehu, E. (2007) Health Enhancing Foods Country Case Studies of China and India. Available at http://siteresources.worldbank.org/INTARD/Resources/healthfoodCandI.pdf (accessed November 2013).
Gorbach, S.L. & Goldin, B.R. (inventors) (1989) *Lactobacillus* strains and methods of selection. New England Medical Center Hospitals, IBM assignee patent no. 4839281.
Health and Consumers Directorate of the European Commission (2008a) Summary report of the Working Groups of the Advisory Group on the Food Chain and Animal and Plant Health on Nutrient Profiles held in Brussels on 8 July and 28 November 2008. Available at http://ec.europa.eu/food/committees/advisory/sum_wg_11072008.pdf (accessed November 2013).
Health and Consumers Directorate of the European Commission (2008b) Working document on the setting of nutrient profiles. Available at http://www.reading.ac.uk/foodlaw/pdf/eu-08024-nutrient-profiles-oct-08.pdf (accessed November 2013).
Health and Consumers Directorate of the European Commission (2009) Working document on the setting of nutrient profiles. Available at http://www.reading.ac.uk/foodlaw/pdf/eu-09012-nutrient-profiles-dec-08.pdf (accessed November 2013).
Health Canada (1998) Policy Paper: Nutraceuticals/Functional Foods and Health Claims on Foods. Available at http://www.hc-sc.gc.ca/fn-an/label-etiquet/claims-reclam/nutra-funct_foods-nutra-fonct_aliment-eng.php (accessed November 2013).
Health Canada (2002) Interim Guidance Document. Preparing a Submission for Foods with Health Claims: Incorporating Standards of Evidence for Evaluating Foods with Health Claims. Available

at http://www.hc-sc.gc.ca/fn-an/label-etiquet/claims-reclam/abstract_guidance-orientation_resume-eng.php (accessed November 2013).

Health Canada (2006) Position Paper on Five US Generic Health Claims Considered for Use in Canada. Available at http://www.hc-sc.gc.ca/fn-an/label-etiquet/claims-reclam/position_paper-enonce_position-eng.php (accessed November 2013).

Health Canada (2009a) Guidance document: Preparing a submission for food health claims. Available at http://www.hc-sc.gc.ca/fn-an/alt_formats/hpfb-dgpsa/pdf/legislation/health-claims_guidance-orientation_allegations-sante-eng.pdf (accessed October 2013).

Health Canada (2009b) Guidance document: The use of probiotic microorganisms in food. Available at http://www.hc-sc.gc.ca/fn-an/alt_formats/hpfb-dgpsa/pdf/legislation/probiotics_guidance-orientation_probiotiques-eng.pdf (accessed October 2013).

Health Canada (2010) Classification of products at the food–natural health product interface: Products in food formats. Version 2.0. Available at http://www.hc-sc.gc.ca/dhp-mps/prodnatur/legislation/docs/food-nhp-aliments-psn-guide-eng.php (accessed November 2013).

Ichikawa, T. (1994) Functional foods in Japan. In *Functional Foods, Designer Foods, Pharmafoods, Nutraceuticals* (ed. Goldberg, I.), pp. 453–467. London: Chapman & Hall.

Information Center of SFDA (2011) SFDA State Food and Drug Administration of the PR China. Available at http://eng.sfda.gov.cn/WS03/CL0756/ (accessed October 2013).

Japanese Ministry of Health Labour and Welfare (2010) List of approved FOSHU (excel file in Japanese only). Available at http://www.mhlw.go.jp/topics/bukyoku/iyaku/syoku-anzen/hokenkinou/xls/hyouziseido-1a.xls (accessed October 2013).

Japanese Ministry of Health Labour and Welfare (2011a) Food for Specified Health Uses (FOSHU). Available at http://www.mhlw.go.jp/english/topics/foodsafety/fhc/02.html (accessed October 2013).

Japanese Ministry of Health Labour and Welfare (2011b) Food with Nutrient Function Claims. Available at http://www.mhlw.go.jp/english/topics/foodsafety/fhc/01.html (accessed October 2013).

Jukes, D. (2009) CLAIMS: Agency update following discussions in Commission Working Group on implementing the Nutrition and Health Claims Regulation 1924/2006. Available at http://www.reading.ac.uk/foodlaw/ (accessed October 2013).

Kan, X. (1996) Legal aspects of functional foods: the situation in China. *Nutrition Reviews*, **54**(11), S162–S166.

L'Abbé, M.R., Dumais, L., Chao, E. & Junkins, E. (2008) Health claims on foods in Canada. *Journal of Nutrition*, **138**(6), 1221S–1227S.

Lähteenmäki-Uutela, A. (2009a) Chinese law on foods and medicines. *Nordic Journal of Commercial Law*, **2**, 1–41.

Lähteenmäki-Uutela, A. (2009b) Foodstuffs and medicines as legal categories in the EU and China. Functional foods as a borderline case. Ph.D. thesis, University of Turku, Finland.

Mellentin, J. (2010) Marketing with - and beyond - health claims. Presentation at Marketing Conference at IDF World Dairy Summit 2010. Auckland, New Zealand.

Metchnikoff, É. (1908) Lactic acid as inhibiting intestinal putrefaction (English translation). In *The Prolongation of Life: Optimistic Studies* (ed. Mitchell, P.C.), pp. 161–183. New York, London: Putnam's Sons.

Ministry of Health of China (1995) Food Hygiene Law of the People's Republic of China. Available at http://www.chinafdc-law.com/laws/detail_156.html (accessed October 2013).

Ministry of Health of China (2001) Drug Administration Law of the People's Republic of China (2001 Revision). Available at http://www.chinafdc-law.com/laws/detail_80.html (accessed October 2013).

Minister of Justice Canada (2011) Food and Drug Regulations CRC c. 870. Available at http://laws.justice.gc.ca/en/showtdm/cr/C.R.C.-c.870 (accessed October 2013).

Mitsuda, H. (1958) Studies on enriched food. *Bulletin of the Institute for Chemical Research, Kyoto University*, **35**(5–6), 140–148.

National People's Congress of China (2009) Food Safety Law of the People's Republic of China. Available at http://www.usdachina.org/info_details1.asp?id=2431 (accessed October 2013).

National People's Congress of China (1995) Food Hygiene Law of the People's Republic of China. Available at http://www.chinafdc-law.com/laws/detail_156.html (accessed November 2013).

National People's Congress of the People's Republic of China (2011) National People's Congress. Available at http://www.npc.gov.cn/englishnpc/Organization/node_2846.htm (Accessed October 2013).

O'Rourke, R. (1999) The international context. In *European Food Law with 1999 Update*, pp. 148–160. Bembridge, Isle of Wight: Palladian Law Publishing Ltd.

Patel, D., Dufour, Y. & Domigan, N. (2008) Functional food and nutraceutical registration processes in Japan and China: Similarities and Differences. *Journal of Pharmaceutical Sciences*, **11**(4), 1–11.

Platzman, A. (1999) Functional Foods: Figuring Out the Facts. Available at http://www.foodproductdesign.com/articles/1999/11/functional-foods.aspx (accessed November 2013).

Porter, D.V. & Earl, R.O. (1992) Contextual factors affecting the regulation of misbranded food. In *Food Labeling: Toward National Uniformity* (eds Porter, D.V. & Earl, R.O.), pp. 35–62. Washington DC: National Academy Press.

Preston, C. & Lawrence, M. (1996) Regulatory and legal aspects of functional foods: the Australian perspective. *Nutrition Reviews*, **54**(11), S156–S161.

Randell, A. & Race, J. (1996) Regulatory and legal aspects of functional foods: an international perspective. *Nutrition Reviews*, **54**(11), 152–155.

RedFern Associates (2010) The China Health Food Market Overview. New Zealand Trade & Enterprise, Shanghai, pp. 1–34. Available at http://www.prclive.com/pdf/uploads/Health%20food%20in%20China%20-%20September%202010%20[by%20REDFERN].pdf (accessed November 2013).

RJS Medical Technology (2011) 27 Categories of health functions claims approved by SFDA. Available at http://www.sfdachina.com/info/86-1.htm (accessed November 2013).

Salminen, S. (1996) Functional dairy foods with *Lactobacillis* strain GG. *Nutrition Reviews*, **54**(11), 99–101.

Sanders, M.E. (2003) Probiotics: considerations for human health. *Nutrition Reviews*, **61**(3), 91–99.

Swann, J.P. (2008) History of the FDA. Available at http://www.fda.gov/AboutFDA/WhatWeDo/History/Origin/ucm124403.htm (accessed November 2013).

Tapsell, L.C. (2008) Evidence for health claims: a perspective from the Australia-New Zealand region. *Journal of Nutrition*, **138**(6), 1206S–1209S.

Tissier, H. (1906) Traitement des infections intestinales par la méthode de la transformation de la flore bactérienne de l'intestin. *Comptes Rendus des Séances et Mémoires de la Société de Biologie*, **60**, 359–361.

Tomita, F. (2007) Japanese Legislation on Nutrition and Health Products: The Regulatory Issues and Products on Japanese Market. Available at http://congress.utu.fi/NHC2007/presentations.php (accessed October 2013).

US Congress (1938) Federal Food Drug and Cosmetic Act of 1938 (Pub. L. 75–717). Available at http://www.fda.gov/RegulatoryInformation/Legislation/FederalFoodDrugandCosmeticActFDCAct/default.htm (accessed October 2013).

US Congress (1990) Nutrition Labeling and Education Act (NLEA) of 1990 (Pub. L 101–535). Available at http://thomas.loc.gov/home/LegislativeData.php?&n=BillText (accessed November 2013).

US Congress (1994) Dietary Supplement Health and Education Act of 1994. Available at http://thomas.loc.gov/home/LegislativeData.php?&n=BillText (accessed November 2013).

US Congress (1997) Food and Drug Administration Modernization Act of 1997. (Pub. L 105–115). Available at http://www.fda.gov/RegulatoryInformation/Legislation/FederalFoodDrugandCosmeticActFDCAct/SignificantAmendmentstotheFDCAct/FDAMA/FullTextofFDAMAlaw/default.htm (accessed October 2013).

Weng, W. & Chen, J. (1996) The Eastern perspective on functional foods based on traditional Chinese medicine. *Nutrition Reviews*, **54**(11), S11–S16.

Yamada, K., Sato-Mito, N., Nagata, J. & Umegaki, K. (2008) Health claim evidence requirements in Japan. *Journal of Nutrition*, **138**(6), 1192S–1198S.

Yamaguchi, P. (2004) FOSHU Approval: Is It Worth The Price? Available at http://newhope360.com/supply-news-amp-analysis/foshu-approval-it-worth-price (accessed November 2013).

Yang, Y. (2008) Scientific substantiation of functional food health claims in China. *Journal of Nutrition*, **138**(6), 1199S–1205S.

Index

Note: page numbers in *italics* refer to figures and in **bold** refer to tables.

Milk and Dairy Products as Functional Foods, First Edition. Edited by A. Kanekanian.